M. Cresti P. Gori E. Pacini (Eds.)

Sexual Reproduction in Higher Plants

Proceedings of the Tenth International Symposium on the
Sexual Reproduction in Higher Plants, 30 May – 4 June 1988
University of Siena, Siena, Italy

With 259 Figures

Springer-Verlag
Berlin Heidelberg New York
London Paris Tokyo

Mauro Cresti
Paolo Gori
Ettore Pacini

Dipartimento di Biologia Ambientale
Università degli Studi di Siena
Via P. A. Mattioli n. 4
Siena 53100, Italy

ISBN 3-540-18673-5 Springer-Verlag Berlin Heidelberg New York
ISBN 0-387-18673-5 Springer-Verlag New York Berlin Heidelberg

Printing and binding: Druckhaus Beltz, Hemsbach
2131/3130-543210 – Printed on acid-free paper

Introduction

From the 30th of May to the 4th of June, 1988, a symposium on "Sexual Reproduction in Higher Plants" was held at the University of Siena, Italy. It was the tenth of a series which started in 1968 in Paris as an informal colloquium, organized by Prof. Favre Duchartre. It was also the second time that the symposium was located in Siena, since the second meeting of the series was organized by the late Prof. Sarfatti in 1972.

During these biennial symposia researchers operating in the wide field of sexual plant reproduction show the progress achieved in their special fields of interest, and discuss them with their colleagues.

Through this, the symposium is a place of exchange of ideas; a place of inspiration for the participants; and a place where new approaches are initiated. Furthermore, it gives an excellent opportunity to make contacts, to organize international cooperations and to prepare future research programmes.

Actually, during the 10th symposium a project was started to organize an international society or association of scientists working in the field of sexual reproduction in higher plants.

The contributions enclosed in this book are those presented at the meeting and ordered according to the symposium program. The content of each contribution is under the responsability of the authors. The symposium comprised 9 sessions of lectures and 3 poster sessions. Seven sessions were devoted to the male gametophyte and its interactions with the female counterpart, one to ovule and seed development and one to taxonomy and embryology. The meeting started with 3 introductory lectures which highlighted some recent and actual topics of plant reproduction research and at the end there was a conspectus of the congress. Compared to the previous symposia there appeared to be an increasing interest in applied aspects, together with a reinvestigation of classical topics by modern techniques and a regression of strictly cyto-embryological researches which dominated during the earliest editions of this symposium series.

This symposium, even if not all aspects of sexual reproduction were equally included (i.e., the female part and the postzygotic aspects were underepresented), it was found very useful and stimulating by the participants and we look forward to the next meeting in Leningrad in 1990.

Acknowledgments

First of all, the organizers would like to thank all the participants for their contributions, because the success and impact of a scientific meeting essentially depends on the quality of papers, the presentation of lectures and posters, and the discussions and exchange of ideas during the meeting.

We want to express special thanks to Ercole Ottaviano, John Heslop-Harrison FRS and David L. Mulcahy, who delivered the introductory invited lectures and to Hans F. Linskens for giving the overview of the proceedings and outlook into the future.

We thank the following participants for acting as chairpersons of different sessions: G. Bergamini Mulcahy; A.E. Cocucci; J. Heslop-Harrison; Y. Heslop-Harrison; M. Hesse; F.A. Hoekstra; J.F. Jackson; H.F. Linskens; C. Nitsch; P.L. Pfahler; M. Sari Gorla; J.A.M. Schrauwen; K.R. Shivanna; J.L. Van Went; M.T.M. Willemse.

Several Institutions (Università di Siena, Ministero della Pubblica Istruzione, Consiglio Nazionale delle Ricerche (C.N.R.), Regione Toscana, Monte dei Paschi di Siena, Banca Toscana) provided essential financial support. Additional support came from Industries (Leitz Italiana, Zeiss Italia, Bio Rad, Boooiliohi). Furthermore, the actual success of the meeting was, to a very great extent due to the staff of the Dipartimento di Biologia Ambientale, Università di Siena. They include Biondi G., Borghi L., Casini S., Ciampolini F., Faleri C., Marchetti P., Milanesi C., Perini C., Romi F., Rosi E., Rossi L., Silvietti A., Silvietti A., Tanganelli A., Valentini M., Vanni F.

We want to express special thanks to Prof. J.L. Van Went, who during his sabbatical in Siena, assisted in the organization of the meeting and to Prof. H.F. Linskens for very useful advice during the editing of the Proceedings.

To each of the Institution, Industries and Colleagues alike, we extend many thanks and congratulations.

The Editors

Contents

FROM POLLINATION TO FERTILIZATION

FROM OVULE TO SEED

Gene-Expression and Transcription

Differential Gene-Expression During Microsporogenesis with *Nicotiana tabacum*

J.A.M. Schrauwen,   M.W.M. Derks,   P.F.M. de Groot,   W.H. Reijnen,
M.M.A. van Herpen & G.J. Wullems
Department of Experimental Botany
Research Group Molecular Plant Physiology
University of Nijmegen
Toernooiveld
NL-6525 ED Nijmegen
The Netherlands

INTRODUCTION

The changes in gene-expression during the development of the plant result in the formation of new transcription and translation products and in formation of other cell types.

Pollen, which plays a key roll in the fertilization process, is a typical product of differential gene- expression occuring in a mature plant. The formation of pollen from vegetative tissue is concomitant with a series of processes by which type specific gene products, like mRNAs, are formed. These transcription products can be distinguished by specificity and function in relation to moment of activity and phase development.

Mature pollen contains at least 10% pollen specific mRNA (Mascarenhas et al, 1984; Stinson et al 1987) that may have a function during the processes of germination and pollen tubegrowth. The latter processes require protein synthesis to occur, right from the start (Linskens et al, 1970), while transcription is not neseccary during the first hours of pollentube growth (Capkova et al, 1983).

During pollen development several pollen specific processes take place, with features like meiosis, mitosis, synthesis of compounds for pollen-wall and -coat, appearance and disappearance of several compounds like starch (Kyo and Harada,1986), phytic acid (Lin and Dickenson, 1984) and enzymes (Linskens, 1966, Zarsky et al, 1985). These pollen specific processes need specific transcription produts

(Stinson et al,1987; Goldberg, 1987) , which may be present only temporaryly during a certain developmental phase.

In this paper we describe the formation of transcription products, mRNAs, during microsporogenesis in *Nicotiana tabacum*. This analyses is carried out to relate morphological and physiological development during microsporogenesis on the one hand and the pattern of gene expression on the other hand.

RESULTS AND DISCUSSION

Morphological development

To relate physiological and morphological development with occurence of specific transcription products one requires a good discription and definition of the various developmental phases. To do so we chose parameters that were representative for these phases and could be measured quickly. The microspore phase of *Nicotiana tabacum* c.v.Petit Havanna was characterised by the sizes of anthers and flowerbuds together with the position and shape of the vegetative and generative nuclei. The nuclei stage was established by staining with 4,6-Diamino-2-phenylindole (DAPI) (Vergne et al, 1987).

With these parameters we could demonstrate a correlation between bud-length and nuclei development in the microspores. However, the population of microspores present in a single anther or filament was not homogenous. The maximum number of cells in th same stage, present in five anthers of a single bud, varies between 42 and 88% (fig 1) depending on the phase of development.

Microspore-gathering

Between 100 -250 depending anthers were collected from buds of equal size and gentle homogenised in 4 ml mannitol 0.3M in DEP 0.1%. The suspension was filtered over nylon sieve (120 µm), centrifuged (150 x g, 1 min) and washed once more. The pellet contained the

M I C R O S P O R O G E N E S I S   I N   *N I C O T I A N A   T A B A C U M*

morphological figure	description	length flower bud	anthers	% of cell: in stage
	POLLEN MOTHERCELLS	8 - 9	2	74,3
meiosis				
	TETRADES	10 - 11	2½	46,9
	MICROSPORES young, with integuments	12 - 16	3½	58,6
	polar nucleus	18 - 22	4	42,6
mitosis				
	two polar nuclei	23 - 24	4½	67,8
	two central nuclei	26 - 36	4½ - 5	59,1
	generative nucleus with moon shape	40 - 48	4½ - 5	64,3
	complete developed nuclei	50 - 52	4½ - 5	88,0

Fig.1: The relation between the sizes of anthers and flower buds and the morphological microspore development of *Nicotiana tabacum*.

microspores and was free from vegetative tissue. A homogeneous microspore suspension was obtained by centrifugation (450 x g, 5 min) of the microspore-population over a discontinuous percoll (50-70%) gradient (Kyo and Harada, 1986).

Transcription product analyses

The microspores were immediately frozen in liquid nitrogen. After standard RNA extraction procedures, the translation capacity was measured in a rabbit reticulocyte system with [35-S]-methionine as a precusor.

The analyses of the transcription and translation products present in microspores from the different phases demonstrated qualitatively and quantitatively differences with respect to the synthesized proteins. The total amount of RNA present in microspores increased from 0.8 to 5 µg per bud from young to completelky developed microspores. The quantitative incorporation of [35-S]-methionine per µg RNA remained almost constant during the early phases and doubled during the development of the generative nucleus. This suggestes an increase in the amount of total and mRNA during microspore development .

The qualitative differences, estimated by two dimensional gelelectrophoreses, revealed a tremendous increase in the number of translation products after mitosis and appearance of certain new proteins at several stages of development. These results indicate continuous RNA syntheses and in addition differential mRNA synthesis during anther development of *Nicotiana tabacum*. These mRNAs may have a function with the development of the microspores and in the mature pollen during the fertilization processes.

REFERENCES

Capková V, Ríhová L, Hrabetová E & Tupy J (1983) The effect of actinomycin D on the gorwth of tobacco pollen tubes. In: Fertilization and embryogenesis in ovulated plants. VEDA, Bratislava, Czechoslovakia, pp 137-139

Goldberg RB (1987) Emerging patterns of plant development. Cell 49:298-300

Kyo M & Harada H (1986) Control of the developmental pathway of tobacco pollen *in vitro*. Planta 168:427-432

Lin JJ & Dickinson DB (1984) Ability of pollen to germinate prior to anthesis and effect of desiccation germination. Plant Physiol 74:746-748

Linskens HF (1966) Die Änderung des Protein- und Enzym-Musters während der Pollenmeiose und Pollenentwicklung. Physiologische Untersuchungen zur Reifeteilung. Planta 69:79-91

Linskens HF, Schrauwen JAM & Konings RNH (1970) Cell-free protein synthesis with polysomes from germinating *Petunia* pollen grains. Planta 90:153-162

Mascarenhas NT, Bashe D, Eisenberg A, Willing RP, Xiao CM, Mascarenhas JP (1984) Messenger RNAs in corn pollen and protein synthesis during germination and pollen tube growth. Theor Appl Genet 68:323-326

Stinson JR, Eisenberg AJ, Willing RP, Pe ME, Hanson DD & Mascarenhas

JP   (1987) Genes expressed in the male gametophyte of   flowering
    plants and their isolation. Plant Physiol 83:442-447
Vergne P, Delvallee I & Dumas C (1987) Rapid assessment of microspore
    and   pollen development stage in wheat in wheat and maize   using
    dapi and membrane permeabilization. Stain Technology 62:299-304
Zársky V, Capková V, Harbetová E & Tupy J (1985) Protein changes
    during pollen development in *Nicotiana tabacum* L. Biologia
    Plantarum 27:438-444

Variations of Nucleolar Ultrastructure in Relation to Transcriptional Activity During G_1, S, and G_2 Periods of Microspore Interphase

M.C. Risueño, P.S. Testillano and M.A. Sánchez-Pina
Centro de Investigaciones Biológicas.- C.S.I.C.
C/Velázquez 144. 28006 Madrid. SPAIN

INTRODUCTION

The nucleolus is a highly dinamic organelle. Variations in the synthesis and/or processing of rRNA produces redistribution of the structural nucleolar components, in both physiological and experimental conditions (Goessens 1984, Hadjiolov 1985, Olmedilla et al. 1987, Risueño and Medina 1986).

Plant germinal cells undergo large variations of activity during the gametogenic process (Mascarenhas 1975). Changes in the fine structure of the nucleolus, at some stages of both macro and microgametogenesis, have been reported by our laboratory in Pisum, Onion and Scilla (Medina et al. 1983a, Sánchez-Pina 1983, Sánchez-Pina et al. 1980, Fakan and Hernández-Verdun 1986, Risueño and Medina 1986). Ribosomal chromatin in plants is localized in both fibrillar centres (FCs) and the dense fibrillar component (DFC); the transcription occurs in the latter while FCs contain the non-transcribing rDNA (Risueño and Medina 1986). DFC contains also growing transcripts and primary transcripts and it is formed by 10 nm fibrils. The granular component (GC) contains 15-20 nm granules which area preribosomal particles.

Other nucleolar components are the vacuoles. They are seen as zones of low electron density but depending on the transcriptional activity, they seem to have different funtcions (Risueño and Medina 1986, Fakan and Hernández-Verdun 1986, Moreno Díaz de la Espina et al. 1980). There are two principal types of FCs in plants nucleoli, related to nucleolar activity: Homogeneous (Hom FC) and Heterogeneous (Het FC). Hom FCs are present in active nucleoli. They are seen as clear areas with 10 nm fibres of decondensed chromatin and they are small and numerous. Het FCs are found in inactive nucleoli. They have large inclusions of condensed chromatin and are bigger and less numerous than the Hom FCs (Medina et al. 1983a, Risueño et al. 1982). Intermediate FCs have been described as the evolution from one type to another, in intermediate active states of the nucleolus. They have condensed chromatin inclusions of variable size inmersed in similar material to those of the Hom FCs (Fakan and Hernández-Verdun 1986, Medina et al. 1983b).

A little data was reported about the behaviour of the plant nucleolus in relation to the interphase of the cell cycle (Arquiaga 1985) but this has not been performed during gametogenesis. However, it would be of very relevant interest to know the changes of nucleolar structure related to the G_1, S and G_2 periods of interphase. Consecuently this work deals with the study of the nucleolar evolution during interphase microspore. We have also analyzed the evolution of the nucleolar chromatin organization.

MATERIAL AND METHODS

The materials used were anthers of Scilla peruviana selected during microspore interphase (from tetrad until the prophasic stage). They were submitted for conventional electron microscopy (glutaraldehyde-osmium fixing and counterstained with uranyl and lead), and for several citochemistry techniques: EDTA-regressive stain (Bernhard 1969) for ribonucleoproteins, osmium ammine (Gautier 1976) for DNA, acetylation in block and uranyl stained sections (Wassef et al. 1979) for nucleic acids and condensed chromatin, EDTA after acetylation for granular and fibrillar RNP visualization, and Ag-NOR technique (Medina et al. 1983c) for argyrophilic proteins of NOR chromatin (DFC and FCs).

RESULTS AND DISCUSSION

In previous cytophotometric studies based on the analysis of variations in DNA amount, we have determined during interphase microspore: a short G_1 followed by a long S period which is superimposed with G_2 period at the end. S period has been divided in S_1, S_2 and S_3 because the DNA synthesis occurs in two moments: at the beginning, and at the end of this period (Risueño et al. 1985).

Evolution of the nucleolus and FCs in relation to transcriptional activity

The microspore nucleolus reactivates during interphase undergoing striking changes on its fine structure, which are representated below:

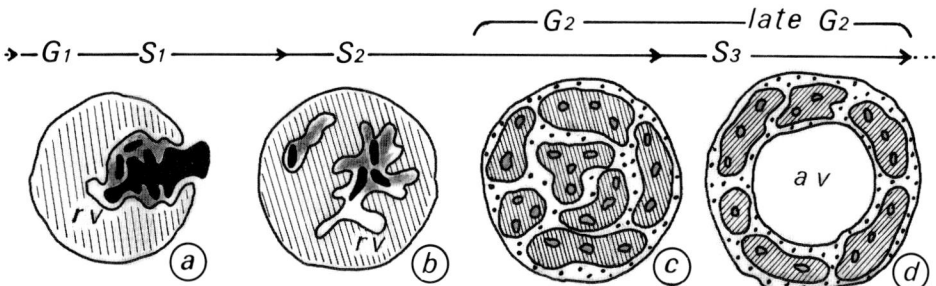

Fig. 1 : Nucleolar behaviour in relation to rRNA activity.
a) Het FCs of inactive nucleoli during G1 and S1 periods; b) Intermediate FCs in connection with vacuoles in reactived nucleoli during S2; c) Hom FCs in active nucleolus; d) Central vacuole of activity not related with the FCs, in highly active state, late G2; DFC; GC; Condensed chromatin inclusions; Decondensed chromatin fibres; rv Reactivation vacuoles; av Active vacuoles.

In a very early G_1, the nucleolus is in a reorganizing state and is formed by irregular fibrillar masses surrounding the chromatin of the nucleolar organizing region (NOR) which shows a structure of het type. During middle G_1 the nucleolus is already formed and it is only composed by DFC showing a great Het FC with large condensed chromatin inclusions (Fig. 1a) inside a medium-electron dense material of 10 nm decondensed chromatin fibres. These Het FCs are seen in connection with low-electron zones, called nucleolar reactivation vacuoles in somatic cells (Risueño and Medina 1986, Fakan and Hernández-Verdun 1986), which at this phase show perichromatin-like-granules (plg) in theirs interiors (Fig. 2). During S_2 period the reactivation vacuoles increase and the FCs are always connected to them. The FCs become of intermediate type having great size and condensed chromatin inclusions of variable measurements (Fig. 1b).

At the end of S_2 period reactivation vacuoles and intermediate FCs become smaller, and they contain few chromatin inclusions while some Hom FCs begin to be seen (Fig. 3). In G_2 period the nucleolus displays a typical active morphology with many Hom FCs and abundant GC (Fig. 1c). At the end of G_2 period the nucleolus is larger and highly active with a central vacuole (Fig. 1d) and small and numerous Hom FCs, showing a patent GC (Fig. 4). The fine structure of this vacuole of activity seems similar to that of reactivation one. However, active vacuoles are never seen in connection with FCs. Similar phenomena were reported in other plant cells (Fakan and Hernández-Verdun 1986, Moreno Díaz de la Espina et al. 1980). After Ag-NOR technique the DFC, Hom FCs and the fibrillar material of Intermediate FCs show silver precipitate while the condensed chromatin inclusions and nucleolar vacuoles are devoid of silver (Fig. 5).

After our data it can undoubtedly be assumed that in the microspore, when the nucleolar activity increases, the number of FCs also increases, while their size becomes smaller being discrete entities named Hom FCs. However, in low active states, intermediate FCs are not discrete but connected by a vacuole. These changes are dependent on the activity and cannot be attributed to the effects of fixatives as was claimed by Motte et al. (1988).

Localization of chromatin in the nucleolus

After the acetylation method that enhances condensed chromatin, no other inclusions than those of Het FCs and intermediate FCs can be seen in the microspore nucleolus. 20-25 nm packed chromatin fibres appear very contrasted (Fig. 6). No other intranucleolar chromatin has been reported in the nucleoli of plants, as occurs in animal cells (Fakan and Hernández-Verdun 1986). This chromatin appears as dense intranucleolar clumps and is not considered as NOR chromatin because it does not contain rDNA.

Different condensation states of rDNA chromatin

After osmium ammine reaction specific for DNA, several states of chromatin condensation can be observed in the FCs (Fig. 7). Packed chromatin fibres of different sizes as well as 3 nm decondensed chromatin can be seen. These fibres would be in a non-nucleosomal structure but in an inactive state. This is a peculiar state of ribosomal chromatin that is in an extended state but inactive in transcription (Derenzini et al. 1983, Fakan and Hernández-Verdun 1986, Risueño and Medina 1986). The protein(s) responsable for this extended state would be acid proteins and with silver affinity (Fig. 5) (Medina et al. 1983c, Medina et al. 1986).We have determined the changes in microspore nucleolar activity are accompanied by variations in nucleolar chromatin organization.

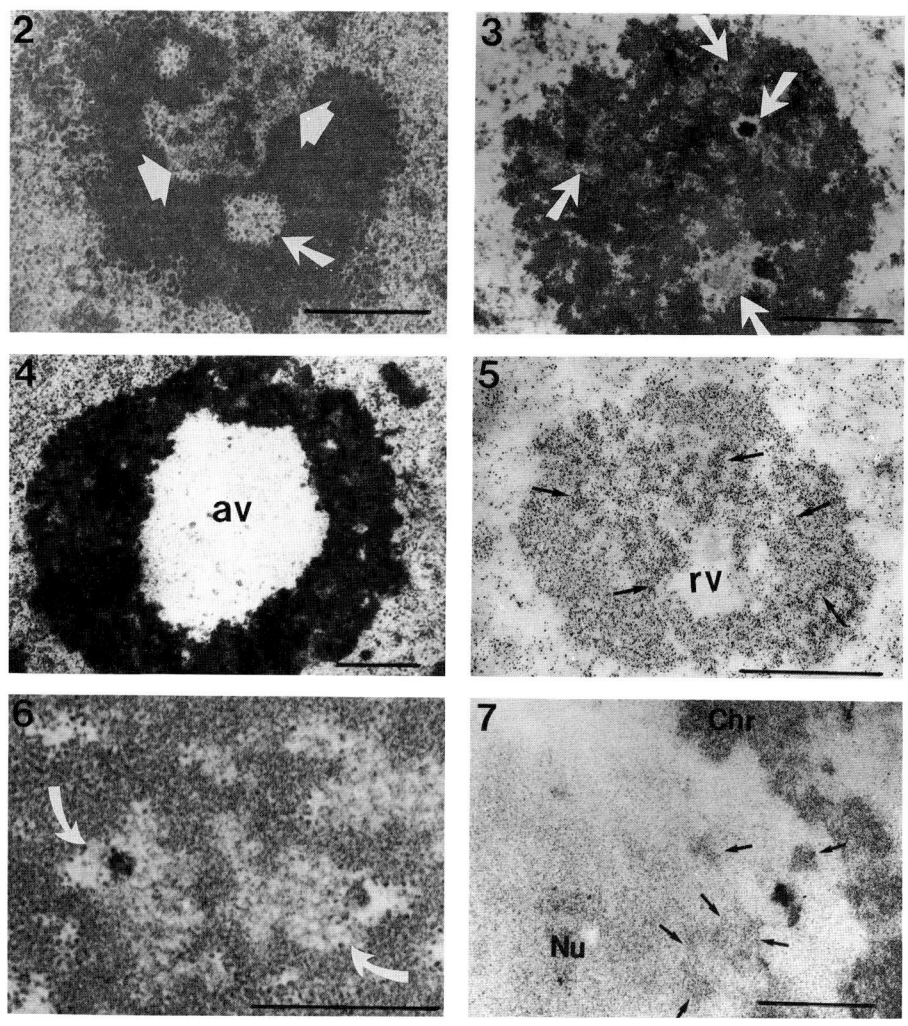

Figs. 2 to 7: Nucleolus during microspore interphase.
Fig.2: Nucleolus at Gl-Sl with Het FCs (arrowheads) in connection
with rv (arrows), see plg in its interior. Gluta-Os fixation.
Fig.3: Middle S2 period. Intermediate FCs (arrows) and rv become
smaller, as well as the chromatin inclusions, which are enhanced
after acetylation and uranyl staining. Fig.4: Late G2 period.
Highly active nucleolus with a central vacuole (av). Fig.5: Middle
S2 period. Ag-NOR technique. DFC and Hom FCs (thin arrows) are
preferentially contrasted. Fig.6: Intermediate FCs show 8-11 nm
chromatin fibres (arrows). After acetylation and uranyl staining
their inclusions are the only intranucleolar condensed chromatin.
Fig.7: Intermediate FCs after osmium ammine reaction. They show
contrasted chromatin inclusions containing differently packed DNA
fibres (arrows). Bars in figs. 2-5: 1 μm; in figs. 6-7: 0,5 μm.

ACKNOWLEDGMENTS
We thank Mr. C Almarza and Miss T. Cortezón their skilful technical
assistance. This work has received finantial support from the
project nº 1-179-2 ID181 CSIC/CAICYT

REFERENCES

Arquiaga MC (1985) Análisis citofotométrico y ultraestructural de la interfase de la microspora de Hyacinthoides non-scripta. Tesina licenciatura. Fac. CC.Biológicas Univ. Complutense Madrid

Bernhard W (1969) A new staining procedure for electron microscopical cytology. J Ultr Res 27: 250-265

Derenzini M, Pession A, Betts-Eusebi CM, Novello F (1983) Relationship between the extended non nucleosomal intranucleolar chromatin "in situ" and ribosomal RNA synthesis. Exp Cell Res 145: 127-143

Fakan S, Hernández-Verdun D (1986) The nucleolus and the nucleolar organizer regions. Part 4B and 5. Biol Cell 56: 189-206

Gautier A (1976) Ultrastructural localization of DNA in ultrathin tissue sections. Inter Rev of Cytol 44: 114-170

Goessens G (1984) Nucleolar structure. Int Review Cytol 87: 107-158

Hadjiolov AA (1985) The nucleolus and ribosome biogenesis. In: Cell Biology Monographs 12. Sgringer Verlag Wien NY

Mascarenhas JP (1975) The biochemistry of angiosperm pollen development. The Botanical Review 41: 259-314

Medina FJ, Risueño MC, Rodíguez-García MI, Sánchez-Pina MA (1983a) The nucleolar organizer (NOR) and fibrillar centres during plant gametogenesis. J Ultr Res 85: 300-310

Medina FJ, Risueño MC, Moreno-Díaz de la Espina S (1983b) 3-D reconstruction and morphometry of fibrillar centres in plant cells in relation to nucleolar activity. Biol Cell 48: 31-38

Medina FJ, Risueño MC, Sánchez-Pina MA, Fernández-Gómez ME (1983c) A study on nucleolar silver staining in plant cells. The role of argyrophilic proteins in nucleolar physiology. Chromosoma 88:149-155

Medina FJ, Solanilla EL, Sánchez-Pina MA, Fernández-Gómez ME, Risueño MC (1986) Cytological approach to the nucleolar functions detected by silver staining. Chromosoma (Berl) 94: 259-266

Moreno-Díaz de la Espina S, Medina FJ, Risueño MC (1980) Correlation of nucleolar activity and nucleolar vacuolation in plant cells. Eur J Cell Biol 22: 724-729

Motte P, Deltour R, Mosen H, Bronchart R (1988) Three-dimensional electron microscopy of the nucleolus and nucleolus-associated chromatin (NAC) during early germination of Zea mays L. Biol Cell 62: 65-81

Olmedilla A, Risueño MC, Fernández-Gómez ME (1987) Action of α-amanitin, ethidium bromide, and cycloheximide on rRNA metabolism in plant cells. Biol Cell 60: 183-192

Risueño MC, Arquiaga C, Sánchez-Pina MA (1985) Determination of the microspore interphase of Hyacinthoides non-scripta. Rev Biol Cel S:29

Risueño MC, Medina FJ (1986) RBC. Cell biol Rev 7: 1-163. Barberá-Guillem (ed) Springer Verlag

Risueño MC, Medina FJ, Moreno-Díaz de la Espina S (1982) Nucleolar fibrillar centres in plant meristematic cells: ultrastructure, cytochemistry and autoradiography. J Cell Sci 58: 313-329

Sánchez-Pina MA (1983) Estudio de las proteínas argirófilas durante el ciclo nucleolar de las células reproductoras masculinas en Liliáceas. Tesis doctoral. Fac. CC. Biológicas. Univ. Complutense. Madrid

Sánchez-Pina MA, Risueño MC, Rodríguez-García MI (1980) Localization of the acid phosphatasic activity in plant cells nucleoli. Cell Biol Int Rep 4: 93-104

Wassef M, Burglen J, Bernhard W (1979) A new method for visualization of preribosomal granules in the nucleolus after acetylation. Biol. Cell 34: 153-158

Pre and Post Anthesis Gene Expression in Maize Pollen

Carla Frova
Dipartimento di Genetica e di Biologia dei Microrganismi
Università di Milano
Via Celoria 26
20133 Milano
ITALY

Introduction

The male gametophytic generation of higher plants can be divided into two distinct phases: 1) pollen formation (microsporogenesis), which takes place within the anther, and 2) pollen function, which begins with the deposition of the grains on the stigma surface and ends with fertilization. Anthesis marks the transition from phase 1 to 2. Microsporogenesis is clearly the more complex of the two phases in terms of development and functions: between the end of meiosis and anther dehiscence the microspore undergoes profound changes which include 1 or 2 mitotic divisions and the synthesis of the cell wall. Pollen function is simpler: apparently all the grain has to do is to put out a tube which grows through the stylar tissues in order to reach the embryo sac and discharge the sperm cells. During both phases pollen is in intimate contact with a sporophytic tissue (the tapetal cells and the style in phases 1 and 2 respectively) and its correct functioning is very dependent on interactions with them. Considered in toto the male gametophytic phase presents a series of differentiation steps: meiosis, mitosis, tube growth. Yet developmentally it is a very simple system as compared with the sporophyte and hence offers unique opportunities to study developmental regulation of gene expression and the functional interactions with quite different tissues. With regard to the first point a large body of evidence shows that post meiotic gene expression is quantitatively considerable. Moreover a large fraction of these genes are also expressed in the sporophyte. The phenomenon, referred to as "haplo-diploid transcription" has important implications with regard to the efficacy of gametophytic selection and its role in the high evolutionary rate of higher plants (for an exhaustive review of these aspects see Ottaviano et al., this volume). However, most of the evidence has been obtained from the analysis of mature pollen only, while the developmental regulation of gene expression has been the subject of relatively little research. Nevertheless, the following indications, which are summarized in Fig.1, allow at least a broad distinction to be made between pollen formation and pollen function in terms of gene expression: i) data on RNA and protein synthesis (Mascarenhas 1975, Hoeckstra and Bruisna

16

1979) indicate that the overall pollen metabolism is more pronounced
during microsporogenesis than during tube growth; ii) induction of
mutations in mature pollen does not disturb its function; however, in
the next generation mutated gametes do not complete their formation
and are eliminated (Rick and Kush 1969); iii) isozyme and mRNA
analysis, which prove extensive gametophytic gene expression, are
referred to microsporogenesis. However :

iv) results from low temperature tolerance selection experiments in
tomato (Zamir et al. 82, Zamir and Vellejos 83) show that the genetic
factors involved in ''tolerance'' are expressed primarily during
pollen function; v) in corn, studies on pollen competitive ability
in terms of tube growth rate show a wide variability of haploid origin
of this trait (Sari Gorla et al. 1975); vi) gametophytic (Ga) factors
exert their influence and interact with the stylar genotype in post
pollination stages (Bianchi e Lorenzoni 1975); vii) a study of
mutants affecting endosperm development and expressed also in the male
haploid phase (de-ga) has shown that some act before, some after
anther dehiscence and some in both phases (Ottaviano et al 1988).

Thus, even though most of the evidence indicates that pollen
formation is the phase at which the majority of the gametophytic genes
are active, a silent role of the haploid genome during pollen function
is to be excluded. However, apart from these general conclusions,
almost nothing is known about the modulation of gene expression which
is expected in relation to specific stages. In particular, are some
genes activated only in one stage (uninucleate, bi or trinucleate)?
Are there differences in gene expression between early and advanced
tube growth stages ? Here an approach serving to provide an answer to
these questions is described and preliminary results are illustrated.

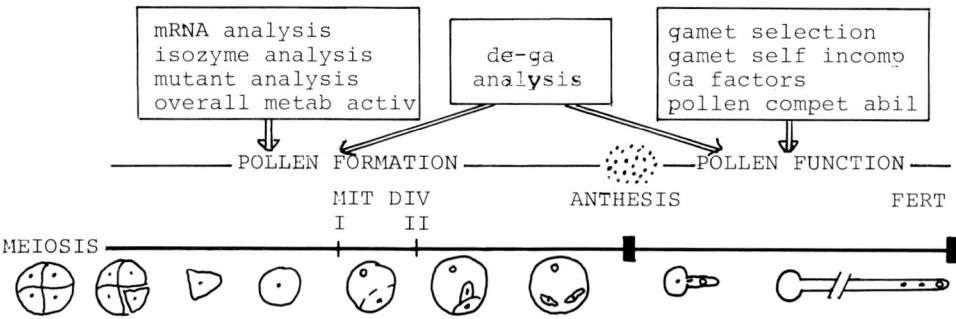

Fig. 1 : Maize pollen life-cycle. Arrows indicate the stage to which
data reported in boxes are referred.

Experimental approach

Microsporogenesis. Microspores can be isolated from the surrounding anther tissues with virtually no sporophytic contamination (Frova et al. 1987). Stages are easily recognizable at the microscope and enough material can be collected even at the very immature stages (uninucleate) for enzymatic or protein analyses. The amount (2-5 mg) however, is insufficient for poly(A)RNA analysis. An alternative approach is the analysis of multimeric enzymes, where the haploid or diploid genetic control of the enzymatic activity of the gametophyte can be established on the basis of the electrophoretic pattern exhibited by pollen (Sari et al. 1986),or of proteins coded by genes transcribed only in response to specific stimuli (light, temperature, chemicals etc.). Among these, genes encoding for heat shock proteins (HSPs) are of great interest for the dissection of the male gametophytic phase, in that they are apparently silent in mature pollen of higher plants (Mascarenhas and Altschuler 1983, Schrauwen et al. 1986, Cooper et al. 1984, Frova et al. 1988), but at least some of them are inducible in developing pollen (Frova et al. 1987). In the following section data regarding genes coding for ADH-1, GOT-1, CAT-1, B-GLU-2 and HSPs will be discussed.

Pollen function. In this phase a direct analysis of isolated pollen is feasible only on *in vitro* germinating grains. This procedure, however, is not satisfactory for two reasons: a) even in the best growth conditions tube length of maize pollen *in vitro* hardly reaches a few millimeters as compared to 20-25 cm. *in vivo*; b) pollen-style interactions, which play a fundamental role in tube elongation, would be completely excluded. Therefore, since pollen tubes cannot be isolated from the stylar tissues they are growing through, an indirect approach was developed, based on the comparative analysis of pollinated and non pollinated silks. *In vitro* germinating grains were analyzed for the early auxotrophic tube growth stages. Three enzymatic activities, ADH-1, GOT-1, CAT-1 we're analyzed and the appropriate electrophoretic variants used as female and pollen source, so that the tube and stylar contribution to activities found in pollinated vs. non pollinated silks could be discriminated. Silks are cut 0.5 cm. above emergence from the husks, and heavily pollinated. 3.5 hours are then left for germination and tube growth. Since maize pollen germinates within 30 min. from the deposition on the silk and tube growth rate is constant and approx. 8 mm per hour, it is expected that after 3.5 hours tube tips have reached a distance of 2-2.5 cm under the silk emergence level. This region of the silks is analyzed. The portion of

the silks where the pollen grains are deposited is analyzed
separately. In Table 1 the two portions are indicated as "base" and
"tip" respectively.

Pre-anthesis gene expression

Even within the small sample of genes considered, there is
variability in the timing of gene expression. In particular, of the
six genes coding for enzymes, two, Adh-1 and B-Glu-2, are transcribed
in all stages, from uninucleate microspore to mature grain. The
others, Got-1 and three Cat genes, are expressed only in some stages.
In the case of catalase a shift of expression between an as yet
unidentified, possibly Cat-3, gene and Cat-1 is observed at first
microspore mitosis (Frova C. manuscript in preparation). This pattern
closely resembles the behaviour of Cat-1 and Cat-2 during the early
development of germinating seeds (Tsaftaris and Scandalios 1983). A
third catalase gene, Cat-4, is weakly expressed from the uninucleate
stage until shortly after second microspore mitosis. Got-1 pattern is
peculiar: a pronounced enzymatic activity is detected in all pollen
formation stages; however, the gametophytic Got-1 gene is expressed
only after first microspore mitosis. Prior to this stage the genetic
control of this isozyme is clearly sporophytic, as indicated by the
presence of a sharp heterodimeric band (Frova et al. 1987).

HSP expression pattern during the male haploid phase in maize has
been described in detail by Frova et al. (1988). To the purpose of
this discussion the main points are: i) before anthesis heat-shock
genes are switched off. No HSPs are inducible in mature and in in
vitro germinating pollen. ii) during microsporogenesis several HS
genes are expressed. Of these only two (HSPs 84 and 72kD) are active
in all stages. The others are stage specific: most (94, 74, 58, 46 kD)
are expressed before the first mitotic division, while one (HSP 66kD)
is expressed after.

Considering all data together (see Table 1) a few conclusions can
be drawn. First, gene expression appears to be related to
microsporogenesis stages. Roughly 70% of the genes considered here
show some stage specificity. This is not unexpected when the diversity
of processes occurring during pollen formation and the different
functional needs of the developing grain are taken into account.
Second, first microspore mitosis seems to be a crucial time for gene
reprogramming: eight out of ten of the stage specific genes are
switched on or off in the transition from uninucleate to binucleate
microspore. The exact meaning of this is not clear at this point.

Table 1: Summary scheme showing enzyme and hsp gene expression during maize pollen life-cycle. Full bars indicate the stages at which the genes are expressed. Dotted bar indicates sporophytic genetic control

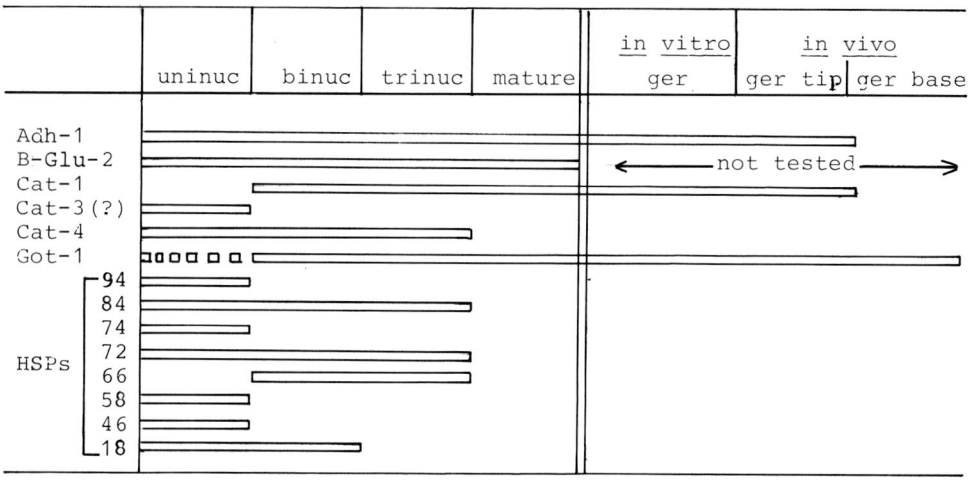

However, it is a fact that the first mitotic division corresponds to a fundamental step in microsporogenesis, that is the end of a relatively totipotent state in uninucleate microspores and the establishment of different developmental programs for the vegetative and the generative cells. It is unlikely that such a differentiation process could take place without significative changes in gene expression. Finally, the GOT-1 case suggests that, even though gametophytic independence is a well documented reality, pollen-tapetum interactions should not be underestimated expecially in the early post meiotic stages.

Post anthesis gene expression

A pronounced activity of the enzymes analyzed (ADH-1, GOT-1, CAT-1) is detectable in mature non germinating pollen and during early stages of both _in vitro_ and _in vivo_ (pollinated silks, tip region) germinating grains. However, when the base region of pollinated silks is analyzed, different results are obtained. In the case of ADH-1 and CAT-1 no gametophytic activity is detectable, while a clear band corresponding to the pollen activity region is visible for GOT-1 (Frova manuscript in preparation). These very preliminary results seem to indicate that gene expression during tube growth is less than in developing microspores or mature grains, even though data on more genes are needed to draw meaningful conclusions. What is reported above shows also that the analysis of in vitro germinating pollen may be misleading and that gene expression during pollen function has to be examined _in vivo_. Finally a short comment on the methodological approach . It is premature to conclude that the Got-1 gene is expressed in elongating pollen tubes, since the presence of activity could be due to migration of the preformed enzyme into the tube growing region. Nevertheless the approach here described appears promising and susceptible of further applications also for the study of pollen-style interactions.

<u>Literature</u>

Bianchi A, Lorenzoni C (1975) Gametophyte factors in Zea mays. In: Mulcahy DL (ed) Gamete Competition in Plants and Animal. North Holland, American Elsevier, p 257

Cooper P, Ho TD, Hauptmann RM (1984) Tissue specificity of the heat shock response in maize. Plant Physiol 75: 431-441

Frova C, Taramino G, Binelli G (1988) Heat shock proteins during pollen development in maize. (Submitted)

Frova C, Binelli G, Ottaviano E (1987) Isozyme and hsp gene expression during male gametophyte development in maize. In: Rattazzi MC, Scandalios JG, Whitt GS (eds) Isozymes: Current Topics in Biological and Medical Research, vol 15. Alan R Liss, New York, p 97

Hoeckstra FA, Bruinsma J (1979) Protein synthesis of binucleate and trinucleate pollen and its relationship to tube emergence and growth. Planta 146: 559-566

Mascarenhas JP (1975) The biochemistry of angiosperm pollen development. Bot Rev 41: 259-314

Mascarenhas JP, Altschuler M (1983) The response of pollen to high temperatures and its potential applications. In: Mulcahy DL, Ottaviano E (eds) Pollen: Biology and Implications for Plant Breeding, Elsevier, New York, p 3

Ottaviano E, Sari Gorla M, Frova C, Pe' E (1988) Male Gametophytic Selection in Higher Plants. This volume p

Ottaviano E, Petroni D, Pe' E (1988) Gametophytic expression of genes controlling endosperm development in maize. Theor Appl Genet 75: 252-258

Rick CM, Khush GS (1969) Cytogenetic explorations in the tomato genome. In: Bogart R (ed) Genetics Lectures 1 p 45

Sari Gorla M, Ottaviano E, Faini D (1975) Genetic variability of gametophyte growth rate in maize. Theor Appl Genet 46: 289-295

Sari Gorla M, Frova C, Binelli G, Ottaviano E (1986) The extent of gametophytic-sporophytic gene expression in maize. Theor Appl Genet 72: 42-47

Schrauwen JAM, Reijnen WH, Deleeuw HCGM, Van Herpen MMA (1986) Response of pollen to heat stress. Acta Bot Neer 35: 321-327

Tsaftaris AS, Scandalios JG (1983) The multi-locus catalase gene-enzyme system of maize. In: Rattazzi MC, Scandalios JG, Whitt GS (eds) Isozymes: Current Topics in Biological and Medical Research, vol 7. Alan R Liss, New York, p 59

Zamir D, Tanksley SD, Jones RA (1982) Haploid selection for low temperature tolerance of tomato pollen. Genetics 101: 129-137

Zamir D, Vellejos EC (1983) Temperature effects on haploid selection of tomato microspores and pollen grains. In: Mulcahy DL, Ottaviano E (eds) Pollen: Biology and Implications for Plant Breeding. Elsevier, New York, p335

Testing the Genomic Instability During the Haploid Life Cycle

G. Kovács - B.Barnabás
Agricultural Research Institute of
the Hungarian Academy of Sciences,
2462, Martonvásár, Hungary

SUMMARY

In this paper, we investigated the bz monogenic marker. The question
was, whether stress treatments during pollen selection can induce
transposon activity in mature pollen. The bz markers are expressed
both in the sporophytic and gametophytic life cycle. The changes in
their autofluorescency and the new genotypes in the offsprings suggest
that transposon activity can be induced during the gametophytic life
cycle.

INTRODUCTION

Recent studies in molecular genetics of pollen have demonstrated, that
approximately 70% of the genomic genes are also expressed during both
the sporophytic and the gametophytic life cycle (Tanksley et al. 1986,
Willing and Mascarenhas 1984, Mascarenhas et al. 1986, Sari-Gorla et
al. 1986). These results provide a firm basis for elaborating methods
of pollen selection. Research in pollen selection has shown so far,
that in different species, one can select for more resistance factors
against environmental stress at the same time, also on pollen level
(e.g. low/high temperature, salinity, agrochemicals, pathotoxins, havy
metals, etc.) (Zamir et al. 1982, Zamir and Vallejos 1983, Sacher et
al. 1983, Smith and Moser 1984, Searcy and Mulcahy 1985, Frova et al.
1986). According to these results, pollen selection can be combined
effectively with traditional breeding methods, and that the so-called
"double selection" is definitely more effective than the traditional
breeding procedures (Pfahler 1983, Barnabás and Kovács 1988).
In our maize pollen selection experiments, we have observed some unexpected results indicating mutation during pollen treatments (Barna-

bás and Kovács 1988), driving our attention to transposable elements.
Since the basic discoveries of McClintock (1964), we have a firm and
reliable body of information concerning different transposons on
sporophytic level (Fendoroff 1983, Nevers et al. 1986). On the other
hand, little is konwn about the gametophytic activities of transposons,
and those published are mostly concerned with one of the more interesting ones, the functioning of the Mutator (Robertson 1985). In this
paper, we will concentrate on the question, whether stress treatments
during pollen selection can induce transposon acitivity in mature
pollen grains.

MATERIALS AND METHODS

As a model species, maize (Zea mays L.) was used in the experiment.
The following genotypes were tested:
1206; 76-1605-2 inbred line genotype: c sh bz wx: A A R y
1202; 108o B-9 x 1088 C-2, containing Ac/Ds transposable elements.
Type of genetic changes on sporophyte level: sh bz Bz
I207; the MF-18 line of the s5 generation was developed in our laboratory. The line contains some unidentified transposons, and on
sporophyte level, bz - Bz genetic changes can be experienced. In the
experiment, we concentrated only on bz from among more markers.
The plants were raised in the field. Pollen was collected at the time
of shedding (about 10 a.m.). After collection, the pollen moisture
content was calculated at the fresh weight basis, then the pollen
samples were dried to 17% water content by a flotation method (Barnabás
and Rajki 1981). They were stored at $4^{o}C$ for 1 day in sealed ampoulas.
The autofluorescence of the bz genotype pollen grains was examined by
means of fluorescent microscope (OPTON, Ultraphot III), after which
the same sample was stained with fluorescein diacetate, in order to
determine the number of viable pollen grains (Heslop-harrison 1970).
The total number of pollen grains was counted in the same sample, by
means of light microscope.
5 ears of the 1206 line were pollinated both with treated and fresh
pollen of line 1207. After harvest, we examined the seed set percentage
and the number of bz and Bz genotype seeds and their percentage, respectively. The results were evaluated by analysis of variances.

RESULTS AND DISCUSSION

The Study of Pollen Populations

The effect of the chosen monogenic marker (bz) can be directly tested
in pollen populations, as the bz gene is also expressed during the
life cycle of the pollen. Expressed bz genes are autofluorescent,
whereas pollens with a normal Bz gene are not. Thus, both different
genotypes can be well distinguished, and therefore genetic changes
in the pollen population can be tested. Table 1. demonstrates changes
of the bz gene, induced by stress treatment of pollens grains.
Table 1.
Stability of the bz gene after stress treatment of pollen grains

Genotype	Pollen grains Total No	living %	bz %	Bz %
1. 1201				
controll	12080	93,8	100	-
treated	1001	51,9^{XXX}	100	-
2. 1202 (Ac/Ds)				
controll	2070	95,7	87,3	12,6
treated	2130	57,2^{XXX}	68,9^{XXX}	31,1^{XXX}
3. 1207				
controll	2020	97,5	92,4	7,6
treated	2130	46,2^{XXX}	61,2^{XXX}	38,8^{XXX}

These results unambiguously prove, that pollen treatment definitely
reduces the number of viable pollen grains, in all genotypes, because
a high percentage of the pollen population died during the treatment.
If the genetic material being tested did not contain active transposon,
no genetic changes could be observed in surviving pollen grains, not
even at the theoretically expected spontaneous mutation rate. In the
case however, when the genotypes contained active transposons, the
frequency of genetic changes has significantly increased. On the genotype with indentified Ac/Ds movable elements, a relatively high spontaneous mutation rate was observed, and following the treatment, the
ratio of Bz genotype pollen grains has dramatically increased. During
the test of our new line with unidentified transposons, we have received
similar results.

Testibility of Changes on the Offspring Generation

Both for evolution and practically, genetic changes are important
only if they are manifest in offspring generations. For this reason,
we have tested the frequency of major genetic alterations experienced
in pollen populations, on offspring generations following pollination.
Our results in Table 2 show that applied pollen treatment definitely
reduces the percentage of seed setting.

Table 2.
New genotypes following pollination, induced by pollen treatment

Genotype	seed-set %	seed No	bz %	Bz %
1206 x 1207				
controll	87,1	1005	99,8	0,2
treated	35,6^{XXX}	554	83,4^{XXX}	16,6^{XXX}

Among the seeds resulting from pollination by untreated pollen, the
ratio of altered genotypes was significatnly low. Higher spontaneous
mutation rates observed in the pollen population were not realized
in the grain yield. On the other hand, pollination with treated pollen
resulted in a number of seeds with an altered genotype. This value is
only about 50% of the frequencies of observed genetic changes in pollen
populations. It is probable that pollen grains with altered genotypes
have a cometitive disadvantage in comparison to normal ones.
Summing up our results, we have demonstrated that transposon activity
can be unduced also during the gametophytic life cycle. Both the
understanding of their functioning and the potentials of practical
application need further investigations on phenotypic and DNA level.

REFERENCES

Barnabás B, Kovács G (1988) Perspectives of pollen and male gamete selection in cereals. In: Wilms HJ, Keijzer CJ (eds) Plant sperm cells. PUDOC, Wageningen, in press

Barnabás B, Rajki E (1981) Fertility of deep-frozen maize (Zea mays L.) pollen. Ann Bot 48:861-864

Fedoroff N (1983) Controlling elements in maize. In Saphiro JA (ed) Mobile genetic elements. Academic Press, New York, pp 1-57

Heslop-Harrison J, Heslop-Harrison Y. (1970) Evaluation of pollen viability by enzymatically induced fluorescence, intracellular hydrolysis of fluorescein diacetate. Stain Technol 45:115-120

Mascarenhas JP, Stinson JR, Willing RP, Pe' ME (1986) Genes and their expression in the male gametophyte of flowering plants. In Mulcahy DL, Bergamini-Mulcahy G, Ottaviano E (eds) Biotechnology and ecology of pollen. Springer, New York, pp 39-44

Nevers P, Shepherd NS, Seadler H (1986) Plant transposable elements. Adv Bot Res 12:103-203

Pfahler PL (1983) Comparative effectiveness of pollen genotype selection in higher plants. In Mulcahy DL, Ottaviano E (eds) Pollen: Biology and implication for plant breeding. Elsevier, New York, pp 366-367

Robertson DS. (1985) Evidente for Mutator activity in the male gametopyte. Maize Genet Coop Newsletter 59:14-15

Sacher RF, Mulcahy DL, Staples RC (1983) Developmental selection during self pollination of Lycopersicon x Solanum F1 for salt tolerance of F2. In: Mulcahy DL, Ottaviano E (eds) Pollen: Biology and implication for plant breeding. Elsevier, New York, pp 329-334

Sari-Gorla M, Frova C, Binelli G, Ottaviano E (1986) The extent of gametophytic-sporophytic gene expression in maize. Theor Apll Genet 72:42-47

Searcy KB, Mulcahy DL (1985) Pollen selection and the gametophytic expression of metal tolerance in Silene dioica (Caryophyllaceae) and Mimulus guttatus (Scrophulariaceae). Am J Bot 72: 1700-1706

Smith GA, Moser HS (1984) In vitro sporophytic screening and gametophytic verification of herbicide tolerance in sugarbeet. Agronomy Abstracts 88

Tanksley SD, Zamir D, Rick CM (1981) Evidence for extensive overlap of sporophytic and gametophytic gene expression in Licopersicon esculentum. Science 213:453-455

Zamir D, Tanksley SD, Jones RA (1982) Haploid selection for low temperature tolerance of tomato pollen. Genetics 101:129-173

Zamir D, Tanksley SD, Jones RA (1981) Low temperature effect on selective fertilization by pollen micture of wild cultivated tomato species. Theor Appl Genet 59:235-238

Zamir D, Vallejos EC (1983) Temperature effects on haploid selection of tomato microspores and pollen grains. In: Mulcahy DL, Ottaviano E (eds) Pollen: Biology and implication for plant breeding. Elsevier, New York, pp 335-341

Willing RP, Mascarenhas JP (1984) Analysis of complexity and diversity of mRNAs from pollen and shoot of Tradescantia. Plant Physiol 75: 865-868

Method Development for Applying Pollen Selection in Cucumber Breeding Programmes: Effects of Centrifugation on Pollen Competence

R.J. Bino[1], J. Franken[1] and E. van der Zeeuw[2]

[1] Institute for Horticultural Plant Breeding (IVT), P.O.B 16, 6700 AA Wageningen, The Netherlands

[2] Department of Genetics, Agricultural University, Gen. Foulkesweg 53, 6703 BM Wageningen, The Netherlands

INTRODUCTION

The possibility to use large populations of haploid genomes makes pollen a potentially efficient plant system to select for sporophytes bearing favorable traits. Screening pollen for sporophytic features is only possible if genes responsible for the particular traits are also expressed during microgametophyte development. Tanksley et al. (1981) and Willing and Mascarenhas (1984) showed that about 60 % of the genes are correspondingly expressed in the gametophytic and sporophytic life cycle of the plant. Overlap in gene expression is also evident from the parallel responses in pollen and plants for tolerances and sensitivities for a wide range of agents (e.g., Mulcahy et al., 1985; Bino et al., 1987). In addition, pathogenesis related processes and mechanisms involved in disease resistance are expressed in both vegetative and generative tissues (Laughnan and Cabay, 1973; Bino et al, 1988). The potential of pollen selection in plant breeding programmes, however, depends on the development of methods for pollen manipulation. As discussed by Bino and Stephenson (1988), selection conditions and methods for separation and concentration of selected from non-selected pollen have to be optimized, while techniques for the application (pollination) of manipulated microgametophytes on pistils insuring fertilization need to be improved.

Methods either based on micro-manipulation, filtration, or centrifugation may possibly be used for the separation of pollen. Germinated pollen may be separated from non-germinated pollen, or, preferably, pollen with short tubes from long tubed pollen. In the present paper we report effects of centrifugation on cucumber pollen.

MATERIAL AND METHODS

Plant material. Cucumber (<u>Cucumis sativus</u> L.) plants were grown in greenhouses at 19 - 24 °C, under summer daylight conditions.

Pollen centrifugation. Pollen was collected from ten flowers at anthesis and germinated for 5 - 20 min in liquid media (5 ml) containing 3 mM H_3BO_3, 1.7 mM $Ca(NO_3)_2$, and 14 % sucrose. At different time intervals, 1 ml samples were centrifugated for 0 - 60 sec at 80 or for 0 - 20 sec at 320 x g in 1.5 ml test tubes using an Eppendorf centrifuge. The supernatant was carefully removed with a pipette. Experiments were executed in triplicate. Viability of pollen in the supernatant and the pellet was determined by observing the streaming of plasma in pollen tubes. Germination percentages and tube lengths were quantified after staining and fixation with aceto-carmine (2%).

RESULTS AND DISCUSSION

<u>Cucumber pollen characteristics</u>.
One male cucumber flower contained about 5000 mature pollen grains, the diameter of the pollen was ± 70 μm and the weight approximately 50 ng. Pollen of cucumber commenced to germinate 5 min after the start of incubation (Fig 1). Speed of tube growth during the first stages of germination was about 30 μm/min.

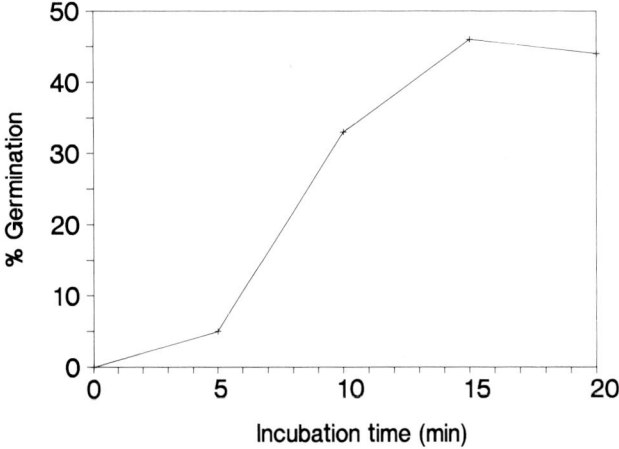

Fig. 1. Effects of incubation time on pollen germination percentages.

Pollen centrifugation.

Pollen samples were incubated for 10 min and subsequently centrifugated for 0
- 60 sec at 80 x g (Table 1). An incubation period of 10 min was optionally
chosen because at that time the sample contained both ungerminated pollen and
pollen with various tube lengths (± 200 μm). By centrifugation it was attempted
to separate germinated from ungerminated pollen. A centrifugation time of 15-
45 sec increased the amount of germinated pollen in the pellet as compared with
the control.   Apparently, at this spinning force,   more germinated than
ungerminated pollen were spun down and the centrifugation procedure enriched
the pellet with tubed pollen. When longer centrifugation times were used,
ungerminated and germinated pollen were pelleted in similar amounts, and the
percentage of germinated pollen in the sample was reduced to the control value.
Centrifugation intervals of 15 - 45 sec did not influence pollen viability, as
compared with the control. At 60 sec, however, 17 % of the germinated pollen in
the pellet was bursted.

Table 1. Effects of different periods of centrifugation at 80 x g on
percentages germination and bursting of pelleted cucumber pollen, 0 sec was the
non-centrifugated control.

Centrifugation time (sec)	% Germination	% Bursting
0	29	6
15	46	5
30	45	8
45	47	3
60	26	17

The use of a higher centrifugal force (320 x g) for 5 - 20 sec increased the
percentage of germinated pollen in the pellet (Table 2).   Correction for the
control value gave a maximal concentration of germinated pollen in the pellet
of 27 % (64 - 37 %). The higher centrifugal force, however, did had a
detrimental effect on pollen viability. Apart from high amounts of bursted
pollen,   some tubes became deformed, a phenomenon best characterized as
'twisted' (Fig. 2).

Table 2. Effects of different periods of centrifugation at 320 x g on percentages germination, bursting and twisting of pelleted cucumber pollen, 0 sec was the non-centrifugated control.

Centrifugation time (sec)	% Germination	% Bursting	% Twisting
0	37	3	0
5	64	19	0
10	57	23	0
15	53	23	1
20	55	18	1

To increase the proportion of germinated pollen, the pellet was re-suspended and re-centrifugated (Table 3). Although high amounts of tubed pollen were recovered, viability was severely reduced as no plasma streaming was observed in tubes of bursted or twisted pollen.

Table 3. Pollen pelleted for different time intervals at 80 x g was re-suspended and re-centrifugated for 15 sec at 320 x g.

1^e Centrifugation time (sec)	% Germination	% Bursting	% Twisting
0	37	3	0
5	56	25	6
10	67	18	4
15	56	25	2
20	48	31	4

Pollinations with centrifugated pollen.

Pollen was incubated for 10 min and centrifugated for 30 sec at 80 x g, subsequently the pellet was resuspended in germination medium and a droplet of this medium was placed upon the stigma of a female cucumber flower. The droplet contained about 100 pollen grains of which ± 68 % was germinated. After 2 - 3 days, pollen tube growth in pistils was examined using aniline blue fluorescence (Kho and Baer, 1968). Few (3 - 4) pollen tubes were found in the style, a single one in the ovarium, but none penetrated the ovules. Apparently, pollen competence was reduced and the number of viable pollen in the droplet was to low for a prosperous fertilization.

CONCLUSIONS

- Germinated and ungerminated cucumber pollen may be partly separated by centrifugation. Best results were obtained using a centrifugal force of 80 x g for 15 - 45 sec. Longer centrifugation periods increased the proportion of ungerminated pollen in the pellet while higher speeds induced bursting and twisting.

- Complete partition of germinated and ungerminated pollen was not possible with the applied methods. Better results may possibly be obtained employing methods as developed for the fractionation of tobacco pollen grains for microspore culture. For this, Kyo and Harada (1985) used a discontinuous Percoll gradient and acquired uniform pollen populations, without adverse effects on viability. Currently, we are attempting similar methods for separating pollen samples on basis of tube lengths.

- In cucumber, production of each mature seed requires at least 10 pollen grains. Moreover, a threshold number of developing seeds is needed for fruit maturation. As challenging agents and separation procedures will reduce the viability and competence of selected microgametophytes, even a larger surplus of manipulated pollen is required to accomplish selection at the gametophyte level. Therefore, before pollen selection may be applied in cucumber breeding programmes, pollen performance and young plant development need to be promoted, for instance by using in vitro pollination and embryo rescue techniques.

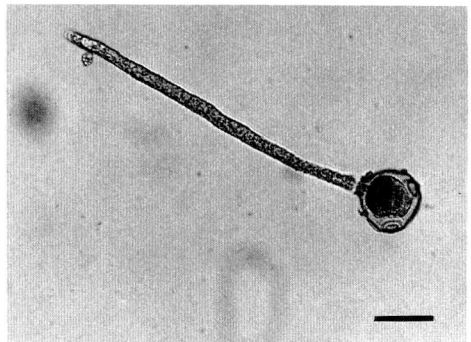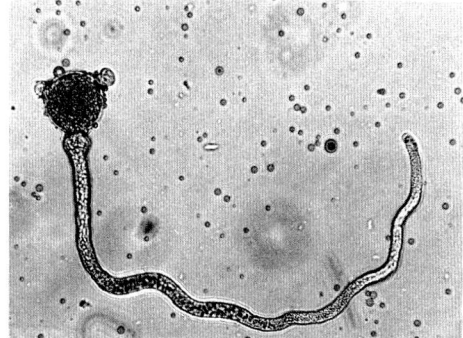

Fig. 2. Cucumber pollen, incubated for 10 min in germination medium, centrifugated for 10 sec at 80 g, and subsequently re-suspended and re-centrifugated for 15 sec at 320 g. Normal pollen tube (left) and 'twisted' pollen tube (right). Bar = 0.1 mm.

ACKNOWLEDGMENT

We thank A.G. Stephenson for stimulating discussion and advice.

REFERENCES

Bino RJ, Franken J, Witsenboer HMA, Hille J, Dons JJM (1988) Effects of
    Alternaria alternata f.sp. lycopersici toxins on pollen. Theor Appl Genet
    (in press)
Bino RJ, Hille J, Franken J (1987) Kanamycin resistance during in vitro
    development of pollen from transgenic tomato plants. Plant Cell Reports 6:
    333-336
Bino RJ, Stephenson AG (1988) Selection and manipulation of pollen and sperm
    cells. In: Wilms HJ, Keijzer CJ (eds) Plant sperm cells as emerging tools
    for crop biotechnology. Pudoc Wageningen, The Netherlands (in press)
Kho YO, Baer J (1968) Observing pollen tubes by means of fluorescence.
    Euphytica 17: 298-302
Kyo M, Harada H (1985) Studies on conditions for cell division and
    embryogenesis in isolated pollen culture of Nicotiana rustica. Plant
    Physiol 79: 90-94
Laughnan JR, Gabay SJ (1973) Reaction of germinating maize pollen to Helmin-
    thosporium maydis pathotoxins. Crop Sci 13: 681-684
Mulcahy DL, Bergamini Mulcahy G, Ottaviano E (eds) (1985) Biotechnology and
    ecology of pollen. Springer-Verlag New York Berlin Heidelberg Tokyo, pp
    528
Tanksley SD, Zamir D, Rick CM (1981) Evidence for extensive overlap of
    sporophytic and gametophytic gene expression in Lycopersicon esculentum.
    Science 213: 453-455
Willing RP, Mascarenhas JP (1984) Analysis of the complexity and diversity of
    mRNAs from pollen and shoots of Tradescantia palludosa. Plant Physiol 78:
    887-890

Pollen Quality and Selection

Male Gametophytic Selection in Higher Plants

E.Ottaviano,   M.Sari   Gorla,   C.Frova,   E.Pè
Dipartimento   di   Genetica   e   di   Biologia   dei   Microrganismi
University   of   Milan
Via   Celoria   26
20133   Milano
I t a l y

Population   Structure   and   Evolution

Gametophytic   selection   refers   to   the   differential   gene transmission   during   the   haploid   phase   in   higher   plants,   which   begins with   the   meiosis   product   and   ends   with   fertilization.   The   special features   of   the   male   gametophyte   (large   population   size,   independence of   the   maternal   plant,   direct   exposure   to   environmental   stresses, competition   within   the   same   style)   suggest   that   selection   in   this   phase may   be   expected   to   act   with   greater   intensity   then   in   the   female gametophyte.

The   phenomenon   was   envisaged   by   Mendel   as   a   mechanisms   which could   produce   distorted   segregations   and   indicated   by   Buchholz   (1922) as   a   factor   increasing   evolution   rate.   However,   the   classical   view,   in considering   the   manifold   positive   aspects   of   diploidy   in   higher   plant genetic   systems,   was   that   of   a   gradual   suppression   of   the   haploid generation,   leading   to   the   establishment   of   an   independent   genetic domain   controlling   specific   pollen   functions.   The   topic   was   elegantly discussed   by   Heslop-Harrison   (1979),   who   pointed   out   that   this expectation   has   not   been   fulfilled   in   the   angiosperms.

Mulcahy   (1979),   in   view   of   the   emerging   information   pointing   to a   large   extent   of   gene   expression   in   the   male   gametophytic   phase,   the results   showing   correlations   between   male   gametophytic   and   sporophytic traits,   and   the   effects   of   male   gametophytic   selection   (MGS)   on   the sporophytic   generation,   proposed   that   this   type   of   selection   played   an important   role   in   the   history   of   higher   plants   by   allowing   a   high   rate of   evolution.   The   high   cost   this   rate   implies   (Haldane   1957)   would   be compatible   with   the   male   gametophytic   generation,   because   of   the   large population   size.

A   significant   role   of   gametophytic   selection   would   also   account for   the   evolutionary   trend   of   the   male   gametophyte   in   plants,   towards an   increase   of   the   efficiency   of   mechanisms   favouring   pollen   selection. As   proposed   by   Hoekstra   (1983),   one   of   these   mechanisms   consists   in the   reduction   of   the   lag   period   of   germination   and   the   increase   of   the pollen   tube   growth   rate.   Species   having   these   properties,   which   maximize pollen competition in the style, are in fact found in the most advanced angiosperms.

Finally, the genetical structure of populations can be regulated by gametophytic selection. Kimura (1959) and Hiraizumi (1964) have shown that a slight prezygotic selection can increase the genetic load (amount of genetic variability in fitness) and a stable equilibrium be attained. Ottaviano et al. (1988b) have proposed a special role of MGS in the regulation of the amount of the genetic load for complex genetic characters in the case of positive correlations between sporophytic and gametophytic selection. In fact, for these types of traits, it is expected that in each generation of polymorphic populations a large amount of genetic load will be produced by the effect of genetic recombination. The sporophytic mean fitness can be held at a stable equilibrium if a large portion of the unfitted genetic combinations are removed in the prezygotic phase, where the cost (loss of pollen genotypes) is compatible with the size of the male gametophyte populations.

Quantitative evaluations of MGS with regard to the evolution rate and population structure require models which take into account the population size and the consequences of the haploid state of the gametophytic generation. Large population size implies the possibility of high intensity of selection; the haploid state increases the probability of selecting complex allele combinations (number of genotypic combinations for n genes is much lower in haploids then in diploids, i.e., 2^n vs. 4^n) and allows efficient selection against deleterious recessive alleles. Models to analyze gametophytic selection under different mating systems (Harding, 1975), in different combinations of gametic-zygotic selection and for comparison of the effectiveness of MGS vs. sporophytic selection (Ottaviano and Sari Gorla 1979; Pfalher 1983) all show that under MGS the number of generations required to obtain a given effect on gene frequencies is greatly reduced and a maximum rate of evolution in the sporophyte is obtained with a positive correlation between gametic and zygotic selection.

Gene expression in pollen

A high evolution rate and regulation of the genetical load in populations require that the following assumptions are satisfied: i) a large extent of gene expression is present in the male gametophytic generation; ii) genes expressed in the gametophytic generation are also expressed in the sporophytic generation; iii) the genes showing gametophytic-sporophytic expression have a significant effect on male gametophytic fitness.

Ottaviano and Mulcahy (1988) in reviewing this topic have showed that a large amount of data concerning inheritance of chromosomal deficiencies, distorted segregations, genetical analyses of single pollen grains based on specific staining techniques or protein electrophoresis, are in accordance with this expectation. However, the most extensive studies testing the first two assumptions have been carried out by means of isozyme and mRNA analyses and the study of gametophytic expression of genes affecting endosperm development.

In tomato Tanksley et al. (1981) assayed 28 structural genes and estimated that 62% of these were expressed in pollen, 58% in both pollen and sporophytic tissues (genetic overlap) and 3% only in pollen. In maize, where 15 enzymes for 34 structural genes have been analyzed (Sari-Gorla et al. 1986), the estimates were 86%, 72% and 6%, respectively. In Populus species, where 15 systems for 45 to 51 genes were analyzed, the results reveal similar values: genetic overlap was from 74 to 80% and genes expressed only in pollen between 11 to 17% (Rajora and Zuffa, 1986). Indications of a large extent of genetic overlap have also been obtained in Hordeum species (Pedersen et al. 1987). In Tradescantia paludosa Willing and Mascarenhas (1984) found that gametophytic and sporophytic tissues contain 20,000 and 30,000 different mRNA's respectively. Heterologous hybridization between cDNA from sporophytic tissues and poly(A)RNA from pollen and the reciprocal hybridization allowed it to be estimated that 60% of the sequences analyzed are found in pollen, 54% are expressed in both gametophytic and sporophytic tissues and 4% are specific to pollen. In maize, colony hybridization involving cloned cDNA and poly(A)RNA from roots and stocks revealed 65% of genetic overlap (Willing et al. 1984). A different class of genes has been studied in maize; Ottaviano et al. (1988a) analyzed a set of 32 endosperm defective (de) mutants representing 32 different genes controlling endosperm development. The segregation pattern of de phenotypes on the ears revealed that 22 (60%) of these genes affect microspore development or pollen tube growth.

Apart from these endosperm mutant data, most of these studies relate to mature pollen (isozyme analysis) or germinating pollen (mRNA analysis). On the other hand, Frova et al. (1986) have shown that pollen gene expression can be larger than indicated by these estimates. In fact there are genes expressed during microspore development, the products of which (enzymes, heat stock proteins) are not found in mature pollen.

Fitness   Variability   within   Experimental   and   Natural   Populations

The   studies   referred   to   in   the   previous   section   clearly   indicate that   gametophytic-sporophytic   gene   expression   involves   a   large   portion of   genes   and   support   the   hypothesis   that   it   is   a   general   phenomenon   in higher   plants.   However,   to   satisfy   all   the   assumptions   required   for   a significant   effect   of   MGS   on   population   structure   and   evolution,   it   is necessary   to   show   that   the   phenomenon   involves   gametophytic   fitness variability   within   populations.

A   number   of   studies   report   that   genetical   differences   for   plant traits   can   be   revealed   by   pollen   assays.   It   has   been   proved   for   growth rate   in   maize,   tolerance   to   herbicide   ethofumesate   in   Beta   vulgaris, resistance   to   pathotoxine   produced   by   Helmynthosporium   maydis, tolerance   to   heavy   metals   in   Silene   dioica and Silene   alba and   to   low temperature   in   tomato,   for   resistance   to   kanamicin   in   transgenic   tomato plants   (see   Ottaviano   1988   for   a   review).

While   this   type   of   association   does   not   discriminate   between sporophytic   or   gametophytic   control   of   the   pollen   character,   the importance   of   the   genome   of   the   haploid   phase   has   been   proved   by gametophytic   selection   experiments.   Results   reported   in   the   literature have   been   produced   using   two   different   criteria: 1)   selection   for   pollen competitive   ability,   obtained   by   applying   high   intensity   of   selection in   the   gametophytic   population,   either   by   the   increase   of   pollination density   (number   of   pollen   grains   per   ovule)   or   by   increasing   the distance   that   competing   pollen   tubes   have   to   cover   in   the   style;   2) selection   for   tolerance   to   environmental   stresses   applied   during   pollen development   in   the   anthers   or   during   pollen   tube   growth   in   the   style. The   use   of   the   methods   based   on   the   first   criterion   produced   positive responses   for   plant   characters   expressing   vigour   and   fertility   in   Vigna sinensis,   wheat,   cotton,   maize,   Turnera   ulmifolia,   Curcubita   pepo, Lotus   corniculatus and   Cassia   fasciculata.   Responses   to   MGS   for tolerance   to   environmental   stresses   have   been   reported   for   low temperature   tolerance   in   tomato   species,   salt   tolerance   in   Solanum species   and   heavy   metal   tolerance   in   Silene   dioica ad Mimulus   guttatus (see   Ottaviano   and   Mulcahy,   1988,   for   a   review).

These   studies   have   proved   that   a   large   set   of   characters,   which are   generally   expressed   at   cellular   level,   can   be   efficiently   assayed and   selected   in   the   gametophytic   generation.   However,   most   of   them   did not   provide   a   population   analysis.   In   fact   the   material   used   was generally   highly   heterozygous F_1's   obtained   by   crossing   divergent selected   genotypes   or   related   species.   Moreover,   the   response   was detected   only   as   distorted   segregation   in   the   sporophytic   generation and information about components of gametophytic fitness was not obtained.

The estimation of fitness values and variability is one of the most difficult tasks in experimental population genetics. It is most easily approached by studying fitness components, such as viability, competitive ability and fertility. The problem has additional difficulties when referred to the gametophytic generation, because the reproductive success of pollen grains depends on both sporophytic and gametophytic control mechanisms.

Information in this regard has been obtained by means of comparisons of gametophytic and sporophytic gene frequencies in multiple census experiments. This approach allowed the estimation of pollen selective values and parameters describing the mating system in Clarkia exilis, Clarkia temporalis (Vasek and Harding, 1976), Phaseolus lunatus and Zea mays (Harding and Tucker, 1969), and barley (Clegg et al. 1978). In all populations analyzed the values of the gametophytic parameter indicate that a significant amount of the net fitness variability relates to that found in the gametophytic generation. However, none of these studies discriminated between gametophytic and sporophytic control.

An experimental approach to evaluate the importance of post-meiotic gene expression in the determination of pollen fitness can be based on the partitioning of the main character into component traits. These include pre-shedding components, i.,e., pollen grain number per plant, pollen variability, pollen competition within the anther during development, and post-shedding components, i.e., differential rate of pollen germination time (lag between pollination and tube emission), pollen tube growth rate and selective fertilization. Analysis of distorted segregations in maize show that selection can act both on the pre-shedding phase, affecting pollen viability and pollen competition within the anther, and on the post-shedding components (Ottaviano et al. 1988a).

The most comprehensive analysis of the genetical control of pollen fitness components was carried out in maize. The research has been focused on parameters describing pollen competitive ability in the early stage of the post-shedding phase and during pollen tube growth in the style (Ottaviano et al. 1988b). The study relates to two sets of families derived from an highly heterozygous synthetic population: one set obtained under high selection intensity and the other under low selection intensity. The analysis of the S3 families derived after two cycles of gametophytic recurrent selection revealed that: i) both characters show a large amount of variability, which is largely under genotypic control (high heritability); ii) a significant response to selection was detected only for the pollen tube growth rate, indicating for this character a significant portion of genetic variance due to post-meiotic gene expression. Heritability values were 0.88, 0.21 and 0.67 between families, between high and low intensity of selection and between families within sets, respectively; iii) a significant correlated response was detected for sporophytic traits, proving genetic overlap for the control of pollen tube growth and vigour of sporophytic tissues. Taking into account that pollen competition, although at lower intensity than in this experiment, is found in normal corn field (about 13 pollen grains load on an individual silk) it is possible to conclude that MGS operates in maize corn fields.

Evidence of MGS in natural population is still fragmentary. Most of the information relevant to this issue derives from survey analysis of pollen load per stigma and from the effect of pollen load on seed set per fruit. Data on 25 different species summarized by Snow (1986) indicate that in only a few cases is seed set limited by pollen availability and that, consequently, post-anthesis pollen competition is a frequent phenomenon in wild populations. In Geranium maculatum it has also been shown that pollen competition occurs even between pollen reaching the stigma at different times (Mulcahy et al. 1983). In this situation, or when the first pollen release is not sufficient for the complete seed set, a very high intensity of selection is reached for tubes from later impollination, because of the reduced number of ovules available. Ramstetter (1987) confirmed the relationship between pollen load and seed set in Aureolaria flava, showing also that the reduced seed set under low pollen load produces effects on seedling survival and other seedling traits. However, apart from the work on the Clarkia species referred to above, data showing fitness genetical variability in the gametophytic populations are not available for natural populations. Therefore more work is needed in this field. For the estimation of sporophytic and gametophytic variability of pollen fitness components a fruitful genetic approach could be based on multiple census experiments with controlled pollinations to discriminate between pre- and post-meiotic gene expression.

Literature

Buchholz J T (1922) Developmental selection in vascular plants. Bot Gaz 73: 249-286.

Clegg M T, Kahler A L, Allard L W (1978) Estimation of life cycle components of selection in an experimental plant population. Genetics 89: 765-792

Frova C, Binelli G, Ottaviano E (1986) Male gametophyte response to high temperature in maize. In Biotechnology and Ecology of Pollen.

Mulcahy   D   L,   Bergamini   Mulcahy   G,   Ottaviano   E   (eds)   pp   33-38
Springer-Verlag   New   York

Haldane   J   B   (1957)   The   cost   of   natural   selection.   J   Genet   55:
511-524.

Harding   J,   Tucker   C   L   (1969)   Quantitative   studies   on   mating   system.
II   Method   for   the   estimation   of   male   gametophytic   selective   values   and
differential   outcrossing   rates.   Evolution   23:85-95.

Harding   J   (1975)   Models   for   gamete   competition   and   self
fertilization   as   components   of   natural   selection   in   populations   of
higher   plants.   In   Gamete   competition   in   plants   and   animals   Mulcahy   D
L   (ed)   North-Holland   Amsterdam

Heslop-Harrison   J   (1979)   The   forgotten   generation: some   thoughts
on   the   genetics   and   physiology   of   Angiosperm   gametophytes.   In   The   plant
genome.   Davies   D   R,   Hopwood   D   A   (eds).   4th   Innes   Symposium   Norwich

Hiraizumi   Y   (1964)   Prezygotic   selection   as   a   factor   in   the
maintenance   of   variability.   Harbor   Symp   Quant   Biol   29:   51-60

Hoekstra   F   A   (1983)   Physiological   evolution   in   angiosperm   pollen:
Possible   role   of   pollen   vigor.   In   Pollen: Biology   and   implications   for
plant   breeding.   Mulcahy   D   L,   Ottaviano   E   (eds)   pp   35-42   Elsevier
Biomedical   New   York

Kimura   M   (1959)   Genetic   load   of   a   population   and   its   significance
in   evolution   (in   japanese   with   english   summary).   Jap   J   Genet   35:   7-33.

Mulcahy   D   L   (1979)   The   rise   of   the   Angiosperms: a   genecological
factor.   Science   206:   20-23

Mulcahy   D   L,   Curtis   P   S,   Snow   A   A   (1983)   Pollen   competition   in   a
natural   population.   In   Handbook   of   experimental   pollination   biology.
Jones   C   E,   Little   R   J   (eds)   pp   330-338   Scient   and   Acad   Edit

Ottaviano   E   (1988)   Selection   pressure   on   pollen   and   its   relevance
to   plant   breeding.   In   Proc   Int   Congress   of   Plant   Physiology.   New   Delhi.

Ottaviano   E,   Sari   Gorla   M   (1979)   Genetic   variability   of   male
gametophyte   in   maize.   Pollen   genotype   and   pollen-style   interaction.
In   Israeli-Italian   joint   meeting   on   genetics   and   breeding   of   crop
plants.   Monogr   Genet   Agraria   IV   Roma

Ottaviano   E,   Petroni   D,   Pe'   M   E   (1988a)   Gametophytic   expression
of   genes   controlling   endorsperm   development   in   maize.   Theor   Appl   Genet
75:   252-258

Ottaviano   E,   Sari   Gorla   M,   Villa   M   (1988b)   Components   of   the   male
gametophyte   fitness   in   maize.   Genetic   variability   and   correlation   with
sporophytic   traits.   Theor   Appl   Genet   (in   press)

Ottaviano   E,   Mulcahy   D   L   (1988)   Genetics   of   Angiosperm   pollen.
proposed   to   Advances   in   Genetics

Pedersen   S,   Simonsen   V,   Loeschcke   V   (1987)   Overlap   of   gametophytic
and   sporophytic   gene   expression   in   barley.   Theor   Appl   Genet   75:   200-206

Pfahler   P   L   (1983)   Comparative   effectiveness   of   pollen   genotype
selection   in   higher   plants.   In   Pollen: biology   and   implication   for
plant   breeding.   Mulcahy   D   L,   Ottaviano   E   (eds)   Elsevier   Biomedical   New
York

Rajora   O   P,   Zsuffa   L   (1986)   Sporophytic   and   gametophytic   gene
expression   in   Populus   deltoides   marsh.,   P   nigra   L,   and   P   maximowiczii
henry.   Can   J   Genet   Cytol   28:   476-482

Ramstetter   J.   (1987)   Pollen   competition   in   Aureolaria.   Ph   D
Dissertation   Univ.   of   Massachusetts   Amherst

Sari-Gorla   M,   Frova   C,   Binelli   G,   Ottaviano   E   (1986)   The   extent
of   gametophytic-sporophytic   gene   expression   in   maize.   Theor   Appl   Genet
72:   42-47

Snow   A   (1986)   Pollination   dinamics   in   Epilobium   canum
(Onagraceae).   Consequences   for   gametophytic   selection.   Amer   J   Bot   73:
139-151

Tanksley   S   D,   Zamir   D,   Rick   C   M   (1981)   Evidence   for   extensive
overlap   of   sporophytic   and   gametophytic   gene   expression   in   Lycopersicon
esculentum.   Science   213:   453-455

Vasek   F   C,   Harding   J   (1976)   Outcrossing   in   natural   populations.
V.   Analysis   of   outcrossing,   inbreeding   and   selection   in   Clarkia exilis
and   Clarkia tembloriensis.   Evolution   30:   403-411

Willing R P, Mascarenhas J P (1984) Analysis of complexity and diversity of mRNAs from pollen shoots of <u>Tradescantia</u>. Plant Physiol 75: 865-868

Willing R P, Eisenberg A, Mascarenhas J P (1984) Genes active during pollen development and the construction of cloned cDNA libraries to mRNAs from pollen. Plant Cell Incomp Newslett 16: 11-12

Pollen Selection for Stress Tolerance or the Advantage of Selecting Before Pollination

DL Mulcahy[+], GB Mulcahy[+], R Popp[+], N Fong[*], N Pallais[**], A Kalinowski[**], JN Marien[***]

[+]Department of Botany
University of Massachusetts
Amherst, MA 01003 USA

A series of studies has indicated that 60 - 75% of the structural genes which are expressed in the sporophyte of both angiosperms and gymnosperms are expressed, and thus subject to selection, also in the pollen (see Ottaviano and Mulcahy, in press, for review). The consequences of this overlap between sporophytic and gametophytic genomes were first indicated by Ter-Avanesian (1949, 1978), later by others (see Ottaviano and Mulcahy, in press), and more recently by Winsor, et al. 1987). Although the first studies of pollen selection have been conducted with cultivated plants, because of the technical advantages that these provide, the efficacy of pollen selection is not limited to cultivars. In *Aureolaria flava* (Scrophulariaceae), seedlings from the highest level of pollen tube competition in natural populations exhibited significantly larger rosettes and were more likely to produce flowering stems in the first season than were seedlings produced with little or no pollen tube competition (Ramstetter and Mulcahy, in prep.) These studies indicate that, in natural populations and in cultivars, pollen competition can be used to select for increased vigor (see Ottaviano, this volume) and for increased stress tolerance, the subject of this article.

We have been considering if pollen selection could increase resistance to osmotic stress. To this end, we exposed pollen of Lycopersicon esculentum to germination medium which had been supplemented with different concentrations of melibiose (See Gibbs and Greenway, 1986). Table 1 presents data on the two most tolerant (see Jaworski et al. 1987) and two least tolerant pollen types identified in a survey of 11 different accessions.

*   International Potato Center (CIP) Lima, PERU
**   Polish Acad. of Science, Inst. of Genetics, Posnan
*** AFOCEL, Cugnaux, FRANCE

Table 1.   Ability of pollen to germinate in solutions of
different osmotic strength.   (Osmotic strength is given as the
equivalent of sucrose solution.)   All pollen sources are
Lycopersicon esculentum.   Each pollen germination solution contained
15% (0.439 M) sucrose, 1.27 mM Ca(NO$_3$)$_2$, 1.62 mM H$_3$BO$_3$, (see
Brewbaker and Kwack, 1963) plus appropriate additions of melibiose
to adjust osmoticum to the specific sucrose equivalent.   The data
are percentage of control pollen germination (control contained 25%
sucrose, plus Ca & B).

SUCROSE EQUIVALENT OF POLLEN GERMINATION MEDIUM

	33	36	39
Pollen Source			
GA 1095	87.5%	43.6%	7.5%
GA 219	46.0%	3.5%	0.0%
New Yorker	2.1%	0.0%	0.0%
Beefsteak	0.9%	0.0%	0.0%

Sporophytes of these four were screened for osmoticum tolerance
at the seedling stage (see Bouslama and Schapaugh, 1984).   Two week
old seedlings, grown in aerated nutrient solutions, and selected for
uniformity, were exposed to polyethylene glycol 500, at - 0.30 and
at - 0.45 MPa.   PEG 500 was chosen as the osmoticum since the large
volume of solution needed prohibited the use of melibiose.

Table 2.   Response of tomato seedlings to osmoticum stress.   Values
are shoot weight after growth in -0.3 MPa PEG, expressed as
percentage of unstressed control.   Six plants per replica.

Source	DF	MS	F	Prob.
Mean	1	116859	6299.00	.008
Rep.	1	19	0.28	.604
Variety	3	134	11.85	.036
Rep. x Var.	3	11	0.17	.914
Remainder	16	66		

These data indicate that the varieties which produce osmoticum tolerant pollen are themselves tolerant to osmotic stress.

As would be predicted from evidence of overlap between gametophyte and sporophyte genomes, it seems that tolerance in one phase often correlates with that in the other. We now consider how to best utilize pollen to select for stress resistant genotypes. In order to answer this question, we must first review the relative influences of three determinants of pollen tube growth.

Influence of the sporophytic pollen source.

In *Zea mays*, pollen from F_1 hybrids frequently outcompetes that from inbred lines, suggesting that the vigor of the sporophytic pollen source has a significant effect upon the performance of the individual gametophytes. (Yamada and Murakami, 1983) Similarly, studying *Zea mays*, Kumar and Sarkar (1980) found a statistically significant correlation between pollen diameter and pollen tube length after 3 hours of *in vitro* growth and, if we compare pollen diameters from F_1 individuals and those of successive inbred generations, we see a steady decline of pollen diameters (Johnson et al. 1976) These and related studies demonstrate that the quality of the sporophytic pollen source has a highly significant effect upon pollen quality.

Influence of the Pollen Genotype.

Recent demonstrations of extensive gene expression in pollen explain earlier reports that variance of *in vitro* pollen tube growth rates is related to the genetic heterogeneity among the individual pollen grains. The interaction between diploid and haploid influences on pollen was demonstrated by the finding that, in progressively inbred lines of *Zea mays*, pollen diameter exhibits decreases in both mean and coefficient of variance. The change in mean reflects the reduction in sporophytic vigor, and the reduction in coefficient of variance is related to increasing homozygosity and resultant convergence of individual haploid genotypes (Johnson et al. 1976)

Influence of the Stylar Genotype.

Stephenson and Bertin (1983, p. 123) reviewed studies of correlations between *in vitro* and *in vivo* pollen tube growth rates and found several cases in which the two were independent of each other, but others in which they were significantly correlated. We can reasonably conclude that, since the style is the normal environment for pollen tube growth, *in vitro* and *in vivo* performance will tend to be uncorrelated. However, exceptions will be found whenever the germination medium effectively mimics the stylar environment (Lafleur and Mascarenhas, 1978) or if the *in vitro* selective pressure is so great that it overwhelms the difference between *in vitro* and *in vivo* environments. This latter possibility is demonstrated in the correlation between *in vitro* and *in vivo* tolerance of pollen to cold (Zamir, et al. 1982)

Granting that the style has a significant effect upon pollen tube growth, we must ask if this effect varies from one pollen type to another, *i.e.*, is the style selective in its influence. One obvious case is self-incompatibility. Studies of *defective endosperm (de)* markers in *Zea mays* also indicate that the outcome of pollen tube competition is influenced by the stylar genotype (Jones, 1928; Pe and Ottaviano, 1988). Another example, less well known, is what might be called "self-advantage", a phenomenon in which, within a mixture of pollen types, self pollen outcompetes non-self pollen, as in *Zea mays* (Jones, 1928) and in *Lycopersicon esculentum* (Hornby and Li, 1975). With these last two studies, it is not known whether the competition between self and non-self males occurs before or after fertilization although recent investigations by Zamir (unpub. and poster of this meeting) may suggest a post-fertilization phenomenon in the case of *Lycopersicon esculentum*, but this is unlikely to be the explanation for *Zea mays*. In two other studies, one of *Zea mays* (Ottaviano et al. 1980) and one of *Raphanus sativus* (Marshall and Ellstrand, 1986) there was no evidence of statistically significant interaction between pollen tube growth rate and the stylar genotype. Some pollen types did grow faster than did others and some styles were more effective in supporting pollen tube growth than were others However, these differences were generally consistent across styles or across different pollen types. Beyond these, there are no other examples known to us in which the style exhibits a <u>selective</u> effect on pollen tubes of different genotypes.

The influence of the style on the pollen makes it possible for even severely weakened pollen to function effectively. For example, stored pollen which exhibits no *in vitro* germination capacity will nevertheless germinate *in vivo* (Linskens, pers. com.). Furthermore, pollen which has been pretreated with germination medium containing either 0.001% or 0.0005% eosin, will not germinate *in vitro*, but will do so *in vivo* (G. Bergamini Mulcahy, unpublished). Searcy and Mulcahy (1985a) also found that, even when styles of copper tolerant plants contained enough copper to inhibit the growth of pollen tubes *in vitro*, pollen tube growth in those styles did not differ from that in control styles. Apparently, the style has a protective influence upon pollen tube growth, although this can be overwhelmed by extreme conditions or those, such as cold, against which the style can apparently provide little protection (see Zamir, et al., 1982)

Given the above information, consider the different methods of selecting for stress tolerance in pollen. Combining pollen from two or more sporophytes will introduce a significant degree of sporophytic influence and thus complicate analysis. Selection during pollen function (in the style) was demonstrated by Zamir, et al. (1982) to be an effective means of selecting for cold tolerance in *Lycopersicon spp.* However, selection during pollen function was not effective in selecting for tolerance to copper (Searcy and Mulcahy, 1985a). Selection for cold tolerance during pollen development (in the anther) was ineffective (Zamir, et al., 1982). In contrast, pollen segregating for copper tolerance was highly responsive to selection during pollen development (Searcy and Mulcahy, 1985a). Selection during development would very likely be effective using a method first described by Rowley and Dunbar (1970). With this, anthers are cultured on selective media and surviving pollen used to make crosses.

Pollen selection *in vitro* for tolerance to cold (Zamir, et al. 1982) or to copper or to zinc (Searcy and Mulcahy, 1985b) is highly effective. To this list, we may now add tolerance to osmoticum, discussed earlier. It may be that, using pollen from one plant, and *in vitro* selection, free from protective systems of anther or style, will be the most effective form of pollen selection.

How then shall we select for, and recover from *in vitro* systems, pollen grains which are stress tolerant? The following section describes the use of density gradients in separating tolerant and sensitive pollen grains.

Density Gradients To Separate Pollen Grains.

Table 3. indicates that, as pollen grains germinate, their density decreases. Furthermore, the longer the pollen tubes, the lower the density. *In vitro*, tubes often burst and density becomes lower still.

Table 3. Buoyant density distribution of Lycopersicon hirsutum (LA 1777) pollen after 2 hours incubation in germination medium. Values are ± standard error of the mean.

Avg. Gradient Position (mm to base)	Percentage of Grains Germinated	Mean Pollen Tube Length	Percentage of Tubes Broken
39	2.0 ± 0.9	0.95 ± 0.12	97.5 ± 1.1
37	66.0 ± 3.5	0.66 ± 0.08	31.0 ± 3.5
33	46.0 ± 6.5	0.33 ± 0.05	3.7 ± 0.9
26	21.3 ± 1.8	0.27 ± 0.03	1.7 ± 0.9

These facts suggest that it should be possible to select for pollen genotypes which germinate under stress by separating these from ungerminated grains.

A variety of centrifugation media are available for the production of liquid density columns. Ludox (colloidal silica particles), although inexpensive, is toxic to pollen and also unstable at physiological pH. More expensive Percoll consists of colloidal silica particles coated with polyvinylpyrrolidone (PVP)), but it too is toxic to pollen, as is Nycodenz, a triiodobenzene ring which is linked to a number of hydrophilic groups. Sucrose should probably not be used since, at least with discontinuous gradients, abrupt osmotic discontinuities may generate artifacts among apparent buoyant densities of pollen grains.

In preliminary studies, Ludox or Percoll are useful in providing indications of which densities can be useful. These two are useful since Percoll (65% v/v) and Ludox (25% v/v), made up in pollen germination medium (see Table 1), will generate continuous density gradients when centrifuged at 26,000 g for 2 hours at an angle $\geq 45^{\circ}$. Addition of density calibration beads (Pharmacia) and hydrated pollen to these gradients and 20 min at 1000 g in a swinging bucket centrifuge head will the range of useful densities.

We have found that Ficoll (dialyzed), a polymer of sucrose, is a suitable centrifugation medium for pollen. Unfortunately it will not self-generate continuous gradients during centrifugation, thus requiring the use of a gradient maker. Loading chambers of a gradient maker with 10% and 40% Ficoll (always in germination medium) will produce an iso-osmotic gradient of 25 - 40% Ficoll. This is appropriate for separating pollen of *Lycopersicon spp.* Ficoll is sufficiently nontoxic to allow its inclusion in pollen germination medium.

The data presented in table 3 show also that, among ungerminated pollen grains, there exists a significant degree of buoyant density heterogeneity. Thus, within each specific density, germinated grains are invariably accompanied by some ungerminated pollen grains. In order to remove all ungerminated grains, we have used a two stage separation. The first is applied to pollen samples immediately after suspension in the germination medium, the second after some pollen germination has occurred.

The next question is whether or not pregerminated pollen will effect fertilization once placed on a stigma. Our preliminary studies have indicated that washed pollen of *Nicotiana alata* will produce a normal complement of seeds and that of *Lycopersicon esculentum* will function normally on stigmas although the pollen tubes grow the equivalent of several pollen diameters before penetrating the stigma. We have not harvested fruits of the latter species so seed set is still unknown. However, these observations indicate that we will be able to utilize pregerminated pollen.

Acknowledgements -- We thank Prof. Sharad C. Phatak, University of Georgia, for kindly providing the GA accessions of Lycopersicon esculentum and both U.S.A.I.D and the Massachusetts Center for Excellence in Biotechnology for financial support.

References

Bouslama, M Schapaugh WT, (1984) Stress tolerance in soybeans. I. Evaluation of three screening techniques for heat and drought tolerance. Crop Sci 24:933-937.

Gibbs A, Greenway B, (1986) Melibiose, a suitable non-permeating osmoticum for suspension of cultured tobacco cells. Jour. Exp. Bot. 37:1079-1089.

Hornby CA, Li Shin-Chai (1975) Some effects of multiparental pollination in tomato plants. Can. Jour. Plant Sci. 55:127-132.

Jaworski CA, Phatak, SC and others. (1987) Ga 1565-2-4 bwt, bacterial wilt-tolerant tomato. HortScience 22:324-325.

Johnson CM, Mulcahy DL, Galinat WC (1976) Male gametophytic in maize:Influences of the gametophytic genotype. Theor. Appl. Genet. 48:299-303.

Jones DF (1927) <u>Selective Fertilization</u>. Univ. Chicago Press. Chicago.

Kumar D, Sarkar KR (1980) Correlation between pollen diameter and rate of pollen tube growth in maize (*Zea mays* L.). Indian Jour. of Exp. Botany 18:1242-1244.

Ottaviano E, Sari Gorla M, Mulcahy DL (1980) Pollen tube growth rate in *Zea mays*:implications for genetic improvement of crops.Science 210:437-438.

Ottaviano E, Mulcahy, D *in press*. Genetics of Angiosperm Pollen. Adv. in Genetics.

Rowley JR, Dunbar A (1970) Transfer of colloidal iron from sporophyte to gametophyte. POLLEN ET SPORES 12:305-325.

Searcy KB, Mulcahy DL (1985) Pollen selection and the gametophytic expression of metal tolerance in *Silene dioica* (Caryophyllaceae)and *Mimulus guttatus* (Scrophulariaceae). Amer. J. Bot. 72:1700-

Searcy KB, Mulcahy DL (1985) The parallel expression of metal tolerance in pollen and sporophytes of *Silene dioica* (L.) Clairv. S. alba(Mill.) Krause and *Mimulus guttatus* DC. Theor. Appl. Genet. 69:597-602.

Ter-Avanesian DV (1949) The role of the number of pollen grains per flower in plant breeding. Bull. Appl. Bot. Plant Breed. (Russian) 28:19-33.

Ter-Avanesian DV (1978) The effect of varying the number of pollen grains used in fertilization. Theor.Appl. Genet. 52:77-79.

Winsor JA, Davis LE, Stephenson AG (1987) The relationship between pollen load and fruit maturation and the effect of pollen load in offspring vigor in *Cucurbita pepo*. Amer. Nat. 129:643-656.

Yamada M, Murakami K (1983) Superiority in gamete competition of pollen derived from F_1 plant in maize. In Pollen:biology and implications for plant breeding, ed. Mulcahy D.L. Ottaviano E. New York:Elsevier Biomedical.

Zamir D, Tanksley S, Jones, J (1982) Haploid selection for low temperature tolerance of tomato pollen. GENETICS 101:129-137.

Breeding for Frost Resistance in *Eucalyptus* Using Pollen Selection: Screening on Viability Using Flow-Cytometry

J.N MARIEN
Association Forêt Cellulose
AFOCEL Région Sud
98, Route de Tournefeuille
31270 CUGNAUX
FRANCE

INTRODUCTION

The extraordinary genetic variation existing in the genus Eucalyptus is a very promising source of breeding potential for a lot of characters, and particulary for adaptation to different ecological areas. Within these adaptative possibilities is the resistance to cold winters.
AFOCEL began to work on Eucalyptus in 1972 and developed a breeding program for successfull plantations in cool temperate climate of Southern France. The winters are generally mild, but sometimes, the temperatures may drop until -15°C to -20°C. Such minima damage many plantations and frost resistance has become the more important criteria in our selection (Marien, 1988).
The different levels of variability have been explored and the best genotypes obtained, at the species, provenances and individual (clonal) levels.
It appeared that the ultimate level of variability was the cellular one, either sporophytic or gametophytic. Studies have been undertaken, using protoplasts (Teulieres, 1985) or pollen (Marien, 1988) to verify the existence of a genetic variability for frost resistance.
The idea of modifying the laws of panmixy, by changing the selecting pressure of one or gamete is not new ; but this theory was developped since some years, as pollen appeared to be an interesting tool for various selection criteria.
Pollen behaviour with high or low temperatures was described by Mulcahy (1984-1985), and on some species as for example, Zea mais (Frova, 1985), Shorgum vulgare (Salgarolo, 1986), Lycopersicon peruvianum (Den Nijs, 1985 ; Maisonneuve, 1986 ; Jones 1982 ; Zamir, 1987) Rheo discolor (Mascarenhas, 1982 ; Souvré, 1981) Oryza sativa (Sakate, 1987)...
In this paper, we describe a possibility to screen pollen after a cold shock, without lost of viability. This point is very important as it allows the use of selected pollen in a breeding strategy.

MATERIALS AND METHODS

Pollen of Eucalyptus gunnii was collected on adult trees, at the anthesis stage, in november 1987, dried and stored at low temperature (-18°C) in vacuum containers (Cauvin, 1984).
Cold treatments were applied according to previous results (Marien, 1988).
Pollen was soaked in a solution of saccharose (20 %) during 16 hours

at +5˚C. It was then introduced in a climatic chamber and submitted
to a controlled freezing program from 0˚C to -6˚C at the rate of -
1˚C/30 minutes. Pollen population was then stained by FDA technique
as described, for example by Salgarolo (1986).
Analysis and screening of the pollen population are made by flow-
cytometry, using an EPICS.C (Coultronics). FDA labelled pollen is
presented in front of a laser ray, providing an exitation to the
marker. Viable cells (corresponding to fluorescent ones) are
deviated and separated from the non fluorescent ones. This technique
also allows a morphologic and quantitative analysis of the pollen
without any marker (Chapel, 1987).
The different subpopulations obtained are scored for fluorescence by
microscope observations and for in vitro germination on an agar
medium (2 g/l) added with saccharose (200 g/l) and boric acid (150
ppm).
Furthermore, artificial pollinations are made in vitro with treated
pollen. The styles are excised after two days. Pollen tube
penetration is observed by staining with aniline blue (Martins,
1959).

RESULTS

Cold treatment (-6˚C) induces a loss of in vitro germination of the
pollen, compared with a control. The mean germination rate of stored
Eucalyptus pollen is variable and the percentage observed here with
the control is 9,5 %. It decreases to 3,5 % after the cold treatment.

The compared analysis of pollen population using flow-cytometry
before and after FDA staining gives a very good idea of their
distributions as shown in table 1.

Table 1 Influence of the size on the distribution of pollen and
relative distribution of fluorescence.

Pollen size (artificial scale)	Bulk population (%)	% fluorescence (FDA test) 100 % population	per size class
> 10	< 1	–	–
9 - 10	2	6	89
7 - 8	5		
5 - 6	10	5	53
3 - 4	20	2	8
0 - 2	63	0	0

The size of the pollen is variable. Only 37 % can be considered as
well developped according to their relative size. The fluorescent
pollen is not evenly distributed. Its percentage decreases from 89 %
to 0 % depending on the size of pollen.
This correspond to a total proportion of 13 % of fluorescent cells
in all the population, and can easely be compared with the 9,5 % of
germination observed.
The structure of the pollen population after a cold treatment is
modified, especially for fluorescence and germination as shown in
table 2.

Table 2   Evolution of in vitro germination of E.gunnii pollen
after cold treatment and flow cytometry screening.

	Percentages compared with :		In vitro germination (%)
	Subpopulation	All population	
* Not screened pollen			
- hydrated	–	–	9,5
- hydrated + cold shock	–	–	3,5
* Pollen screened by flow-cytometry (after a cold shock)			
- all the population	100	100	–
- all sizes - No FDA (fragments exepted)	85% of 100%	85	4,2
- big size only - No FDA	11% of 85%	9	15,4
- all sizes - with FDA (fragments exepted)	12% of 85%	10	24,5
- big size only with FDA	66% of 9%	6	34,5

In this case, 85 % of the whole population is scored. Then 12 % of
the grains appear to be fluorescent. When we consider only the
biggest pollen grains (11 % of the population), then, the percentage
of fluorescence raises to 66 %.
Pollen germination after a cold treatment is not significantly
different for the whole population, with or without flow-cytometry
screening (3,5 to 4,2 %). These percentages raise dramatically when
the screening is done on FDA fluorescence. They can reach up to 34,5
% if big sizes are also used as a screening criteria. Unfortunately,
the number of grains in this case is reduced as only 6 % of the
initial population is concerned.

The penetration of the pollen tubes in the pistil have been observed
and preliminary results are shown in table 3.

Table 3   Pollen load on stigma and pollen tube penetration for
various cold treatments after flow-cytometry screening.

Pollen treatments	Pollen load on stigma	Growth of pollen tube in styles
- Dry	* * *	* * *
- Hydrated (20˚C and cell sorter)	* *	* *
- Hydrated (0˚C and cell sorter)	*	*
- Hydrated (-2,5˚C and cell sorter)	*	*
- Hydrated (-7˚C and cell sorter)	*	*

Number of events :	*	1 to 10
	* *	10 to 50
	* * *	> to 50

Despite the few cold treated and screened pollen grains set down on the stigma, compared to those used in classical artificial hybridizations, it appears clearly that pollen screened by flow-cytometry after a cold shock is enable to germinate and produce a pollen tube going down to the ovary level.

CONCLUSION

E.gunnii pollen stressed by a cold shock shows a very strong decrease in its viability. This fact, added with other observations let us think that we can confirm the existence of a variability for frost tolerance at various genetic levels (including the individual one).
The efficiency of the screening would not be of a great interest in a breeding program if it was impossible to use the pollen after screening.
Flow cytometry appear to be a very important tool in the strategy of pollen selection . We could demonstrate that viability and ability for screened cold treated pollen to germinate in vitro or in the pistil after artificial fecondation was not lost. We could also get separated fractions of pollen population. This appear to be very important as it gives us the possibility to work on large quantities of vegetal material and to concentrate at high levels the surviving cells.

REFERENCES

Cauvin B (1984) Eucalyptus : hybridations contrôlées, premiers résultats In Annales Afocel 1983: 85-117
Cauvin B et al (1987) Eucalyptus : hybridation artificielle - barrières et hérédité des caractères In Annales Afocel 1986: 255-303
Chapel M (1987) Hybridation somatique par électrofusion de protoplastes chez Nicotiana In Thèse Doctorat Université Toulouse, Nov 1987.
Den Nijs A P M et al (1985) Pollen selection in breeding glasshouse tomatoes for low energy conditions In Springer Verlag (ed), Biotech. and ecol. of pollen: 125-130.
Frova C et al (1985) Male gametophyte response to high temperatures in maize In Springer Verlag (ed), Biotech. and ecol. of pollen: 33-38
Jones RA (1982) Pollen selection In California Agriculture, 36: 26-27
Maisonneuve B et al (1984) In vitro pollen germination and tube growth of tomato and its relation with plant growth In Euphytica 33: 833-840
Marien JN (1988) Action de basses températures sur le pollen d'Eucalyptus gunnii In Annales Afocel (à paraître).
Marien JN (1988) Eucalyptus breeding for frost resistance in proceeding of the Australian Forestry Conference - AFDI vol.5.
Martin FW (1959) Staining and observing pollen tubes in the style by mean of fluorescence In Stain Technology 34: 125-128.
Mascarenhas JP et al (1983) The response of pollen of high temperatures and its potential applications In Elsevier medical (ed), Pollen : Biology and implication for plant breeding: 3-8.
Mulcahy DL et al (1975) The inflence of gametophytic competition on sporophytic quality in dianthus chinensis In Theor. Appl. Genet.46: 277-280.
Satake T et al (1987) Male sterility caused by cooling treatment at theyoung microspore stage in rice plants in Japan. Jour. Crop

Science Vol 56/3: 405-410.

Salgarolo P (1986) Etude des stérilités physiologiques chez le sorgho et de l'effet inducteur des abaissements noturnes de température In Institut National Polytechnique de Toulouse.

Souvré A (1981) Etude cytophysiologique et ultrastructurale de l'action du froid sur les microsporocytes et le tapis de Rheo discolor In Thèse – Université P.Sabatier Toulouse.

Teulière C, Alibert G, Marien JN et Boudet AM (1985) Isolement de protoplastes d'Eucalyptus. Relation entre leur survie à basse température et la résistance au froid en conditions naturelles de différents clones In Annales Afocel: 89-103.

Zamir D et al (1987) Pollen selection for low temperatures adaptation in tomato In Theor. Appl. Genet. 74: 545-548.

In Vitro Pollen Selection in *Brassica napus* L.

T. Hodgkin
Scottish Crop Research Institute
Invergowrie
Dundee DD2 5DA
United Kingdom

Introduction

The possibility that pollen selection could be used to bring about changes in the frequencies of genes controlling characters of agronomic significance has been explored by a number of workers (Ottaviano, Sari Gorla & Pe, 1982; Pfahler, 1983; Ottaviano & Mulcahy, 1986). The evidence indicates that 50-60% of the isoenzyme genes expressed in the sporophyte are active in the male gametophyte (Tanksley, Zamir & Rick, 1981; Pedersen, Simonsen & Loeschcke, 1987). Further, Willing & Mascarenhas (1984) have estimated that some 64% of pollen genes in Tradescantia paludosa L. are also expressed in shoots.

Evidence of expression in both sporophyte and gametophyte of genes of economic significance is much more limited. There are examples of functional genetic correlations such as resistance in Zea mays L. to the pathotoxin from Helminthosporium maydis Nisik and Mayake race T (Laughnan & Gabay, 1973). A number of selection experiments have also been described which suggest that gametophyte expression of genes controlling the particular traits occurs. Thus Zamir & Gadish (1987) have shown that pollinations carried out at low temperatures favoured the transmission of Lycopersicon hirsutum Humb. & Bonpl. genes for cold tolerance in the F_2 and backcross from a cross between L. hirsutum and the less tolerant L. esculentum Mill.

Until now pollen selection has been carried out in vivo, either by maintaining the female parent in a potentially selective environment (Zamir, Tanksley & Jones, 1982), or by applying selective agents to the style (Simon & Sanford, 1986), or by investigating the effect on the sporophyte of selection for a character such as pollen competitive ability (Ottaviano, Sari Gorla & Pe, 1982). In vitro selection, applying selection pressure to pollen populations prior to pollination, would clearly extend the potential use of pollen selection to a much greater range of characters.

Pollen of Brassica napus L germinates well in vitro (Hodgkin & MacDonald, 1986) and was used in thin layer chromatographic bioassay to investigate the effect of phytotoxic compounds present in culture filtrates of the fungal pathogen _Alternaria brassicicola_ (Schw.) Wilts. (Hodgkin & MacDonald, 1986). Those fractions of the culture filtrate extract which killed sporophytic secondary embryoids also inhibited pollen germination in such assays. A procedure by which incubated pollen of _B. oleracea_ L. could be used for seed production was also developed (Hodgkin, 1987). This paper describes experiments in which pollen, incubated in the presence of a toxic culture filtrate extract of _A. brassicicola_, was used for seed production to determine whether _in vitro_ pollen selection would result in plants with improved resistance to the toxic extract.

Material and Methods

For the selection experiments phytotoxic extracts of _A. brassicicola_ culture filtrates were prepared initially as described by MacDonald & Ingram (1986). These extracts were further purified by partitioning three times with ethyl acetate, drying the water soluble fraction and redissolving in water to give the desired final concentrations of 10 mg ml^{-1} or 20 mg ml^{-1} in germination medium of the partially purified toxic extract.

Pollen was collected from freshly opened flowers of glasshouse grown plants of the cultivars Herkules (oil seed rape) and Arran (forage rape). Cv. Herkules has some resistance to the toxic extract from _A. brassicicola_ (MacDonald pers. comm.) and cv. Appin is partially resistant to the pathogen. The pollen was incubated in shake culture for 1 h either in germination medium (Hodgkin & Lyon, 1986) or in germination medium containing 20 mg ml^{-1} of the water soluble fraction of the _A. brassicicola_ culture filtrate. Following incubation, the pollen was centrifuged, rinsed twice, recentrifuged and used to pollinate decapitated pistils of flowers of glasshouse grown plants of cv. Primor (an oilseed rape cultivar susceptible to _A. brassicicola_ and its toxic extract) which had been emasculated prior to anthesis.

Seeds harvested from pollinations with incubated pollen, together with seed from control pollinations with unincubated pollen of cvs. Appin and Herkules, were sown under glass, vernalised for 8 weeks at 6°C and flowered in an unheated glasshouse. Shortly after germination the sexual origin of the seeds was determined from their cotyledon acid phosphatase isoenzyme profile, using polyacrylamide gel electrophoresis (Wills, Fyfe & Wiseman, 1979), for which the parental cultivars differed. Before vernalisation the effect of the water soluble extract of the _A. brassicicola_ culture filtrate on the progeny seedlings was tested by a leaf disc bioassay (MacDonald & Ingram, 1986).

Pollen germination of the progeny plants was determined in standard germination medium and in medium containing 10 mg ml^{-1} or 20 mg ml^{-1} of the toxic extract. Germination tests comprised at least 2 replicate hanging drops and counts were made after incubation for 4 h at 20°C of at least 300 grains per hanging drop from random fields at 60x magnification.

Results

Incubated pollen of the three B. napus cultivars germinated freely, both in hanging drops (mean 68.9%) and in shake culture, although germination rates were somewhat reduced in the latter. Preliminary experiments confirmed the results obtained by Hodgkin (1987) that incubated pollen could be centrifuged, rinsed, recentrifuged and resuspended with little loss of viability and that pollination of decapitated pistils with such pollen would result in seed set.

Pollen germination of all three cultivars was markedly reduced in the presence of 10 mg ml^{-1} of the water soluble fraction of the toxic extract. Pollen from cv. Primor germinated least well in the presence of the toxic extract with germination reduced to 21% of its control while germination of pollen from cvs. Herkules and Arran was only reduced to 38% and 36% respectively of their controls. Pollen-tube lengths of all three cultivars were reduced to approx. 38% of that of the controls. At 20 mg ml^{-1} of the toxic extract only occasional malformed pollen-tubes were observed in all of the cultivars and tests with fluorescein diacetate showed that over 90% of the pollen of each cultivar failed to stain after incubation for 1 h.

Table 1. Experimental data for pollinations of decapitated pistils of B. napus cv. Primor with pollen from cvs. Appin or Herkules incubated in germination medium, in medium containing 20 mg ml^{-1} A. brassicicola toxic extract, or left unincubated (control).

Pollen parent	Pollen treatment	Flowers pollinated	Seeds set	Seeds sown	% germinated	% maternal
Appin	incubated + toxin	163	81	40	41	25
	incubated - toxin	142	85	42	48	0
	control	129	554	50	53	0
Herkules	incubated + toxin	251	241	120	59	31
	incubated - toxin	209	237	120	56	6
	control	92	250	50	48	0

Surprisingly, seed set per pollinated flower from pollen of both cultivars incubated in the presence of the toxic extract was only slightly lower than that from pollen incubated in control germination medium (Table 1). However, unincubated pollen of both cultivars gave much higher seed sets than incubated pollen (4.29 seeds/flower for cv. Appin and 2.72 for cv. Herkules, Table 1).

Half the seeds obtained from each of the incubated pollen treatments and 50 seeds from each of the control pollinations were sown. Seed germination was poor (approx 50%) and tests to determine the sexual origin of the seedlings showed that a significant proportion (30%) of those from pollinations using pollen incubated in the presence of the toxic extract did not result from cross-pollination (Table 1). In contrast, only 5% of those from pollination with incubated unselected pollen were not from cross pollinations and none were found in progeny from the control pollinations. It is likely that seedlings with the maternal isoenzyme pattern arose as apomictic maternals (Mackay, 1972).

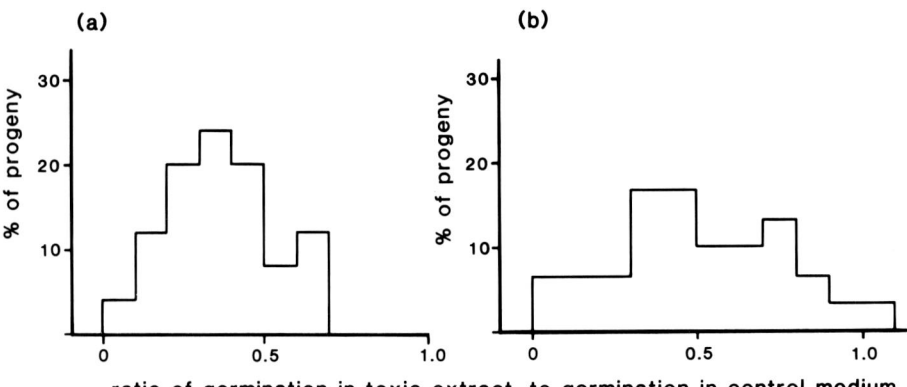

Fig 1. Proportions of B. napus progeny plants from pollinations with (a) unselected pollen and (b) pollen incubated in 20 mg ml^{-1} toxic extract grouped by ratio of germination in medium containing 10 mg ml^{-1} toxic extract to germination in control germination medium.

Pollen from the F_1 progeny was germinated in control medium and in media containing 10 mg ml^{-1} and 20 mg ml^{-1} of the toxic extract. Germination of pollen from individual plants ranged from 21.7 - 90.7% (mean = 59.1%) in the control and from 0 - 71.8% (mean = 24.2%) in 10 mg ml^{-1} while no germination was obtained in 20 mg ml^{-1} of the toxic extract. The ratio of germination in 10 mg ml^{-1} toxic extract to that in control medium ranged from 0 - 0.66 for progeny from pollinations with unincubated pollen (Fig 1). The progeny from pollinations with incubated unselected pollen gave a similar range (0 - 0.67) except for one plant. Pollen from this plant germinated almost as well in 10 mg ml^{-1} toxic extract (66.5%) as in the control medium (67.7%). The progeny from pollinations with pollen incubated in toxic extract showed a wider range of ratios (0 - 1.0, Fig 1) and 8 plants exceeded the germination ratios of progeny from unincubated pollinations. Additional tests on these 8 plants and a random sample of 8 other plants confirmed these results.

Discussion

The results obtained demonstrate that B. napus pollen can be incubated in vitro, subjected to a selection procedure and used successfully in pollinations. The techniques used in the experiments were simple and employed no expensive or sophisticated equipment and it is likely that similar techniques could be used with a wide range of species. However, in Brassica spp., where apomictic maternals can occur, the results also show that some genetic test of the hybrid nature of the derived progeny is essential and it would be wise to include such a test in any experiments of this nature.

The results suggest that pollen from eight plants (23%) of the progeny derived from pollination with selected pollen and one plant derived from unselected incubated pollen have enhanced resistance to the toxic extract. The F_2 progeny from these plants will be tested and compared with control F_2 progenies to determine the inheritance of this character. The tests will be done on both sporophyte and pollen to determine whether the character is expressed in both. The mode of action of the resistance shown to the toxic extract must also be investigated. It may involve some property unique to the gametophyte which is not expressed in the sporophyte.

The development of techniques for in vitro pollen selection considerably enhances the potential of gametophyte selection in crop improvement. Attractive characters for pollen selection experiments in Brassica spp. include cold tolerance where this depends on membrane integrity, resistance to phytotoxins from Phoma lingam (Sacristan, 1982), herbicide resistance or resistance to antibiotics (Bino, Hille & Franken, 1987). However, in each case, it will be necessary to determine that gametophytic and sporophytic gene expression overlap (or that a useable gametophytic gene is linked to the sporophytic gene of interest), that no other gametophytic genes act to modify or mask the effect of the genes of interest and that an effective selection system can be devised which operates within a sufficiently short time (1-2h).

Where these criteria are fulfilled, pollen selection may offer advantages over other in vitro systems acting on the sporophyte. Thus, as with cell selection

large numbers of individuals are exposed to selection pressure and, as with
microspore selection, selection is combined with gene segregation and recombination. In addition, difficulties in plant regeneration that occur in other in
vitro systems are avoided and interference with plant breeding programmes can
be minimised.

References

Bino RJ, Hille J, Franken J (1987) Kanamycin resistance during in vitro develop-
    ment of pollen from transgenic tomato plants. Plant Cell Reports 6: 333-336
Hodgkin T (1987) A procedure suitable for in vitro pollen selection in Brassica
    oleracea. Euphytica 36: 153-159
Hodgkin T, Lyon GD (1986) The effect of Brassica oleracea stigma extracts on
    the germination of B. oleracea pollen in a thin layer chromatographic bio-
    assay. J Exp Bot 37: 406-411
Hodgkin T, MacDonald MV (1986) The effect of a phytotoxin from Alternaria bras-
    sicicola on brassica pollen. New Phyt 104: 631-636
Laughnan JR, Gabay SJ (1973) Reaction of germinating maize pollen to Helmintho-
    sporium maydis pathotoxins. Crop Sci 13: 681-684
MacDonald MV, Ingram DS (1986) Towards the selection in vitro for resistance
    to Alternaria brassicicola in Brassica napus ssp. oleifera, (Metzg.) Sinsk.,
    winter oilseed rape. New Phyt 104: 621-629
Mackay GR (1972) On the genetic status of maternals induced by pollination of
    B. oleracea L. with B. campestris L. Euphytica 21: 71-77
Ottaviano E, Mulcahy DL (1986) Gametophytic selection as a factor of crop
    plant evolution. In: Barigozzi C (ed) The Origin and Domestication of Culti-
    vated Plants. Elsevier, Amsterdam Oxford New York Tokyo p 101
Ottaviano E, Sari Gorla M, Pe E (1982) Male gametophyte selection in maize.
    Theor Appl Genet 63: 249-254
Pedersen S, Simonsen V, Loeschcke V (1987) Overlap of gametophytic and sporo-
    phytic gene expression in barley. Theor Appl Genet 75: 200-206
Pfahler PL (1983) Comparative effectiveness of pollen genotype selection in
    higher plants. In: Mulcahy DL, Ottaviano E (eds) Pollen: Biology and Impli-
    cations for Breeding. Elsevier, New York Amsterdam Oxford p 361
Sacristan MD (1982) Resistance responses to Phoma lingam of plants regenerated
    from selected cell and embryogenic cultures of haploid Brassica napus.
    Theor Appl Genet 61: 193-200
Simon CJ, Sanford JC (1986) Induction of gametic selection in situ by stylar
    application of selective agents. In: Mulcahy DL, Mulcahy B, Ottaviano E
    (eds) Biotechnology and Ecology of Pollen. Springer, New York Berlin Heidel-
    berg Tokyo, p 107
Tanksley DS, Zamir D, Rick CM (1981) Evidence for extensive overlap of sporo-
    phtyic and gametophytic gene expression in tomato. Science 213: 453-455
Willing RP, Mascarenhas JP (1984) Analysis of the complexity and diversity of
    mRNAs from pollen and shoots of Tradescantia. Plant Phys 75: 865-868
Wills AB, Fyfe SK, Wiseman EM (1979) Testing F$_1$ hybrids of Brassica oleracea
    for sibs by seed isoenzyme analysis. Ann Appl Biol 91: 263-270
Zamir D, Gadish I (1987) Pollen selection for low temperature adaptation in to-
    mato. Theor Appl Genet 74: 545-548
Zamir D, Tanksley SD, Jones RA (1982) Haploid selection for low temperature to-
    lerance of tomato pollen. Genetics 101: 129-137

Possibility of Expanding Genetic Variation by Limited Pollination Based on the Reproductive Success Rate (RSR) of Pollen Grains Deposited on Stigma

H.Namai and R.Ohsawa

Institute of Agriculture and Forestry

University of Tsukuba

Tsukuba, Ibaraki, 305

Japan

1. Introduction

We have already proposed the new concept of "Reproductive Success Rate" (RSR) pertaining to the pollen grains deposited on a stigma and the ovules in a pollinated flower (Namai and Ohsawa 1986). The RSR of pollen grains deposited on a stigma (RSR-P) and the RSR of ovules in a pollinated flower (RSR-O, i.e. ordinary seed set percentage) are calculated from the following equations:

RSR-P(%) = (No. of seeds obtained) × 100 / (No. of pollen grains deposited)

RSR-O(%) = (No. of seeds obtained) × 100 / (No. of ovules in pollinated flower)

We also described the possible effect of the number of pollen grains deposited on a stigma to RSR-P and RSR-O such that the highest RSR-P is accomplished under limited pollination with about 1:1 ratio of pollen grains deposited to ovules in pollinated flower. Such limited pollination is presumed to give rise to the lowest intensity of male gametic competition in post pollination.

According to our subsequent mathematical analyses, the highest intensity of transferring progenitor genes and genotypic variation in progeny can be obtained under limited pollination. Therefore, the possibility of expanding genetic variation in the progeny through limited pollination is strongly suggested.

Based on the concept of RSR-P and RSR-O, the wide variations among plants derived from less pollen grains deposition (Ter-Avanesian 1978) as well as the vigorous growth of plants derived from excessive pollen grains deposition (Mulcahy and Mulcahy 1975; Mulcahy et al. 1975) can be well explicated.

2. Relationships between the number of pollen grains deposited on a stigma and RSR P and
   RSR-O in some cultivated plants

The following are the brief results of our pollination studies to determine the
relationships between the number of pollen grains deposited on a stigma and RSR-P and
RSR-O in three species (Namai and Ohsawa 1986; Namai and Kato 1987). These data
explained graphically the changes of RSR-P and RSR-O with different number of pollen
grains deposited on a stigma. Similar phenomena were also observed in Geranium
maculatum and Mirabilis jalapa (Mulcahy et al.1983, Cruden 1977, Namai and Ohsawa 1986).

(1) Brassica juncea

Under field condition with many insect pollinators, about 500 to 1,000 pollen grains are
deposited on a stigma of Brassica juncea flower with about 20 to 25 ovules (Ohsawa and
Namai 1987). As shown in Fig.1, RSR-P increased abruptly from 0% in flowers with 1 to
4 selfed pollen grains on stigma to 12% in those with 10 pollen grains and approximately
25% in those with 20 pollen grains which is about the same number as the number of
ovules per flower. However, RSR-P decreased rapidly in the flowers with more than 20
to 50 pollen grains deposited on a stigma and decreased slowly from 6% in those with
250 grains to 3% in those with 500 pollen grains. On the other hand, RSR-O increased
from 3% in flowers with more than 4 selfed pollen grains to 80% in those with 500
pollen grains (Namai and Ohsawa 1986).

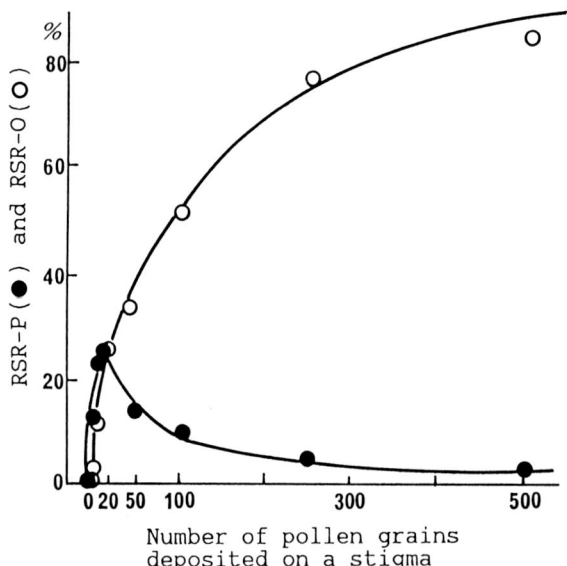

Fig.1.   Correlation
between the number
of pollen grains
deposited on a stigma
and RSR-P and RSR-O.

(Adapted from Namai
and Ohsawa 1986)

(2) Fagopyrum esculentum

About 10 compatible pollen grains are deposited on a stigma in Fagopyrum esculentum (buckwheat) with one ovule per flower (Namai 1986).   As shown in Fig.2, both RSR-P and RSR-O were 40% in flowers with only one compatible pollen grain.   In flowers with 10 compatible pollen grains RSR-O increased to 90%, though RSR-P decreased to 9%. Therefore, the highest intensity of RSR-P appeared in the flowers with approximately 1:1 pollen grain to ovule ratio (Namai and Ohsawa 1986).

(3) Oryza sativa

In improved cultivars of rice (Oryza sativa) with one ovule per flower, dozens to hundreds of selfed pollen grains are deposited gravitationally, though only a few crossed pollen grains (mean of 5.4 pollen grains) are deposited by wind in potted male sterile plants exposed to air-borne pollen in paddy field.   As shown in Fig.3, the highest intensity of RSR-P was also observed in flowers with 1:1 pollen grain to ovule ratio (Namai and Kato 1987).

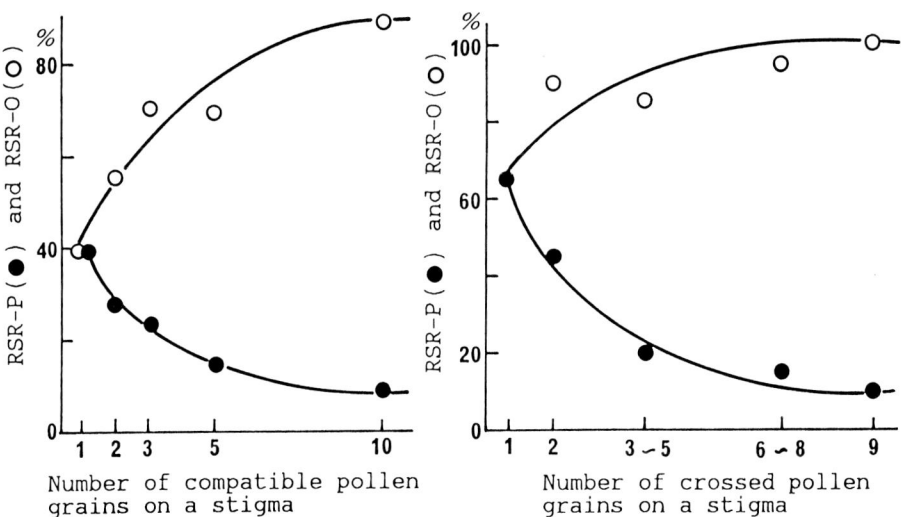

Fig.2.   Correlation between the the number of compatible pollen grains on a stigma and RSR-P and RSR in Fagopyrum esculentum. (Adapted from Namai & Ohsawa 1986)

Fig.3.   Correlation between the number of crossed pollen grains on a stigma and RSR-P and RSR-O in Oryza sativa. (Adapted from Namai and Kato 1987)

3. Evaluation of the possibility of expanding genetic variation through limited
   pollination

In order to use various plant genetic resources efficiently, it is very important to
expand genetic variations and evaluate completely different valuable traits by simple
technique.   For these purposes, limited pollination wherein approximately 1:1 ratio of
the number of pollen grains deposited on stigma to the number of ovules in pollinated
flower should be a practical method due to production of a lot of seeds with wide
genetic variations.

(1) Efficient utilization of 1,000 pollen grains for producing a lot of seeds

Table 1 shows the expected number of seeds obtained from artificial pollination with
different number of pollen grains per stigma in Brassica juncea.   Pollination of
approximately 500-1,000 pollen grains onto a flower with about 20 ovules (pollen-ovule
ratio is approximately 25-50 : 1) produces only 16-20 seeds.   However, limited
pollination of approximately 20 pollen grains (pollen-ovule ratio is approximately 1:1)
produces more than 200 seeds in all.   Therefore, the most number of seeds are obtained
under such limited pollination (Namai and Ohsawa 1986).

Table 1. Variation of expected number of seeds
          obtained from 1,000 pollen grains by
          self-pollination with different number
          of pollen grains per stigma

Number of pollen grains per stigma	Number of flower pollinated	Expected number of seeds obtained
4	250	0.0
5	200	124.0
10	100	236.0
20	50	245.0
25	40	200.8
50	20	132.4
100	10	103.2
250	4	62.0
500	2	32.6
1,000	1	20.0

(Adapted from Namai and Ohsawa 1986)

(2) Evaluation of the total intensity of transferring progenitor genes and the intensity
    of genotypic variation in progeny on the basis of RSR-P and RSR-O

The total intensity of transferring progenitor genes (Tr) can be evaluated using RSR-P
and RSR-O, and corresponds to the possibility that fertilization will occur.   This is
obtained by the following equation:

$$Tr(\%) = \frac{2\times (\text{Number of seeds obtained})}{(\text{Number of pollen grains deposited}) + (\text{Number of ovules per flower})} \times 100$$

The intensity of genotypic variation in progeny (Gv) means the ratio of the number of practical fertilizing gamete combinations producing seeds to the total number of possible fertilizing combinations between the pollen grains deposited and the ovules in a pollinated flower. This is obtained by the following equation:

$$Gv = (RSR\text{-}P) \times (RSR\text{-}O)$$

Table 2 shows the expected RSR-P and RSR-O in pollination with different number of pollen grains deposited per stigma and the corresponding Tr and Gv values in B. juncea. The limited pollination onto many flowers by dividing a certain number of pollen grains in the pollen-ovule ratio of approximately 1:1 is presumed to give rise to the highest intensity of Tr(24.5) and Gv(600.3). This increases the possibility of fertilization as well as the intensity of genetic variation. In F. esculentum and O. sativa, such limited pollination is also presumed to give rise to the highest Tr and Gv.

Table 2. Variation of expected total intensity of transferring progenitor genes (Tr) and intensity of genotypic variation in progeny (Gv) in Brassica juncea

No.of pollen grains/stigma	No.of flowers pollinated	RSR-P (%)	RSR-O (%)	Tr (%)	Gv
5	200	12.4	3.1	5.0	38.4
10	100	23.5	11.8	15.7	277.3
20	50	24.5	24.5	24.5	600.3
25	40	20.1	25.1	22.3	504.5
50	20	13.2	33.1	18.9	436.9
100	10	10.3	51.6	17.2	531.4
250	4	6.2	77.5	11.5	480.5
500	2	3.3	81.5	6.3	269.0
1,000	1	2.0	100.0	3.9	200.0

4. Conclusion

Our studies on RSR-P and RSR-O revealed the pathway for expanding genetic variation by limited pollination as follows:
① The highest RSR-P will be accomplished under limited pollination with 1:1 pollen to ovule ratio,
② The lowest intensity of male gametic competition in post pollination will appear due to the highest RSR-P,

③ The highest frequency of transferring progenitor genes (Tr) will perform due to the highest total intensity of RSR-P and RSR-O in pollination,

④ The widest genetic variation will occur due to the highest intensity of genotypic variation (Gv).

Therefore, limited pollination must be effective for expanding genetic variation whereas excessive mixed pollination or sufficient number of insect pollinators must be needed for enhancing random mating within a seed growing population of entomophilous plant (cf. Ohsawa and Namai 1988) as well as for contracting genetic variation.

References

Cruden RW (1977) Pollen-ovule ratio: A conservative indicator of breeding systems in flowering plants. Evolution 31: 32-46

Mulcahy DL, Mulcahy GB (1975) The influence of gametophytic competition on sporophytic quality in Dianthus chinensis. Theor Appl Genet 46: 277-280

Mulcahy DL, Mulcahy GB, Ottaviano E (1975) Sporophytic expression of gametophytic competition in Petunia hybrida. In: Mulcahy DL (ed) Gamete Competition in Plants and Animals. North-Holland Publ, Amsterdam. pp227-232

Mulcahy DL, Curtis PS, Snow AA (1983) Pollen competition in a natural population. In: Jones CE, Little RJ (ed) Handbook of Experimental Pollination Biology. Scientific and Academic Editions, New York. pp330-337

Namai H (1986) Pollination biology and seed multiplication method of buckwheat genetic resources. In: Inst Soil Sci & Plant Cul (ed) Buckwheat Research 1986. The Organizing Committee of 3rd Intl Symp Buckwheat, Plaway Poland. pp180-186

Namai H (in press) Inducing cytogenetical alterations by means of interspecific and intergeneric hybridization in brassica crops. Proc Gamma Field Symp 26

Namai H, Kato H (1987) The number of pollen grains deposited upon pistil assuring seed setting of male sterile seed parent in rice (Oryza sativa L.). Japan J Breed 37: 98-102

Namai H, Ohsawa R (1986) Variation of reproductive success rates of ovule and pollen deposited upon stigmas according to the different number of pollen on a stigma in Angiosperm. In: Mulcahy DL, Mulcahy GB, Ottaviano E (ed) Biotechnology and Ecology of Pollen. Springer-Verlag, New York Berlin Heidelberg Tokyo. pp423-428

Ohsawa R, Namai H (1987) The effect of insect pollinators on pollination and seed setting in Brassica campestris cv. Nozawana and Brassica juncea cv. Kikarashina. Japan J Breed 37: 453-463

Ohsawa R, Namai H (1988) Cross-pollination efficiency of insect pollinators (Shimahanaabu, Eristalis cerealis) in rapeseed, Brassica napus L. Japan J Breed 38: 91-102

Ter-Avanesian D V (1978) The effect of varying the number of pollen grains used in fertilization. Theor Appl Genet 52: 77-79

^{31}P and ^{1}H NMR as a Non-Destructive Method for Measuring Pollen Viability

J. A. R. Ladyman[1] and R. E. Taylor[2]

Shell Agricultural Chemical Co.

Modesto, California, U.S.A.

<u>Introduction.</u>

Most techniques that provide information about the viability of pollen, e.g. staining with vital dyes or germinability tests, are destructive and thus of limited value where pollen supplies are small and the pollen is required for pollination or destructive biochemical experiments. Dumas et al. (1982) have reported using proton (^{1}H) nuclear magnetic resonance (NMR) to correlate water content with pollen viability. Another method for estimating pollen viability would be to use phosphorus (^{31}P) NMR which provides information as to the amounts of organic phosphorus (e.g., ATP) available for metabolism.

These experiments show some limitations to using ^{1}H NMR to determine pollen viability, but ^{31}P NMR appears to be a very promising technique.

<u>Experimental Procedures and Results</u>

^{1}H NMR Experiments. Experiments were performed at ambient temperature with a Bruker WM-360 spectrometer. The decoupling coil of a 5mm ^{13}C probe was used because it reduced ringdown giving a better baseline for observing broad NMR lines. Resonances with full widths at half maximum of 10 kHz could easily be detected. Data were aquired with quadrature detection with unlocked magnetic field. A typical aquisition consisted of 32 scans in a

---

Present address: 1. Plant Cell Research Institute, Inc., 6560 Trinity Court, Dublin, CA 94568, USA. 2. Bruker Instruments Inc., Billerica, MA 01821, USA

total time of approximately 5.3 minutes on 0.5-1 mg pollen. Proton NMR could be used to follow moisture loss from the same sample of pollen over time, e.g. Table 1. Germination of pollen (rye, *Secale cereale*, and corn, *Zea Mays*) was on an agar-based medium. Fresh pollen routinely germinated >75%. Pollen grains were judged to have germinated when the length of the tube was >1.5x pollen grain diameter.

Table 1. ^{1}H NMR data of desiccating rye pollen.

Time (hr)	% Original area under signal	FWHM (Hz)	% germination
0	---	---	77
0.25	100	696	71
1.25	97	708	47
2.25	89	713	<9
3.25	77	740	0

The resonance that is attributed to moisture declines as the germinability of pollen declines. However, in terms of determining viability, rehydration after desiccation confounds interpretation (Figure 1).

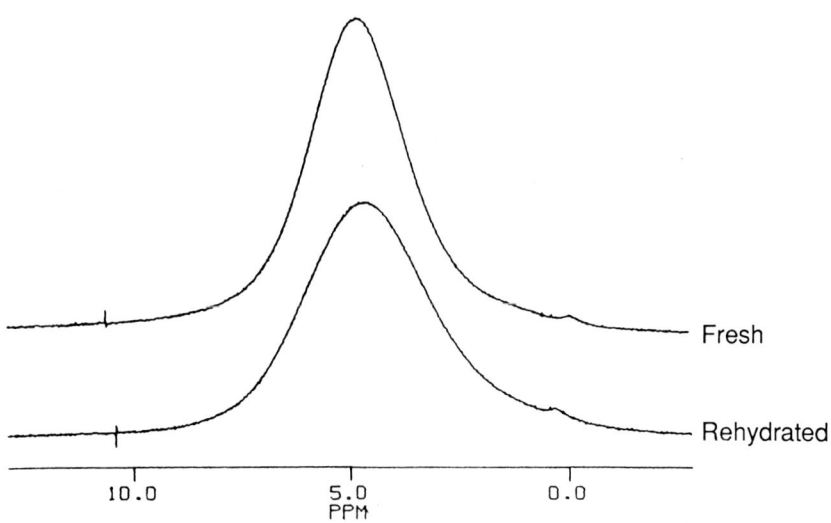

Figure 1. ^{1}H NMR spectra of fresh and rehydrated rye pollen.

The ^{1}H linewith is about 1 Hz for liquid water and about 45 kHz for rigid molecules in ice. In our studies there was <u>no</u> indication that two clearly different types of water, i.e. bound and free, were present in cereal pollen.

It has been suggested (Duplan and Dumas, 1984) that viability of pollen may best be measured using the Carr-Purcell-Meiboom-Gill (CPMG) sequence and measuring the relaxation time (T2) of the ^{1}H signal. Under the CPMG sequence not all the resonances of the ^{1}H spectrum relaxed at the same rate (Figure 2).

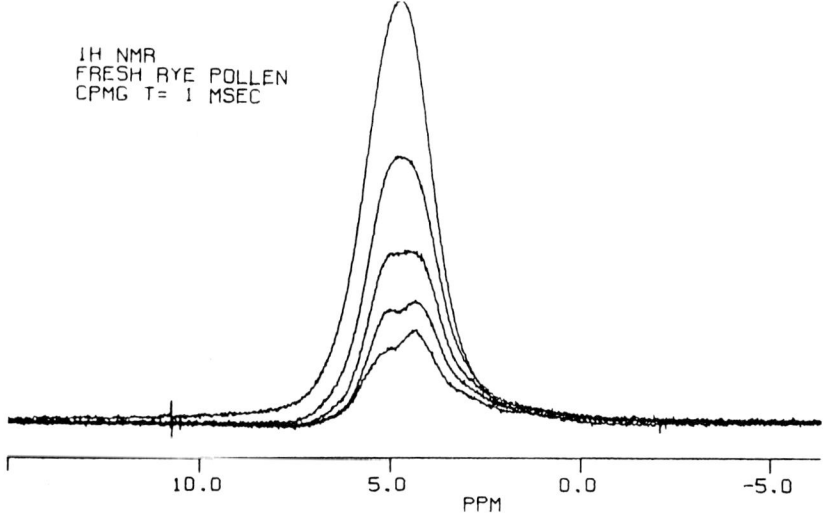

Figure 2. Spectrum of the ^{1}H NMR decay rate under the Carr-Purcell-Meiboom-Gill (CPMG) sequence 15 minutes after rye pollen collection.

This may have been due to the different signals from water and lipids in the pollen. In the 360 MHz ^{1}H spectra an upfield shoulder that is associated with lipid components was present (Priestly and Kruijff, 1982). The prominence of the upfield shoulder was dependent upon species, e.g it was larger in corn than in rye pollen. The CPMG relaxation rates, as a function of time, did not show a monotonic relationship as pollen lost water (e.g. Table 2). However, accurate measurements of chemical shifts and linewidths are difficult because of bulk magnetic susceptibility effects in this relatively immobile

Table 2.   Relaxation times of the ^1H signal from rye pollen using the CPMG sequence.   The CPMG experiment is performed on the WM-360 with the limitation $\tau > 0.6$msec.

Time 0	T2 (msec): Experiment 1	T2 (msec): Experiment 2
Plus 4 minutes	9.56	8.91
Plus 1 hour	14.68	10.00
Plus 2 hour	10.21	10.80
Plus 3 hour	10.99	9.97
Plus 4 hour	12.48	16.60

system. Duplan and Dumas used an 80 MHz instrument, which may account for differences observed between their work and the present study.   The line widths in these studies were generally larger than those reported by Duplan and Dumas.   Since chemical shift dispersion and magnetic susceptibility increase with higher magnetic fields, the differences probably can be ascribed to the difference in magnetic field strength between the two instruments.

31P NMR experiments.   Experiments were performed with the Bruker WM-360 NMR spectrometer using a 10 mm broad band VSP probe.   Typical aquisitions were 4500 scans/15.38 minutes. ^{31}P NMR spectra   reflected germinability more clearly. The quality of the spectra from pollen was dependent upon the plant species and the quantity of pollen.   The amounts of fresh pollen required for obtaining acceptable spectra ranged from 100mg (cotton, *Gossypium hirsutum*) to 400mg (rye and corn).   This may be a function of size, shape and mobility of the pollen.   There was generally a lower signal to noise ratio for corn than for rye pollen, but the behaviour of the peaks relative to germinability was similar.   The γ, α and β phosphate (^{31}P) resonances of ATP are at -5, -10, and -18 ppm, respectively.   The ^{31}P resonances from UDPG, NADP and the α phosphate from ADP are also at -10 ppm.   The ^{31}P resonance at -5 ppm also includes that from the β phosphate of ADP.   The inorganic and sugar phosphates appear at about 3 ppm.   For the following discussion the term organic phosphates excludes sugar phosphates.   Introduction of pollen

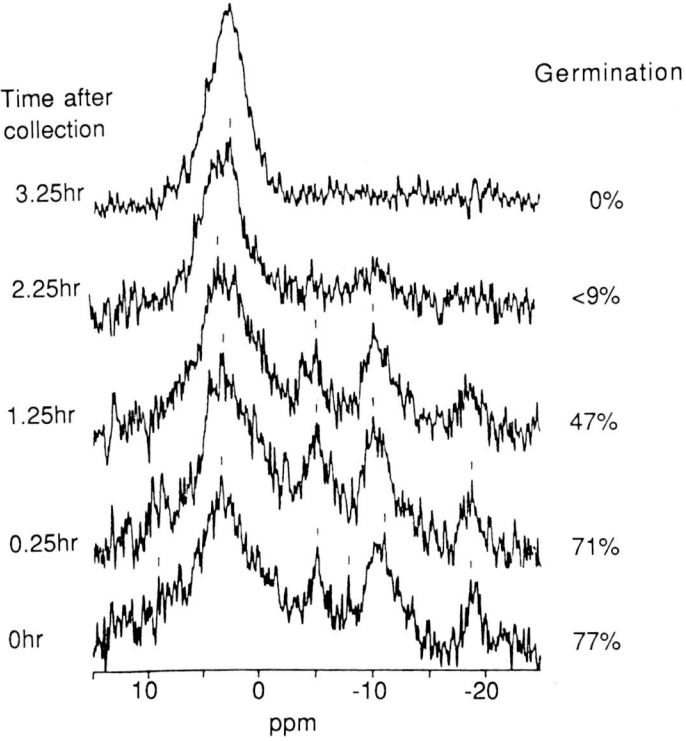

Figure 3. ^{31}P NMR spectra of rye pollen with different germination (%).

into an anaerobic environment or water led to an immediate disappearance of organic ^{31}P peaks for all species of pollen tested (data not shown).

Germinability and the size of the organic ^{31}P peaks, especially the one corresponding to the β phosphate group of ATP, correlated well with each other for several species. Data for rye are shown in Figure 3 and can be directly compared to the data in Table 1. Figure 4 shows the ^{31}P NMR spectra of corn pollen before and after drying and after rehydration. By solid state NMR, inorganic phosphorus was shown to be present in the dry sample. On rehydration, this peak, unlike the organic phosphorus peaks, reappeared.

Conclusion.

Our experiments have shown that for measuring the water loss from a sample of pollen, ^{1}H NMR can be used easily. The technique is also useful for investigating pollen composition. However, for estimation of pollen viability ^{1}H NMR can not be recommended. ^{31}P NMR is a very promising technique for measuring viability and has the advantage that biochemical events can be monitored non-invasively. Rehydration after dying is not a complication in the interpretation of ^{31}P results as it is with ^{1}H NMR (compare Figure 1 to Figure 4). A disadvantage of ^{31}P NMR - with the instrument used in these experiments - is the relatively large sample size. There is great potential in using both ^{1}H and ^{31}P NMR for investigating the effects of chemicals, the environment, and *in vitro* manipulation, on pollen.

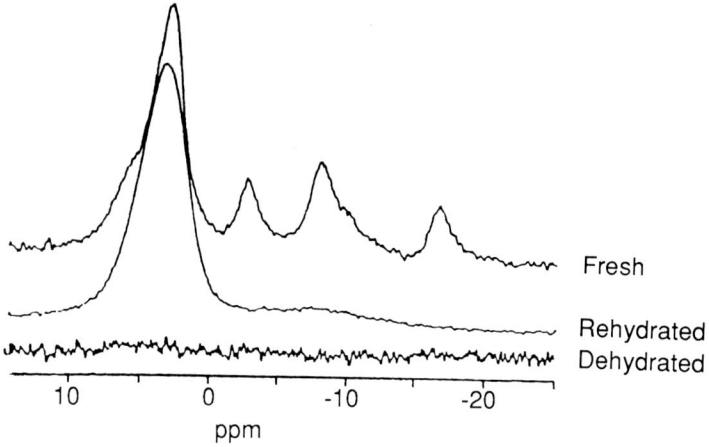

Figure 4. ^{31}P NMR spectra of fresh, dry and rehydrated corn pollen.

References.

Dumas C, Duplan J, Said C, Soulier J (1983) [1]H Nuclear magnetic resonance to correlate water content and pollen viability. In: Mulcahy DL, Ottoviano E (Eds) Pollen: Biology and Implications for Plant Breeding. Elsevier, Amsterdam, p15

Duplan J, Dumas C (1984) Viabilite Pollinique et Conservation du Pollen. In: Herve Y, Dumas C (Eds) Incompatibilite Pollinique et Amelioration des Plantes. Le Departement de Formation Continue de l'Ecole Nationale Superieure Agronomique, Rennes Cedex, France, p 40

Priestly DA, de Kruijff B (1982) Phospholipid motional characteristics in a dry biological system. Plant Physiol. 70: 1075-1078

The Effect of Pollen Load on Pollen Tube Performance in Apple, Pear and Rose Styles

T. VISSER[1]), RENATA SNIEZKO[2]) & CLARA M. MARCUCCI[3])

1) Institute for Horticultural Plant Breeding
Mansholtlaan 16
6708 PA Wageningen
The Netherlands

INTRODUCTION

In vitro, pollen density has been shown to markedly influence the germination of pollen of many plant species (VISSER, 1955; BREWBAKER & MAJUMBER, 1961). In vivo, WILLIAMS & MAIER (1977) observed that tube length of self-incompatible pollen in apple styles (insignificantly) increased with increasing numbers of pollen grains on the stigma. LEE (1980), working with plum pollen, found significant correlations between the number of grains on the stigma and pollen tube growth in the style in 5 out of 7 instances, but these correlations lack causality. In view of the scant information on an in-vivo effect, it was thought useful to investigate if and to what extent pollen tube growth in the style depends on the 'pollen load', in casu the number of (germinated) pollen grains on the stigma.

METHODS AND MATERIAL

Ways and means of experimentation with apple and pear were largely the same as earlier described by VISSER & OOST (1982). In the spring of 1983, branches were cut and put in a 'Chrysal' solution. The flowers were depetalled and emasculated in the balloon stage and thereafter hand-pollinated (once) with fresh pollen. Four trials on the apple cvs 'Golden Delicious' and 'James Grieve' and the pear cvs 'Doyenne du Comice' and 'Bonne Louise' were carried out at a controlled temperature of 18°C at Wageningen; two trials on the apple 'Starkrimson' and the pear 'Abate Fetel' were done at ambient temperature of 16°-20°C at Bologna. Apart from compatible pollen, also self-incompatible and incongruent (apple x pear,

2) Institute of Biology, University Marie Curie Slodowska, Lublin, Poland
3) Istituto Coltivazioni Arboree, Universita di Bologna, Italy

pear x apple) pollen was used. Samples for studying germination and tube growth, consisting of 30 styles (6-7 flowers)/cultivar/ pollen type) were collected after 8, 14 or 16, 24, 40 and/or 48, 64 or 72 hours after pollination. Per style we determined the number of germinating grains on the stigma and the number of pollen tubes which were present at one third of the style below the stigma (self-incompatible and incongruent tubes ususally did not penetrate any further: MARCUCCI & VISSER, 1988). The percentage pollen tube growth of a sample consisted of the latter number expressed as a percentage of the former.

In 1986, the rose trial (on 'Sonia') was similarly carried out on cut flowers kept in a 'Chrysal' solution at 22°C constant. The flowers were emasculated before anthesis and after two days pollinated once or twice (interval one day) with viable compatible pollen. Tube penetration into the ovary was ascertained 4, 5 and 6 days after the first pollination. Per time and treatment 1-3 flowers, each with 50-70 styles, were sampled. All flowers were temporarily kept in a fixation liquid and prior to maceration rinsed thrice with water. The apple and pear styles were subsequently kept in 1N NaOH at 70°C for 5-6 hours and, after rinsing again, coloured in 0.1 N K3PO4 + 0.1N anilin blue for several hours; the rose styles were kept for 30 min in 0.1N NaOH and stained likewise (after KHO & BAER, 1968). The squash preparations were studied under a UV microscope.

RESULTS

Apple and pear. It was found that 8-14 hours after pollination on average (of all pollen types) some 40% of the apple and pear stigmas were devoid of pollen; this proportion diminished to about 10% later on. Apparently, on many stigmas the pollen had initially not germinated as yet and was rinsed off during the preparation of the material. By the time it had germinated and penetrated the stigma surface it was unaffected by rinsing; the remaining 10% of the stigmas without pollen was presumably unpollinated (see MARCUCCI & VISSER, 1988).

Within samples (of some 30 styles/pollination) the number of germinated pollen grains per stigma greatly varied, which variation allowed arranging the styles in 9 classes (3-6 styles/class) of 1<10, 10<20,......, 70<80 and > 80 <90 grains per stigma. As in the latter class the 'maximum' percentage pollen tube growth was already attained (MARCUCCI & VISSER, 1988), higher counts were not used in this paper. To compare the data irrespective of cultivar and assessment time, we determined comparative pollen tube growth which stands for the ratio - 3 - of the percentage pollen tube growth of a given class and that of the 'maximum' class (>80 grains) in the same sample. For example, if in classes <10, 10<20 and 20<30 these percentages were 25, 35 and 40 and for the maximum class 50%, comparative tube growth of the classes would be 25/50 = 0.50, 35/50 = 0.70 and 40/50 = 0.80 respectively. In this way we assessed for each of the three pollen types with pear and apple the class means (of 2-8 samples) of the 9 classes (<10,.....>80 grains/stigma).

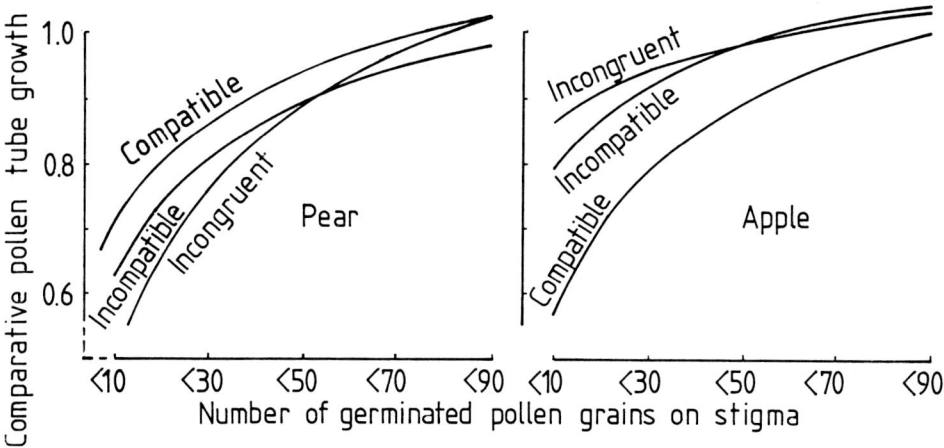

Figure 1. Comparative pollen tube growth (y) of three pollen types in the apical part of pear and apple styles as a function of the pollen load (x).

Pollen type	Pear	Apple
Compatible	y=0.34 logx+0.36(r=0.91)3	y=0.46 logx+0.11(r=0.93)3
Self-incompatible	y=0.38 logx+0.25(r=0.93)3	y=0.27 logx+0.52(r=0.95)3
Incongruent	y=0.57 logx-0.09(r=0.98)3	y=0.18 logx+0.78(r=0.75)1

1) P=0.05, 3) P=0.001; n=9

There was an asymptotic relation between pollen load and comparative tube growth. It was calculated from the regressions (linear on log basis, see Fig.1) that in five instances 40-50 grains per stigma produced tube growth not differing significantly (P=0.05) from that of the maximum (> 80 <90 grains), in the case of the incongruent pollen with apple this number was 26 grains per stigma only.

Rose. The flowers of the hybrid tea-rose 'Sonia' had up to 70 pistils of which the ovary usually contained one ovule only (SNIEZKO et al., 1989). Instead of the pollen tube growth in the style, tube penetration in the ovary was observed in relation to three pollen load classes: < 10, 10 < 20 and > 20 germinated grains on the stigma. Among the tubes penetrating the ovary, a distinction was made between normal tubes and abnormal tubes, which grew erratically, often forming loops and knots (SNIEZKO et al., 1989). Maximal penetration was obtained five days after pollination (Table 1). The percentage total penetration was the same for single and double pollination, but the proportion abnormal tubes was always significantly lower and effective penetration thus slightly higher in the latter case. The percentage abnormal tubes increased with time after pollination in both cases, apparently because it takes time for the abnormal growth to show up.

With respect to pollen load (Table 2), after single pollination the total percentage tubes penetrating the ovary clearly increased with it. After pollinating twice only the lowest pollen load had a depressing effect. At all pollen loads double pollination resulted in fewer abnormal tubes than single and,

Table 1. Effect of single and double pollination on the total percentage of ovaries penetrated by a pollen tube (normal plus abnormal) and the percentage effective penetration (normal tubes only) as related to time after pollination.

After polli- nation	Pollinated once			Pollinated twice		
	No of pistils	% Penetration Total*	Effective	No of pistils	% penetration Total*	Effective
4 Days	104	37(24)**	28**	76	43 (16)	37
5 "	220	70(28)	50	232	67 (19)	54
6 "	135	67(32)	46	243	67 (22)	52

*)of which the % abnormal tubes is given in brackets; **)estimated.

Table 2. Effect of the number of germinated grains per stigma on the total percentage of ovaries penetrated by a pollen tube (normal plus abnormal) and the percentage effective penetration (normal tubes only) in relation to single and double pollination (means of data of 5 and 6 days after pollination).

Number of grains per stigma	Pollinated once			Pollinated twice		
	No of pistils	% Penetration Total*	Effective	No of pistils	% penetration Total*	Effective
1-10	49	51 (20)	41	156	62 (15)	53
11-20	72	65 (34)	43	39	72 (21)	56
> 20	215	80 (30)	56	266	73 (23)	56

*)of which the % abnormal tubes is given in brackets.

except for the highest pollen load, produced better effective penetration. The fact that the lowest pollen load was accompanied by the lowest percentage abnormal pollen tubes is probably attributable to their slower growth, leading to a slower development of abnormalities, as can be derived from the lower percentage at four than at five or six days after pollination (Table 1).

DISCUSSION AND CONCLUSIONS

Lacking adherence to the stigma (GAUDE & DUMAS, 1982), ungerminated pollen is washed off during preparation of the material for examination. Accordingly, our data refer to the number of germinated grains, which number is supposedly close, if not equal, to the number of *viable* grains brought on the stigma. Taking this into account, the data show a significant relation between pollen load on the stigma and tube growth in the style, inferring that after single pollination in rose on average at least 20 and in apple and pear 40 germinated grains are needed for 'optimal' tube growth. Such average amounts are normal for hand pollination, which may vary between a few and 200 or more grains per stigma, as observed in this paper and by others for apple, pear, cherry and plum stigmas (MODLIBOWSKA,

1945; SCHMADLAK, 1961; ANVARI, 1973; LEE, 1980). In apple and pear the five styles cater each for two, sometimes more, ovules (ANVARI & STOSSER, 1984; VISSER & VERHAEGH, 1987) in cherry and plum the one style serves one ovule and in rose each of the many styles is connected usually with only one ovule (SNIEZKO, et al., 1989). Therefore, the pollen loads mentioned far exceed the theoretical requirement of one or two viable grains to fertilize the one or two ovules. It is of interest that, just as the compatible pollen, tube growth of the self-incompatible and incongruent pollen in the (upper part of the) style depends on the pollen load too. It may be further questioned whether or not the in-vitro and in-vivo effect of pollen density are the same phenomena. As to that, pollen of many plant species appear to have sufficient reserves to germinate and form tubes in vitro, apart from frequently needing boron for germination (VISSER, 1955) and, according to BREWBAKER & MAJUMBER (1961), calcium to counter the negative effect of low pollen density. In vivo, it may be presumed that the stigma provides a priori the proper conditions for the germination of whatever number of pollen grains deposited. It has been earlier suggested (VISSER & MARCUCCI, 1983) that the excess pollen serves in vivo as 'mentor pollen', stimulating the performance of the rest. That is to say, a portion of the pollen, through initiating tubes, initiates processes in the style apex promoting tube growth of the pollen subsequently germinating (see also VISSER & VERHAEGH, 1980). If so, the conclusion is, that the in vitro and in vivo effects of pollen density are dissimilar phenomena.

Eventually it is noteworthy that in the rose trial, irrespective of pollen load, double pollination consistently resulted in fewer abnormal pollen tubes than single. This fits the theory that in single pollination the 'mentor' role is played by the earlier-, or 'stronger' rather than the slower-germinating, or 'weaker' pollen (producing abnormal tubes). In a double pollination the first pollen, called 'pioneer pollen', promotes both the stronger and weaker pollen in the second dose to the disadvantage of the latter, which thus leads to fewer abnormalities. It has been suggested that in this way pollen load plays a role in pollen competition (see SNIEZKO, et al., 1989; VISSER & VERHAEGH, 1989).

REFERENCES

ANVARI SF (1977) Untersuchungen uber das Pollenschlauchwachstums und die Entwicklung der Samenanlagen in Beziehung der Fruchtansatz bei Sauer kirschen (Prunus cerasus L.). Diss. Un. Hohenheim (from LEE, 1980b)

ANVARI SF, STOSSER M (1984) Das Wachstum der Pollenslauche im Kernhaus bereich beim Apfel nach Bestaubung einzelner Narben. Mitt. Klosterneuburg 34: 221-225

BREWBAKER JL, MAJUMBER SK (1961) Cultural studies of the pollen population effect and the self-incompatibility inhibition. Amer. J. Bot. 48: 457-464

GAUDE T, DUMAS C (1982) Stigma-pollen recognition: Importance of the adhesion and hydration of the pollen grain in the incompatibility response. Incomp. Newsletter 14: 7-16

KHO YO, BAER J (1968) Observing pollen tubes by means of influorescence. Euphytica 17: 298-302

LEE CL (1980) Pollenkeimung, Pollenschlauchwachstums und Befruchtungsverhaltnisse bei Prunus domestica. II Pollenschauchwachstums im Griffel. Gartenbauwiss. 45: 241-248

MARCUCCI CLARA M, VISSER T (1988) Pollen tube growth in apple and pear styles in relation to self-incompatibility, incongruity and pollen load. (in press)

MODLIBOWSKA IRENA (1945) Pollen tube growth and embryo-sac development in apples and pears. J. of Pom. 21: 57-89

SCHMADLAK J (1961) Untersuchung des Pollenschlauchwachstums in Apfelgriffeln. Ber. Deutsche Bot. Ges. 74: 337-342

SNIEZKO RENATA, PIJNACKER HORDIJK JP, DUBOIS LIDWIEN AM, VISSER T (1989) Pollen tube growth as affected by double pollination and pollen load of the hybrid tea rose cv. 'Sonia' (in press)

VISSER T (1955) Germination and storage of pollen. Meded. Landbouwhogeschool, Wageningen 55: 1-68

VISSER T, MARCUCCI CLARA M (1983) Pollen and pollination experiments. IX The pioneer pollen effect in apple and pear related to the interval between pollination and temperature. Euphytica 32: 703-709

VISSER T, OOST EH (1982) Pollen and pollination experiments. V An empirical basis for a mentor effect observed on the growth of incompatible pollen tubes in pear. Euphytica 31: 305-312

VISSER T, VERHAEGH JJ (1980) Pollen and pollination experiments. II The influence of the first pollination on the effectiveness of the second one in apple. Euphytica 29: 385-390

VISSER T, VERHAEGH JJ (1987) The dependance of fruit and seed set of pear and apple on the number of styles pollinated. Gartenbauwiss. 52: 13-16

VISSER T, VERHAEGH JJ (1988) The influence of double pollination and pollen load on seed set and seedling vigour of apple and pear. 'Sexual Reproduction of Plants', Siena 1988, Springer Verlag:

WILLIAMS RR, MAIER MARIA (1977) Pseudocompatibility after self-pollination of the apple Cox's Orange Pippin. J. Hort. Sci. 52: 475-483

DNA Repair in *Petunia hybrida* Pollen

J. F. Jackson
Department of Agricultural Biochemistry
University of Adelaide
Waite Agricultural Research Institute
Glen Osmond
South Australia 5064
AUSTRALIA

INTRODUCTION

According to Charnov (1979), Bateman's principle (Bateman, 1948) as applied to plants, suggests that there is considerable advantage in not allowing pollen grains to be rendered inviable. Sperm competition may have played an important part in the evolution of hermaphroditism, male reproductive success being the selective force; once the pollen grain is on the stigma, it has no other option but to attempt fertilization. If this is so, then DNA repair is necessary to maintain pollen fitness, since interaction with the environment and the passage of time can lead to damage to sperm DNA. Exposure to ultraviolet irradiation particularly could be damaging (Jackson and Linskens, 1978) especially as pollen may be in a more vulnerable dehydrated state, a point which will be explored further in this communication. It is possible that if pollen DNA is also in a dehydrated state, it could have a higher spontaneous rate of damage due to base deamination and could bind mutagens more easily (Jackson, 1987), making a DNA repair system in pollen an important factor in male reproductive success. The DNA repair capabilities of pollen will be discussed in this communciation as well as the implications of sperm DNA being in a dehydrated form and the development of a means for testing changed conformation of DNA under conditions of low water activity (i.e. dehydration).

Comparison of DNA Repair in Pollen Grains With Repair in Mammalian Cells

The detection of DNA repair in most pollen grains is relatively simple since there is little or no background of DNA replication to interfere with the interpretation of experimental results (Jackson and Linskens, 1978). This generalization holds good for most dicotyledonous plants. However, amongst the monocotyledons, including Zea mays for example, it seems that radioactively labelled thymidine used to measure DNA repair as an unscheduled DNA synthesis (Jackson and Linskens,

1978), is not taken up by germinating pollen (Jackson, 1987). DNA repair is not, therefore, easily measured in monocotyledonous pollen. Gymnosperm pollen, on the other hand, does take up thymidine but it seems that DNA replication is taking place in this pollen, and gives a high background against which DNA repair has to be measured (Jackson, 1987). Nevertheless, it is possible to estimate DNA repair in, for example, pollen from the gymnosperm Pinus mugo, and there appears to be considerable DNA repair activity in this pollen. DNA repair in mature pollen is best understood in Petunia hydrida, where the transport of thymidine into pollen occurs readily (Kamboj and Jackson, 1984; 1985), and where there is little or no DNA replication to complicate repair measurement. DNA repair as gleaned from Jackson and Linskens (1978; 1979; 1980) and from Jackson (1987), in response to the interaction of pollen with various mutagenic agents is compared in Table 1 with the DNA repair response of human HeLa cells in culture to the same mutagens. While it is difficult to make precise comparison because conditions of culture are somewhat different, and we do not know the size of the thymidine pool in either cell which could have relatively large effects on the measure of DNA repair, nevertheless it can be seen from Table 1 that the estimation of the magnitude of DNA repair in Petunia hybrida pollen compares more than favorably with that in HeLa cells, normally considered a "metabolically active" mammalian cell. Pollen could therefore be considered to give a strong repair response to mutagen interaction. The response of human cells to ultraviolet irradiation is difficult to compare with pollen under the conditions used, and therefore there is no entry for the effect of UV on human cells in Table 1; however, the response of pollen cells to UV is listed so as to provide a comparison of the effects of UV irradiation with other direct-acting mutagens.

Table 1 also shows that the "direct-acting" electrophilic mutagens are far more effective in bringing about a DNA repair response than mutagens requiring metabolism in both pollen and human cells. Further inspection of Table 1 reveals also that pollen and human cells respond to a similar degree to each of the mutagens--thus a mutagen giving a large response in pollen will also give a high response in human cells, and vice versa.

Table 1. Comparison of unscheduled DNA synthesis (UDS) brought about in Petunia hybrida pollen with that in human HeLa cells in culture after exposure of each to various mutagens. Pollen was cultured in the presence of mutagen, ^{3}H-thymidine and 10% sucrose for 2 h at 25° (Jackson and Linskens, 1980), while HeLa was cultured for 1 h at 37° in dilute buffer (Martin et al., 1978). Both systems utilized ^{3}H-thymidine of specific activity 26 C_i/mmole. N.A. indicates results not available.

Mutagen	(mM) Concentration	UDS (max. dpm/μg DNA) Pollen	HeLa
Direct-acting			
4-nitroquinoline-1-oxide	0.001-0.1	5,950	1.850
N-methyl-N'-nitro-N-nitrosoguanidine	0.001-0.1	8,012	1,208
Methylmethanesulfphonate	0.1-10	176	436
Ethylmethanesulphonate	0.1-10	13	22
Cyclophosphamide	0.001-1.0	3	116
Azaserine	0.01-10	1,702	N.A.
UV (2,537 Å)	0-10,000 J/m^2	345	N.A.
Requiring Metabolism			
Aflotoxin B_1	0.001-0.1	28	368
2-acetylaminofluorene	0.1-25	41	60
Dimethyl-p-aminobenzene	0.001-25	20	44
Diethylnitrosamine	0.1-100	50	33

DNA Repair is Localized in the Generative Nucleus

When Petunia hybrida pollen was treated with the mutagen 4-nitroquinoline-1-oxide in the presence of ^{3}H-thymidine, and subsequently subjected to autoradiography, microscopic examination showed that tritium labelling was confined to the generative nucleus. The larger vegetative nucleus had no more than the background level of silver grains. No significant labelling of mitochondria above the background level was detected.

Metal Ions and Pollen DNA Repair

One of the unusual effects given by Petunia pollen is the DNA repair response to multivalent metal ions. As shown in Table 2, Al^{3+}, Fe^{3+}, V^{2+} and Be^{2+} are the most effective in this regard, while Ca^{2+}, Zn^{2+}, Mn^{2+}, Cd^{2+}, Ni^{2+} and Cr^{3+} all give significant responses. Comparison with responses in human cells is not available, as such experiments have not been (or cannot be) carried out with HeLa cells. When this metal ion effect was first reported (Jackson and Linskens, 1982), it was pointed out that perturbation in fidelity of DNA synthesis in _vitro_ (Sirover and Loeb, 1976) is brought about by the same metal ions which also give unscheduled DNA synthesis in pollen cells.

Table 2. Unscheduled DNA synthesis response given by metal ions when added to cultured pollen of _Petunia_ _hybrida_. Metal ions were added to 2 mM, otherwise conditions of experiment were as described by Jackson and Linskens (1982).

Magnitude of Response	Metal Ion	Unscheduled DNA Synthesis (dpm/μg DNA)
Strong	Al^{3+}	9,375
	Fe^{3+}	8,125
	V^{2+}	7,500
	Be^{2+}	6,875
Medium	Ca^{2+}	2,437
	Zn^{2+}	2,125
	Mn^{2+}	1,875
	Cd^{2+}	1,687
	Ni^{2-}	687
	Cr^{3+}	625
Weak	Sr^{2+}, Cu^{2+}, Co^{2+} Sn^{2+}, Pb^{2+}, Sb^{2+}	20-300

It should be noted that many metal ions are capable of complexing strongly with DNA in the Z conformation (Jovin and Soumpasis, 1987), just as are some of the mutagens listed in Table 1 (see Jovin and Soumpasis, 1987). It is tempting therefore to seek an answer to the relatively high repair response to metal ions in the conformation of DNA in pollen.

State of Hydration of Pollen

Like bacterial spores (Jackson, 1988), pollen from dicotyledonous plants when shed from the anthers, is often in a dehydrated state (Jackson, 1987). One may therefore learn more about the implications of this dehydrated state in pollen from earlier work with bacterial spores. When _Bacillus_ _subtilis_ spores are irradiated with ultraviolet light, one obtains a special type of thymine dimer often known as "spore product," where adjacent thymine bases of DNA are covalently linked in a different way to the cyclobutane-type thymine dimer obtained when the freshly hydrated vegetative _B_. _subtilis_ is irradiated (Varghese, 1970). The same "spore-product' is generated when a sample of purified DNA is dehydrated prior to irradiation (Smith and Hanawalt, 1969). It seems reasonable to assume then that spore DNA is in a dehydrated state (Jackson, 1987). Under conditions of high water activity, the B conformation of DNA prevails, while at low water activity (dehydration) the A conformation or even the Z conformation is favored (Saenger, Hunter and Kennard, 1986). It is possible then that at least some of the pollen DNA is also in the Z conformation, and that the high repair response to multivalent metal ions is a result of the high affinity of Z-DNA for metal ions, giving distorted regions of DNA requiring repair. Alternatively, Z-DNA itself may also be recognized as requiring repair since on hydration the metal ions may stabilize the DNA in the Z form (Jovin and Soumpasis, 1987), instead of the DNA assuming the normal B conformation as water activity increases during the initial stages of pollen germination.

Discussion

How then can we test the hypothesis that the DNA of dry, mature pollen is substantially in the Z conformation. An approach similar to that used in the spore experiments can be applied; however, it is not yet certain what conformation the dehydrated DNA has assumed when the so-called spore product is formed. Certainly the amount of Z-DNA must be increased with dehydration, but we do not know that "spore product" type thymine dimers result from irradiation of Z-DNA unequivocally. Another approach could be to make use of the antibodies to Z-DNA as developed by Rich et al. (1981). Such an approach has begun in this laboratory, using goat anti-Z DNA antibodies. Polyd(G-C)·d(G-C), treated with Mn^{2+} at 55°, has been used in these studies since it is known that under these conditions the synthetic polymer assumes a Z-conformation (van de Sande, McIntosh and Jovin, 1982). After reaction with the goat antibody, further reaction with mouse anti-goat antibody containing electron dense gold spheres of 10 nm diameter was carried out (Birrel et al., 1987). After spreading as described by Inman (1982), inspection under the electron microscope showed DNA-like strands with numerous gold spheres attached. The method will be further refined to be used as a probe for Z-DNA. It is hoped that this approach can help determine how much pollen DNA is in the Z conformation under a variety of conditions.

References

Bateman AJ (1948) Intra-sexual selection in Drosophila. Heredity 2:349-368

Birrel AB, Hedberg KK, Griffiths, OH (1987) Pitfalls of immunogold labelling: analysis by light microscopy, transmission electron microscopy, and photoelectron microscopy. J Histochem Cytochem 35:843-853

Charnov EL (1979) Simultaneous hermaphroditism and sexual selection. Proc Natl Acad Sci (US) 76:2480-2484

Inman RB (1982) Electron microscopy in biology. Vol 2 (Griffith JD, ed), pp 237-271, John Wiley & Sons, New York

Jackson JF (1987) DNA repair in pollen--a review. Mut Res 181:17-29

Jackson JF (1988) Dehydration and rehydration in plant reproductive tissue. Modern Methods of Plant Analysis (Linskens HF and Jackson JF, eds), Vol 9, in press

Jackson JF (1978) Evidence for DNA repair after ultraviolet irradiation of Petunia hybrida pollen. Mol Gen Genet 161:117-120

Jackson JF and Linskens HF (1979) Pollen DNA repair after treatment with the mutagens 4-nitro-quinoline-1-oxide, ultraviolet and near ultraviolet irradiation, and boron dependence of repair. Mol Gen Genet 176:11-16

Jackson JF, Linskens HF (1980) DNA repair in pollen: range of mutagens inducing repair, effect of replication inhibitors and changes in thymidine nucleotide metabolism during repair. Mol Gen Genet 180:517-522

Jackson JF, Linskens HF (1982) Metal ion induced unscheduled DNA synthesis in Petunia pollen. Mol Gen Genet 187:112-115

Jovin TM, Soumpasis DM (1987) The transition between B-DNA and Z-DNA. Ann Rev Phys Chem 38:521-560

Kamboj RK, Jackson JF (1984) Divergent transport mechanisms for pyrimidine nucleosides in Petunia pollen. Plant Physiol 75:499-501

Kamboj RK, Jackson JF (1985) Pyrimidine nucleoside uptake by Petunia pollen: specificity and inhibitor studies on the carrier-mediated transport. Plant Physiol 79:801-805

Martin CN, McDermid AC, Garner RC (1978) Testing of known carcinogens and noncarcinogens for their ability to induce unscheduled DNA synthesis in HeLa cells. Canc Res 38:2621-2627

Saenger W, Hunter WN, Kennard O (1986) DNA conformation is determined by economics in the hydration of phosphate groups. Nature 24:385-388

Sirover M, Loeb LA (1976) Infidelity of DNA synthesis in vitro: screening for potential metal mutagens or carcinogens. Science 194:1434-1436

Smith KC, Hanawalt PC (1969) Molecular Photobiology, Academic Press, New York

van de Sande JH, McIntosh LP, Jovin TM (1982) EMBO J 1:115-120

Varghese AJ (1970) 5-Thyminyl-5,6-dihydrothymine from DNA irradiated with ultraviolet light. Biochem Biophys Res Commun 38:484-490

Gametophyte and Sporophyte Transformation and Manipulation

Transformation, T-DNA and Sexual Reproduction in Higher Plants

George J. Wullems
Department of Experimental Botany
Research Group Molecular Plant Physiology
University of Nijmegen
Toernooiveld
6525 ED Nijmegen
The Netherlands

INTRODUCTION

Each organism develops from one single cell and the development follows different ways. The initial cell is a somatic cell or a sexual cell which can either directly initiate a new organism or functions as zygote, after sexual fertilization with another sexual cell, as the initial cell.
The development of a multicellular organism is the result of differentiation, growth and organization that occur at every biological level, from molecular- through cellular structures, organelles, cells, tissues and organs to the whole organisms. Finally all cells composing an organism are different from the common initial cell, whereas the genetic constancy is maintained in the germ line. The process resulting in new cell types, other tissues, functional organs is called 'Differentiation'. Cells are the smallest, independent elements in an organism and therefore it is natural to focus the study of differentiation, growth and organization on the level of the cell. At a certain stage, a cell is determined to develop into one direction. At that stage development in another direction is blocked and the direction into which it will differentiate has been fixed.
Compared with animals, higher plants have unique developmental processes. In plants spore-forming and gamete-forming generations are alternating and are separated temporally and spatially during the life cycle. The male and female gametophytes reside in distinct sporophytic organ systems and contain the sperm and egg cells respectively. Due to differentiation processes gametophytic cells are functionally distinct from one another. In addition, unlike animal cells, plants contain meristematic cells capable of differentiating into the sporophytic organ system throughout the life cycle. Another difference with animal cells is that the reproductive organ developes at the end of the life cycle. This offers the possibility to study aspects of differentiation in relation to sexual reproduction independent from embryogenic differentiation.
Some aspects of the cellular and molecular processes that control plant development are slowly becoming to be understood with the aid of the modern molecular biological approaches. However, the processes responsible for differentiation of meristematic cells into the mature sporophytic organ systems are unknown, although it is generally accepted that certain growth substances, called 'phytohormones' play a crucial role in plant cell, growth and development. Two types of phytohormones are necessary for growth of plant cells *in vitro*: auxins and cytokinins. The signal function of these phytohormones in differentiation becomes more evident when we see that when their relative concentrations in the culture medium are altered it is possible to manipulate the

respons of the tissue in such a way that either unorganized callus or roots, shoots or both forms of differentiation occur. The classical experiments of Skoog and Miller (1957) are a good illustration of this phenomenon. Other examples demonstrating the role of phytohormones are 2,4D dependent regeneration of shoots in tissue cultures of certain monocots (Vasil, 1983) and the development of callus, vegetative buds or flower buds in thin layer cultures of tobacco under influence of the hormone balance (Tran Thanh Van, 1973).

The experimental systems described above are appropriate as model systems to study the physiological basis of differentiation but has not yet showed up with answers concerning the molecular basis of differentiation and development. Obviously differentiation and development concerns regulation of gene expression as a result of auxin levels, cytokinin levels and auxin/cytokinin balance in plant cells. Since cellular genes responsible for phytohormone synthesis have not been identified and isolated, the *Agrobacterium* system might be a good alternative system since it induces developmental abnormalities like, crown gall and hairy roots, that can be considered as developmental mutants with an impact both on vegetative and on generative developmental processes in plants.

Agrobacterium tumefaciens and crown gall

In *Agrobacterium tumefaciens* three genes, analogue to cellular phytohormone synthesis genes have been identified. These are bacterial genes with eukaryotic expression signals and hence can be expressed inside plant cells. For a recent overview see Schell, 1987. Two of these genes code for auxin synthesis; another gene codes for synthesis of cytokinin. These three genes are located on a piece of bacterial DNA, the T-region which is part of a large plasmid inside *Agrobacterium*. This T-region can be transferred to wounded plant cells where it becomes integrated as T-DNA in the plant genome. The expression of these genes, which are called *onc*-genes, results in elevated levels of auxin and cytokinin leading to tumor formation at the side of infection. More evidence for developmental control of these genes came from studies with mutations in either the auxin or cytokinin genes. If the auxin genes are inactivated there will be no extra auxin produced, leaving the tissue with overproduction of cytokinin resulting in shoot induction on the tumour. If the cytokinin gene is inactivated, the tissue will develop roots as a result of the overproduction of auxin. These results show that the expression of the three T-DNA *onc*-genes mimic fairly accurately *in planta* developmental processes as a result of variation in endogenous hormone concentrations. Furthermore transgenic shoot or root tissue that arises on tumours as a consequence of inactive *onc*-genes can be grown in tissue culture as organ cultures or even as plants. They therefore provide an elegant model system for studying the molecular basis of differentiation of plants both at the level of organ differentiation and of sexual reproduction.

There is a second application of *A. tumefaciens* in which the *onc*-genes are deleterious. In this case the bacterium is used as a tool to introduce 'desired genes' into plant cells. The aim is then to obtain healthy, well differentiated plants with newly introduced genes. To achieve this, *Agrobacterium* strains have been constructed with artificial T-regions, which contain marker genes and single cloning sites for inserting desired genes. Desired genes may be of commercial value (plant improvement) or of scientific value (promoters to study regulation of gene expression).

Initially, transformation of plant cells was developed with *Agrobacterium* on the level of single plant cells (protoplasts) (Marton and Wullems, 1979; Krens *et al.*, 1982), however, with the disadvantage of having to regenerate plants through the whole tissue culture procedure. This was followed by the development of transformation procedures starting with tissue explants (leaf discs) (Horsch *et al.*, 1985). Although these methods have shown elegant applications, they have the disadvantage of the necessity to need tissue culture and regeneration procedures, which are not available to all species. Monocot plants are recalcitrant in this respect. In addition, tissue culture induces genetic instabilities. To overcome these problems people are searching for procedures to achieve genetic manipulation through transformation via the reproductive process (gametes, embryo's).

T-DNA AND SEXUAL REPRODUCTION

As a result of transformation experiments with mutated *Agrobacterium* strains it was possible to regenerate transgenic plants which contained T-DNA on which the auxin locus was inactive (deleted) (Wullems *et al.*, 1981). The main characteristics have been described elsewhere. In the scope of these proceedings I will summarize the main properties of these tissues and illustrate the effect of the presence of T-DNA, in particular the T-DNA gene 4, on pollen development and male sterility.

A shooty crown gall line that was studied was SR1-4013-3. The origin and the characterization of this line has extensively been described (Wullems *et al.*, 1981, Peerbolte *et al.*, 1987). The line was derived from co-cultivation of tobacco protoplasts with a wild type *Agrobacterium tumefaciens* strain. The shoots that were obtained from the transformed callus were octopine synthase positive (Ocs$^+$), rooting deficient (Rod$^+$), had thick fleshy leaves (Tfl$^+$), showed reduced apical dominance (Rad$^+$).

In order to allow the shoots to reach maturity they were grafted onto non-transformed SR1 tobacco rootstocks. The transformed grafts retained a reduced apical dominance but after subsequent graftings the plants were able to flower. However, flowers showed hyperstyly (Hst$^+$) and hardly produced any pollen. In fact the plants showed to be male sterile (Mst$^+$) (Fig. 1).

To study the inheritance of the properties of the transgenic plants they were cross pollinated with pollen from healthy tobacco plants. Of the resulting offspring plants, 50% showed the crown gall phenotype, being Ocs$^+$, Rod$^+$, Tfl$^+$, Rad$^+$, Hst$^+$ and Mst$^+$. The seggregation of these markers is according to Mendellian laws. Surprisingly however, the other 50% of the seedlings, which had no tumour markers, also showed the maternal characteristics hyperstyly (Hst$^+$) and male sterility (Mst$^+$). Apparently , these two traits had been inherited by 100% of the seedlings.

The molecular basis of the occurrence of the male sterility is still unknown. In accordance to other forms of male sterility which are associated with the mitochondrial genome, like in maize, we have looked at restriction fragment length polymorphisms (RFLP's) in mitochondria of transgenic plants. We observed slight aberations in the restriction patterns between Mst$^+$ and Mst$^-$ tobacco plants. In addition to these DNA patterns, studies on translation products obtained after *in vitro* translation of mitochondrial RNA from Mst$^+$ and Mst$^-$ plants showed different banding patterns, indicating that presumably the cytochrome oxidase II (CoxII) protein in the Mst$^+$ plants is modified (Dr. F. van

92

der Markt, personal communication). Taken together these data,
this form of male sterility, which presumably is induced by the
overproduction of cytokinins in the transgenic tumorous plants,
resembles the forms of male sterility in other plants in such a
way that the mitochondrial genome is involved. How the T-cyt gene
of the T-DNA influences the pollen development at the level of
gene expression is currently under investigation, in order to gain
information about the molecular basis of male sterility.

Fig. 1.   A: Grafted, mature, flowering transgenic tobacco plant
           expressing the T-cyt gene of *Agrobacterium tumefaciens*.
           B: Detail of a male sterile (Mst⁺) transgenic flower
           showing hyperstyly (Hst⁺) and reduced pollen formation
           (Part of the corolla is removed).

Horsch RB, Fry JE, Hoffman ML, Eichholtz O, Rogers SG & Frayley RT
    (1985) A simple and general method for transferring genes
    into plants. Science 227:1229-1231
Krens FA, Molendijk L, Wullems GJ & Schilperoort RA () *In vitro*
    transformation of plant protoplasts with Ti plasmid DNA.
    Nature 296:72-74

Marton L, Wullems GJ, Molendijk L & Schilperoort RA (1979) *In vitro* transformation of cultured cells from *Nicotiana tabacum* by *Agrobacterium tumefaciens*. Nature 277:129-131

Peerbolte R, Floor M, Ruigrok P, Hoge JHC, Wullems GJ & Schilperoort RA (1987) T-DNA stability and expression in F1 tobacco transformants studied at various stages of differentiation. Planta 172:448-462

Schell JS (1987) Transgenic plants as tools to study the molecular organization of plant genes. Science 237:1176-1183

Skoog F, Miller CO (1957) Chemical regulation of growth and organ formation in plant tissues cultured *in vitro*. Symp Soc Exp Bot 11:118-140

Tran Thanh Van K (1973) *In vitro* control of *de novo*, flower, bud, root, and callus differentiation from excised epidermal tissues. Nature 246:44-45

Vasil IK (1983) Regeneration of plants from single cells of cereals and grasses. In: Genetic engineering in Eukaryotes. Eds. PF Lurguin, A Kleinhofs, NATO ASI Series A Life Sciences Vol 61 pp 233-252. Plenum Press

Wullems GJ, Molendijk L, Ooms G & Schilperoort RA (1981) Differential expression of crown gall tumor markers in transformants obtained after *in vitro A. tumefaciens* induced transformation of cell wall regenerating protoplasts from *N. tabacum*. PNAS USA 78:4344-4348

Wide Hybridization in Crop Brassicas

P.B.A. Nanda Kumar, Shyam Prakash* and K. R. Shivanna
Department of Botany
University of Delhi
Delhi 110 007
India

A wide gene pool exists in wild species of most of the crop plants. In recent years, there has been a greater emphasis on the use of wild species in plant breeding, especially for transferring genes resistant to diseases and pests. Considerable number of examples in which desirable genes have been transferred from wild species to the cultivars are available (see Harlan 1984; Goodman et al. 1987).

Brassica is one of the most important oil yielding plants. In India, the important cultivated species are B. juncea and B. campestris. Of late, plant breeders have been showing interest in B. napus as it is resistant to white rust. Attempts are being made to breed varieties which show early maturity with non-shattering fruits. Most of the cultivars of both B. juncea and B. campestris are devoid of resistant genes to major diseases, particularly white rust and alternaria blight, and pests. Wild species of Brassicas constitute a good source of genes imparting resistance to many diseases and pests. Many of the wild species are capable of growing under moisture and saline stress conditions. Although Brassicas are known to exhibit hybrid vigour, only a few cytoplasmic male sterile (CMS) systems are available in cultivated species which include the synthetic alloplasmic lines obtained by combining cytoplasm of the wild species with the nuclear genomes of crop species (Hinata and Konno 1979; Prakash and Chopra 1988). Thus, there is considerable scope for wide hybridization in Brassicas both for gene introgression and for inducing CMS lines. Because of the presence of crossability barriers, it has been difficult to obtain wide hybrids in most of the combinations (Harberd 1975). We have attempted to raise interspecific hybrids of Brassica using three cultivated species (B. campestris ssp. oleifera (2n = 20, AA), B. juncea (2n = 36, AA BB) and B.napus (2n = 38, AA CC) and two wild species (B. fruticulosa (2n = 16, FF) and B. gravinae (2n=4x=40,GG).

*Division of Genetics, Indian Agricultural Research Institute,
New Delhi 110 012, India

This paper presents details of three successful crosses (B. fruticulosa x B. campestris, B. juncea x B. gravinae and B.napus x B. gravinae).

Materials and Methods

Plants of all species were grown under field conditions. Controlled pollinations were carried out on emasculated buds and bagged. Bags were removed 4-8 days after pollination by which time the stigma had dried. Some of the pollinated flowers were left on the plant until they dried or developed into mature fruits. The remaining flowers were used to culture ovaries (4-5 days after pollination) and ovules (8-15 days after pollination).

Murashige and Skoog's (1962) medium (MS) containing 3-5% sucrose, 0.8% agar and various combinations of growth hormones and casein hydrolysate (CH 500 mg/l) were used to culture ovaries and ovules (Nanda Kumar et al. 1986, 1988a). Cultures were maintained at 25±2°C under continuous illumination (light energy of approx. 23 watts/metre^2). Cultured ovaries were maintained until fruits reached maturity, and seeds were harvested. Ovules from some of the cultured ovaries were isolated (8-22 days after culture) and sown on fresh medium. Cultured ovules which continued development, germinated in situ and gave rise to seedlings.

Harvested seeds (from cultured ovaries as well as from field pollinations) were sown in soil as well as on culture media. Seedlings raised in vitro were transferred to soil and grown to flowering. Some of the in vitro grown seedlings were used to multiply hybrids through shoot tip and single node segments (Nanda Kumar et al. 1988b). Cytological analysis of hybrids were carried out in acetocarmine squash preparations.

Results

Although pollinations were carried out in all reciprocal combinations, some seeds were obtained only in three of the crosses (Table 1). Very few of these seeds germinated.

In the reciprocal crosses of B. fruticulosa x B. campestris and B. campestris x B. fruticulosa, cultured ovaries yielded marginally better fruit and seed set when compared to flowers maintained on plants (Table 2). In other crosses, ovary cultures did not yield any seeds.

Table 1: Results of field pollinations

Cross	No. of pollinations	No. of pods formed (% pod set)	No. of seeds/pollination
B.fruticulosa x B.campestris	956	367 (38.4)	0.5
B.campestris x B.fruticulosa	266	5 (1.9)	0.03
B.gravinae x B.juncea	57	0	-
B.juncea x B.gravinae	104	8 (7.7)	0.13
B.gravinae x B.napus	61	0	-
B.napus x B.gravinae	109	0	-

Table 2: Responses of Cultured Ovaries

	No. of ovaries cultured	No. of ovaries developed into fruits (% fruit set)	No. of seeds obtained	No. of seeds per ovary cultured
B.fruticulosa x B.campestris	348	172(49.43)	213	0.62
B.campestris x B.fruticulosa	60	4(6.67)	4	0.07
B.gravinae x B.juncea	68	0	-	-
B.juncea x B.gravinae	25	0	-	-
B.gravinae x B.napus	35	0	-	-
B.napus x B.gravinae	24	0	-	-

Responses of ovule cultures especially those raised from in vitro cultured ovaries was much better (Table 3). This was the only effective method to raise hybrids in the cross B.napus x B.gravinae.

The hybrids obtained in vitro were multiplied by culturing shoot tips and single node segments on MS medium containing 6-benzylaminopurine. Proliferated shoots were rooted on a medium containing NAA, transferred to soil and grown to flowering in the field (Nanda Kumar et al. 1988b).

Germination of hybrid seeds and seedling establishment was very poor when sown in soil (Table 4). Response of seeds was much better when seeds (particularly those obtained from cultured ovaries) were cultured on the medium.

Table 3: Responses of cultured ovules

Cross	No. of ovules cultured	No. of seed-lings obtained	% cultured ovules developed into seedlings	No. of hybrids grown to flowering*
B.fruticulosa x B.campestris from pistils maintained on plants	255	9	3.5	
from cultured ovaries	61	34	55.7	91
B.juncea x B.gravinae from cultured ovaries	21	5	23.8	8
B.napus x B.gravinae from cultured ovaries	4	1	25	13

*includes hybrids multiplied in vitro

Table 4: Responses of hybrid seeds

Seed source	No. of seeds used for germi-nation	No. of established hybrid seedlings	% seed-lings estab-lished
B.fruticulosa x B.campestris Developed on plants			
Sown in pots	132	4	3.03
Cultured on the medium	58	18	31.03
Developed in cultured ovaries			
Sown in pots	67	2	2.99
Cultured on the medium	104	44	42.31
B.juncea x B.gravinae Developed on plants			
Sown in pots	14	3	21.43
B.campestris x B.fruticulosa Developed in cultured ovaries			
Cultured on the medium	4	1	25.0

Hybridity was confirmed by morphological and cytological studies (Figures 1-3). Hybrids showed characters of both the parents in vegetative as well as floral characters.

Meiotic analysis of the hybrid B.fruticulosa x B.campestris (2n = 18) showed mostly 18 univalents and in some cells 0-5 bivalents. They were completely pollen sterile and no seed was obtained on being allowed to open pollinate or in back crosses with both the parents.

Meiotic analysis of B. juncea x B. gravinae (2n = 38) hybrid revealed 10 II + 18 I; some cells, however, showed up to 16 bivalents. The hybrids showed about 6% pollen fertility. Fifteen seeds developed after self- as well as open-pollinations.

B. napus x B. gravinae (2n = 39) hybrid showed 10 II + 19 I in a majority of the PMC's. However, as many as 19 II were observed in some cells. Pollen fertility was about 9%. Ten seeds were harvested following self- and open-pollinations.

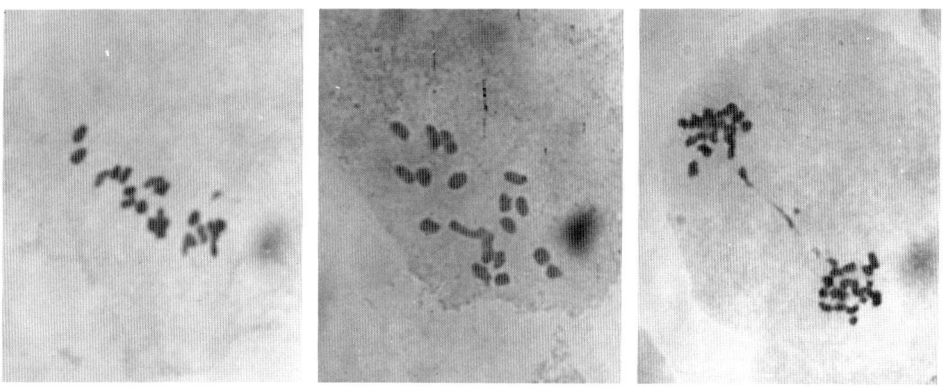

Fig.1-3. Meiosis of the F₁ hybrids. 1. Metaphase I of B. juncea x B. gravinae showing 2 I + 18 II. 2. Metaphase I of B. napus x B. gravinae showing 18 II + 1 III. 3. Anaphase I of B. napus x B. gravinae showing a chromosome bridge and a fragment.

Investigations are in progress to induce amphidiploidy and to raise F₂ and back crosses.

Acknowledgements

Award of a research fellowship by the Centre of Advanced Studies, Department of Botany to one of us (NK) is gratefully acknowledged.

References

Goodman RM, Hauptli H, Crossway A, Knauf VC (1987) Gene transfer in crop improvement. Science 236: 48-54
Harberd DJ (1975) Cytotaxonomic studies of Brassica and related genera. In: Vaughan JG, Macleod AJ, Jones BMJ (eds) The

biology and chemistry of the Cruciferae, Academic Press, London, p 47

Harlan JR (1984) Evaluation of wild relatives of crop plants. In: Holden JHW, Williams JT (eds) Crop genetic resources: conservation and evaluation, George Allen and Unwin, London, p 212

Hinata K, Konno N (1979) Studies on a male sterile strain having the B. campestris nucleus and the Diplotaxis muralis cytoplasm. I. On the breeding procedure and some characteristic of the male sterile strain. Jpn J Breed 29: 305-311

Murashige T, Skoog F (1962) A revised medium for rapid growth and bioassays with tobacco tissue cultures. Physiol Plant 15: 473-497

Nanda Kumar PBA, Shivanna KR (1986) Interspecific hybridization between Brassica fruticulosa and B. campestris. Cruciferae Newslett 11: 18

Nanda Kumar PBA, Shivanna KR, Prakash S (1988a) Wide hybridization in Brassica: crossability barriers and studies on hybrid and synthetic amphidiploid of B. fruticulosa x B. campestris. Theor Appl Genet (communicated)

Nanda Kumar PBA, Shivanna KR, Prakash S (1988b) In vitro propagation of Brassica interspecific hybrids. Plant Cell Tissue and Organ Culture (communicated)

Prakash S, Chopra VL (1988) Synthesis of alloplasmic Brassica campestris and induction of cytoplasmic male sterility. Plant Breeding (in press)

Preparation of Sporoplasts for Studies of Pollen Physiology

Bruce G. Baldi[1], John D. Everard and Frank A. Loewus
Institute of Biological Chemistry
Washington State University
Pullman, Washington 99164-6340 USA.

ABSTRACT

Sporoplasts (pollen protoplasts) are produced in quantity from pollen of _Lilium longiflorum_ Thunb. by exposure to 4-methylmorpholine N-oxide (MMNO). Treatment for 1 h at 25°C with aqueous MMNO, 15% w/v, in buffered solution, pH 5.3, containing 0.5% cellulase, 0.25% pectinase and 0.1% bovine serum albumin loosens the attachment of the sporoplast to exine but fails to effect release. Displacement from exine is achieved by a series of three treatments. First, MMNO-treated pollen is suspended in a buffer containing 0.5 mM EDTA, 0.6 M mannitol and 10 mM MES, pH 5.5 which effects release of sporoplasts from exines (25-30%). Then it is suspended in a buffer containing 5 mM $CaCl_2$, 0.6 M mannitol and 5 mM MES, pH 5.5. Finally, it is suspended in 5 mM $CaCl_2$, 0.3 M pentaerythritol and 3 mM MES, pH 5.3. Separation of sporoplasts from exines and undigested pollen is accomplished on a discontinuous sucrose density gradient. Purified sporoplasts are readily ruptured by passing the suspension through a 15 um^3 nylon mesh. Data is presented on cell-free distribution within self-generated Percoll gradients of phytase and a phytate-rich component.

INTRODUCTION

Production of pollen protoplasts from mature pollen is largely limited by the presence of an outer wall, the exine, which resists the action of reagents normally used for preparing vegetative plant protoplasts. To circumvent this barrier, most methods have selected pollen at stages of development preceding encasement by exine or have chosen germinating pollen from which pollen tube protoplasts can be isolated (Bajaj, 1974; Bhojwani and Cocking, 1973; Imamura and Potrykus, 1983; Ito, 1973; Maeda et al., 1979; Power, 1973, Rajasekhar, 1973; Tanaka et al., 1987; Zhu et al., 1984). Over the past four years we have developed a simple reliable method for release of large quantities of pollen protoplasts (which we have termed sporoplasts) from mature pollen through a solvolytic process involving use of 4-methylmorpholine N-

---

[1]Present address: USDA-ARS Plant Hormone Lab, Beltsville Agricultural Research Center, Beltsville, MD 20705 USA.

oxide (MMNO) (Baldi et al., 1986; Baldi et al., 1987b; Loewus et al., 1985). Apparently, this reagent breaks hydrogen-bonded elements which maintain attachment of intine to exine, permitting the sporoplast to escape its exine shell, possibly through a weakened aperture. The precise mode of action of MMNO requires further study. Treatment with MMNO under mild conditions (Baldi et al., 1987b) yields sporoplasts which are biologically competent as judged by histochemical stains, recovery of enzymatic activities, and metabolic functions. The exine fraction from MMNO treated pollen has been useful in examining the structure of sporopollenin (Espelie et al., 1988; Given et al., 1988). Although no attention has been given to the composition of intine derived from MMNO treated pollen, that possibility exists since the sporoplasts emerge with their intine layer largely intact.

Currently, our interest in phytic acid metabolism in pollen (Baldi et al., 1987a; Baldi et al., 1988; Scott and Loewus, 1986a,b) has led to efforts to improve the methodology of recovery of sporoplasts and to examine their subcellular components for evidence of localization of phytic acid and related phytases. Here, we report on the role of calcium in maintaining plasma membrane integrity of sporoplasts and the distribution of phytate and phytases in self-generating Percoll gradients of sporoplastic organelles.

MATERIALS AND METHODS

Pollen from Lilium longiflorum Thunb., cv. Nellie White which had been harvested during the 1986 and 1987 seasons was used (Baldi et al., 1987b). Sporoplasts were prepared according to Method III with a slight modification. One-half g of pollen was suspended in 7.5 ml of buffer consisting of 3 mM MES (2-N-morpholino-ethanesulfonic acid), 0.3 M pentaerythritol and 5 mM $CaCl_2$, pH 5.3 for 30-45 min to allow hydration. MMNO (60% aqueous), to a final concentration of 15% w/v, was added to the buffered pollen suspension and then Cellulysin, Macerase (Boerhinger-Calbiochem), and bovine serum albumin to final concentrations of 0.5, 0.25 and 0.1%, respectively. The suspension was incubated for 1 h in a shaking water bath at 74 osc/min, 25°C. Following enzymatic digestion, the pollen which had as yet not released its sporoplasts, was transferred to a 1.8 x 15 cm test tube and pelleted by centrifugation at 180g for 5 min. The pellet was resuspended in 20 ml of buffer consisting of 0.5 mM EDTA, 0.6 M mannitol, and 10 mM MES, pH 5.5 and gently shaken on a Burrell wrist shaker for 10 min. After pelletizing again, the procedure was repeated, first in 20 ml of buffer composed of 5 mM $CaCl_2$, 0.6 M mannitol, and 10 mM MES, pH 5.5 and lastly, in 20 ml of buffer composed of 5 mM $CaCl_2$, 0.3 M pentaerythritol, and 3 mM MES, pH 5.3. Free sporoplasts were separated from exine and incompletely digested pollen by suspending the final pellet in 10 ml of buffer #1 composed of 2 mM $CaCl_2$, 0.4 M sucrose, 9.1% dextran (35-50 x 10^3 mol wt), and 10 mM HEPES, pH 7.0. This suspension was overlaid with 5 ml each of buffer #2 composed of 2 mM $CaCl_2$, 0.4 M sucrose, 10 mM HEPES, pH 7.0, buffer #3 composed of 2 mM $CaCl_2$, 0.4 M mannitol and 10 mM HEPES, pH 7, mixed in a ratio 1.5 of buffer #3 to 2.5 of buffer #2, and buffer #4 composed of 2 mM $CaCl_2$, 75 mM NaCl, 0.15 M mannitol and 0.1 M HEPES, pH 7.0. The gradient was spun at 320g for 20 min. Highly enriched sporoplast fractions were recovered at the interfaces between buffers #2/#3 and #3/#4. Partially digested pollen was recovered at the interface between buffers #1/#2. Pure exine was recovered in the pellet fraction.

Sporoplasts were ruptured by forcing the suspension through a 15 um^2 nylon mesh (TETKO Inc., Elmford, NY) fitted over a 10 ml plastic syringe. This treatment expelled sporoplast contents through the mesh while retaining exine, pollen grains and intine/plasma membrane fragments. Preliminary examination of sporoplastic contents was made on a discontinuous gradient consisting of 2 ml of organellar suspension overlaid with 1 ml each of 5, 10, 15, 20, and 25% Ficoll that was spun at 10g for 1 h, 4°C. One ml fractions were collected and stored at -20°C until analyzed.

Sporoplastic contents were also fractionated in a modified osmocentrifugational cell of the type described by Nunes and Galembeck (1985). Here, one chamber of the special cell, fitted with a Nuclepore SN:MF, 0.1 um membrane, Lot 712, was filled with 12 ml of a solution composed of 50% Percoll, 0.3 M pentaerythritol, 2 mM $CaCl_2$, and 3 mM MES, pH 5.3 while the adjoining chamber was filled with a similar buffer lacking Percoll. Two ml of sporoplast suspension was layered on top of the Percoll solution and the cell was spun at 650g for 30 min, 4°C. One ml fractions were removed serially from the top of the gradient. The refractive index of each was measured and corresponding densities determined by extrapolation from a standard curve of refractive index versus gravimetrically determined density.

Ficoll and Percoll gradient fractions were analyzed for phytic acid (Latta and Eskin, 1980), pH 5 and pH 8 phytases, myo-inositol monophosphatase (Ficoll gradient only), and total protein (Scott and Loewus, 1986b).

RESULTS

Although 1 h of digestion of pollen with MMNO/carbohydrase failed to effect complete separation of sporoplast from exine, it did result in partial emergence of intine-clad sporoplast through the aperture in about 90% of the pollen grains; an indication that loosening of exine/intine bonding had occurred. Subsequent treatment with EDTA-containing, buffered osmoticum provided sporoplasts in 25-30% yield with about 30% contamination from exines and partially digested pollen (Table 1). When 0.5 mM EGTA or 5 mM CaCl2 replaced EDTA, the yield of free sporoplasts was much lower and was accompanied by a greater degree of contamination.

Table 1. Influence of the composition of initial wash buffer on sporoplast yield and purity in MMNO/carbohydrase-treated lily pollen.

Composition of initial wash[1]		Sporoplasts[2] (# x 10^3)	Exines (# x 10^3)	Contamination (%)	Yield (%)
+ 0.5 mM EDTA	(7)	310 $\pm$ 40	85 $\pm$ 10	28 $\pm$ 3	26 $\pm$ 4
+ 0.5 mM EGTA	(3)	178 $\pm$ 63	72 $\pm$ 30	39 $\pm$ 3	14 $\pm$ 5
+ 5.0 mM CaCl$_2$	(3)	50 $\pm$ 11	64 $\pm$ 17	123 $\pm$ 3	4 $\pm$ 1

[1] 0.6 M Mannitol, 10 mM MES, pH 5.5. Values in parentheses are the number of replicates examined. EDTA, (ethylenedinitrilo)-tetraacetate; EGTA, ethylene-bis(oxyethylenenitrilo)tetraacetate.

[2] The total number of pollen grains in 5 ml of pollen digest was 1,188,000 $\pm$ 19,000.

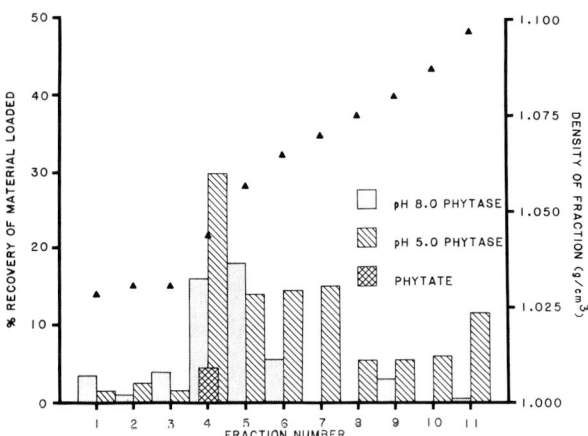

Figure 1. Distribution of phytate and pH 5 and pH 8 phytase activities in a self-generated Percoll density gradient. Fraction volume was 1 ml. A profile of the density gradient is indicated by triangular symbols.

Fractionation of contents from ruptured sporoplasts on Ficoll or Percoll gradients revealed a close association between pH 5 and pH 8 phytases and phytate reserves. In discontinuous Ficoll gradients, the bulk of the protein and phosphatase activity appeared between 5-10% Ficoll. Phytate was distributed in increasing concentration through the gradient with its highest value in 25% Ficoll. Both phytases migrated into denser regions of the gradient with highest specific activities in 20% Ficoll but on a total activity basis, substantial phytase remained in low density fractions. In Percoll gradients (Fig. 1), phytate and the bulk of the pH 5 and pH 8 phytase activities appeared in buffer fractions #4 and #5 (mean densities: 1.035 and 1.048 g/cm3, respectively). No phytate was detected in other fractions but pH 5 phytase activity appeared throughout the gradient. Only 4.1% of the phytate initially loaded on the gradient was recovered after centrifugation.

DISCUSSION

The observation that EDTA greatly increases release of sporoplasts from MMNO/carbohydrase-treated pollen while Ca^{2+} represses such release suggests the presence of at least three types of exine/intine bonding, one linked to hydrogen bonds, a second involving covalently-bound structures subject to carbohydrase attack, and a third involving ionic linkages. The fact that EGTA is less effective than EDTA might indicate an involvement by ionic linkages other than Ca^{2+}, possibly Mg^{2+}, at the exine/intine interface. Clearly, a more detailed study of sporoplast release is indicated. Such an effort will reveal structural details regarding the association of exine and intine layers in the pollen wall that have, until now, been largely ignored.

Access to sporoplasts of mature ungerminated pollen provides the pollen physiologist with a wealth of new opportunities to examine the properties of subcellular structures and to explore their biochemical and physiological roles. The non-invasive nature of MMNO solvolysis which promotes the loosening of exine from intine and leads to release of free sporoplasts has now been modified to produce sporoplasts which exhibit the same potential for experimental research found in vegetative protoplasts. Our studies on the localization of phytate accumulation and associated phosphohydrolyases in mature pollen is but one aspect of an increasingly promising new approach to pollen physiology.

ACKNOWLEDGEMENTS

Supported in part by NSF grants DMB-8404157 and DMB-8715482. A scientific paper in Project 0266, College of Agriculture and Home Economics Research Center, Washington State university, Pullman, WA 99164. The authors thank the Oregon Lily Company, Brookings, OR for permission to harvest lily pollen.

LITERATURE CITED

Baldi, B.G., V.R. Franceschi and F.A. Loewus, 1986. Dissolution of pollen intine and release of sporoplasts. Mulcahy, G. Mulcahy-Bergamini and E. Ottaviano, eds., Biotechnology and ecology of pollen. Springer, New York. pp 77-82.

_____ 1987a. Localization of phosphorus and cation reserves in Lilium longiflorum pollen. Plant Physiol. 85:1018-1021.

_____ 1987b. Preparation and properties of pollen sporoplasts. Protoplasma 141:47-55.

_____ J.J. Scott, J.D. Everard and F.A. Loewus 1988. Localization of constitutive phytases in lily pollen and properties of the pH 8 form. Plant Sci. (in press).

Bajaj, Y.P.S. 1974. Isolation and culture studies on pollen tetrad and pollen mother cell protoplasts. Plant Sci. Lett. 3:93-99.

Bhojwani, S.S. and E.C. Cocking 1972. Isolation of protoplasts from pollen tetrads. Nature New Biol. 239:29-30.

Espelie, K., F.A. Loewus, R.J. Pugmire, B.G. Baldi, and P.H.Given. Structural studies of Lilium longiflorum sporopollenin by ^{13}C-NMR. (submitted for publication).

Given, P. H., N. J. Ryan-Gray, G. Davidonis, P. C. Painter and A. Traverse 1988. The chemical structure of sporopollenin, the precursor of sporinite macerals. Org. Geochem. in press.

Imamura, J. and I. Potrykus 1983. Isolated tetrad protoplasts develop to the binucleate stage in tobacco (Nicotiana tabacum cv. Havana) Sixth Inter. Protoplast Symposium, poster proceedings, Protoplasts 1983, pp 15.

Ito, M. 1973. Studies on the behavior of meiotic protoplasts I. Isolation from microsporocytes of Liliaceous plants. Bot. Mag. Tokyo 86:133-141.

Latta, M. and M. Eskin, 1980. A simple and rapid colorimetric method for phytate determination. J. Agric. Food Chem. 28:1313-1317.

Loewus, F.A., B.G. Baldi, V.R. Franceschi, L.D. Meinert and J.J. McCollum, 1985. Pollen sporoplasts: Dissolution of pollen walls. Plant Physiol. 78:652-654.

Maeda, M., M. Yoshioka and M. Ito, 1979. Studies on the behavior of meiotic protoplasts IV. Protoplasts isolated from micro-sporocytes of Liliaceous plants. Bot. Mag. Tokyo 92:111-121.

Nunes, S.P. and F. Galemback 1985. Percoll and Ficoll self-generated density gradients by low speed osmocentrifugation. Anal Biochem. 146:48-51.

Power, J.B., 1973. Isolation of mature tobacco pollen protoplasts. in: H.E. Street, ed., Plant tissue and cell culture. Blackwell Sci. Publ., Oxford. pp 118-119.

Rajasekhar, E.W., 1973. Nuclear divisions in protoplasts isolated from pollen tetrads of Datura metel. Nature 246:223-224.

Scott, J.J. and F.A. Loewus, 1986a. Phytate metabolism in plants. in: E. Graf, ed., Phytic Acid: Chemistry and Applications. Pilatus Press, Minneapolis. pp23-42.

_____, 1986b. A calcium-activated phytase from pollen of Lilium longiflorum. Plant Physiol. 82:333-335.

Tanaka, I., C. Kitazume and M. Ito, 1987. The isolation and culture of lily pollen protoplasts. Plant Sci. 50:205-211.

Zhu, C., Y. Xie and S. Hu, 1983. Isolation and cultural behavior of pollen tube subprotoplasts in Antirrhinum majus L. Acta Bot. Sinica 26:459-465.

Structure and Variability of Embryos, Endosperm and Perisperm During in Vitro Culture of Sugar Beet, *Beta vulgaris* Ovules

P. Olesen
Biotechnology Section
A/S De Danske Sukkerfabrikker
P.O. Box 17, DK-1001 Copenhagen K
Denmark

E. Buck & B. Keimer
Maribo Seed
14, Højbygårdsvej
P.O. Box 29, DK-4960 Holeby
Denmark

Introduction

In recent years much progress has been made in the induction of haploid plants in economic crop species (e.g. Hu & Yang 1986) and doubled haploid lines are frequently used as a source of homozygosity in breeding programmes as well as in basic genetic studies. By far the most popular and widespread technique for haploid induction has been to exploit the androgenic pathway from in vitro cultured anthers or microspores (Hu & Yang 1986). However, androgenic methods have not been developed in all economic plants and sugar beet (*Beta vulgaris*) is one of the more important crop species where haploid induction through anther culture has proven unsuccessful until now (Rogozinska et al. 1977; Van Geyt et al. 1985). Instead, a very successful system for the production of gynogenic haploids through in vitro culture of unpollinated beet ovules (Hosemans & Bossoutrot 1983, 1985; Bossoutrot & Hosemans 1985; Van Geyt et al. 1987) and their use in practical breeding progammes (D'Halluin & Keimer 1986) has been established.

In practise, high frequencies of ovule response (20-30%) have been obtained in sugar beet (D'Halluin & Keimer 1986) but compared to the more efficient and less expensive androgenic pathway operating in other species, a number of problems remain to be solved. Among those are pronounced genotype effects (inter and intravarietal), polyembryony and secondary embryony and low efficiencies of chromosome doubling as well as final production of dihaploids. In order to throw some light on the pronounced variations in haploid embryo morphology and physiology, a detailed anatomical analysis of in vitro cultured pollinated and unpollinated beet ovules as well as in situ embryo development was initiated. Hopefully, the results of this study will give some indications on how to improve the overall efficiency of the gynogenic haploid induction in sugar beet.

The present communication presents some preliminary data on the cellular origin of the haploid embryos as well as the structural variability of embryos, endosperm and perisperm during in vitro culture.

Materials and Methods

Plant material was DDS-Maribo Seed proprietary lines grown under controlled greenhouse or growth chamber conditions (D'Halluin & Keimer 1986). Ovules from selffertile donor plants (OT and 2nMM) were isolated from flower buds before anthesis to avoid self pollination (ovule length 0.3 - 0.6 mm). For comparative purposes, male sterile lines were included in some experiments and a few studies were made also on in sita developed ovules directly from the plants or after some culture periods. For in vitro culture conditions, see D'Halluin & Keimer (1986).

For microscopy ovules were fixed in p-formaldehyde-glutaraldehyde (2% - 2.5%) in phosphate buffer (0.05 M, pH 7) for 2-4 hours at room temperature before being rinsed four times over 1 - 2 hours with buffer. In most cases a few drops of detergent (Tween 80) was added to both fixatives and rinses to improve penetration and embedding. Specimens were embedded in glycol methacrylate plastic (Feder & O'Brien 1968). 3.0 μm sections were stained with periodic acid - Schiff (PAS) for insoluble polysaccharides and Aniline Blue Black for proteins. Sections were studied and photographed with a Reichert-Jung Polyvar microscope. Photomicrographs were made from Kodak Ektachrome 50 or 160 diapositives via contact copies onto Ilford Pan F black and white negatives.

Results and Discussion

In general a pattern for the time course of in vitro haploid embryo development could not be determined because of very extensive variation between both varieties and individual ovules. Figs. 1 - 4 illustrate the general morphology of haploid embryo development which, until release from the cultured ovules (Fig. 4), follow the same pattern as in zygotic embryos. Main deviations in the haploid pattern were: (1) the suspensor being broader, containing less starch and showing a tendency to callus-like proliferation (Fig. 2); (2) more variable morphology and (3) the cotyledons being relatively under-developed (Fig. 4).

The induced haploid embryos develop through precocious germination without a dormancy stage into embryos being released or growing out of the ovules. Typically a medium shift (subculture) was needed for development of shoots and plantlets (see also D'Halluin & Keimer 1986). Although cellular differentiation of the perisperm (cf., Artschwager 1927) appeared normal during haploid development, the extensive starch accumulation typical for normal diploid development was generally not seen (Figs. 3-4). Only one haploid embryo was developed per

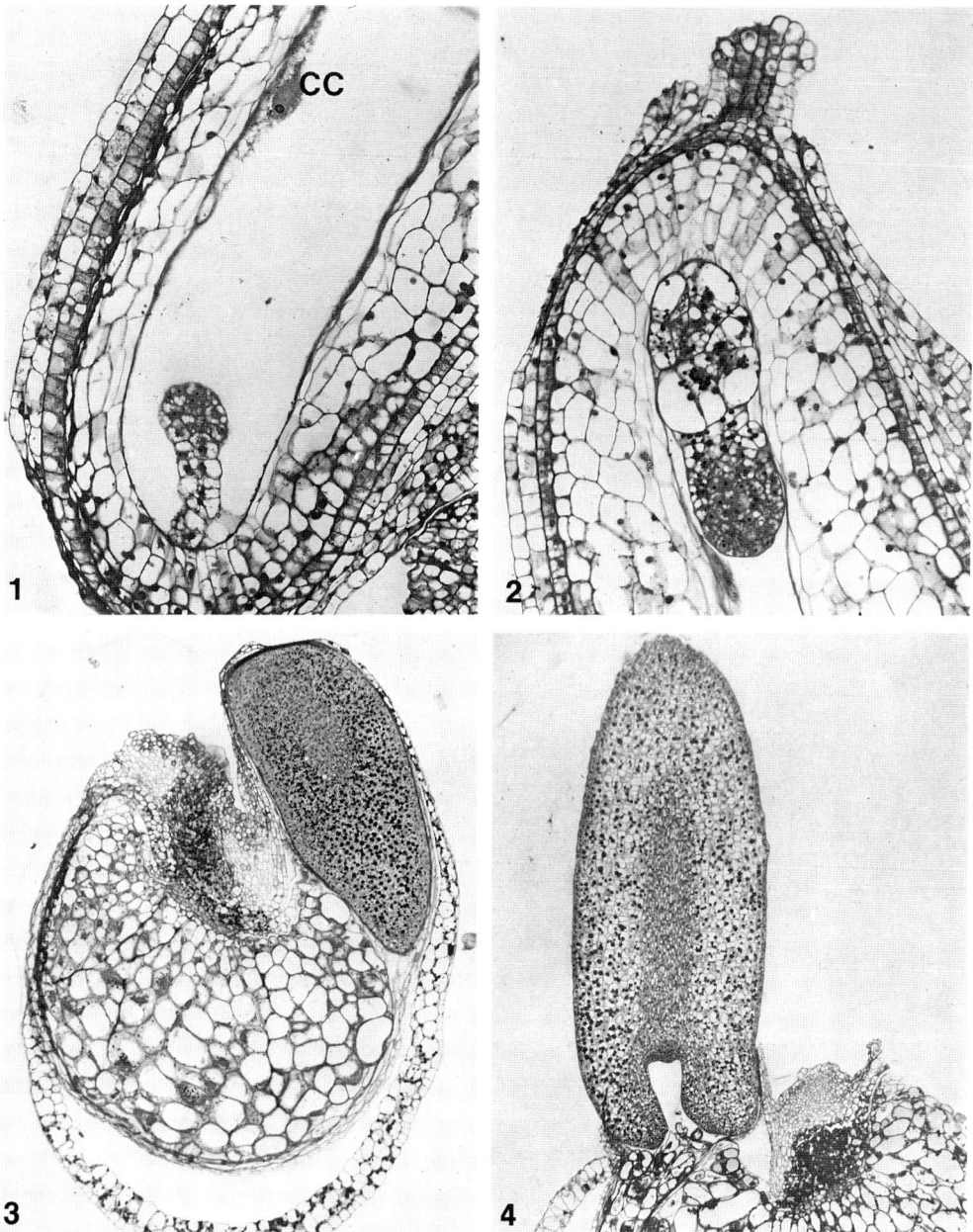

Figs. 1 - 4. Typical development of haploid embryos in cultured sugar beet ovules. 1. Young embryo stage and its suspensor originating from the unfertilized
egg cell. Unfertilized central cell (CC). x 200. 2. Later development of haploid embryo with some callus-like proliferation of the suspensor region. x 200.
3. Mid-stage haploid embryo in ovule with collapsed nucellus and non-starchy perisperm (P). x 80. 4. Release of haploid embryo from ovule with well-developed
perisperm without starch accumulation. Cytoledons relative small compared to
root and hypocotyl region. x 50.

ovule (although in one single case a twin was observed) and release happened
either through the micropyle or sidewards by breakage of the integuments (cp.,
Bossoutrot & Hosemans 1985; Bornman 1985 and Van Geyt et al. 1987).

The exact cellular origin of the single haploid embryo always seemed to be the
egg cell which is placed laterally to the synergids (Bruun 1987). Although the
first divisions of the egg cell proved difficult to observe with certainty, the
relative position of the synergids during embryo initiation could always be ascertained by their conspicuous filiform apparatus which persisted long after
their cellular degeneration. The egg cell origin is in accordance with results
of both Borman (1985) and Bossoutrot & Hosemans (1985) but the suggestion of the
latter that haploid embryo might originate also from antipodal cells could not
be verified. In only one single case an abnormality in embryo induction was observed (fig. 9). Here, in a cultured ovule with an empty and collapsed embryo
sac, a row of globular, multicellular structures resembling proembryoids were
seen extending from the embryo sac to the outer nucellus in the micropylar region. Probably these structures were formed by induced division in the outer
cell layers of the nucellus which often proliferate (increase in cell size and
stainability of cytoplasm) during embryo development.

Although careful precautions were taken to avoid self pollination, a significant
proportion of the responding ovules showed a conspicuous endosperm development
(Figs. 5-8). Principally, this might happen by three different pathways: (1)
double induction of both egg cell and central cell to develop without being fer-
tilised, (2) accidental pollination and fertilisation before culture initiation
or (3) apomictic development from unreduced female gametophyte. At present it is
impossible to ascertain which pathway would be the most plausible. Jassem (1973)
showed some plasticity in endosperm development in different beet cross-combina-
tions but, on the other hand, endosperm development without embryo development
was never observed in the present study.

In those cases where endosperm developed in intentionally unpollinated ovules,
it mainly followed the pattern from diploid or tetraploid development after fertilisation (present study; Jassem 1973; Artschwager 1927). In several cases an
extreme differentiation of the outermost cell layer of the endosperm took place
(Fig. 7). These cells were extremely large with huge nuclei and typically filled
with starch and protein bodies. In many respects these cells show morphological
features common to aleurone cells in cereals. Fig. 6 is an example of an ovule
where such cells did not differentiate. In one case a very conspicuous callus
formation (fig. 8) was initiated from the aleurone-like cells.

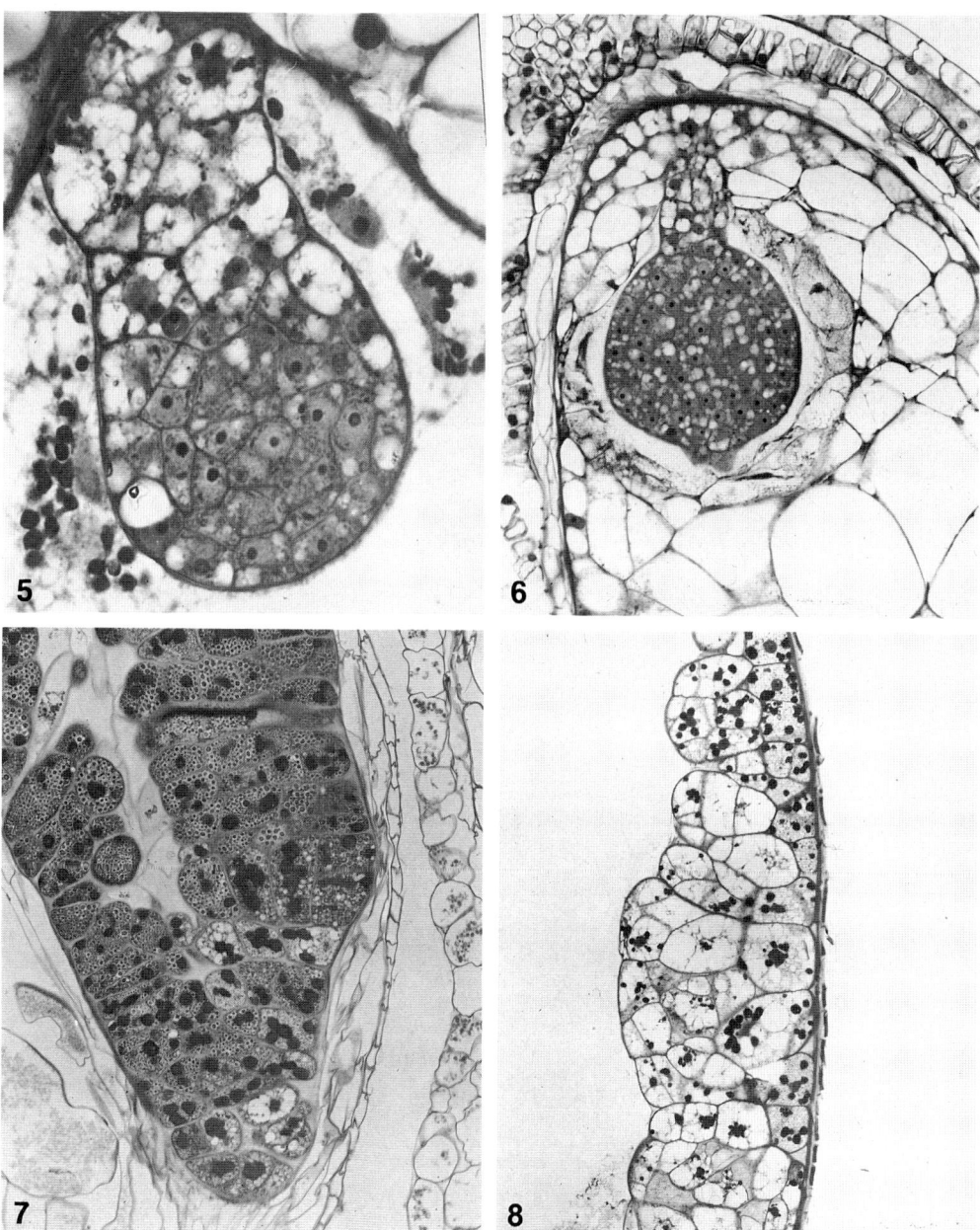

Figs. 5 - 8. Anatomical variations associated with embryo formation in cultured unpollinated beet ovules. 5. Young embryo with broad suspensor surrounded by starchy cytoplasm and free nuclei presumably indicating endosperm initiation. x 800. 6. Well-developed cellular endosperm containing a growing globular embryo surrounded by zone of lysed cells. x 320. 7. Tangential section of endosperm with conspicious aleuron-like outer cells. x 200. 8. Empty ovule after release of embryo: callus formation from the remaining aleuron-like outer cell layer of the endosperm. x 200.

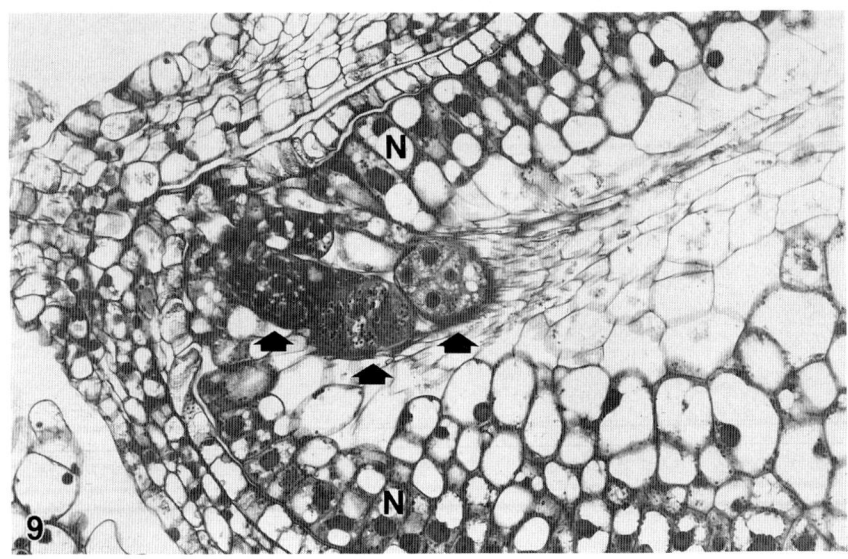

Fig. 9. Oblique row of proembryoid complexes (arrows) located in the micropylar nucellus between the empty embryo sac and outer proliferating nucellus cells (N). No embryo or endosperm structures were found in this ovule. x 320.

Acknowledgements

The authors want to thank the personnel at the Maribo Seed tissue culture facilities for handling of plant material and ovule preparation, and Lone Bruun (M.Sc.) and H. Elsted Jensen, Botanical Laboratory, University of Copenhagen, for valuable discussions and photographic work, respectively.

References

Artschwager E (1927) J Agr Res 34: 1-25
Bornman CH (1985) In Vitro 21: 36A
Bossoutrot D, Hosemans D (1985) Plant Cell Reports 4: 300-303
Bruun L (1987) Nord J Bot 7: 543-551
D'Halluin K, Keimer B (1986) In: Horn W et al. (eds) Genetic manipulation in plant breeding. W de Greuter & Co., Berlin New York
Hosemans D, Bossoutrot D (1983) Z Pflanzenzücht 91: 74-77
Hosemans D, Bossoutrot D (1985) In: Chapman GP et al. (eds) Experimental manipulation of ovule tissues. Longman Inc., New York
Hu H, Yang H (eds) (1986) Haploids of higher plants in vitro. Springer, Berlin Heidelberg New York
Jassem M (1973) Genetica Polonica 14: 295-303
Rogozinska JH, Goska M, Kuzdowicz A (1977) Acta Soc Bot Pol 46: 471-479
Van Geyt JPC, D'Halluin K, Jacobs M (1985) Z Pflanzenzücht 95: 325-335
Van Geyt JPC, Speckmann Jr. GJ, D'Halluin K, Jacobs M (1987) Theor Appl Genet 73: 920-925

Cytological Aspects of in Vitro Androgenesis in Cereals

B.Barnabás, É,Szakács, K.Liszt[+]
Agricultural Research Institute of
the Hungarian Academy of Sciences,
2462, Martonvásár, Hungary

INTRODUCTION

In the Gramineae family many aspect of microspore division and differentiation in vitro were studied (Sun 1978, Miao et al. 1978, Zheng and Ouyang 1980, Idzikowska et al. 1982, Huang 1986). It has been discovered, that in the cereal species the same series of developmental pathways of in vitro androgenesis exist as in other plant species. It was observed in maize (Miao et al. 1978, Barnabás et al. 1987) that mainly the vegetative cell contributes to the multicellular pollen grain (MPG) formation. In wheat the equal division of microspore nucleus occured more frequently in anther culture (Zheng and Ouyang 1980). Still there have been only a few publications on the ultrastructural aspects of androgenesis in vitro (for a survey see Huang 1986, Barnabás et al. 1987). The aim of the present study was to characterize the different types of the MPGs formed in wheat anther culture as a result of equal or unequal divisions.

MATERIALS AND METHODS

Triticum aestivum L.cv., Ciano was used as the experimental material. Spikes were collected from field grown plants and the anthers containing microspores in late uninucleate stage were inoculated onto P 2 medium (Chuang et al. 1978) in Petri dishes under sterile conditions. The cultures were then incubated at 29°C in the dark. In general, 5-5 anthers were randomly chosen from the spikes and from the Petri

[+]Department of Plant Anatomy
Eötvös Loránd University
1088, Budapest,
Múzeum krt. 4/a
Hungary

dishes after 1, 7, 9, 11 and 14 days in culture. Five anthers from
each sampling were used for cytological observations. The microspores
released from the anthers were stained with 1 ug ml^{-1} DAPI (4,6-dia-
midino-2-phenylindole) by the method of Coleman and Goff (1985) and
examined by means of fluorescent microscope (OPTON, Ultraphot-III).
The other 5 anthers were fixed in 2,5% glutaraldehyde in phosphate
buffer (pH 7,2) containing 5% sucrose for 2 h and postfixed in 1% OsO$_4$
solution in the same buffer for another 2 h at room temperature. After
dehydration the anthers were embedded in Araldite. Ultrathin sections
were prepaired using Reichert Ultracut ultramicrotome. The sections
were poststained with uranyl acetate and lead citrate and were examined with Tesla BS 500 transmission electron microscope at 60 kV.

RESULTS AND DISCUSSION

In the late uninuclear stage the wheat microspore (Fig. 1.) contained
a large nucleus (n) and an organellum rich cytoplasm. Microbodies (mb),
proplastids (p) and endoplasmic reticulum (er) could be observed. In
this stage there is a direct contact between the microspores and the
tapetum layer (t) of the anther wall. The Ubisch bodies (Ub) are clearly visible. The intine (i) is beginning to develop. Microspores from
1-day culture have completed the first mitotic division. Two ways of
division : symmetric (Fig. 2a.) and asymmetric (Fig. 2b.) occurred.
In the DAPI-stained microspores the daughter nuclei showed similar or
different intensity of fluorescence on the basis of their DNA content.
The vegetative-like nucleus (vn) seemed to be fainter than the gener-
ative-like (gn) one. In the case of symmetric division the two nuclei
had the same brightness. After one week in culture a number of multicellular pollen grains appeared (Fig. 3.).

By means of electron microscope four different types of the MPGs
could be distinguished between 9 and 14 days in culture (Fig. 4.).
Type 1 : the MPG has heterogenous construction (Fig. 4a.). Some cells
         contains very dense cytoplasm with much lipids (l). These
         cells show degeneration. The other cells seem to be function-
         able, containing heterochromatic nucleus and normal cell or-
         ganelles. There are amyloplasts (ap) in both types of cells.
Type 2 : the constructing cells are also not uniform (Fig. 4b.).

One part of cells are highly vacuolated (v) and the others
contain much reserve materials: starch (st) and lipids (l).

Type 3 : the MPG consists of nearly uniform cells with dehydrated,
dense cytoplasm (Fig. 4c.). The quality of the cytoplasm
and the swollen cell walls (cw) indicates that the cells
might not be unviable, but have a latent condition.

Type 4 : this is an entirely viable construction (Fig. 4d.). It consists of homogeneous cells with large, heterochromatic nuclei.
Each nucleus has two nucleoli (nc). The cytoplasm of the cells
is extremely rich of ribosomes. These cells have meristematic
habit.

In the course of our cytological examinations it has become obvious
that a DNA-specific fluorochrome such as DAPI was suitable to distinquish the different ways of microspore divisions occurred in wheat
anther culture. During the electron microscopical observations we
could find some new types of MPGs. Sunderland (1979) and Huang (1982,
1986) characterised the so cold "partionated" MPGs in barley and wheat.
These structures are consisted of highly vacuolated cells besides some
cells with dense cytoplasm. This seems to be similar to the Type 2
MPG þy our classification. The Type 1 MPG has not been observed before.
We suppose that these heterogeneous structures also developed by the
repeated asymmetric divisons of the microspores. The viable, function-
able part of these MPGs can develop further into primer calli. By our
hypothesis, the globular pollen embryos might be formed from those
MPGs which consist of uniform cells (Type 3 and 4) as a result of
equal divisions.

116

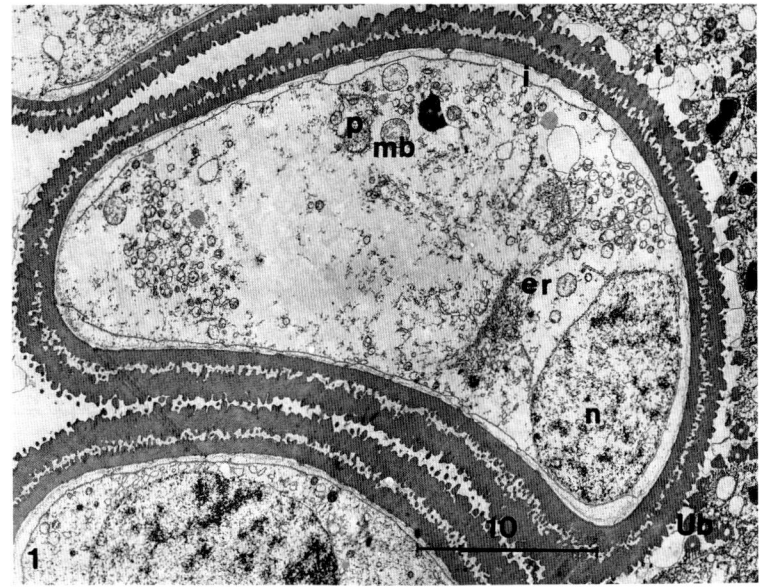

Fig.1. Wheat microspore in the late uninuclear stage

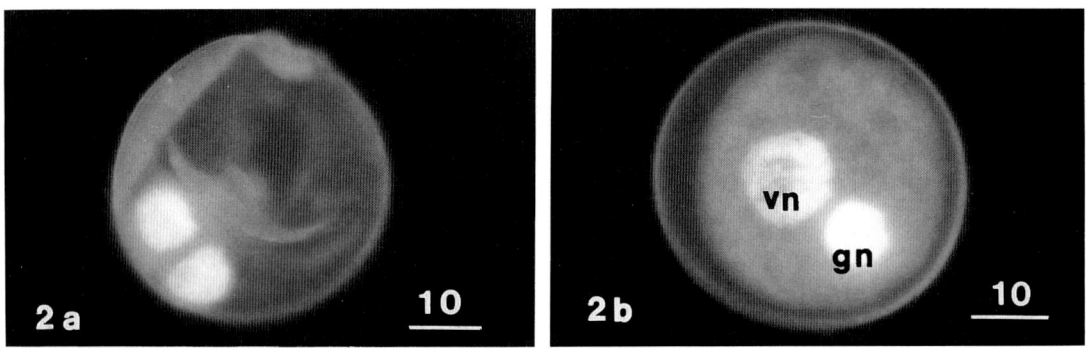

Fig.2. First mitotic division of the microspores

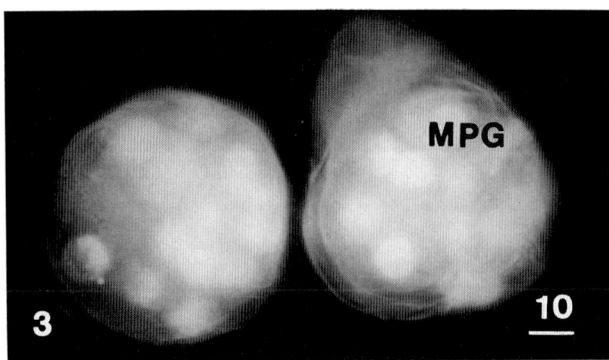

Fig.3. Multicellular pollen grains

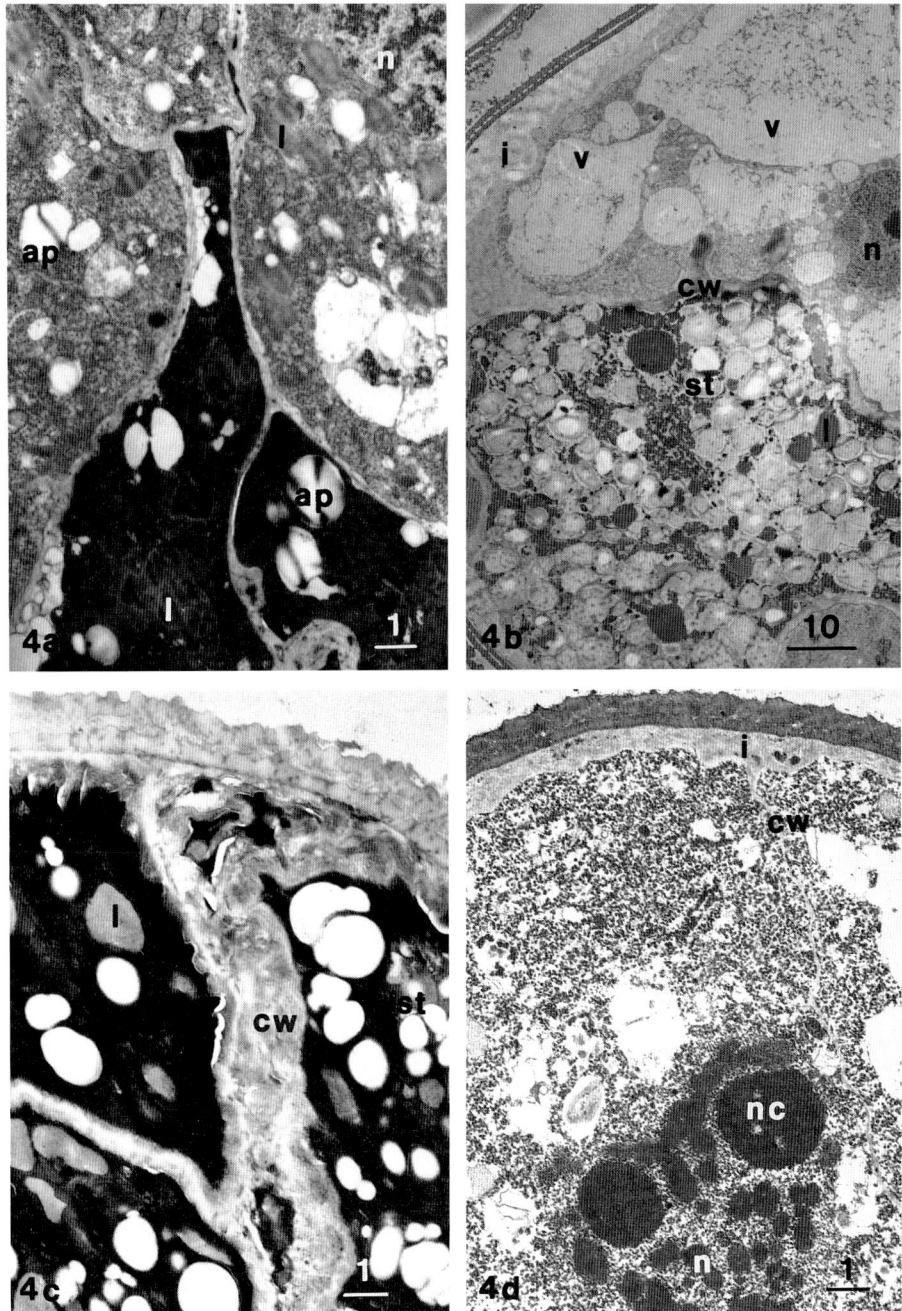

Fig.4. Different types of the multicellular pollen grains

118

REFERENCES

Barnabás B , Fransz P F, Schel J H N (1987) Ultrastructural studies
    on pollen embryogenesis in maize (Zea mays L.) Plant Cell Reports
    6: 212-215
Chuang C C, Ouyang T W, Chia H, Chou S M, Ching C K (1978) A set of
    potato media for wheat anther culture. In: Proc. Symp. Plant Tissue
    Culture. Science Press Beijing 51-56
Coleman A W, Goff L J (1985) Application of fluorochromes to pollen
    biology. I. Mithramycin and 4,6-diamidino-phenylindole as vital
    stains and for quantitation of nuclear DNA. Stain Technol 60: 145-
    154
Huang B (1982) Developmental studies of anther cultures of Hordeum,
    Triticum and Peonia. PhD Thesis Univ East Anglia Norwich UK
Huang B (1986) Ultrastructural Aspect of Pollen Embryogenesis in
    Hordeum, Triticum and Peonia. In: Haploids of Higher Plants in
    vitro.Hu Han and Yang Hongyuan (eds) Springer-Verlag Berlin Hei-
    delberg New York Tokyo 91-118.
Idzikowska K, Ponitka A, Mlodzianowski F (1982) Pollen dimorphism and
    androgenesis in Hordeum vulgare. Acta Soc Poloniae 51: 153-156
Miao S, Kuo C, Kwei Y, Sun A, Ku S, Lu W, Wang Y, Chen N, Wu M, Hang L
    (1978) Induction of pollen plants of maize and observation on their
    progeny. In: Proceedings of Symposium on Plant Tissue Culture.
    Science Press Peking 23-33
Sun C (1978) Androgenesis of cereal crops. In: Proceedings of Sym-
    posium on Plant Tissue Culture Science Press Peking 117-123
Sunderland N, Roberts M, Evans L J, Wildon D C (1979) Multicellular
    pollen formation in cultured barley anthers. I. Independent division
    of the generative and vegetative cells. J Exp Bot 30: 1133-1144
Zheng J, Ouyang J (1980) The early androgenesis in in vitro wheat
    anthers under ordinary and low temperature. Acta Gen Sin 7: 165-175

A Micromanipulation Method for Artificial Fertilization in *Torenia*

C.J. Keijzer, M.C. Reinders and H.B. Leferink-ten Klooster
Department of Plant Cytology and Morphology
Agricultural University
Arboretumlaan 4
6703 BD WAGENINGEN
The Netherlands

Introduction

The present study is partly a consequence of our interspecific cross research programme in Lilium spp. (Van Tuyl et al., 1987). Van Roggen et al. (1988) demonstrated that sperm cells generally do not reach the embryosac following the cut-style pollination method which is necessary for interspecific lily crosses. As a first step to study the fate of sperm cells in a non-specific embryo sac, we developed a system for artificial (intraspecific) sperm cell injection into the embryo sac. For this, Torenia fournieri was used as a model plant, since the embryo sac emerges from the integuments in this species (Van der Pluijm, 1964; Wilms and Keijzer, 1985).

Materials and methods

Pollen grains of Torenia fournieri were germinated in an aqueous solution of 10% sucrose and after 12 hours sperm cells were isolated by osmotically bursting the pollen tubes with water. Bursting of sperm cells was prevented by immediatedly raising the osmotic pressure again with the initially used sucrose solution.

Ovaries were surface-sterilized in 70% ethanol for 10 seconds and mechanically isolated and/or manipulated ovules were cultured in a liquid CC-medium (Mol, 1986), either with or without a little piece of the placenta. To obtain plantlets in vitro, ovules with a piece of the placenta were cultured on a MS-medium containing 3% sucrose and 0.6% agar.

Micropipettes with a diameter smaller than $0.4\mu m$ were made using a Leitz pipette puller. For the injection of sperm cells into the embryo sac, such micropipettes were changed into either micropipettes with a lateral hole and a tip diameter smaller then $0.1\mu m$ or micro-forceps, using a microforge (De Fonbrune, 1949). Micromanipulations were carried out using Leitz micromanipulators and either a Leitz inversion microscope or a Nikon interference microscope.

Observations and discussion

The structure of the embryo sac of Torenia fournieri enables easy
observation and manipulation of this organ (figs 1 and 2). The
base of the synergids contains a prominent filiform apparatus, in
which the pollentube can be easily observed, both in fresh ovaries
under the light microscope and in electron microscope sections
(fig.3). Penetrating the synergids with a micropipette via this filiform apparatus appears to be impossible without seriously damaging
the embryo sac. Since the filiform apparatus is the only site connecting the synergids with the embryo sac wall, artificial injection
of sperm cells into the synergids has to take place in other sites,
i.e. via the central cell. Likewise, despite the presence of a narrow
canal-like extension connecting the egg cell with the embryo sac wall,
the problems in observing this structure using light microscopy
(fig.2) forced us to use the same way for injections into the egg
cell, i.e. also via the central cell.
From 4 days after (natural) fertilization, which occurs 24 hours
after pollination, both spherical-shaped embryos and a multicellular
endosperm can be directly observed in cleared intact ovules (fig.4),
thus offering a quick and easy method to trace successful artificial
fertilization attempts. Also in vitro, ovules develop an embryo and
endosperm, when cultured from 24 hours after pollination (figs 5
and 6). The main structural differences between the vivo and vitro
ovules are the swelling cells of the integument tapetum in the latter
(fig.6). This phenomenon might be due to a too low osmotic value of
the medium used. However, it apparently does not influence the processes for which the culture is carried out, being embryo-, endosperm-
and plantlet formation (fig.7).
Sperm cells, necessary for artificial fertilization, are isolated
from pollen tubes in vitro (fig.8), using the osmotic shock procedure
(fig.9). Addition of water to the germination medium is immediately
followed by bursting of the pollen tubes and accordingly the release
of the sperm cells (fig.9). Once the sperm cells are freed, the
osmotic pressure is raised again to prevent the sperm cells from
bursting. Careful selective sucking the sperm cells into the micropipette can more or less separate the cells from the vegetative
cytoplasm (fig.10). The narrow diameter of the lateral hole in the
injection pipette prevents excessive evaporation, by which the pipette
can be used as a storage tube for isolated sperm cells during the
time needed to prepare the acceptor ovules.
The main problem of sperm cell injection into the embryo sac is the
elastic wall of the latter, permitting the pipette tip only to
penetrate after dramatic inward bending of this wall (figs 11 and 12).
This results into a sudden, too far proceeding penetration of the
injection pipette, which might damage the central cell or other
structures of the ovule (fig.12). Although never leading to rupture
of the embryo sac or one of its cells, the deposition of the sperm
cells and germination medium in the embryo sac cells presumably has
to be carried out carefully, since raising the pressure inside the
embryo sac might cause physiological damage. Retraction of the empty
pipette from the embryo sac should occurr slowly in order to prevent

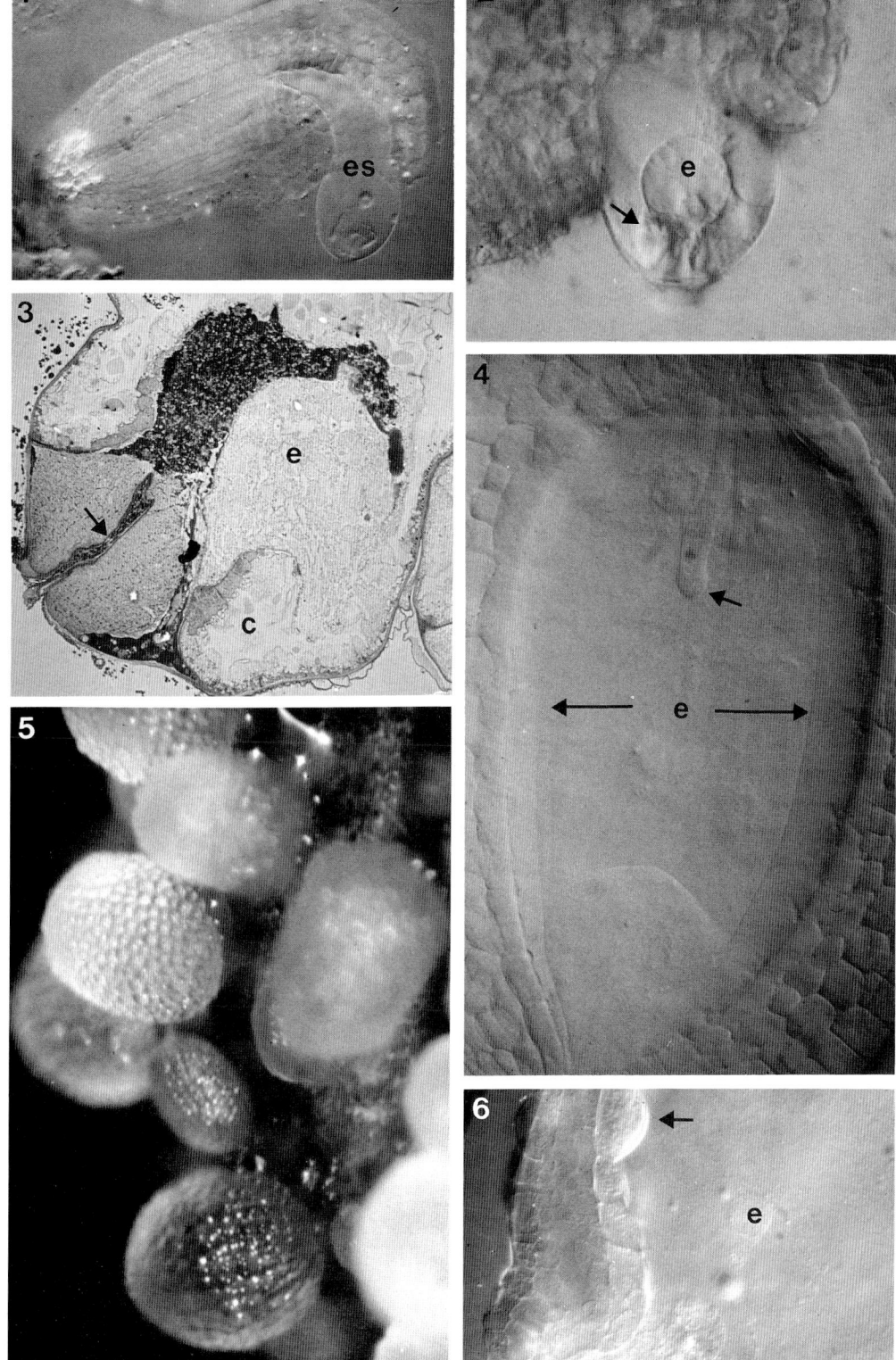

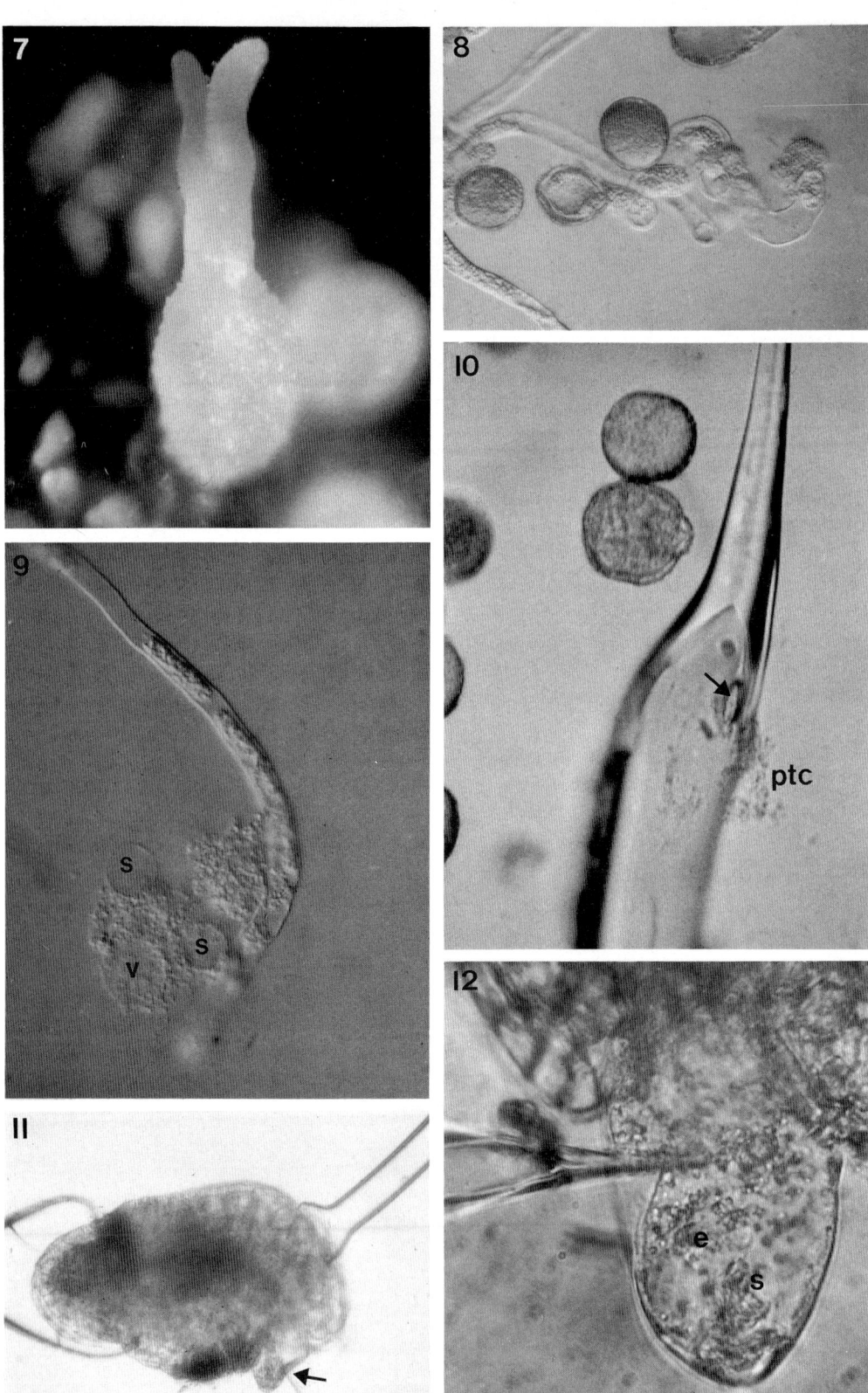

leakage of the latter.

The method presented here has to be improved. Up to now, only a few attempts have led to sperm cell deposition in the embryo sac. Further studies must reveal the fate of the sperm cells in such a situation (fertilization?). Moreover, they can show the possibilities for the many variations that are available using this technique in reproduction processes.

Figure legends

1. The ovule of Torenia after clearing with Herr's solution, showing the emerging embryo sac (es).
2. The emerging part of the embryo sac showing the egg cell (e). The synergids are out of focus (arrow).
3. TEM detail of fig. 2. The filiform apparatus contains a pollen tube (arrow). e=egg cell, c=cental cell.
4. Six days after pollination an embryo (arrow) and endosperm (e) can be observed after clearing with Herr's solution.
5. After 6 days in vitro swollen fertilized ovules can be distinguished from unfertilized smaller ones.
6. The same situation as in fig. 4, after 6 days in vitro. Note the swelling cells of the integument tapetum (arrow). e=embryo.
7. After 4-6 weeks in vitro plantlets emerge from the ovules.
8. After a few hours in the germination medium, pollen grains germinate.
9. Decreasing the osmotic pressure of the germination medium ruptures the pollen tubes and sets both the sperm cells (s) and the vegetative nucleus (v) free.
10. Part of the pollen tube content (ptc) with the sperm cells is sucked into the lateral hole (arrow) of the injection pipette.
11. An ovule, kept in position by microforceps, is injected via the lateral side of the embryosac (arrow).
12. The same situation as in fig. 11, after penetration of the pipette into the central cell. e=egg cell, s=synergid.

Acknowledgements

The authors thank Prof. dr. M.T.M. Willemse for critically reading the text, mr. S. Massalt for preparing the fotographs and miss. N. van Dam and miss. F. Bonebakker for typewriting the manuscript. Part of this research was supported by the Commission of the European Communities under grant number 0202-NL.

References

De Fonbrune P (1949) Technique de micromanipulation, Masson, Paris
Herr JM (1971) A new clearing squash technique for the study of ovule development in angiosperms. American Journal of Botany 58: 785-790

Mol R (1986) Isolation of protoplasts from female gametophytes of
    Torenia fournieri. Plant Cell Reports 3: 202-206
Van der Pluijm JE (1964) An electron microscopic investigation of
    the filiform apparatus in the embryo sac of Torenia fournieri.
    In: Linskens HF (ed.), Pollen Physiology and fertilization.
    North Holland Publishing Company, Amsterdam, pp. 8-16
Van Roggen PM, Keijzer CJ, Wilms HJ, Van Tuyl JM and Stals AWDT
    (1988) A SEM study of pollen tube growth in intra- and inter-
    specific lily crosses. Botanical Gazette, in press
Van Tuyl JM, Kwakkenbos AAM, Keijzer CJ and Wilms HJ (1987)
    Interspecific hybridization between Lilium longiflorum and the
    white asiatic hybrid "Mont Blanc". NALS Lily Yearbook, in press
Wilms HJ and Keijzer CJ (1985) Cytology of pollen tube and embryo
    sac development as possible tools for in vitro plant (re-)produc-
    tion. In: Chapman GP, Mantell SH and Daniels RW (eds.) Experimen-
    tal manipulation of ovule tissues. Longman, New York, London,
    chapter 3

The Isolated Embryo Sac of *Zea mays:* Structural and Ultrastructural Observations

V. T. Wagner, Y. Song, E. Matthys-Rochon, and C. Dumas
Reconnaissance Cellulaire et Amelioration des Plantes
Université Claude Bernard Lyon I
43 Bd. du 11 Novembre 1918
69622 Villeurbanne Cedex - France

SUMMARY

A technical procedure has been described for the enzymatic isolation of the corn embryo sac (Wagner et al., in prep.). Light and electron microscopy were used to assess the condition of the embryo sac. The results show that the female gametes are intact with "active" cytoplasms and only remnants of a cell wall.

INTRODUCTION

In Angiosperms the female gametophyte is often deeply embedded in the tissues of the ovule; consequently, physiological measurements and morphological observations are difficult to obtain. Micromanipulator-aided dissections (Allington, 1985), acid digestion and squashing under a cover slip (Bradley, 1948), partial enzymatic digestion of ovules followed by manual dissection (Forbes, 1960), and recently a strict enzymatic maceration of the ovule (Zhou, 1985; Zhou and Yang, 1985) are some of the several different methods developed for isolating the embryo sac from the obstructing ovular tissue. The enzymatic maceration method is exceptional because there is no physical manipulation and possible impairment of the embryo sac, and relatively large quantities of embryo sacs can be acquired in a short period of time. Until now, enzymatic digestion techniques have not been successfully applied to the cereals, but micromanipulator-aided dissections have been accomplished (Allington, 1985). In Zea mays, embryos have been individually dissected out of the ovule but only after considerable postfertilization enlargement (Mock and Dahmen, 1973; Sheridan et al., 1978). This paper presents preliminary observations on the isolated Zea mays embryo sac.

MATERIALS AND METHODS

Ovules dissected from ovaries were digested in an enzyme solution consisting of sucrose, cytohelicase, pectinase, cellulase and hemicellulase, and pectolyase. The enzyme-ovule mixture was agitated for 1 to 3 hrs. After liberation of the embryo sacs from the ovular tissue, the embryo sacs were purified from debris by filtration through nylon filters then removed by rinsing and resuspension into a sucrose solution (Wagner et al., in prep.). The embryo sacs could then be individually selected using a Pasteur micropipette.

The embryo sacs were observed directly after isolation or were fixed, embedded, and sectioned as follows. Embryo sacs were fixed in 4% gluteraldehyde in phosphate buffer (0.1M, pH 7.0, 7% sucrose) for two hrs. Post-fixation was in 2% osmium tetroxide in the same buffer. The material was then dehydrated in an acetone series and embedded in Spurr's resin (Spurr,1969). Serial thick sections of the embryo sac were cut and observed with phase contrast microscopy. Selected sections were embedded on the tip of a blank epoxy block for subsequent ultrathin sectioning (Mogensen, 1971). The sections for ultrastructural analysis were stained with uranyl acetate and lead citrate or according to the PATAg procedure (Thiery, 1967) and observed with an JEOL 1200 electron microscope.

RESULTS

Phase contrast light microscopy of unfixed material reveal that the isolated embryo sac is pear-shaped and the antipodals are absent (Fig 1). Serial thick sections of several isolated embryo sacs demonstrate that the constitutive cells are intact and in proper orientation inside of the embryo sac. The central cell is highly vacuolar with the majority of its nonvacuolar cytoplasm and paired polar nuclei found near the egg cell (Fig 2). The synergid pair is found next to the egg cell at the micropylar end of the embryo sac. The egg and central cell are often found to be seperated from each other, at the micropyle region (Wagner et al., in prep.). The diagramatic representation of the isolated embryo sac is shown in Fig 3.

Electron microscopy reveals a complete complement of organelles for the egg and central cell cytoplasms. The egg and central cell plasma membranes are intact and organelles are structurally normal: mitochondria have distinct cristae, nuclei are surrounded by a pored envelope, and amorphous plastids are observed (Fig 4). No polysaccharide cell wall was found between or around any of the constitutive cells (Fig 5); however, slight remnants of a fiberous cell wall near the outside of the central cell were observed (Fig 6).

DISCUSSION

The exposure to the enzyme mixture and agitation during the isolation procedure has not caused cellular damage to the isolated embryo sac. Light microscopic observations suggest that the isolated embryo sac remains structurally intact during the isolation procedure and that no cellular rupturing of the constitutive cells has occured. The organization of the constitutive cells is similar to the nonisolated condition. A central cell with coupled polar nuclei and nonvacuolar cytoplasmic polarization toward the micropyle, and a vacuolar egg cell with a centrally located nucleus, are indicative of a normal mature embryo sac (Van Lammeren, 1987). In addition, electron microscopy reveals intact plasma membranes with a full complement of "active" organelles in the central and egg cell.

Although the cellular and nuclear organization of the isolated embryo sac is normal, the enzyme maceration technique has altered the nature of the physical associations between the constitutive cells of the embryo sac. The antipodals are no longer found attached to the central cell; and the egg and central cell are not in close proximity at the micropyle. Perhaps the maceration mixture has partially digested the shared polysaccharide walls between the cells, allowing the individual constitutive cells of the isolated embryo sac to seperate.

128

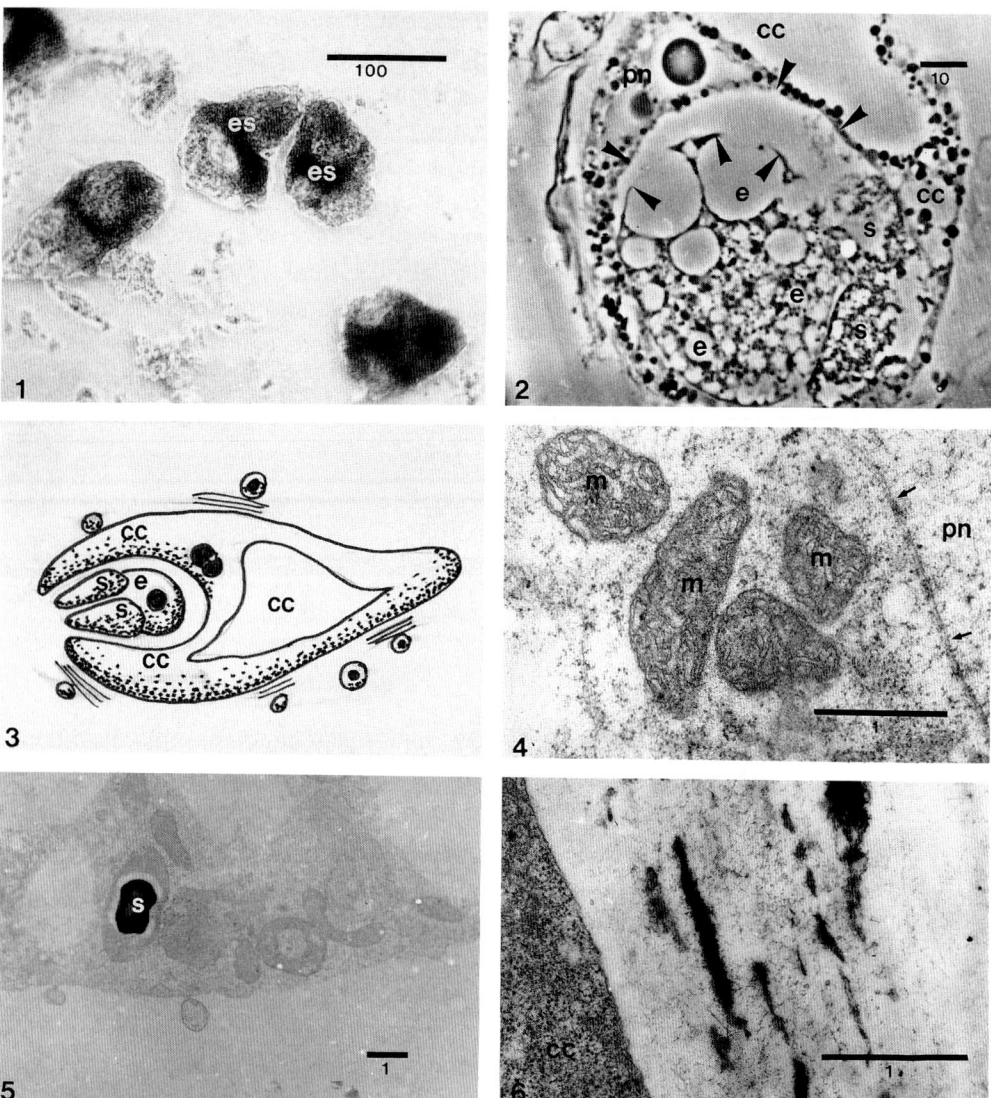

Fig.1   Light microscopic view of several isolated embryo sacs (ES) of corn.

Fig.2   A crosssectional light micrograph showing the constitutive cells of the
embryo sac: vacuolar central cell (CC) with coupled polar nuclei (PN), two
synergids (S), and egg cell (E). Please note the distance between the egg
and central cell (arrows).

Fig.3   Diagramatic representation of the constitutive cell organization in the
isolated embryo sac. Central cell (CC), egg cell (E), and synergids (S)

Fig.4   Electron micrograph of the central cell cytoplasm of the isolated embryo
sac. Please note intact nuclear membranes (arrows) of the polar nuclei (PN)
and the "active" configuration of the mitochondria (M).

Fig.5   Results of the PATAg test for the presence of polysaccharide cell walls
around the egg cell near the micropyle. Please note the positive reaction
of the starch grain (S) in the plastid, but no polysaccharide cell wall is
detectable.

Fig.6   Electron micrograph of the remnant cell wall found near the central cell
(CC).

In conclusion, this new technique opens several possibilities for future investigations. The isolated embryo sac will allow for a precise microinjection of exogenous DNA into the chosen target cell which is not physically or visually hampered by the surrounding ovular tissue. Another research possibility is to use a second maceration mixture to selectively digest the remnant cell wall remaining around the isolated embryo sac, liberating the female gametes. Preliminary studies have produced small numbers of isolated female gametes and a previous report describes a procedure to isolate the male gametes (Dupuis et al., 1987). These techniques are predecessors for the fusion of isolated gametes.

ACKNOWLEDGEMENTS

We greatfully acknowledge research support provided by the Institut National de la Recherche Agronomique, Centre National de Recherche Scientifique, and the Biotechnology Action Programm of the Commission of European Communities (No. BAP-0203-F).

REFERENCES

Allington P M (1985) Micromanipulation of the unfixed cereal embryo sac, in: The experimental manipulation of ovule tissues, Longman, New York: 39-51

Bradley M V (1948) An aceto-carmine squash technique for mature embryo sacs. Stain Technol 23: 29-40

Dupuis I, Roeckel P, Matthys-Rochon E, Dumas C (1987) Procedure to isolate viable sperm cells form the corn (Zea mays L.) pollen grains. Plant Physiol 85 (4): 876-878

Forbes I (1960) A rapid enzyme-smear technique for the detection and study of plural embryo sacs in mature ovaries in several Paspalum species. Agron J 52: 300-301

Mock J J, Dahmen W J (1973) Sterile culture of immature maize seeds and embryos. Crop Sci 13: 764-766.

Spurr A R (1969) A new viscosity resin embedding medium for electron microscopy. J Ultrastruct Res 26: 31-43

Sheridan W F, Neuffer M G, Bendbow E (1978) Rescue of lethal defective endosperm mutants by culturing immature embryos. Maize Coop News Letter 52: 88-90

Mogensen H L (1971) A modified method for re-embedded thick epcxy sections for ultratomy. J Ariz Acad Sci 6: 249-250

Thiery J P (1967) Mise en évidence des polysaccharides sur coupes fines en microscopie electronique. J Microscopie 6: 967-1018

Van Lammeren (1987) Embryogenesis in Zea mays L. A structural approach to maize caryopsis development in vivo and in vitro. U Wageningen

Zhou C, Yang H Y (1984) The enzymatic isolation of embryo sacs from fixed and fresh ovules of Antirrhinum majus L. Acta Biol Exp Sin 17: 141-147

Zhou C, Yang H Y (1985) Observations on enzymatically isolated, living and fixed embryo sacs in several angiosperm species. Planta 165: 225-231

Potential of Unreduced Pollen for Breeding Tetraploid Perennial Ryegrass

A.P.M. den Nijs and A.G. Stephenson*

Foundation for Agricultural Plant Breeding (SVP)

POB 117, 6700 AC WAGENINGEN, The Netherlands

ABSTRACT

Tetraploid cultivars of perennial ryegrass are slowly gaining popularity for fodder and pasture in north-western Europe. Breeding of new cultivars relies on colchicine treatment of young diploid seedlings, selection of tetraploid non-chimaeric individuals, intercrossing, evaluation, and finally selection of half-sib families as a basis for new synthetics.

Pollinations of tetraploid plants by selected diploid clones producing appreciable percentages of unreduced (2n) pollen in addition to the normal, reduced pollen, would facilitate enlargement of the tetraploid gene pool. This approach is already in wide use, e.g. with potato and alfalfa. In order to evaluate this approach of ploidy breeding for perennial ryegrass, we examined as a first step the presence of unreduced pollen in samples of mature pollen from field-grown plants. A search of over 900 plants belonging to many diploid cultivars and inbred populations thereof revealed some 20 plants which produced 5-26% large pollen (>48 μm) in addition to the normal pollen (30-45 μm diameter). Stainability of the pollen varied widely among genotypes and was not related to the percentage of large pollen. Clonal ramets of three selected genotypes grown in temperature-controlled glasshouses produced most stainable pollen at 14 and 18°C, but the fraction of large pollen was highest at 22 and 26°C.

Seed set on self-incompatible plants of several tetraploid cultivars was low following pollination by selected diploids producing large pollen, and most of the offspring were tetraploids. Although contamination by rare inbred offspring due to selfing cannot be ruled out, these preliminary results indicate that especially at elevated temperatures sexual polyploidization can be used to enrich the tetraploid population of perennial ryegrass.

---

* This contribution is partly based on collaborative research during a sabbatical of A.G.S. with SVP. Present address of A.G.S. is Dept. of Biology, Penn. State University, University Park, Pennsylvania 16802, U.S.A.

INTRODUCTION

Tetraploidy is prevalent in a number of forage crops, and not surprisingly breeders have introduced tetraploidy in the ryegrasses (x=7) as well. The first 4x cultivars of ryegrasses were introduced in the Netherlands in the sixties based on early work by Wit at SVP (Wit, 1959). Tetraploid cultivars of perennial ryegrass excel in palatability and resistance to several diseases, but due to a less pronounced tillering they form a rather open sward, are weak competitors with diploids and are not very resistant to grazing. Early 4x cultivars were only fit for fodder production, but continued breeding and selection has yielded new cultivars with improved grazing tolerance. These cultivars are slowly gaining popularity as fodder and pasture crops in north-western Europe, both in mixtures with diploids and in tetraploid mixes. The creation of new tetraploid cultivars is, however, very timeconsuming as it is entirely based on colchicine-doubling of germinating diploid seeds followed by a number of cycles of selection for non-chimaeric stable tetraploidy in addition to the desired agricultural characters. During this process fertility needs to be improved as well (Wit, 1959). In short, the mitotic chromosome doubling method entails a number of disadvantages which must be overcome.

In other species a different method of polyploidization has also been employed, which relies on numerically unreduced gametes resulting from meiotic irregularities. Interploidy crosses yield thus new, sexually originated polyploids. This method of sexual polyploidization has certain inherent advantages for ploidy breeding of heterozygous crops such as perennial rye grass, reason why we decided to evaluate the method as a new breeding tool for the grass breeder.

MITOTIC DOUBLING

Ryegrasses are normally strongly allogamous due to a well developed self-incompatibility system. This is usually accompanied by a high level of heterozygosity and heterogeneity within cultivars. The choice of seedlings for colchicine treatment is therefore important for success of the tetraploidized population. Such a preselection is however impossible because germinating seedlings are used for the treatment. This leads to a very variable starting material for the c-treatment. The treatment itself typically yields a few percent doubled plants, as the meristem of many plants is not 'hit', and further mostly mixoploids or chimaeras are obtained. This, and the low fertility of the surviving individuals require several breeding cycles to provide a 4x source population for cultivar breeding. The low fertility must for a large part be attributed to the inherent inbreeding due to the colchicine treatment. Ideally, the total genome has been completely doubled, leading to increased homozygosity.

Breeding companies have managed to cope with these problems and create 4x cultivars. However, a recent study of the effects of colchicine treatment using a set of highly inbred lines by Hague and Jones (1987) revealed, that the colchicine treatment per sé induced a range of morphological variation which was independant of the doubling of the chromosomes. This variation appeared to be non-random and showed a strong genotypic effect, and thus interferes with the purpose of truefully doubling the chromosome number of selected parents.

The basis of this induction of variation is as yet not understood, but the authors put the use of colchicine for elevating the ploidy level in a doubtful light, suggesting that the mutant diploid lines obtained as a byproduct of the C-treatment surpass the tetraploids as breeding material (Hague and Jones, 1987).

SEXUAL POLYPLOIDIZATION

Polyploid species evolved in nature mainly by sexual polyploidization (review e.g. Harlan and De Wet, 1975), and the same principle has been applied to ploidy breeding of a number of crops (Mendiburu and Peloquin, 1977; Jahr et al., 1963).

The sexual polyploidization can avoid the inbreeding effect associated with mitotic doubling and therefore generally results in higher fertility of the polyploids, as was emphasized by Skiebe (1966). No mixoploids and few aneuploids are expected, and no mutagenic treatments are required, so that a clean, simple method emerges, provided that genotypes can be discovered which produce 2n gametes in addition to the normal gametes. Usually interploidy crosses are employed so an initial tetraploid is also necessary. Between n=x and 2n=2x pollen tubes inside the 2n=4x stylar canal competition may occur, in which the 2n pollen tubes are at an advantage due to heterotic effects. Besides, also selection is possible within the 2n pollen population resulting in extra vigourous 4x offspring. Lastly, outstanding heterotic diploid genotypes can be transferred largely intact to the tetraploid level.

SEARCH FOR 2N POLLEN PRODUCERS

The feasibility of polyploidization by 4x x 2x crosses requires the occurrence
of unreduced 2n pollen in reasonable numbers. The capacity to produce 2n
pollen in addition to normal pollen depends in a number of species both on
genotype and environment (review Kaul and Murthy, 1985). At SVP, a search for
genotypes producing at least 5% extra large pollen among over 900 spaced
plants of diverse 2x cultivars and inbred populations thereof during 1986 and
1987 yielded over 20 plants with 5-26% large pollen. For simplicity this
pollen will be referred to as 2n pollen although proof of this still depends
on the outcome of crossing experiments. Pollen slides were sometimes made
directly in the field, but more often immature inflorescenses were brought
into the lab, and pollen was dusted onto slides after anther dehiscence.
Stainability in acetocarmine varied from 40 to 90 percent, and this percentage
was not related to the fraction 2n pollen. The genotypes with some 2n pollen
were mostly found within inbred populations, two of which contained five and
four 2n pollen producing plants among 45 and 57 progeny, respectively. The
highest percentage 2n pollen, 26%, occurred in a single plant among 71 inbred
progeny of the New Zealand cultivar A236.

TEMPERATURE EFFECTS

In order to examine the influence of temperature on the fraction of extra
large pollen, three genotypes with an appreciable fraction of extra large
pollen were cloned, vernalized and from about tillering on grown at four
different constant temperatures in climatized glasshouses of the Phytotron of
the Institute for Horticultural Plant Breeding. Another two genotypes with low
percentages large pollen and two without any were added. The temperatures were
14, 18, 22 and 26 °C. There were six clonal ramets per temperature for most
genotypes. Plants were kept upright and treated regularly with insecticides to
combat aphid infestations.
The high temperature regimes tended to devernalize the plants, especially of
the two genotypes with a low percentage of large pollen. Of the flowering
plants pollen samples were examined for stainability and percentage of extra
large pollen. For most genotypes data were collected on at least three days
for each temperature. Pollen was stained with acetocarmine in glycerin and the
number of viable (intensely stained and regularly shaped) grains was
determined from a random sample of 200 grains per slide. From the viable
grains in turn the percentage large was calculated. Data were analyzed with

the General linear model of the SAS statistical package which is especially robust for unbalanced data sets. The percentages were arcsine square root transformed before analysis, to meet the requirement for normality. Both viability and percentage of large pollen depended on genotype as well as temperature, and there were interactions as well.

The temperature effect was most pronounced with respect to the percentage of large pollen, and the higher temperatures (22 and 26 $^\circ$C) permitted much more formation of large pollen grains than the low ones.

CROSSING EXPERIMENTS

Two diploid clones with 2n pollen were chosen for crossing experiments with three tetraploid cultivars. Potted plans were placed together in an unheated glasshouse and immature inflorescences of both parents were bagged in together with a sleeve of parchment closed at top and bottom by a cotton swab: to one inflorescence of the tetraploid there were about 10 spikes of the diploid male parent. In this way conditions for cross-fertilization were optimized. The bags were shaken periodically. Seed was only harvested on the tetraploid plants. As a check for pseudo-selfcompatibility, groups of 10 spikes of the tetraploids were bagged.

Selfing of the three tetraploids met with little success, and seed set in the interploidy crosses was also rather low (Table 1). In a normal cross in tetraploid perennial rye some 50-80 seeds per spike are expected, but it should be kept in mind that glasshouse conditions in a Dutch summer are not ideal for maximum seed yields. Emergence of the obtained seeds was rather normal relative to the conditions of seed set. The ploidy level was determined via root tip chromosome counts which enables a classification in 3x (=21) and 4x (=28) categories, but does not establish the actual number of chromosomes accurately.

The majority of the counted offspring were classified as tetraploids.

Both in the (small) selfed offspring and in the crosses a few triploids were identified, indicating that the so called 'triploid block' precluding development of triploid embryos may be strong in this grass, but not absolute. Nevertheless, for practical purposes the preponderance of 4x offspring is useful.

The plants had not been emasculated, so a few 4x plants may derive from self pollination rather than the 4x x 2x cross via functional 2n pollen. Self incompatibility is usually strong in perennial rye and is not at all weakened by chromosome doubling. The very low seed set in the bagged spikes points in this same direction.

We conclude, that these preliminary data on 2n pollen distribution and crossing experiments hold promise for the use of sexual polyploidization in perennial rye, and warrant further study of genetic and temperature effects on 2n pollen formation, cytological examination of the pollen development, crossing experiments and critical evaluation of sexually derived tetraploids.

Table 1. Number of seeds per spike and ploidy level of offspring after selfing or crosses with diploids.

Family	Spikes harvested	Seeds/spike	Ploidy level (Chromosome numbers)	
			3x (21-25)	4x (26-29)
Citadel x 3-8	2	18	5	19
Modus x 3-8	2	16	1	17
Madera x 7-2	3	7	3	4
Citadel self	10	0.9	-	-
Modus self	20	0.4	1	3
Madera self	10	1	1	3

REFERENCES

Hague, L.M. and R.N. Jones, 1987. Cytogenetics of Lolium perenne. 4. Colchicine induced variation in diploids. TAG 74: 233-241.
Harlan, J.R. and J.M.J. de Wet, 1975. On O. Winge and a prayer: the origins of polyploidy. The Botanical Review 41: 361-389.
Jahr, W., K. Skiebe and M. Stein, 1963. Bedeutung von Valenzkreuzungen fuer die Polyploidiezuechtung. Z. Pflanzenzuecht. 50: 26-33.
Kaul, M.L.H. and T.G.K. Murthy, 1985. Mutant genes affecting higher plant meiosis. TAG 70: 449-466.
Mendiburu, A.O. and S.J. Peloquin, 1977. The significance of 2n gametes in potato breeding. TAG 49: 53-61.
Skiebe, K., 1966. Polyploidie und Fertilitaet. Z. Pflanzenzuecht. 56: 301-342.
Wit, F., 1959. Chromosome doubling and the improvement of grasses. Genetica Agraria 11: 97-115.

Sporophytes and Male Gametophytes from in Vitro Cultured, Immature Tobacco Pollen

Rosa Maria Benito Moreno, Florian Macke, Marie-Theres Hauser[*],
Anna Alwen, Erwin Heberle-Bors
Institute of Microbiology and Genetics
University of Vienna
Althanstr. 14
A-1090 Wien
Austria

SPOROPHYTES (HAPLOID PLANTS) FROM POLLEN

Anther culture is one of the few biotechnologies which has found its way into breeding practice. Still however, many species are recalcitrant and yields are low, indicating that the mechanisms by which immature male gametophytes change their development pathway to form a sporophyte directly and asexually, are not well understood.

Working with the model system tobacco, a number of factors could be identified which allowed a certain degree of generalization of the mechanisms involved in the induction of pollen embryogenesis (Heberle-Bors 1985). Tobacco is a plant displaying pollen dimorphism. The plants contain, in their mature anthers, embryogenic pollen grains which can be separated from the normal gametophytic pollen by density centrifugation (Heberle-Bors and Reinert 1980, Protocol 1 in Fig. 1). These embryogenic pollen grains, also called P-grains, are sterile pollen grains for the donor plants. They are produced by the plants in condition promoting male sterility. These conditions are stress conditions for the tobacco plants, such as short days, low temperature and nitrogen starvation (Heberle-Bors 1982, 1983). Growth substances known to affect male fertility, such as auxins and anti-gibberel-

[*] present address:

Max-Planck-Institut für Biochemie, Martinsried,
D-8033 München, FR Germany

```
PROTOCOL 1:                              PROTOCOL 2:
------------------------------------------------------------------
```

- SPECIFIC GROWTH CONDITIONS,   - PLANTS IN GENERALLY GOOD
  TREATMENTS, GENOTYPES   CONDITION

- FLOWERS WITH NEARLY MATURE POL-   - FLOWERS WITH EARLY
  LEN AND POLLEN DIMORPHISM   BINUCLEATE POLLEN

- POLLEN ISOLATION   - POLLEN ISOLATION

- SEPARATION OF P-GRAINS   - NO POLLEN SEPARATION

- POLLEN CULTURE AT LOW CELL   - POLLEN CULTURE AT LOW CELL
  DENSITY (5.10^4/ml) IN: DENSITY (5.10^4/ml) IN:
    - MILLER MACROSALTS   - MILLER MACROSALTS
    - SUCROSE (0.25 M)   - MANNITOL (0.4 M)
    - pH 7   - pH 7
    - DARKNESS   - DARKNESS

- MEDIUM CHANGE AFTER 2 WEEKS,   - MEDIUM CHANGE AFTER 1 WEEK,
  HIGH CELL DENSITY (2.10^5/ml)   HIGH CELL DENSITY (2.10^5/ml)
                            IN:
                            - MILLER MACROSALTS
                            - MS MICROSALTS
                            - FE-EDTA (10^{-4}M)
                            - SUCROSE (0.25 M)
                            - pH 7
                            - DARKNESS

- "GERMINATION" OF TORPEDO-SHAPED EMBRYOS ON AGARMEDIUM:
                            - NITSCH MACROSALTS
                            - MS MICROSALTS
                            - FE-EDTA (10^{-4}M)
                            - SUCROSE (4 mM)
                            - AGAR (0.8 %)
                            - pH 5.8
                            - LIGHT

- TRANSFER OF YOUNG PLANTLETS TO SOIL

Fig. 1: Protocols for pollen embryo formation in isolated pollen cultures of tobacco. Protocol 1: P-grain induction in vivo, protocol 2: P-grain induction in vitro (starvation).

lins, also promote P-grain formation and haploid production (Heberle-Bors 1983). In wheat, cytoplasmic male sterile lines produce high numbers of P-grains and are more productive in anther culture than fertile lines (Picard et al. 1978, Heberle-Bors and Odenbach 1985, Hadwiger and Heberle-Bors, unpublished results). Potatoe lines, selected for high yields in anther culture, are often male sterile (Johansson, pers. commun., Uhrig, pers. commun.). Gametocides applied on wheat plants also increase pollen embryo formation in wheat anther culture (Schmid and Keller 1986, Picard et al. 1987). All these data point to the assumption, that pollen starvation due to dysfunction of its nutritional tissue, the tapetum, may be a necessary requirement for the induction of repeated cell divisions in the pollen grains. Elegant proof for this hypothesis were experiments by Kyo and Harada (1986). They simulated the conditions causing male sterility and P-grain formation in situ by an in vitro starvation treatment of isolated pollen grains. Own experiments in this direction confirmed these results (Protocol 2 in Fig. 1). Both protocols are highly efficient with respect to the number of pollen grains cultured in vitro. While, however, the procedure as in protocol 1 relies on the induction of pollen dimorphism in situ and the effective separation of P-grains from the normal pollen grains, the procedure as in protocol 2 is simpler and allows induction of pollen embryogenesis in the majority of pollen grains isolated from anthers, the only limitation being pollen viability.

MALE GAMETOPHYTES FROM IN VITRO CULTURED MICROSPORES

If starvation of the pollen leads to a reprogramming of male gametophytic development towards cell division and embryogenesis, then in vitro culture of isolated pollen or microspores in a reverse-type of culture medium, i.e. a rich medium, should lead to normal gametophytic development. In other words, the medium should simulate the tapetum. We developed a medium suited for such an in vitro maturation of microspores and immature pollen (Planta, in press). This medium contains Murashige and Skoog minerals, sucrose in high concentration (0.5 M), coconut water (2 %), lactalbumin hydrolysate (1 g/l), glutamine (3 mM) and inositol (100 mg/l) at pH 7. Within one day, the microspores pass through first pollen mitosis, and after further four days the pollen grains are mature. This period of time is similar to that for tobacco pollen maturation in vivo. Unlike the in vivo development, the microspores start incorporation of starch immediately after culture initiation. The final stages of pollen maturation take place in a nutrient-deprived medium with lower osmolarity (0.25 M sucrose, Kyo and Harada (1986) minerals which lack nitrogen, pH

7). This medium appears to simulate the conditions in the nearly mature anther (mobilization of reserve materials, see Hoekstra et al., this volume) and is required for efficient seed seet, as well as for pollen germination in vitro and pollen tube growth.

SEED SET AFTER POLLINATION WITH IN VITRO MATURED POLLEN

The in vitro matured tobacco pollen can be used to pollinate emasculated flowers in situ. It is spun down and resuspended in a germination medium (Brewbaker and Kwack 1964). Droplets of 4 ul containing about 10.000 pollen grains are transferred to the stigma surface. The flowers set seed and the seeds germinate normally. Crosses with in vitro matured pollen from transgenic plants containing marker genes (KAN[R], NOS) showed Mendelian segregation proving that is was the in vitro matured pollen which performed fertilization. This was the first report that in vitro matured pollen can be used to pollinate and fertilize embryo sacs in situ.

In vitro cultures of plant parts are usually only used to regenerate whole plants. In vitro cultures to study particular phases of normal in vivo development (cytodifferentiation) are rare. The present system is one of the very few in vitro systems that can be used in this way. The better control of development in in vitro cultures should facilitate the study of pollen development on the molecular level. Particularly, the study of phenomena in which the pollen interacts during its development with the surrounding tapetum, such as pollen wall formation (including the pollen allergens), incompatibility (incongruity), and male sterility may profit from the use of this in vitro system. Pollen selection can be simplified by applying selection pressure during in vitro pollen development rather than during pollen germination. This may overcome the problem that in vivo matured pollen invariably has formed pollen tubes after selection which renders the pollen tube less capable of fertilization after pollination.

Seed formation with pollen which has not interacted with the tapetum may be interesting for the breeders in order to overcome fertility barriers, both in male sterile lines, in self-incompatible species, and in hybrids.

EXPERIMENTAL CONTROL OF THE ALTERNATION OF GENERATIONS

Perhaps the most interesting aspect of in vitro pollen development is the efficient control of the switch from gametophytic to sporophytic development. Starting out from pollen grains in the early binucleate stage, the pollen can be directed either towards maturation (rich medium) or towards embryogenesis (starvation treatment followed by embryogenesis medium). The most fundamental developmental decision in plant development, the alternation of generations, is thus controlled by the nutritional status of the cells involved. This is a striking similarity to yeast, where a hunger signal induces the vegetatively growing cells to undergo meiosis and produce spores. It remains to be seen whether a hunger signal is a general trigger for phase change in plants, particularly whether the egg cell experiences such a hunger signal in the ovule. Together with the availability of high numbers of synchronous cells (2.10^5 pollen grains per tobacco flower), this aspect makes the study of pollen development in vitro a promising system to study plant reproductive development on the molecular level.

POLLEN AS A TARGET AND SUPERVECTOR FOR GENE TRANSFER

Finally, the immature pollen grain may be used as a target cell for gene transfer. The transformed pollen can then either be matured in vitro and used as a super vector to transfer the genes into the embryo sac, or induced to form pollen embryos. Both pathways may lead to the formation of transgenic plants in species which cannot yet be transformed because vegetative tissues (protoplasts) are unable to regenerate whole plants. In vitro maturation appears to be particularly interesting in this respect, since germ line (pollen) transformation would circumvent in vitro regeneration completely.

REFERENCES

Benito Moreno RM, Alwen A, Heberle-Bors E (1988) In situ seed production after pollination with in vitro matured, isolated pollen. Planta, in press

Brewbaker JL, Kwack BH (1964) The essential role of calcium ion in pollen germination and pollen tube growth. Amer J Bot 50:859-865

Heberle-Bors E (1982) In vitro pollen embryogenesis in <u>Nicotiana tabacum</u> L. and its relation to floral induction, sex balance, and pollen sterility of the pollen donor plants. Planta 156:396-399

Heberle-Bors E (1983) Induction of embryogenic pollen grains and subsequent embryogenesis in <u>Nicotiana tabacum</u> L. by treatments of the pollen donor plants with feminizing agents. Physiol Plant 59:67-72

Heberle-Bors E (1985) In vitro haploid formation from pollen: a critical review. Theor Appl Genet 71:361-374

Heberle-Bors E, Reinert J (1980) Isolated pollen cultures and pollen dimorphism. Naturwiss 67:311

Heberle-Bors E, Odenbach W (1985) In vitro pollen embryogenesis and cytoplasmic male sterility in <u>Triticum aestivum</u>. Z Pflanzenzücht 95:14-22

Kyo M, Harada H (1986) Control of the developmental pathway of tobacco pollen in vitro. Planta 168:427-432

Picard E, de Buyser J, Henry Y (1978) Technique de production d'haploides de Blé par culture d'anthères in vitro. Le Sélectionneur Francais 26:25-37

Picard E, Hours C, Grégoire S, Phan TH, Meunier JP (1987) Significant improvement of androgenetic haploid and doubled haploid induction from wheat plants treated with a chemical hybridization agent. Theor Appl Genet 74:289-297

Schmid J, Keller ER (1986) Improved androgenetic response in wheat (<u>Triticum aestivum</u>) as a result of gametocide application to anther donor plants. Abstracts of the 4th Int Cong Plant Tissue Cell Culture, 3-8 August, 1986. University of Minnesota, p 146

Microsporogenesis and Pollen Development

Extranuclear Extrusions of Nucleolar Material During Diplotene in Microsporocytes of *Larix decidua* Mill

A. Górska-Brylass, B. Wróbel, A. Narbutt
Department of Plant Cytology and Genetics
Institute of Biology Copernicus University
Gagarina 9
87-100 Toruń
Poland

Our earlier studies on anther meiosis in Larix decidua Mill. have revealed increased metabolic activity in microsporocytes during diplotene. It is reflected, among other things, in the variation of the total protein content, and in the last diplotene phase also in high level of protein synthesis /Chwirot and Górska-Brylass 1981/, in considerable variations in the activity level of the two mitochondrial enzymes, Mg^{2+} ATPase and cytochrome c-oxidase /Chwirot and Górska-Brylass 1987/ and in endogenous auxin content /Górska-Brylass et al. 1981/.

During the long diplotene striking structural changes are also observed both in the cytoplasm and in the nucleus. Among the latter we considered in particular the nuclear bodies, whose number varied from phase to phase of diplotene. The common feature of these structures is their affinity for silver ions, which is regarded as an indicator of presence of nucleolar proteins /Hubble et al. 1979/. To get a better uderderstanding of the nature and function of argyrophilic nuclear bodies /ANBs/ we undertook studies on their structure and on the dynamics of changes in the number of ANBs during diplotene.

Material and methods

The material for study were male cones of Larix decidua Mill. in 12 succesive stages of diplotene collected directly from trees every week or every fortnight from mid November to mid February. The number and volume of ANBs were determined in a light microscope in uncut microsporocytes impregnated with silver nitrate after Howell and Black /1980/. The results were statiscally analysed. For electron microscopy, silver impregnated material was prepared according to the technique of Ploton et al. /1985/ For some material, the standard electron microscopy method /glutaraldehydeosmium-epon/ was used.

Results and discussion

In larch microsporocytes during diplotene there occur, besides the main nucleolus, numerous ANBs 0.3-5.0 μm in diameter / Fig.4/. They include two categories of structures. The first one consists of loose ANBs made up fluffy threads 30-50 nm wide / Fig.3/. Some pictures suggest that grains which can often be seen near the loose body are part of these threads. It is difficult to find out if what takes place here is a process of aggregation or of disintegration. The second category of ANBs are dense bodies whose ultrastructure corresponds to the fibrillar part of the main nucleolus / Fig.6/. These bodies are rather frequently seen close to the nucleolar organizer region.

During diplotene the number ANBs in the microsporocytes under goes considerable variations, from three ANBs per nucleus in middle diplotene to more than thirty at the end of this period / Fig.1/.

The comparison of total volumes of all ANBs in the microsporocytes in successive diplotene phases / Fig.2/ has revealed peculiar sequence of alternate phases: decrease and increase in total volume of ANBs in the cell. Particulary spectacular is the phase of decrease in middle diplotene, when total ANBs volume is reduced to 1/3 its total volume in the preciding phase. In microsporocytes going through the ANBs decrease phase there can be found pictures of extrusion of these bodies into the cytoplasm through evagination of the nuclear envelope / Fig.8/. By this way ANBs, or at least one of their categories, are extruded into the cytoplasm. They become cytoplasmics nucleoloids / Fig.9/, which for some time still show affinity for silver ions. There can be observed in them a certain segregation of argyrophilic material / Fig.10/, which may reflect further processes of maturation of nucleolar ribonucleoproteids.

The cytoplasmic nucleoloids are surrounded by a large number of ribosomes. Ribosomes occur in large number in larch microsporocytes throughout diplotene, but a striking increase in their number is noted in the phase of the highest ANBs loos from the nucleus, which falls on the first two decades of January. In that period the cytoplasm becomes electron-opaque because of the large number of ribosomes.

Our results have demonstrated that in Larix decidua diplotene, like in the whole meiotic prophase of Pinus silvestris /Willemse 1971/ and of Taxus baccata /Pennel and Bell 1986/, there is no elimination of ribosomes from the cytoplasm, which has been noted in some angiosperms, e.g. in Lilium

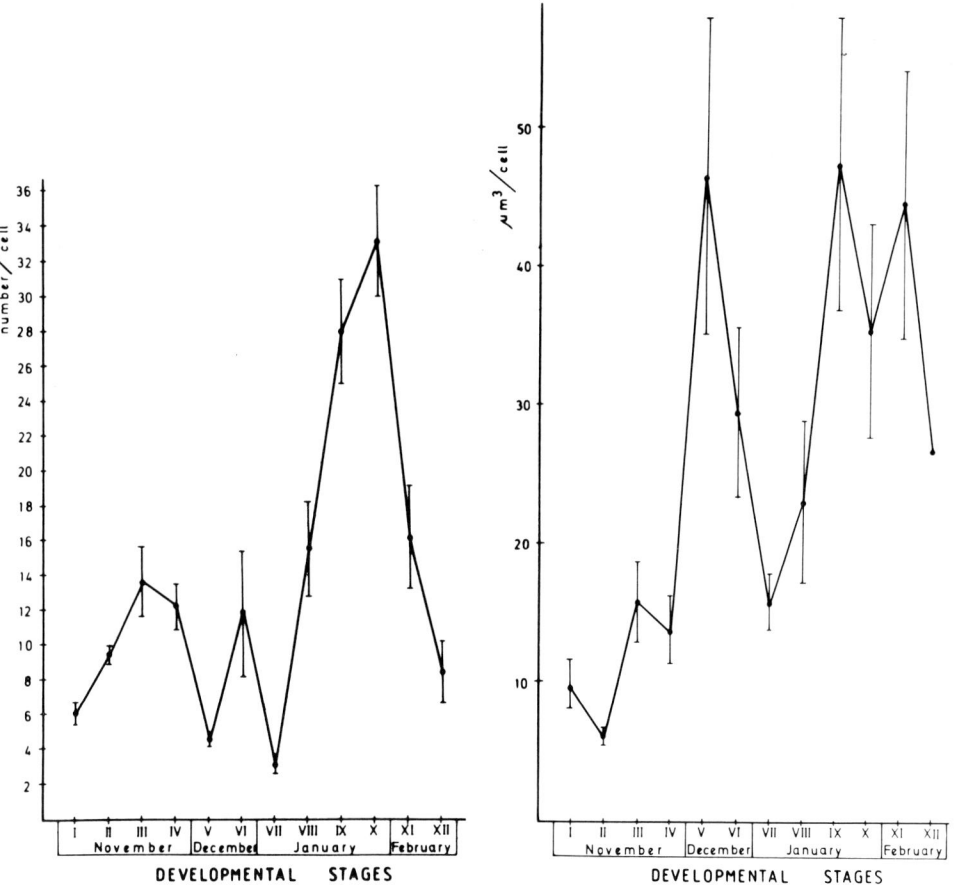

Fig.1 Number of ANBs in microsporocytes of Larix decidua in successive diplotene phases

Fig.2 Total volume of ANBs in microsporocytes of Larix decidua in successive diplotene phases. Standard errors indicated

/Mackenzie et al. 1967/. In Larix decidua there is even a temporary increase in their number. This happens in the situation when the nucleolus has exhibited no transcriptive activity for a long time. In anther meiosis of Larix decidua incorporation of H3 uridine in the nucleolus stops completely before the microsporocytes enter diplotene, and is restored only in the young microspore /Górska-Brylass, Wróbel unpublished/. It seems that during Larix decidua diplotene the mechanism which is responsible for supplying the cell with an adequate number of ribosomes under conditions of prolonged inactivi-

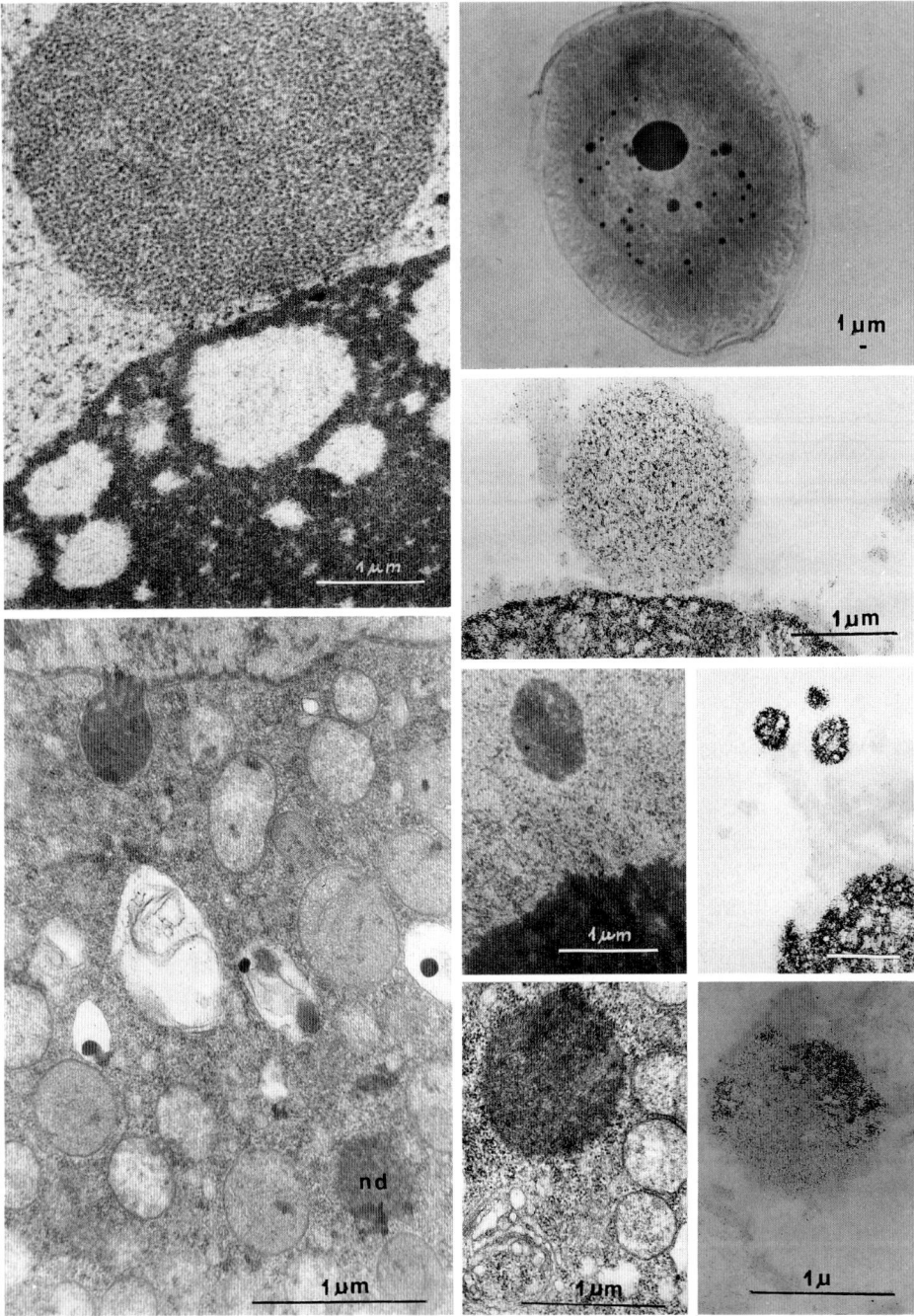

ty of nucleolus is extranuclear extrusion into the cytoplasm of part of the argyrophilic nucleolar material. In the cytoplasm the material in the form nucleoloid presumably goes through further stages of maturation /segregation of part of the argyrophilic material in the nucleoloid/and eventual disintegration forming a population of new ribosomes. The ribosomes formed in this way are active, evidence of which is the high level of protein synthesis in larch microsporocytes in the final phase of diplotene /Chwirot and Górska-Brylass 1981/.

Fig.3 Loose ANB near the nucleolus /N/; Fig.4 Light micrograph of silver-impregnated nucleolus and nuclear bodies /ANBs/ in diplotene microsporocyte; Fig.5 Siver-impregnated material showing silver deposits in both the nucleolus and the loose ANB;
Fig.6 Dense ANB; Fig.7 Silver-impregnated dense ANBs, N-nucleolus;
Fig.8 Evagination of nuclear envelope containing electron-opaque material similar to dense ANB, nd-nucleoloid; Fig.9 Cytoplasmic nucleoloid;
Fig.10 Silver-impregnated cytoplasmic nucleoloid showing segregation of argyrophilic material /unstained with uranyl acetate/.
Bars represented 1 /um in all cases.

This work carried out under Program C.P.B.P. 04.04.9.04. coordinated by the Institute of Dendrology, Polish Academy od Sciences

REFERENCES

Chwirot WB, Górska-Brylass A /1981/ Variations of total protein content and protein synthesis during microsporogenesis in Larix europaea DC. Acta Soc Bot Pol 50:33-38
Chwirot WB, Górska-Brylass A /1987/ Activity of Mg^{2+} ATPase and cytochrome c-oxidase in microsporocytes and anther wall during microsporogenesis in Larix europaea DC. Biologia Plantarum /Praha/ 29:167-174
Górska-Brylass A, Chwirot WB, Michalski L /1981/ IAA-peroxidase relation in the microsporocytes and anther wall during seccessive stages of meiosis in Larix europaea DC. Acta Soc Bot Pol 50:67-73
Howell WM, Black DA /1980/ Controlled silver-staining of nucleolus organizer regions with a protective colloidal develope: 1-step method. Experientia 36:1014-1015
Hubbell HR, Rothblum LJ, Hsu TC /1979/ Identification of a silver binding protein associated with the cytological silver staining of actively transcribing nucleolar regions. Cell Biol Int Rep :615-622
Mackenzie A, Heslop-Harrison J, Dickinson HG /1967/ Elimination of ribosomes during meiotic prophase. Nature 215:997-999
Pennel RJ, Bell PR /1986/ Intracellular RNA during microsporogenesis in plants: Taxus baccata as a model system. Acta Soc Bot Pol 55:11-16
Ploton D, Menager M, Adnet JJ /1985/ Simultaneous ultrastructural localization of Ag-NOR /nucleolar organizer region/ proteins and ribonucleoproteins during mitosis, in human breast cancerous tissues. J Cell Sci 77:239--256
Willemse MThM /1971/ Morphological and quantitative changes in the population of cell organelles during microsporogenesis of Pinus silvestris L. I Morphological changes from zygotene until prometaphase I. Acta Bot Neerl 20:261-274

Evolution of Nuclear Interchromatin Structures During Microspore Interphase Periods

P.S. Testillano and M.C. Risueño
Centro de Investigaciones Biológicas.- C.S.I.C.
C/Velázquez 144. 28006 Madrid. SPAIN

INTRODUCTION

The pollen grain plays an essential role in the sexual reproduction of higher plants, carrying a part of the genetic information of the future embryo, and being one of the first points of genetic selection (Mulcahy and Mulcahy 1983).

A good understanding of the processes that take place during the development of the pollen grain, as well as the stages in which these processes happen, is necessary to carry out genetic modifications in the male gametophyte, which is a way used to obtain improved plants (Mulcahy 1986).

During the pollen grain development, the postmeiotic interphase constitutes a long period in which striking changes of activity occur in order to prepare the cell for its gametophitic function; the synthesis of important molecules takes place during this period (Mascarenhas 1975).

Only a few papers have analyzed this interphase (Gorska-Brylass et al. 1984, Tanaka et al. 1979, Tanaka et al. 1980), but in a fragmentary way. However, a study of microspore nuclei referring to G_1, S and G_2 periods has not been carried out.

The nuclear interchromatin region constitutes the extranucleolar ribonucleoprotein (RNP) network in which pre-mRNA synthesis and maturation take place (Puvion et al. 1984). In this region, the relationship between non-nucleolar RNP structures and the RNA metabolism has not been completely established; but it has been reported in animal cells that changes in synthesis, proccesing and/or transport of RNA involve variations in the density and location of these structures (Puvion and Viron 1981).

So, it has been demonstrate that hn-RNA synthesis takes place around condensed chromatin masses, and perichromatin fibres (PF) are the morphological substrate of these newly-synthesized RNAs. After transcription, these fibres migrate into the interchromatin space, forming interchromatin fibres (IF) which contain pre-mRNA in maturation (Puvion and Viron, 1981). Perichromatin granules (PG) have been described as previous stages in this procces, constituting a step between PF and IF (Puvion-Dutilleul and Puvion 1981). It has been reported that these PGs increase in number under experimental conditions when mRNA transport or processing is blocked (Puvion-Dutilleul and Puvion 1981, Cervera et al. 1985, Dupuy-Coin et al. 1978).

However, the functional significance of interchromatin granules (IG) is less known. In plants, IG have been related to the network of nuclear bodies called micropuffs (Risueño et al. 1978). They contain snRNPs, and several authors have related them to wrongly processed rRNAs or in degradation (Puvion et al. 1984).

In this paper we analyse interchromatin structures in the microspore and their evolution related to the G_1, S and G_2 periods, which have previously been determined in our material (Risueño et al. 1985). These structures can be related to different stages in hnRNA proccesing.

MATERIAL AND METHODS

Anthers of <u>Scilla peruviana</u> were selected during the whole microspore interphase. Anthers in each stage were submitted for: a) Glutaraldehyde fixation, and b) Acetylation in block (Wassef et al. 1979). In both cases, some ultrathin sections were stained with uranyl and lead, and other with EDTA regressive staining (Bernhard 1969).

EDTA stained sections from acetylated samples provided a very good visualization of granular and fibrilar RNPs. Nucleic acids were preferentially stained with uranyl after acetylation, with this method condensed chromatin masses showed more electron density than the RNP structures.

RESULTS AND DISCUSSION

The interphasic periods in our material have previously been
determined with Feulgen preparations of microspore nuclei in which
a quantification of the relative amounts of DNA per area has been
carried out in an automatic image analyzer (Risueño et al. 1985).
The results showed that DNA replication occurs in two steps: at the
beginning of the interphase, when microspores are enclosed in the
tetrad; and at the end, near the first postmeiotic mitosis. These
two increases in DNA amount led us to divide the S period into
three subperiods S_1, S_2 and S_3.

Throughout the interphase, chromatin undergoes different
condensation states. The density of fibril-granular material in the
interchromatin region is related to chromatin state, being this
material very abundant when chromatin is decondensed (Fig. 3) and
scarce in stages with condensed chromatin (Fig. 4).

During the early G_1 period the EDTA regressive staining reveals a
very contrasted fibrilar material around the bleached masses of
chromatin (Fig. 1), this material is PF which represents the
morphological substrate of hnRNA transcription (Fakan and Puvion,
1980). This fact corresponds to the transcription reactivation
which takes place after meiosis. Many RNP granules and fibres also
appear in the interchromatin region (Fig. 1).

From this period of the tetrad till the beginning of the vacuolated
microspore stage (S_1 and S_2 periods), an accumulation of
different-shaped granules is observed (Fig. 2). With the
acetylation method and then staining with uranyl, these structures
show greater electron density in their outline than in their
interior (Fig. 2), and their size is 30-60 nm.

Based on their fine structure these granules could be assimilated
to the PGs, their morphology is similar to grouped granules induced
under viral infections (Dupuy-Coin et al 1978), and they do not
seem to ressembly the IG clusters.

Densely stained 10 nm fibrillar material is also observed with the
EDTA staining (Fig. 3) which seems to correspond with accumulations

of IFs. These fibrils and granules represent different stages of hnRNA processing (Fakan and Puvion 1980), but we cannot determine the morphological secuence at the moment. In the next stages, the accumulations of granules and fibres decrease and, at the end of interphase, the interchromatin region presents less fibril-granular material, showing clear areas (Fig. 4). At this phase chromatin is in a very condensed state which represents an inactivated state of transcription (Puvion and Viron 1981). Several studies carried out in animal cells under experimental conditions demonstrate that increases of PGs and IFs are related to an impairment of hnRNA processing or transport (Cervera et al. 1985, Dupuy-Coin et al. 1978, Puvion and Viron 1981, Puvion et al. 1984). Such events have not been reported in physiological conditions, either in animal cells or in other plant cells.

In the case of the microspore, these accumulations of material in the interchromatin region could represent a way of storage of mRNAs during the interphase for a subsequent transport of them in the cytoplasm. Autoradiographycal studies with incorporation of tritiated uridine in the microspore will allow us to have more complete knowledge of the processes which occur during postmeiotic interphase.

These results could be related to a high protein synthesis which occurs at the end of this interphase (Raghavan 1984, Chwirot and Gorska-Brylass 1981).

These striking changes reported by us in the interchromatin region throughout the interphase, are due to variations in nuclear transcriptional activity.

ACKNOWLEDGEMENTS

We thank Mr C. Almarza and Miss T. Cortezón for their skilful technical assistance. This work has received financial support from the Project n° 1-179-2 ID181 CSIC/CAICYT.

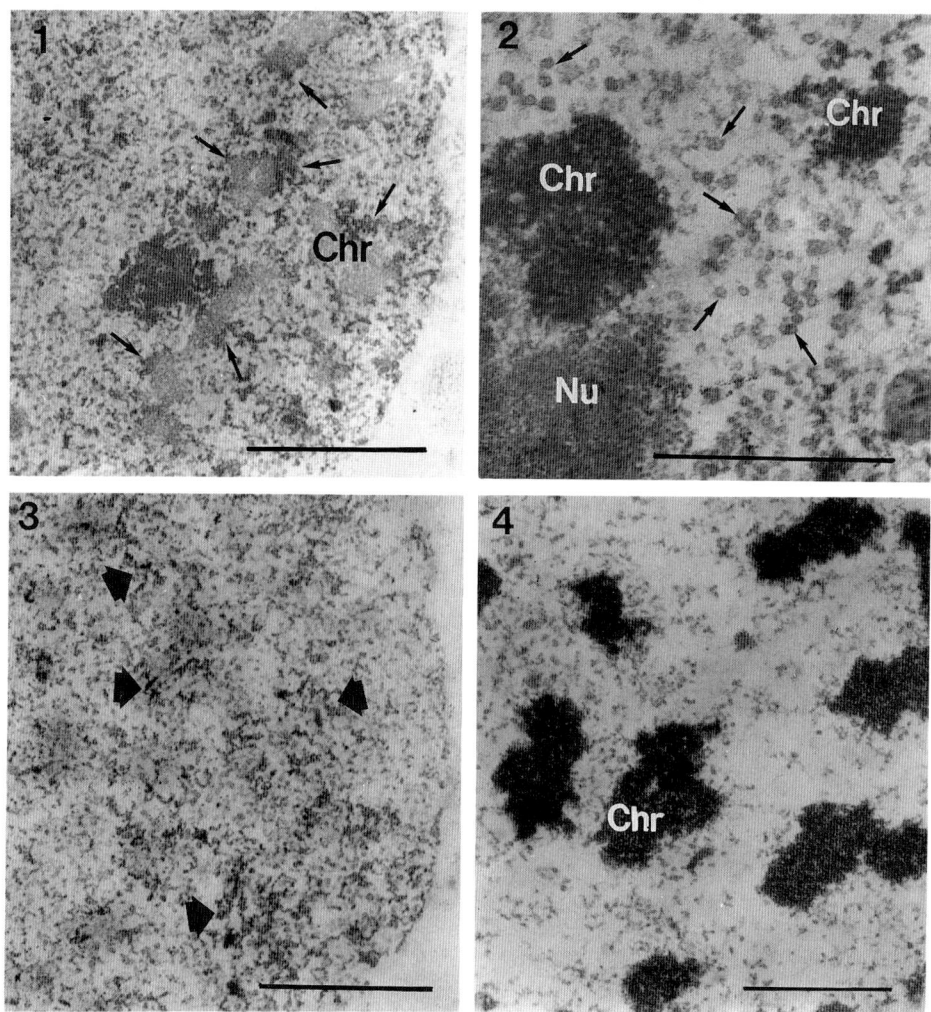

Figs. 1 to 4: Nucleus during microspore interphase.
Fig.1: G1 period. Acetylation in block, EDTA staining. Chromatin
masses appear bleached (Chr) and surrounded by fibrillar RNP
material which is PF (arrows). Interchromatin region shows many
granules and fibres. Fig.2: Early S2 period. Acetylation in block,
uranyl-lead saining. Condensed chromatin masses show more contrast
than RNP structures and nucleolus (Nu). Different-shaped granules
are seen in the interchromatin region, they appear grouped or
individually (thin arrows). Fig.3: Middle S2 period. Acetylation in
block, EDTA staining. Many grouped or individual IF (arrowheads)
appear very contrasted, intermingled with RNP granules. Chromatin
is in a decondensed state. Fig.4: Late S2 period. Acetylation in
block, uranyl-lead staining. Chromatin is very condensed while
interchromatin region presents less fibril-granular material,
showing clear areas.
Bars in figs. 1,3,4: 1 μm; in fig. 2: 0,5 μm.

REFERENCES

Bernhard W (1969) A new staining procedure for electron microscopical cytology. J Ultr Res 27: 250-269

Cervera J, Báguena-Cervellera R, Martínez A (1985) The effects of zinc chloride on the RNP structures in HEp-2 cells: Accumulation of perichromatin granules. J Ultr Res 93: 129-137

Chwirot WB, Gorska-Brylass A (1981) Variations of total protein content and protein synthesis during microsporogenesis in Larix europaea L. Acta Soc Bot Pol 50: 33-38

Dupuy-Coin AM, Arnoult J, Bouteille M (1978) Quantitative correlation of morphological alterations of the nucleus with functional events during "in vitro" infection of glial cells with Herpes simplex hominis (HSV 2). J Ultr Res 65: 60-72

Fakan S, Puvion E (1980) The ultrastructural visualization of nucleolar and extranucleolar RNA synthesis and distribution. Int Rev Cytol 65: 255-299.

Gorska-Brylass A, Chwirot W, Majewska A (1984) Ultrastructural and metabolic transformations of the larch microspore during G1 period of the postmeiotic interphase. Postepy Biologii Komorki 11: 577-580

Mascarenhas JP (1975) The biochemistry of angiosperm pollen development. The Botanical Review 41: 259-313

Mulcahy DL, Mulcahy GB, Ottaviano E (eds) (1986) Biotechnology and Ecology of Pollen. Springer Verlag NY Berlin Heidelberg Tokyo

Mulcahy DL, Ottaviano E (eds) (1983) Pollen Biology and implications for plant breeding. Elsevier Biomedical NY Amsterdam Oxford

Puvion-Dutilleul F, Puvion E (1981) Relationship between chromatin and perichromatin granules in cadmium-treated isolated hepatocytes. J Ultr Res 74: 341-350

Puvion E, Viron A (1981) "In situ" structural and functional relationships between chromatin pattern and RNP structures involved in non-nucleolar chromatin transcription. J Ultr Res 74: 351-360

Puvion E, Viron A, Xu X (1984) High resolution autoradiographical detection of RNA in the interchromatin granules of DRB treated cells. Exp Cell Res 152: 357-367

Raghavan V (1984) Protein synthetic activity during normal pollen development and during induced pollen embryogenesis in Hyoscyamus niger. Can J Bot 62: 2493-2513

Risueño MC, Arquiaga C, Sánchez-Pina MA (1985) Determination of the microspore interphase of Hyacinthoides non-scripta. Cell Biol Rev (RBC) S: 29

Risueño MC, Medina FJ (1986) The nucleolar structure in plant cells. Cell biol Rev (RBC) 7: 1-163

Risueño MC, Moreno Díaz de la Espina S, Fernández-Gómez ME, Giménez-Martín G (1978). Nuclear micropuffs in Allium cepa cells. I. Quantitative, ultrastructural and cytochemical study. Cytobiologie 16: 209-223

Tanaka I, Taguchi T, Ito M (1979) Studies on microspore development in liliaceous plants. I. The duration of the cell cycle and developmental aspects in lily microspores. Bot Mag 92: 291-298

Tanaka I, Taguchi T, Ito M (1980) Studies on microspore development in liliaceous plants. II. The behaviour of explanted microspore of the lily, Lilium longiflorum. Plant Cell Phys 21: 667-676

Venable J, Goggeshall R (1965) A simplified lead citrate stain for use in electron microscopy. J Cell Biol 25: 407-408

Wassef M, Burglen J, Bernhard W (1979) A new method for visualization of preribosomal granules in the nucleolus after acetylation. Biol Cell 34: 153-158.

Effect of Gamma Irradiation on Nucleus DNA Amount and Syntheses of Nucleic Acids in Microsporocytes and Tapetum of *Nicotiana tabacum* L. Var. *Xanthi* Dulieu Issued from Irradiated Seeds (DNA) or from Directly Irradiated Excised Inflorescences (DNA, RNA)

C. Almhana, L. Albertini, A. Souvré
Laboratoire de Cytologie et de Pathologie végétales
E.N.S. Agronomique
145 avenue de Muret
31076 Toulouse Cédex
France

Introduction

According to Evans and Van't Hof (1975), DNA is the primary gamma-radiation target in the proliferative root meristems of seven plant species. These authors have also reported that the relative radiosensivity of the $G_1 \longrightarrow S$ and $G_2 \longrightarrow M$ transition points is characteristic of the species and independent of the amount of DNA per chromosome and (per nucleus) and of the mitotic cycle duration.

Other investigations have shown that :

- a weak or moderate irradiation (from a fraction of Gy to 100 Gy) induces an increase in the mitotic cycle duration and especially in the sensitive S phase (e.g., *Nicotiana*, Michel-Delbos, 1979) in relation with a repair synthesis of DNA at the level of chromatic injuries (Veleminsky and Gichner, 1978) ;

- a medium or strong irradiation (from 100-200 Gy to several thousands of Gy), not only inhibits DNA reduplication and mitosis (Haber and Foard, 1964), but also breaks chromosomes (chromatoclasic effect), leading to the elimination of chromosome fragments and, consequently, to a reduced amount of DNA in the nucleus (Haber, Foard and Perdue, 1969 ; Michel-Delbos, 1979).

The **present paper** reports the effects of gamma irradiation on the nuclear DNA content and on the syntheses of nuclear DNA and cellular RNA in microspore mother cells (MMC) and tapetum of plants of *Nicotiana tabacum* L. var.

xanthi Dulieu issued from irradiated seeds (DNA) or from excised inflorescences directly irradiated (DNA, RNA). Owing to the lack of references concerning the meiocytes and tapetal cells, our results are analysed in the light of the literature relative to the cells of vegetative organs.

Material and methods

The different stages of the microsporogenesis of *N. tabacum* and the nuclear evolution of tapetum cells were previously described by Almhana (1988). The seeds and excised inflorescences were gamma-irradiated using a gamma cell (^{60}Co source). The dose scale applied to the seeds was : 0, 100, 500 and 1000 Gy ; the excised inflorescences received the strong dose of 3000 Gy. The nuclear DNA content of microsporocyte and tapetum cell was determined on thin sections (20 μm) of anthers initially fixed by ethanol-acetic acid 3/1 and embedded in paraffin, after Feulgen staining, using the two-wavelengths cyto-photometric method (MP$_2$ Leitz cytophotometer) : each value in Table 1 is the average of 30-50 nuclear measurements. Inflorescences of plants issued from irradiated seeds were treated, after excision, with [methyl-^{3}H] thymidine (0.74 MBq/ml) during 24 hr according to a method previously described (Albertini, 1970) ; inflorescences having to be directly irradiated after excision were treated during 24 hr either with ^{3}H-thymidine before irradiation, or with ^{3}H-thymidine or [5-^{3}H] uridine (0.74 MBq/ml) after irradiation. The anthers fixed by ethanol-acetic acid 3/1, then embedded in paraffin, were sectionned (8 μm) and treated by Ficq's (1961) radioautographic technique (time of exposure in dark room : 1 month). The total activity of a given cellular structure was determined as being the number of silver grains counted over this structure after allowing for the background (each average grain count was obtained from 30 nuclei, nucleoli or cytoplasms at a given stage). Each result in Tables 2 and 3 is the ratio (%) of the average grain count over an irradiated structure to the average grain count over the corresponding control structure.

Results

Nuclear DNA amount (Tables 1 and 2)

In **tapetum** cells of plants issued from irradiated seeds as well as in the corresponding **MMC**, only the strong 500 Gy treatment induces, at pachytene, a reduction of the nuclear DNA amount which is stronger (28 %) than that determined in the corresponding MMC (20 %). The direct irradiation (3000 Gy) of inflorescences is still more drastic ; in this case, as well as in the tapetum of plants issued from seeds irradiated with 500 Gy, the various categories of nuclei (2C ; 2C —> 4C ; 4C ; 4C —> 8C) seem to be affected by the irradiation as shown by representative histograms (Almhana, 1988).

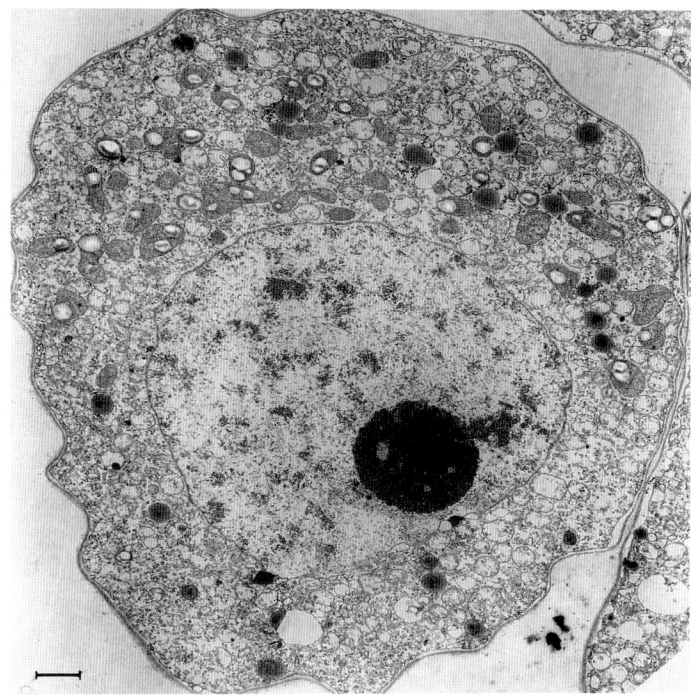

Fig. 6. Microsporocyte of Nymphaea at early prophase I; aggregation
of plastids - with small starch grains - and mitochondria, some mitochondria and many osmiophilic bodies scattered around the nucleus,
bar = 2µm.

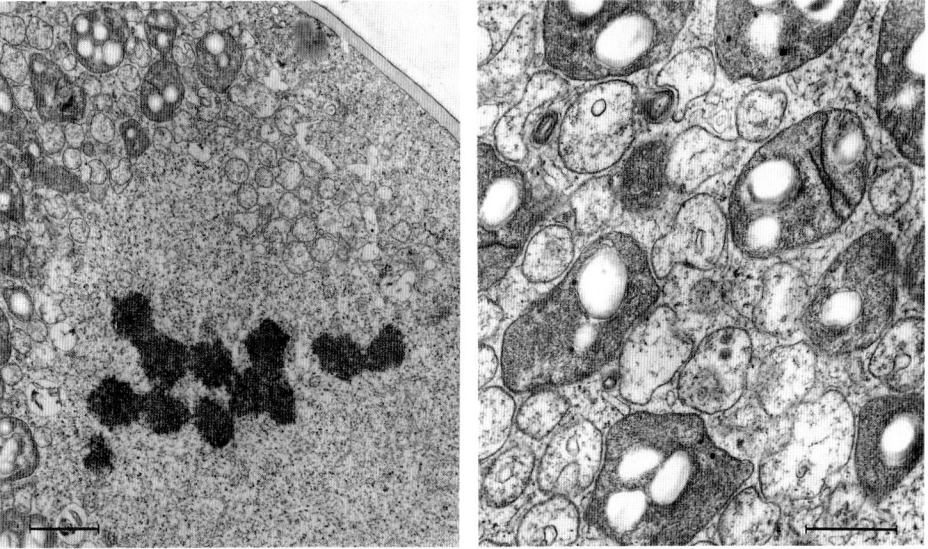

Figs. 7 and 8. Microsporocyte of Nymphaea at metaphase I; chromoso-

178

mes in homogenous area adjacent to aggregation of plastids and mitochondria, bar = 2µm. Fig. 8. Aggregated mitochondria and plastids
with starch grains, bar = 1µm.

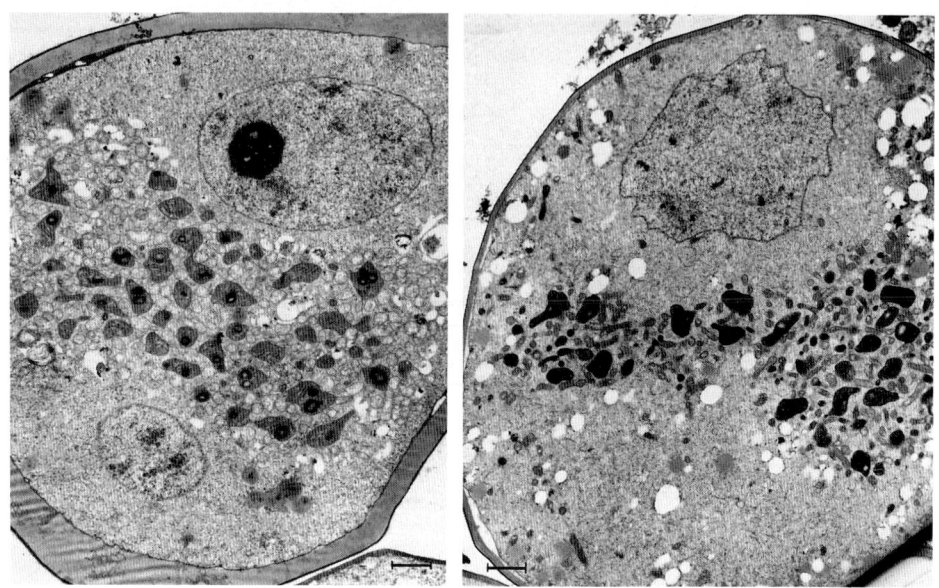

Figs. 9 and 10. Microsporocytes of Nymphaea. Fig. 9. Post-telophase
I, equatorial plate of plastids and mitochondria. Fig. 10. Post-telophase II, a part of complicated organelle aggregation which partitions the cell into four mononucleate parts, bar = 2µm.

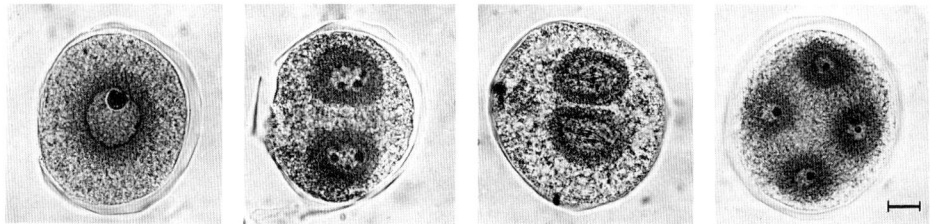

Fig.11. Microsporogenesis in Malva. Organelles in: late prophase I,
dyad, mctaphase II, tetrad before cytokinesis, bar = 10µm.

are constricted, apparently undergoing division (Fig. 2). The starch
grains in the plastids grow and after staining the group of organelles becomes distinctly visible in a light microscope (Fig. 1 inset).
Meanwhile, on the other side of the nucleus many parallel ER cisternae appear (Fig. 3). They fold and extend in some places
(Fig. 4), eventually the whole area is occupied by concentrically
arranged ER cisternae (Fig. 5). They gradually disappear and cyto-

Table 1 - Effect of gamma irradiation on nuclear DNA amount in MMC and tapetal
cells, at pachytene stage (*Nicotiana tabacum* var. *xanthi* Dulieu ;
Feulgen ; cytophotometry).

	Dose	MMC		Tapetum	
Irradiated seeds	Control :	330 bc	311 c	439 a	406 a
	100 Gy :	360 b	324 c	384 ab	397 a
	500 Gy :	217 e	295 c	286 c	324 bc
	1000 Gy :	269 d	247 d	-	-
Directly irradiated inflorescences	Control :	243 b	247 b	286 a	301 a
	3000 Gy :	164 c	174 c	176 c	173 c

For each treatment, measures concern two plants whence two means (in arbitrary
units) by treatment. The test of Student was applied in the two experiments :
average values followed by the same letter are not different at 5 % level.

Table 2 - Effect of gamma irradiation on the DNA amount and labelling (with
[methyl-³H] thymidine) of MMC and tapetum nuclei in *N. tabacum* var.
xanthi Dulieu.

Percent reduction (relatively to control) of nucleus DNA amount (Feulgen ; cytophotometry)			
	Irradiated seeds		Directly irradiated inflorescences
	500 Gy	1000 Gy	3000 Gy
MMC at pachytene	20	20	31
Tapetum (pachytene)	28	-	40

Percent reduction (relatively to control) of nucleus labelling with
³H-thymidine (0.74 MBq/ml) at S phase (historadioautography)

	Irradiated seeds		Directly irradiated inflorescences (3000 Gy)		
	500 Gy	1000 Gy	before radiotracer supply	after radiotracer supply	(synthesis inhibition)
MMC at PRM	38	44	93	74	(19)
M2	33	40	76	35	(41)
Tapetum PRM	13 (NS)	17 (NS)	94	66	(28)
P-D	24	29	88	49	(39)
MAT$_I$	30	25	85	31	(54)
M$_2$	25	45	75	44	(31)

The average percent reductions were calculated from experimental results of
Table 1 (cytophotometry) and from silver grain countings (radioautography).
PRM = premeiosis ; P-D = pachytene-diakinesis ; MAT$_I$ = metaphase I-anaphase I-
telophase I ; M$_2$ = microspores at stage 2.

Nuclear DNA synthesis (Table 2)

MMC and microspores. In MMC and microspores of plants issued from irradiated seeds, only the strong doses induce significant reductions of nuclear DNA synthesis. On inflorescences directly irradiated (3000 Gy), the reductions of nuclear incorporation of [methyl-^{3}H] thymidine, more marked than those above-mentioned, are more important when MMC and microspores are gamma irradiated before the precursor supply than after.

Tapetum. In anthers of plants issued from seeds irradiated with 500 or 1000 Gy, the tapetum nucleus incorporates almost normally [methyl-^{3}H] thymidine at the premeiotic S phase preceding its division (at synizesis) (Table 2). Later, the endopolyploidization process which concerns the two nuclei of the tapetum cell is rather not much disturbed by 500 Gy or 1000 Gy dose (average level reduction : 25 %-30 %) (Table 2). On the other hand, when inflorescences are directly irradiated with 3000 Gy, tapetum nuclei, at premeiotic S phase or during endoreduplication, appear as sensitive as corresponding MMC and microspore nuclei.

RNA synthesis (see Table 3)

Table 3 - Effect of 3000 Gy dose of gamma rays directly applied to excised inflorescences of *N. tabacum* var. *xanthi* Dulieu on subsequent incorporation of [5-^{3}H] uridine (0.74 MBq/ml) into chromatin (no-nu), nucleoli (nu) and cytoplasm (cyt) of MMC and tapetal cells : each number is an average percentage of precursor incorporation inhibition (% Inh.) relatively to its specific control.

Tissue	MMC			Tapetum					
Stage	Leptotene			Pachytene			Microspore stage 2		
Structure	no-nu	nu	cyt	no-nu	nu	cyt	no-nu	nu	cyt
% Inh.	90	83	95	90	83	88	70	40	82

Remarks and discussion

A **strong** initial (500-1000 Gy) or direct (3000 Gy) gamma **irradiation** induces, in MMC at pachytene as well as in the corresponding tapetum cells (with endoreduplicating nuclei), a significant decrease in nuclear DNA amount (from 20 % to 40 %) : these quantitative data, obtained not only on special somatic cells (tapetum cells) but also on meiotic cells, corroborate those reported in the literature ; the latter were usually obtained with an initial irradiation of the seeds (Haber et al., 1969 , Lactuca sativa ; Michel-Delbos, 1979, N. tabacum). The agreement between our radioautographic results (^{3}H-thymidine) and our cytophotometric determinations (Feulgen) can be emphasized : e.g., the decrease in chromatin radioactivity determined at the end of a direct irradiation of inflorescences following the precursor supply is another manner of measuring the DNA loss induced by irradiation (Table 2). Furthermore, it may be remarked that a strong direct irradiation (3000 Gy) of inflorescences induces in MMC, at premeiosis, and in tapetum cells, at the premitotic S phase or, later on, with endoreduplicating nuclear DNA, a relative inactivation of the chromatin remained in situ, as shown by the marked decrease in ^{3}H-thymidine incorporation into this structure (Table 2). In addition to the chromatin loss induced by irradiation, there is an inhibition of DNA biosynthesis affecting the remaining chromatin, as already mentioned by various authors for the somatic cells of various plant species (Haber et al., 1964, 1969 ; Inoue, Hasegawa and Hori, 1975 ; Schaeverbeke-Sacré and Matheron, 1983).

The strong irradiation (500-1000 Gy) of seeds of N. t. var. xanthi Dulieu does not affect DNA synthesis in tapetum nucleus at premeiosis before mitosis (at synizesis), but significantly alters DNA endoreduplication of the two nuclei of the tapetum cell during meiosis and afterwards : contrary to the observation of Callebaut, Van Oostveldt and Van Parijs (1980) in Pisum sativum, DNA endoreduplication, in the tapetum of Tobacco, therefore appears to be more sensitive to initial irradiation than premitotic reduplication.

The direct irradiation (3000 Gy) of inflorescences induces at all the stages studied, in MMC as well as in tapetum cells, marked reductions (usually betweeen 80 % and 90 %) of ^{3}H-uridine incorporation into chromatin, nucleolus and cytoplasm. The above data are consistent with those of Callebaut *et al*. (1980) and Schaeverbeke-Sacré *et al*. (1983) : for the latter authors, who carried out investigations on *Helianthus tuberosus* explants irradiated from 25 to 5000 Gy and then grown *in vitro*, the inhibition of RNA synthesis is directly related to the dose received.

References

Albertini L (1970) Les acides nucléiques et les protéines au cours de la micro-sporogénèse chez le *Rhoeo discolor* Hance. Etude autoradiographique et cyto-photométrique. Thèse Doctorat Etat, Université Paris VI, p 186

Almhana C (1988) Effets du rayonnement gamma sur le développement végétatif, la microsporogénèse et les acides nucléiques microsporocytaires et tapétaux de *Nicotiana tabacum* L.. Etude cytophotométrique (DNA) et radioautographique (DNA, RNA). Thèse Doctorat INP Toulouse, p 142

Callebaut A, Van Oostveldt P, Van Parijs R (1980) Stimulation of endomitotic DNA synthesis and cell elongation by gibberellic acid in epicotyls grown from gamma-irradiated pea seeds. Plant Physiol 65 : 13-16

Evans LS, Van't Hof J (1975) Dose rate, mitotic cycle duration, and sensitivity of cell transitions from $G_1 \longrightarrow$ S and $G_2 \longrightarrow$ M to protracted gamma radiation in root meristems. Radiation research 64 : 331-343

Ficq A (1961) Contribution à l'étude du métabolisme cellulaire au moyen de la méthode autoradiographique. Monographie n° 9, Institut Interuniversitaire des Sciences Nucléaires, Bruxelles

Haber AH, Foard DE (1964) Further studies of gamma-irradiated wheat and their relevance to use of mitotic inhibition for developmental studies. Amer Jour Bot 51 : 151-159

Haber AH, Foard DE, Perdue SW (1969) Actions of gibberellic and abscisic acids on lettuce seed germination without actions on nuclear DNA synthesis. Plant Physiol 44 : 463-467

Inoue M, Hasegawa H, Hori S (1975) Physiological and biochemical changes in gamma irradiated rice. Radiation Botany 15 : 387-395

Michel-Delbos M (1979) Effet du rayonnement gamma sur des bourgeons axillaires de *Nicotiana tabacum* (dose-action, fractionnement de dose) : étude cytologi-que. Thèse Doctorat 3e cycle, Université Paris XI, p. 37

Schaeverbeke-Sacré J, Matheron B (1983) Action des rayons gamma du Cobalt 60 sur la teneur en acides nucléiques et l'histogenèse des tissus de tubercule de topinambour cultivés *in vitro*. Can J Bot 61 : 1448-1455

Veleminsky J, Gichner T (1978) DNA repair in mutagen-injured higher plants. Mut Res 55 : 71-84

Why Do Nuclear Vacuoles Appear in the Prophasic Nucleus of Pollen Mother Cells? Facts and Hypotheses

M.I. Rodríguez-García, A. Majewska-Sawka[1] and M.C. Fernández
Estación Experimental del Zaidín (C.S.I.C.)
Profesor Albareda 1
18008 Granada
Spain

INTRODUCTION

The presence of nuclear vacuoles during meiotic prophase has been shown in both animal (Rasmussen, 1976) and plant cells, including ferns (Sheffield and Bell, 1979; Sheffield, Laird and Bell, 1983), gymnosperms and angiosperms (Sheffield et al., 1979; Karasawa and Ueda, 1983a; Karasawa and Ueda, 1983b; Sangwan, 1986). An intimate relationship between the appearance of nuclear vacuoles and meiotic processes seems to be a common feature of all cells. Although there is no lack of hypotheses put forth by different authors, the whys and wherefores of these structures have yet to be satisfactorily explained.

In a detailed study of meiotic prophase in the silkworm Bombyx mori, Rasmussen (1976) reported the presence of nuclear vacuoles limited by a single membrane, which first appeared in early zygotene as a result of the invagination of the inner membrane of the nuclear envelope toward the karyoplasm. Three years were to pass before the same structures were described in meiotic prophase in plants (Sheffield et al., 1979). Dilations of the perinuclear space had previously been observed in post-meiotic microspores of Podocarpus macrophyllus (Aldrich and Vasil, 1970) as well as in covering tissues of Pisum sativum L. ovules and Allium cepa L. anthers at the beginning of meiosis (Risueño et al., 1975). None of these reports however relate the blebbing of the outer membrane of the nuclear envelope to the formation of nuclear vacuoles. Such nuclear structures may well have gone unnoticed in earlier studies performed on micro- and macrosporogenesis in various plants, or else have been disregarded as artifacts, hence their relative absence in the literature.

[1] A. Majewska-Sawka was the recipient of a grant from the Spanish Ministry of Foreign Relations and the Polish Ministry of Education. Permanent address: Institute for Plant Breeding and Acclimatization, Weyssenhoffa 11, 85-950 Bydgoszcz, Poland.

The present study describes the appearance,　formation and development of nuclear vacuoles during meiotic prophase in the pollen mother cells (PMCs) of two widely different species: Beta vulgaris L. and Olea europaea L. The sequence of morphological and developmental changes displayed by the nuclear vacuoles of both species during meiotic prophase was essentially the same, thus corroborating that such vacuoles show a universal pattern of behavior throughout this particular part of reproduction. Their functional significance has yet to be clarified, making similar studies in other species highly useful　for purposes of comparison as well as to test the various hypotheses of other authors.

INITIATION OF VACUOLIZATION DURING MEIOSIS IN THE PROPHASIC NUCLEUS

At the beginning of meiotic prophase, PMCs look very similar to young somatic cells in terms of nuclear organization and cytoplasmic organelles, which at this point include mitochondria, proplastids, ER, Golgi bodies and a few small vacuoles (fig. 1A). In leptotene, when the chromatin begins to undergo condensation, small dilations are seen in the perinuclear space, as well as noticeable invaginations of the double nuclear membrane, folding it in toward the karyoplasm (fig. 1B). The dilations and invaginations appear to be distributed randomly across the nuclear envelope, without showing any specific localization. The perinuclear dilations also become larger as leptotene progresses (fig. 1D), pushing the inner nuclear membrane toward the karyoplasm until free vacuoles are pinched off around the periphery of the nucleus (fig. 1E). At this time ER cisternae can be seen in direct contact with the perinuclear space. These observations seem to suggest that transport takes place between the cytoplasm and the nucleus via the ER‑nuclear envelope. Sheffield et al. (1979) saw no continuity between the ER and nuclear envelope, and they believe that the former was not involved in formation of the vacuoles. Another type of vacuole also noted at the periphery of the nucleus during the same period is made up of double or multiple membranes surrounding remains of ribosomes (fig. 1C). Our observations concur with those of Karasawa and Ueda (1983a), with whom we share the opinion that the invaginations of the inner and double nuclear membranes reflect two different mechanisms of nuclear vacuole formation. Sangwan (1986) however, observed the second kind of vacuole containing sequestered cytoplasmic structures only during late prophase, and thus believed that two types of nuclear vacuoles existed, each with a different function.

In zygotene - pachytene, which are characterized by the presence of synaptonemal complexes, the nucleus is greater than at any other moment during meiotic prophase.This increase in size is accompanied by an increase in total vacuolar volume,which comes to occupy up to half the volume of the nucleus (fig. 1F). The vacuoles can grow by fusion of two or more smaller ones and /or by accumulating material which may diffuse in from the karyoplasm through the membranes (Karasawa and Ueda, 1983; Sheffield et al., 1979). In pachytene the vacuolar compartment can displace the karyoplasm toward one side, producing images which may represent the morphological expression, at the ultrastructural level, of the so called "synizetic knot" observed under light microscope in solanaceous plants (Cawood and Jones, 1980). The vacuoles are initially electron light, their contents in some cases consisting in electron opaque fibrilo - granular material, which becomes more evident in pachytene (fig. 1F). During meiotic prophase considerable numbers of pores are present in the nuclear membrane.

DECREASE IN NUCLEAR VACUOLES DURING LATE MEIOTIC PROPHASE

In late meiotic prophase during diplotene-diakinesis, stages which are clearly characterized by a marked condensation of the paired chromosomes, nuclear vacuoles are smaller and less numerous (fig. 1G). The few remaining vacuoles are located preferentially at the periphery of the nucleus and between the two membranes of the nuclear envelope (fig. 1H). Evaginations of the outer nuclear membrane protrude into the cytoplasm, possibly to form free vacuoles within it in subsequent moments. Large numbers of small vacuoles are seen in the cytoplasm. The nucleus also becomes smaller. All these obsevations support the possibility that nuclear vacuoles and their contents may be eliminated toward the cytoplasm: if a drop in nuclear volume is observed at the same time as when nuclear vacuoles become scarcer, the implication is that the vacuoles must pass into the cytoplasm. Karasawa and Ueda (1983a) and Sagawan (1986) also noted a decrease in the number of vacuoles in the nucleus during late prophase. In metaphase I, no vacuoles are seen in the area occupied by the chromosomes and spindle, while some small vacuoles are seen interspersed with plastids and mitochondria in more outward portions of the cell.

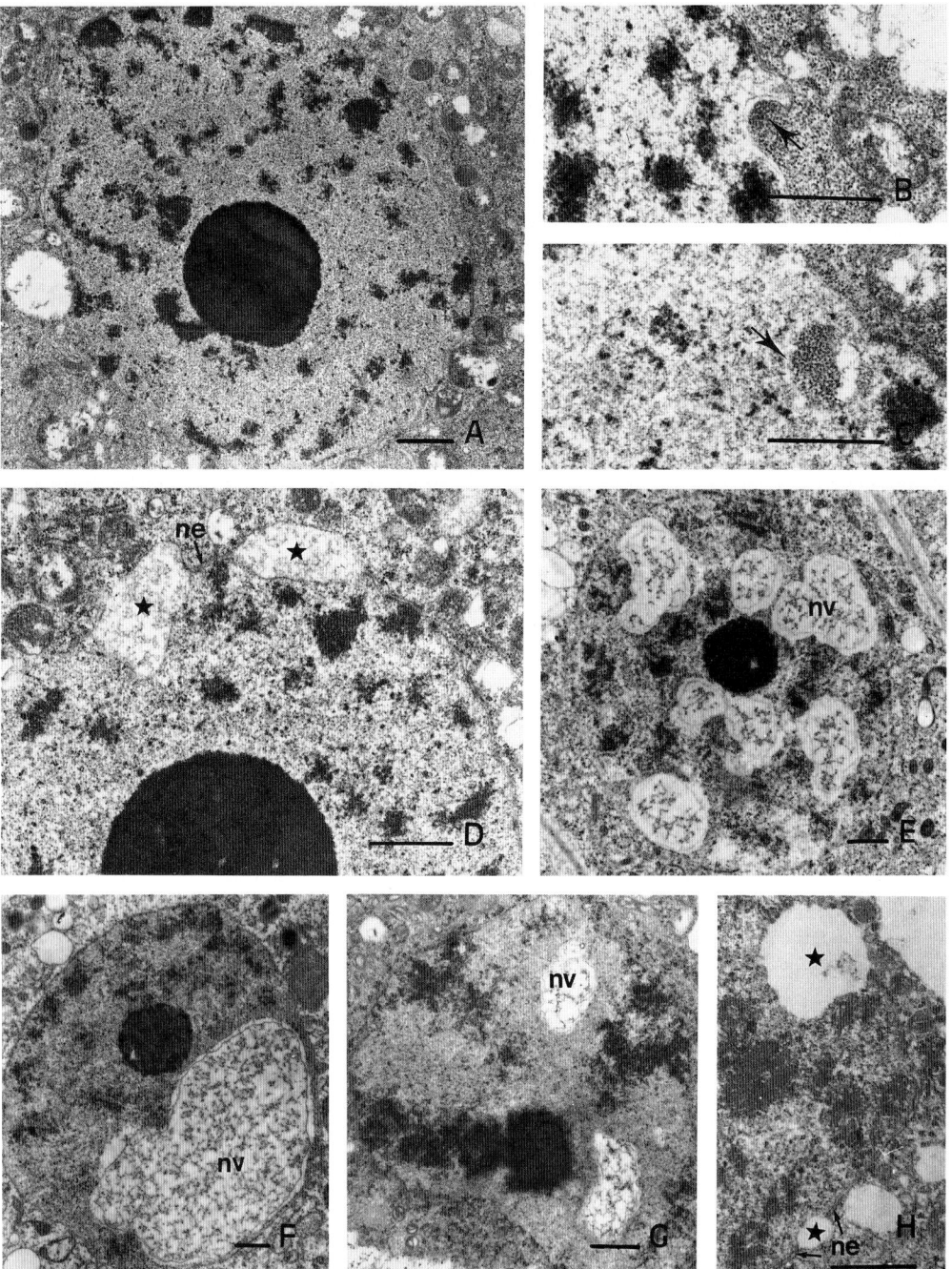

Fig. 1. Formation and development of nuclear vacuoles during meiotic prophase.
A: Early prophase nucleus without any signs of nuclear vacuole formation. B,C:
Invagination of the double nuclear membrane encapsulating a part of the cyto-
plasm in leptotene. D: Dilations of the perinuclear space in leptotene. E,F:
Free lying nuclear vacuoles in the karyoplasm in pachytene. G: Decrease in
nuclear vacuole volume in diplotene/diakinesis. H: Possible elimination of
nuclear vacuoles in diplotene/diakinesis. A,B,C,D,H: Olea europaea; E,F,G:
Beta vulgaris. All bars represent 1 μm.

FUNCTIONAL DYNAMICS OF NUCLEAR VACUOLES

The hypothesis according to which vacuoles are formed in the nucleus to facilitate the pairing of homologous chromosomes (Karasawa and Ueda, 1983) does not appear to be supported by our observations: homologous chromosomes have practically finished pairing off by the time nuclear vacuoles occupy their largest portion of the nucleus. The fact that nuclear vacuoles reach their peak volume just when the nucleus itself is largest make it hard to understand how the vacuoles could favor contraction of the karyoplasm. Our observations in Olea and Beta, together with those in other species (Karasawa and Ueda, 1983; Sangwan, 1986) provide evidence of the cytoplasmic origin of nuclear vacuoles which suggests that a possible transport of cytoplasmic components towards the nucleus may take place. The large number of pores in the nuclear envelope during meiotic prophase points to a high degree of activity and nucleo-cytoplasmic exchange. The requirements of the cell in these moments of meiosis may however outstrip the capacity of the available pores, hence the cell may have to produce invaginations of the nuclear membrane to enhance the passage of materials from the cytoplasm to the nucleus. Sheffield et al. (1979) suggested that vacuoles might act as reserve structures in which metabolites were accumulated in the course of the reorganization of the nucleus during meiotic prophase in preparation for gametophyte generation. This interpretation seems compatible with our hypothesis. The temporary compartmentation of the nucleus also suggests a possible interchange of material between the vacuoles and the karyoplasm. Finally, the eventual drop in nuclear volume noted concomitantly with the rise in the number of small cytoplasmic vacuoles may well be evidence of elimination towards cytoplasm of nuclear metabolic waste products. It seems clear in any case that nuclear vacuoles are changeable, transient structures responding to changes in functional dynamic conditions dependent upon the meiotic events in the prophasic nucleus. Their function may be related to processes of transport, interchange and/or elimination. A more detailed knowledge of the molecular nature of the vacuolar contents will help us to understand the phenomenon of nuclear compartmentation during meiotic prophase.

CONCLUSIONS

Nuclear vacuoles appear to be involved in some as yet unexplained way in meiotic processes during prophase. The nuclear envelope plays a fundamental role in the formation of these vacuoles, giving rise to a considerable increase in membrane synthesis. Vacuoles can be formed in two different ways: either by dilation of the perinuclear space and invagination of the inner membrane toward the karyoplasm until an individual vacuole is pinched off, or by invagination of the double nuclear membrane toward the karyoplasm, encapsulating and sequestering a portion of the cytoplasm which eventually forms a vacuole. The formation of vacuoles in the early prophasic nucleus (leptotene-zygotene) coincides with a considerable increase in nuclear volume. This suggests that the karyoplasm does not really contract, but rather becomes polarized in response to the appearance of nuclear vacuoles. In late meiotic prophase (diplotene-diakinesis) the nucleus becomes notably smaller, while at the same time nuclear vacuoles become scarcer, possibly indicating that vacuoles are eliminated into the cytoplasm.

Future research should set as one of its top priorities the precise identification of the nature and content of these vacuoles, as such information would go a long way toward clarifying the functional significance of these compartments in the meiotic nucleus.

Acknowledgements: This research was supported by the Comisión Asesora para la Investigación Científica y Técnica and C.S.I.C. (Project No. 1-179-2ID 181/906). We thank Ms. M. Garrido for her skilful technical assistance and Ms. K. Shashok for translating the original manuscript.

REFERENCES

Aldrich HC, Vasil IK (1970) Ultrastructure of the postmeiotic nuclear envelope of Podocarpus macrophyllus. J Ultr Res 32: 307-315

Cawood AH, Jones JK (1980) Chromosome behaviour during meiotic prophase in the Solanaceae. Chromosoma 80: 57-68

Karasawa R, Ueda K (1983a) Nuclear vacuoles and synizesis during meiotic prophase in Haplopappus gracilis. Cytologia 48: 819-826

Karasawa R, Ueda K (1983b) Occurrence of nuclear vacuoles in meiotic prophase nuclei in Compositae. Caryologia 36: 145-153

Rasmussen SW (1976) The meiotic prophase in Bombyx mori female analysed by three-dimensional reconstructions of synaptonemal complexes. Chromosoma 54: 245-293

Risueño MC, Galán Cano J, Giménez-Martín G (1975) Ultrastructure of the nuclear envelope in the covering tissues of the ovule and anther beginning of meiosis. Mikroskopie 31: 5-13

Sangwan RS (1986) Formation and cytochemistry of nuclear vacuoles during meiosis in Datura. Eur J Cell Biol 40: 210-218

Sheffield E, Bell PR (1979) Ultrastructural aspects of sporogenesis in a fern, Pteridium aquilinum (L.) Kuhn. Ann Bot 44: 393-405

Sheffield E, Cawood AH, Bell PR, Dickinson HG (1979) The development of nuclear vacuoles during meiosis in plants. Planta 146: 597-601

Sheffield E, Laird S, Bell PR (1983) Ultrastructural aspects of sporogenesis in the apogamous fern Dryopteris borreri. J Cell Sci 63: 125-134

Aspects of Sporopollenin Biosynthesis: Phenols as Integrated Compounds of the Biopolymer

S.Herminghaus, S. Arendt, S. Gubatz, M. Rittscher and R. Wiermann

Botanisches Institut der Universität Münster

Schloßgarten 3, D-4400 Münster/Westf., FRG

INTRODUCTION

The structure and biosynthetic pathway of sporopollenin are largely unknown. In 1971, Brooks and Shaw have postulated that sporopollenin is generated from carotenoids and carotenoid esters. This assumption is based on comparartive studies on the naturally occuring sporopollenin and polymers which were synthesized from carotenoids. Both the natural and the synthetic macromolecules exhibit a high level of resistance to acetolysis; both show similar IR spectra and have similar elemental composition.

As far as higher plants are concerned, this hypothesis has not yet been confirmed by tracer experiments. However these are exactly the kind of experiments which are needed to establish whether isoprenoid metabolism is involved in sporopollenin biosynthesis and whether further, or completely different , metabolic pathways participate.

To remove the confusion about the nature of this important plant material, the following experiments were performed: 1. Highly purified sporopollenin was degraded in order to analyse the resulting low molecular degradation products. These investigations were focussed on the appearance of phenols as degradation products. These results obtained from naturally occuring sporopollenin were compared with those from "synthetic sporopollenin". 2. On the basis of an optimized technique extensive tracer experiments were carried out with a variety of added precursors.

MATERIAL AND METHODS

MATERIAL

Pollen from *Corylus avellana* were collected from various localities around Münster, stored at -20°C and freeze-dried before use. Intact anthers of *Tulipa* cv. Apeldoorn were prepared from young bulbs purchased from Nebelung (Münster).

A polymer synthesized from ß–carotene (= "synthetic sporopollenin") was a kind gift from Dr. Geisert (Mainz, FRG).

METHODS

Purification of the sporopollenin fraction

The purification of the sporopollenin was carried out as described previously (Herminghaus et al., 1988; Rittscher and Wiermann, 1988) either using a gentle method employing hydrolytic enzymes [Corylus and Tulipa (Rittscher and Wiermann, 1988)] or by a treatment for several days with 80% phosphoric acid [Tulipa, Rittscher and Wiermann, 1988].

The application of labeled precursors

The application of labelled precursors to intact anthers was accomplished as described by Rittscher and Wiermann, 1988. After 24 h, the anthers were washed, decapitated, and their contents (= pollen/tapetum fraction) were squeezed out and treated as described above in order to purify the sporopollenin fraction.

Chemical degradation of the sporopollenin fraction

The sporopollenin fraction was degraded by potash–fusion and nitrobenzene oxidation. The degradation products were analysed by TLC, HPLC and GC (for detail see Herminghaus et al., 1988.

RESULTS AND DISCUSSION

1. The results of the degradation experiments

An important prerequisite for these experiments was the isolation and purification of sporopollenin. Instead of using the conventional aggressive methods such as acetolysis or treatment with 80% phosphoric acid, a gentle procedure using various hydrolytic enzymes was developed (Herminghaus et al., 1988).

The highly purified sporopollenin fractions were decomposed by different techniques. Apart from potash-fusion, nitrobenzene oxidation , which is commonly used in lignin research , was employed for the first time. Nitrobenzene oxidation degraded sporopollenin with nearly 100% efficiency. The acidified diethylether extracts following <u>nitrobenzene oxidation</u> were analysed by TLC and HPLC. The results show that phenolic substances like p-hydroxybenzoic acid, vanillic acid, p-hydroxybenzaldehyde, and vanillin were formed. In addition to these compounds, some not yet identified compounds were found. The release of p-coumaric acid, a further degradation product, depends strongly on the conditions under which nitrobenzene oxidation is carried out. (Schulze Osthoff and Wiermann, 1987; Herminghaus et al.,1988).

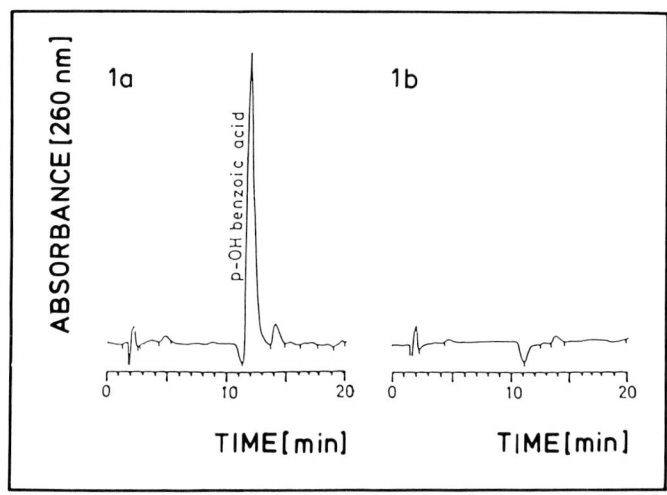

Fig.1: HPLC diagrams obtained with the acidified ether extract following potash-fusion.(1a: *Corylus* sporopollenin; 1b: "synthetic sporopollenin")

Following degradation of sporopollenin by potash-fusion, different compounds were released. In addition to some not yet identified compounds, which appeared without exception at low concentrations , the major degradation product was p-hydroxybenzoic acid.

The appearance of phenols following the degradation of sporopollenin by potash - fusion was interpreted by Brooks and Shaw (1968) as artefacts arising from aromatisation of the carotenoid skeletons under these conditions (see Shaw, 1971 too). In order to test, whether , and if so, to what extent, phenols appear after degradation of a polymer synthesized from carotenoids, the carotenoid-polymer was degraded exactly under the same conditions as natural sporopollenin. The results clearly demonstrate that phenols, if they indeed resulted from the degradation of <u>"synthetic sporopollenin"</u> are generated only in extremely small quantities in comparision to those released after degradation of *Corylus* sporopollenin (Fig.1).

Similar results were obtained after degradation of both polymers by nitrobenzene oxidation (data not shown, see Herminghaus et al.,1988). Thus we conclude that phenols are integral constituents of natural sporopollenin.

2. The results of tracer experiments

The view that phenols are involved in sporopollenin biosynthesis is clearly supported by the results of tracer experiments. The purification of the labeled sporopollenin fraction was accomplished either by a gentle method employing hydrolytic enzymes (pronase, amylase, amyloglucosidase, cellulase, pectinase, lipase) and alkaline hydrolysis, or by the conventional aggressive procedure, where the material was enriched by alkaline hydrolysis and treatment for several days with 80% phosphoric acid (Rittscher and Wiermann, 1988). ^{14}C-labeled mevalonate , glucose, acetate, malonic acid, phenylalanine, tyrosine and p-coumaric acid were applied to anthers of *Tulipa* cv. Apeldoorn.

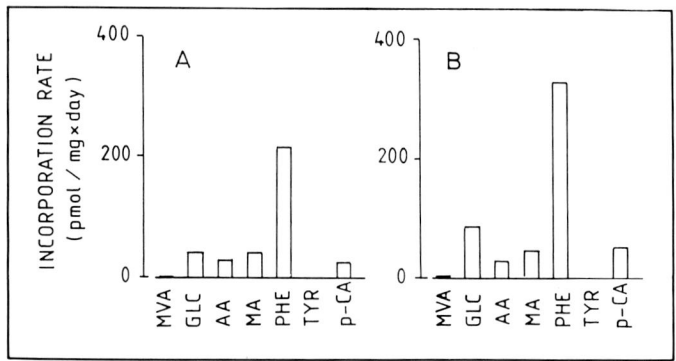

Fig.2: The extent of incorporation of different precursors into sporopollenin (A: gentle method, B: conventional aggressive method; MVA: mevalonic acid, GLC: glucose, AA: acetic acid, MA: malonic acid, PHE: phenyl-alanine, TYR: tyrosine, p.CA: p-coumaric acid)

The results show that a high level of incorporation into the sporopollenin fraction was always seen with $\left[\text{U-}^{14}\text{C}\right]$-phenylalanine, regardless of the method of enrichment used (Fig.2). In comparision with phenylalanine, only a low level of incorporation into sporopollenin was achieved after the application of mevalonic acid.

The level of radioactivity in the sporopollenin fraction following application of phenylalanine was to an extent that degradation products of the labeled polymer could be analyzed. (Fig.3, Tab.1). After potash-fusion protocatechuic acid , m- and p–hydroxybenzoic acid and syringic acid were unambiguously identified by HPLC. Among these compounds, the main degradation component , p–hydroxybenzoic acid, was the most strongly labeled substance. 74% of the radioactivity of ether– soluble acids,which represents 43.1% of the enriched sporopollenin fraction (Tab.1) eluted with this substance , only 4% with peak 2, an as yet unknown product. Because of the overlapping of peaks 2 and 3, the small amount of radioactivity which eluted with peak 2 is probably also attributable to p–hydroxybenzoic acid.

Thus it could be shown that at least a portion of the phenols released by potash–fusion is originally of phenolic nature and does not come from isoprenoid or other metabolic pathways. In contrast to earlier assumptions of Brooks and Shaw (1968, 1971) we conclude that the synthesis of phenolics via phenylalanine ammonia–lyase is definitely an essential part of sporopollenin biosynthesis.

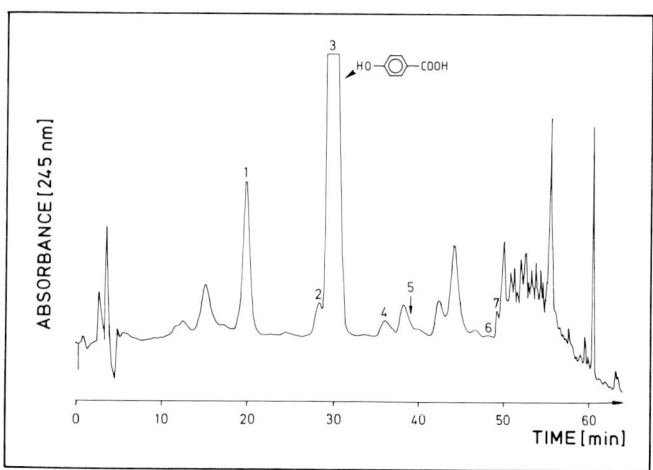

Fig.3: HPLC diagram obtained with ether soluble acids released during degradation of the labeled sporopollenin from *Tulipa* by potash fusion.
(1.: protocatechuic acid; 2.: unknown compound; 3.: p–hydroxybenzoic acid; 4.: m–hydroxybenzoic acid; 5. vanillic acid; 6.: syringic acid; 7.: solvent;) The values of radioactivity eluted from peak 3 are seen in Tab.1)

Tab. 1: The radioactivity of sporopollenin as well as of insoluble residues and
p-hydroxybenzoic acid obtained after the degradation of the sporopollenin
fraction labelled with $[U-^{14}C]$-phenylalanine.

Sporopollenin		insoluble residues		p-hydroxybenzoic acid	
9.63 mg	(=100 %)	1.02 mg	(=16.6 %)	0.36 mg	(= 3.8 %)
10102 dpm	(=100 %)	275 dpm	(= 2.7 %)	4356 dpm	(=43.1 %)
1049 dpm/mg		269 dpm/mg		12030 dpm/mg	

Acknowledgements

These studies were financially supported by the Deutsche Forschungsgemeinschaft and
the Fonds der Chemischen Industrie. The authors are indebted to Dr.Weltring for
revising the English.

References

Brooks J, Shaw G (1968) Chemical structure of the exine of pollen walls and a new
    function for carotenoids in nature. Nature 219: 532-533
Brooks J, Shaw G (1971) Recent developments in the chemistry, biochemistry, geo-
    chemistry and post-tetrad ontogeny of sporopollenins derived from pollen and
    spore exines. In: Heslop-Harrison J.(ed) Pollen development and physiology.
    Butterworths, London, 99-114
Herminghaus S, Gubatz S, Arendt S, Wiermann R (1988) The occurence of phenols as
    degradation products of natural sporopollenin - a comparision with synthetic
    sporopollenin. Z. Naturforsch 43 c: in press
Rittscher M, Wiermann R (1988) Studies on sporopollenin biosynthesis in *Tulipa*
    anthers. -II. Incorporation of precursors and degradation of the radio-
    labelled polymer. Sex Plant Reprod: in press
Schulze Osthoff K, Wiermann R, (1987) Phenols as integrated compounds of sporo-
    pollenin from *Pinus* pollen. J.Plant Physiol 131: 5-15

Organelle Aggregations During Microsporogenesis in *Stangeria, Nymphaea,* and *Malva*

B. Rodkiewicz, E. Duda and K. Kudlicka
Institute of Biology
Maria Curie-Skłodowska University
20 033 Lublin
Poland

Abstract. Microsporocytes of Stangeria and Nymphaea acquire a polarized appearance for a short part of prophase I, when the areas on
the opposite sides of the nucleus differ in their organelle contents.
By the end of prophase I in microsporocytes of Nymphaea plastids and
mitochondria form a group which after telophase I stretches into the
equatorial plane, and after telophase II separates a tetrad into four
regions. During microsporogenesis in Malva organelles form a dense
coat around the late prophase I nucleus and later around each nucleus
in dyads and tetrads.

Key words: microsporogenesis, organelle aggregations, organelles in
meiosis, Stangeria, Nymphaea, Malva.

Introduction
Plastids and mitochondria aggregated into groups were described in
certain stages of microsporogenesis and sporogenesis in several species of higher plants (ref. Rodkiewicz and Duda 1988). These organelles aggregate and disaggregate, following basically similar
schedule, in microsporogenesis with simultaneous cytokinesis, except
for that in Malva described in this communication.

Material and Methods
Microsporangiate cones were collected in October from Stangeria erio-
pus (Kunze) Nash grown in the glasshouse. Microsporangia of Stangeria and anthers of Nymphaea alba L were fixed in 2% glutaraldehyde in 0,05M cacodylate buffer at pH 7, and embedded in Spurr.
Ultrathin sections stained with uranyl acetate and lead citrate were examined with an EM. Anthers of Malva silvestris L fixed in Rega-
ud fluid (formaldehyde and potassium bichromate) were squashed
and stained with 2% toluidyne blue.

Results
Prophase I in microsporocytes of Stangeria - In late leptotene or in
zygotene all plastids and mitochondria are aggregated on one side
of nucleus (Fig. 1). Plastids contain small starch grains; some

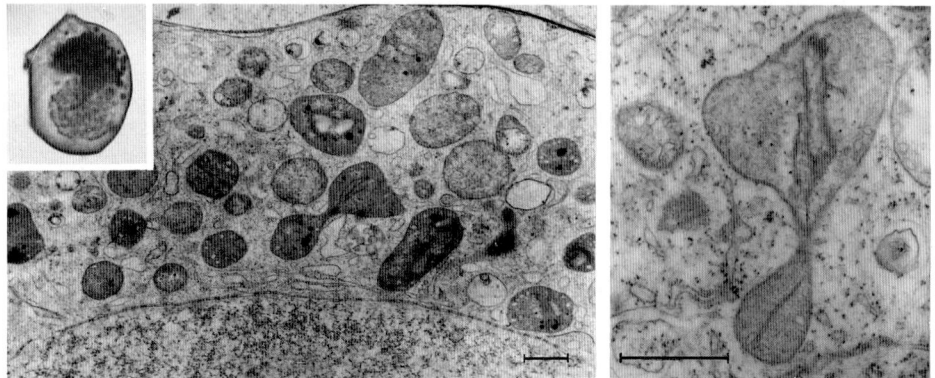

Figs. 1 and 2. Microsporocyte of Stangeria at early prophase I; aggregation of plastids and mitochondria, bar = 1μm; in the inset plastids with starch grains stained with PAS. Fig. 2. Apparently dividing plastids in the aggregation, bar = 1μm.

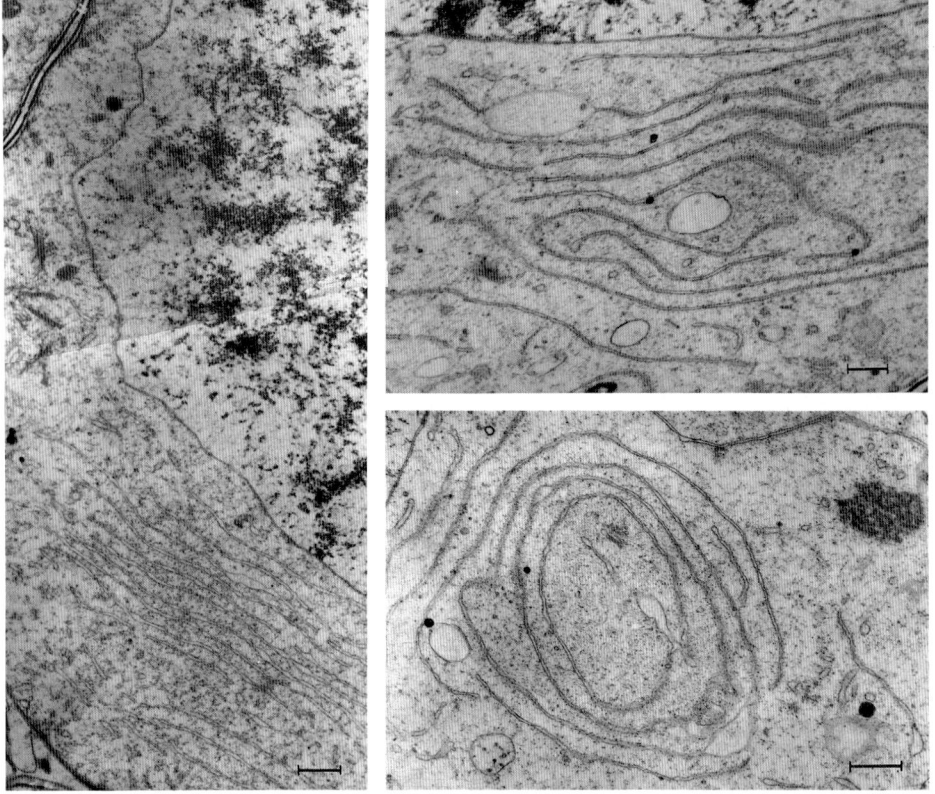

Figs. 3 - 5. Microsporocyte of Stangeria at early prophase I: Fig.3. Opposite side to that in Fig. 1, array of ER, at cisternae ends several dictyosomes, bar = 1μm. Figs. 4 and 5. Slightly later stage, cisternae folded and in a concentric arrangement, bar = 1μm.

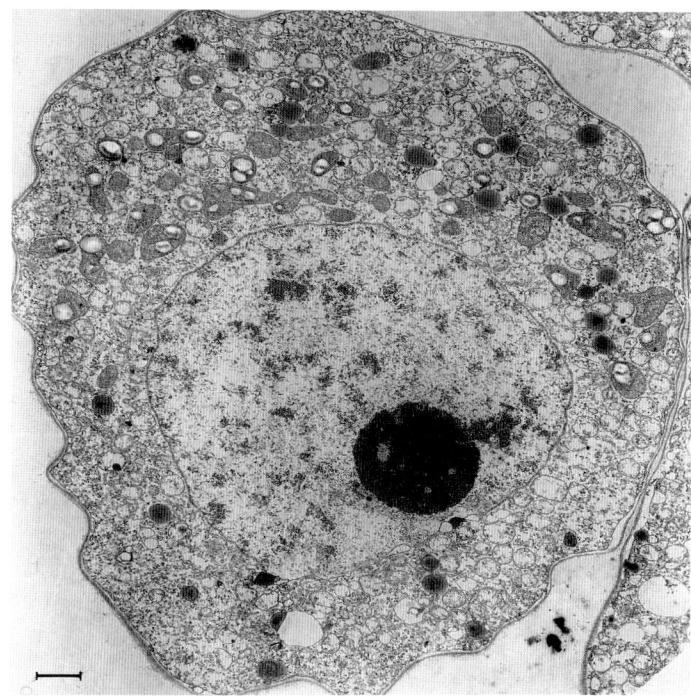

Fig. 6. Microsporocyte of Nymphaea at early prophase I; aggregation of plastids - with small starch grains - and mitochondria, some mitochondria and many osmiophilic bodies scattered around the nucleus, bar = 2µm.

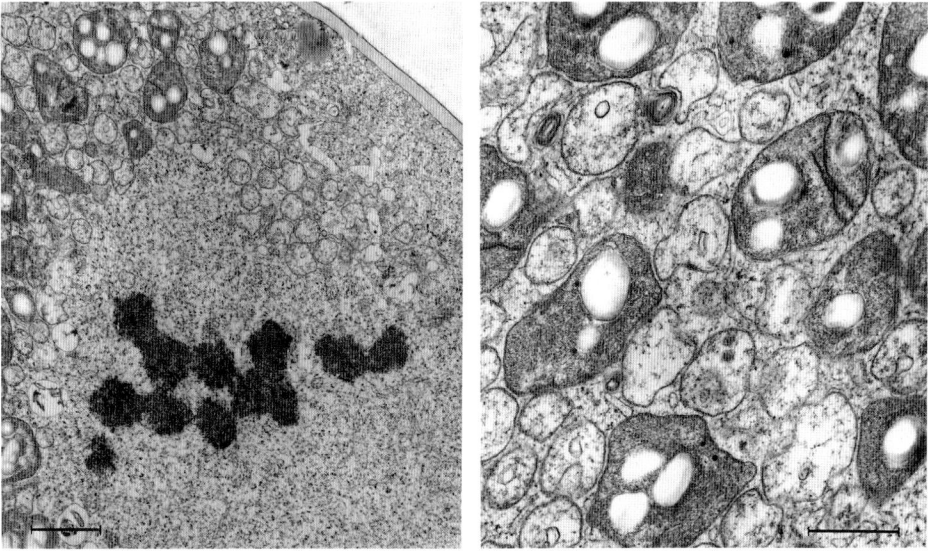

Figs. 7 and 8. Microsporocyte of Nymphaea at metaphase I; chromosomes in homogenous area adjacent to aggregation of plastids and mitochondria, bar = 2μm. Fig. 8. Aggregated mitochondria and plastids with starch grains, bar = 1μm.

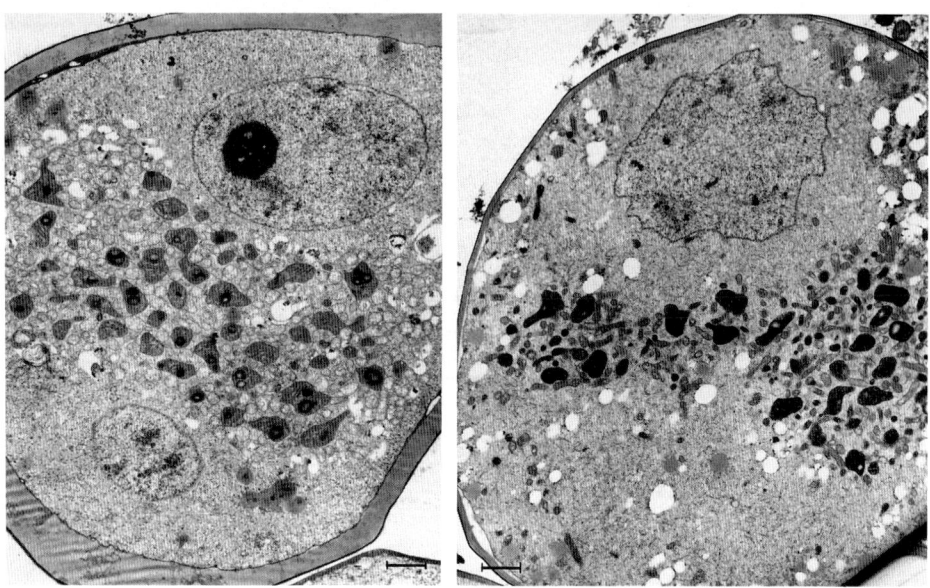

Figs. 9 and 10. Microsporocytes of Nymphaea. Fig. 9. Post-telophase I, equatorial plate of plastids and mitochondria. Fig. 10. Post-telophase II, a part of complicated organelle aggregation which partitions the cell into four mononucleate parts, bar = 2μm.

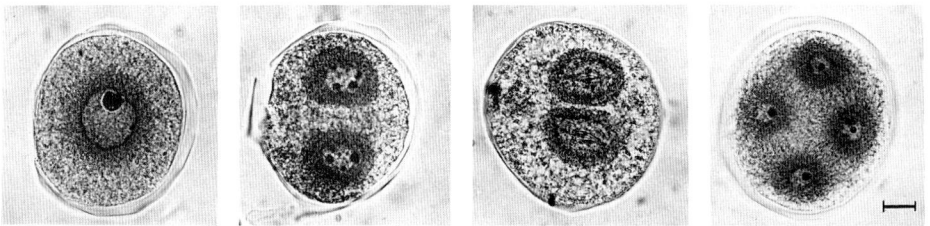

Fig.11. Microsporogenesis in Malva. Organelles in: late prophase I, dyad, metaphase II, tetrad before cytokinesis, bar = 10μm.

are constricted, apparently undergoing division (Fig. 2). The starch grains in the plastids grow and after staining the group of organelles becomes distinctly visible in a light microscope (Fig. 1 inset). Meanwhile, on the other side of the nucleus many parallel ER cisternae appear (Fig. 3). They fold and extend in some places (Fig. 4), eventually the whole area is occupied by concentrically arranged ER cisternae (Fig. 5). They gradually disappear and cytoplasm becomes less dense. Some mitochondria are detached from the organelle group and get scattered around the nucleus.

The microsporocytes of Nymphaea also show similar-positioned orga-nelles (Fig. 6). In later prophase I, the group is dispersed, plas-tids and mitochondria are about evenly distributed in cytoplasm of the microsporocyte. In Nymphaea by the end of prophase I they form again a large group which persist through meiosis I (Figs. 7 and 8). After telophase I this group is reshaped into a plate occupying the equatorial plane of the microsporocyte (Fig. 9). After telophase II (Fig. 10) the plate of organelles is transformed in such a way that it divides a tetrad into four parts which finally are separated by cell plates laid down in simultaneous cytokinesis.

Organelles during microsporogenesis in Malva form a dense coat around the late prophase I nucleus (Fig. 11). Inside this coat the meiotic spindle develops and the first meiotic division is carried out. La-ter each nucleus in a dyad is coated with an organelle layer, again inside this coat the second meiotic division takes place and each resulting nucleus is coated by the organelles. A tetrad divides fi-nally by simultaneous cytokinesis.

Discussion

The behaviour of plastids and mitochondria in prophase I of micro-sporocyte may be assumed by comparing various kinds of organelle disposition described in several plants (ref. Rodkiewicz and Duda 1988).

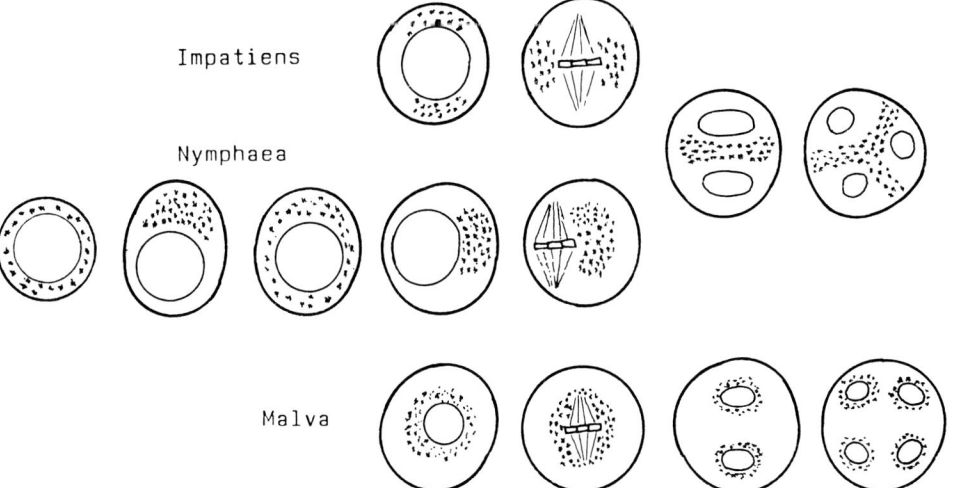

Fig. 12. Mitochondria and plastids during microsporogenesis in Im-patiens (Rodkiewicz et al 1986), Nymphaea and Malva. Positions of organelles before late prophase I were reconstructed by combining the data on Stangeria and Nymphaea microsporocytes.

At early prophase I mitochondria and plastids aggregate in a group. After a short time mitochondria gradually begin to detach from the group which dissapears by the late pachytene. The changing positions of mitochondria and plastids during simultaneous microsporogenesis are shown in Fig. 12. We assume that they display a common pattern of behaviour until late prophase I. There is no explanation of early prophase I organelle aggregation which seems to visualize polarization of a microsporocyte. The second organelle aggregation after telophase I may prevent coalescence of nuclei which during telophase I were kept apart by the phragmoplast. The phragmoplast is dismantled before spindles of meiosis II begin to be formed (Van Lammeren et al 1985, Brown and Lemmon 1987, Hogan 1987). These spindles are also separated by the organelle aggregation which lasts until the end of telophase II. Long preservation of organelle aggregation may facilitate their equal apportionment among the microspores in tetrads which are divided by simultaneous cytokinesis.

Acknowledgements. We thank dr R. P. Singh, NBRI, Lucknow, for discussing the manuscript and dr H. Jakubiec-Werblan, Director of the Botanical Garden of Warsaw University, for Stangeria cones.

References.

Brown RC, Lemmon BE (1987) Division polarity, development and configuration of microtubule arrays in Bryophyte meiosis. II anaphase I to the tetrad. Protoplasma 138: 1-10

Hogan CJ (1987) Microtubule patterns during meiosis in two higher plant species. Protoplasma 138: 126-136

Rodkiewicz B, Bednara J, Mostowska A, Duda E, Stobiecka H (1986) The change in disposition of plastids and mitochondria during microsporogenesis in some higher plants. Acta Bot Neerl 35: 209-215

Rodkiewicz B, Duda E (1988) Aggregation of organelles in meiotic cells of higher plants. Acta Soc Bot Pol (in press)

Van Lammeren AAM, Keijzer CJ, Willemse MTM, Kreft H (1985) Structure and function of the microtubular cytoskeleton during pollen development in Gasteria verrucosa Mill H Duval. Planta 165: 1-11

Amylogenesis and Amylolysis During Pollen Grain Development

PACINI E.* and FRANCHI G.G.**

* Dipartimento di Biologia Ambientale – Sezione Botanica
** Dipartimento Farmaco Chimico Tecnologico – Sezione Botanica Farmaceutica
University of Siena
Via Mattioli 4
53100 Siena
ITALY

INTRODUCTION

In pollen mother cells, the organelles dedifferentiate at the onset of meiosis (Bird et al., 1983). Plastids previously containing starch are completely starch-free at zygotene (Bird et al., 1983; Pacini and Franchi, 1983). Organelle dedifferentiation is accompanied by ribosome reorganization (Bird et al. 1983). After telophase, plastids may differentiate again in the amyloplasts, but their behavioural pattern varies according to the species. At pollen shedding some species have starch and others have not (Baker and Baker, 1979; Franchi and Pacini, 1988).

In this paper we present the differentiation and dedifferentiation of plastids in five Angiosperm species; they are correlated with the physical and chemical properties of starch.

MATERIALS AND METHODS

Anthers of the species listed in Fig. 1 were collected, fixed, embedded, cut, stained and observed by TEM as previously described (Pacini and Juniper, 1979). 1 - 2 micron sections were cut from the same EM blocks and tested for total polysaccharides with the PAS reaction preceeded by aldehyde blockade.

Anthers from flower buds at different stages were squashed on slides and: a)tested with iodine / potassium iodide (IKI test) (Jensen, 1962); b) obseved under polarized light in order to detect starch birefringence.

OBSERVATIONS

The plastids dedifferentiate during early prophase in the species studied (Franchi and Pacini, 1980; Franchi et al., 1984; Pacini and Franchi, 1983; Pacini and Juniper 1979; 1984; Pacini et al., 1985; 1986). They differentiate again at the end of meiosis, but not at the same stage in all species (earlier in Parietaria judaica and Smilax aspera and later in Olea europea and Lycopersicum peruvianum) (Fig. 1). Starch is hydrolyzed during the middle or late microspore stage (Fig. 1) and the plastids are completely undifferentiated before pollen mitosis; they start to differentiate again during the early or middle two-celled stage. In Lycopersicum peruvianum, Smilax aspera, Prunus avium and Olea europaea

182

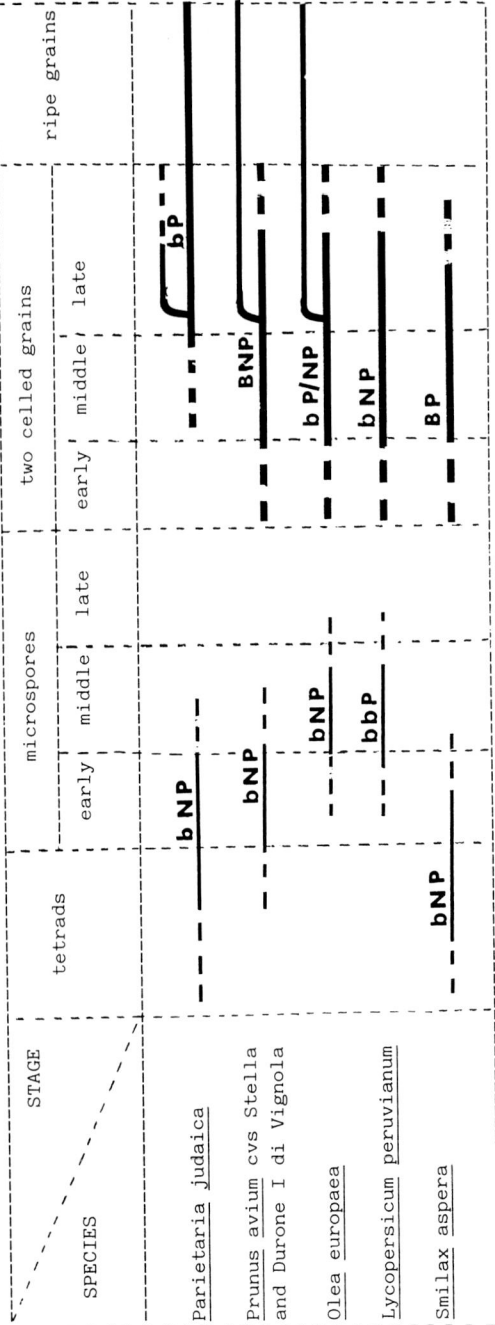

Fig. 1.
Amylogenesis-amylolysis during pollen grain development of five angiosperm species. Two amylogenesis occur after meiosis, the first during microspore stage and the second during the two-celled stage. The amount of starch produced by the first one is less abundant than that of the second, as indicated by the solid line. The morphology of the amyloplasts during both periods of amylogenesis is different, as well as the physical and chemical properties of the starch, as evidentiated by the colour after the IKI test and polarization microscopy. Only few pollen grains of _Prunus avium_ cvs have starch at the mature stage. Some _Olea europaea_ cultivars have starch in ripe pollen grains and some have not. The occurrence of starch in _Parietaria judaica_ pollen grains is a function of the season.

B black ⎫
b brown ⎬ starch colour after IKI test
bb blue ⎭

P polarizable ⎫
⎬ starch with polarization microscopy
NP not polarizable ⎭

starch starts to be hydrolyzed during the late two-celled stage and the process is complete just before anther dehydration. Prunus avium and Olea europaea may contain starchy and starchless pollen grains contemporaneously in the same loculus, in percentages which vary from cultivar to cultivar. In Parietaria judaica the ripe pollen grains contain starch, but there are also a few starchless grains, the percentage of which varies over the long blooming period of this plant, from May to November/December.

The physical and chemical properties of the starch deposited in the two amylogeneses also vary: in fact starch may be polarizable or not, and it has different amounts of amylose and amylopectin according to the IKI test (Fig. 1) (Franchi and Pacini, 1988).

Even the amount of starch produced during the two amylogeneses is different, and always higher during the second. The morphology of the amyloplasts differs as shown in Fig. 2 for Lycopersicum peruvianum, Olea europaea and Parietaria judaica at maximum starch bulk.

Amylogenesis and amylolysis are phenomena shared by all the microspores and pollen grains of a loculus, but they do not behave synchronously; this is particularly evident in squashed preparations stained with IKI.

DISCUSSION

Starch is the only insoluble polysaccharide stored temporarily in plastids during pollen grain development. It derives from the photosynthetic activity of the mother plant: sugars from the flower can directly reach the loculus via the tapetum or can be temporarily polymerized to starch in the exotecium or filament (Pacini and Franchi, 1983; Pacini et al., 1986).

During the late tetrad and early microspore stages the loculus may also act as a temporary storage site (Gori, 1982; Pacini and Franchi, 1983).

Two or more waves of amylogenesis/amylolysis were observed during pollen grain development; in dicotyledons two waves occur after meiosis; in monocotyledons more than two waves occur.

Sugars absorbed by the microspores and bicellular pollen grains may be deposited as starch or rapidly utilized. In both cases they may be used as energy sources for normal spore or gametophyte metabolism, and for the build up of intine, pores and furrows.

In all the five observed species starch hydrolysis of the first amylogenesis occurs at the same time as intine build up.

In some species all the starch is hydrolyzed before the pollen is ripe, but the polysaccharides are not consumed and persist in the cytoplasm (Franchi and Pacini, 1988). The simultaneous presence of starchy and starchless grains in the same anther is reported for species having both short and long blooming periods. The ratio of starchy to starchless grains is almost the same every year in plants having a short blooming period, and might therefore be genetically determined (Pacini et al., 1985; 1986). On the other hand in plants with a long blooming period this ratio fluctuates widely according to environmental conditions (Franchi et al., 1984).

The occurence of polysaccharides in the cytoplasm, ready to be utilized for pollen tube walls and the energy demands of germination does not mean that starchless pollen grains germinate faster. In fact Gramineae starchy pollen grains germinate in a few seconds, and fecundate in a few minutes (Heslop-Harrison, 1979).

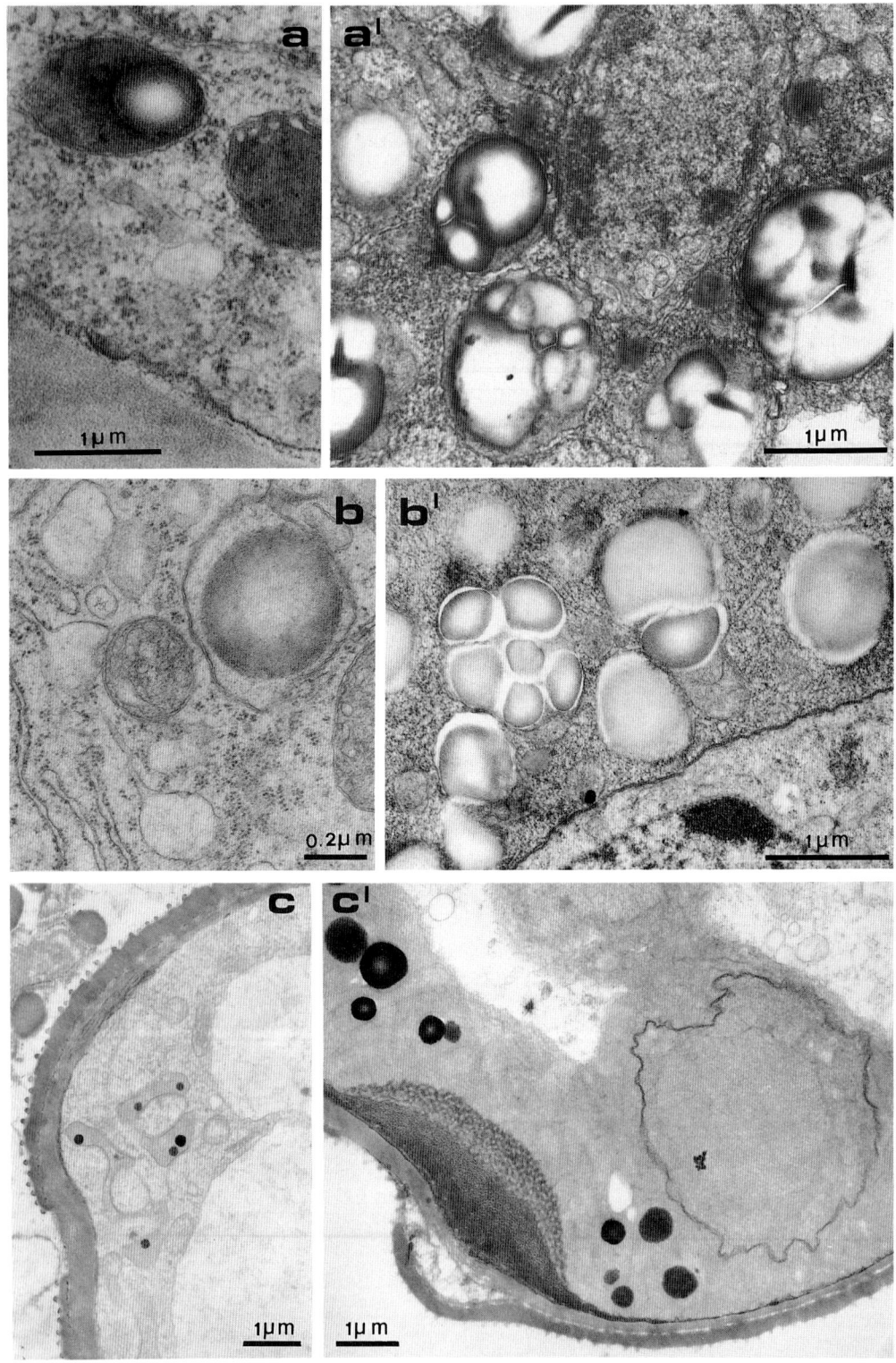

It is possible to appreciate plastid differentiation and dedifferentiation even by light microscopy using the IKI test and polarization microscopy: with these two easy and inexpensive technique it is possible to observe physical and chemical variations in the starch which depend on the ratio between amylose and amylopectin.

The importance of reproductive calendars as the basis for other studies was stressed recently.

The occurrence of starch during pollen grain development is a reproductive calendar parameter (Pacini and Sarfatti, 1978; Pacini and Franchi, 1983; Pacini et al., 1985; 1986).

Quite recently Sangwan and Sangwan Norreel (1987) noticed that plastid differentiation and dedifferentiation is important in the induction of androgenesis. They tested twenty-six species and found that only those which are androgenic have undifferentiated plastids at the right stage for embryo induction. This again shows the importance of plastid differentiation during pollen grain development. Why plastids inside the generative cell do not differentiate while those in the vegetative cell do, is a question which requires investigation.

ACKNOWLEDGMENTS

This research was performed under the program of CNR group: "Biology of Reproduction and Differentiation".

REFERENCES

Baker HG and Baker I (1979), Starch in Angiosperm pollen grains and its evolutionary significance. Amer J Bot 66: 591-600.

Bird J, Porter EK and Dickinson HG (1983), Events in the cytoplasm during male meiosis in Lilium. J Cell Sci 59: 27-42.

Franchi GG and Pacini E (1980), Wall projections in the vegetative cell of Parietaria officinalis pollen. Protoplasma 104: 67-74.

Franchi GG and Pacini E (1988), Pollen polysaccharide reserves in some plants of economic interest. This volume.

Franchi GG, Pacini E and Rottoli P (1984), Pollen grain viability in Parietaria judaica L. during the long blooming period and correlation with meteorological conditions and allergic diseases. Giorn Bot Ital 118: 163-178.

Gori P (1982), Accumulation of polysaccharides in the anther cavity of Allium sativum, clone Piemonte. J Ultrastr Res 81: 158-162.

Heslop-Harrison J (1979), Aspects of the structure, cytochemistry and germination of the pollen of rye (Secale cereale L.). Ann Bot 44 (suppl.): 1-47.

Fig. 2 – Amyloplasts during the first and the second amylogeneses; a, a' Parietaria judaica; b, b' Olea europea, c, c' Lycopersicum peruvianum after PATAg test.

Jensen WA (1962), Botanical histochemistry. WH Freeman and Co, San Francisco.

Pacini E and Franchi GG (1983), Pollen grain development in Smilax aspera L. and possible functions of the loculus. In: Mulcahy DL and Ottaviano E (Eds) "Pollen: biology and implications for plant breeding", pp. 183-190. Elsevier Science Publishing Co, New York.

Pacini E and Juniper BE (1979), The ultrastructure of pollen-grain development in the olive (Olea europaea). 1. Proteins in the pore. New Phytol 83: 157-163.

Pacini E and Juniper BE (1984), The ultrastructure of pollen grain development in Lycopersicum peruvianum. Caryologia 37: 21-50.

Pacini E and Sarfatti G (1978), The reproductive calendar of Lycopersicum peruvianum Mill. Soc Bot Fr, Actualités botaniques 125: 295-299.

Pacini E, Franchi GG and Bellani LM (1985), Pollen grain development in the olive (Olea europaea L.): ultrastructure and anomalies. In: Willemse MTM and Van Went JL (compilers) "Sexual reproduction in seed plants, ferns and mosses", pp. 25-27. PUDOC, Wageningen.

Pacini E, Bellani LM and Lozzi R (1986), Pollen, tapetum and anther development in two cultivars of sweet cherry (Prunus avium). Phytomorphology 36: 197-210.

Sangwan RS and Sangwan Norreel BS (1987), Ultrastructural cytology of plastids in pollen grains of certain androgenic and non-androgenic plants. Protoplasma 138: 11-22.

Histological and Biochemical Analyses of Microsporogenesis in the Normal and Male Sterile, Stamenless-2 Mutant of Tomato

V.K. SAWHNEY and S.K. BHADULA
Department of Biology
University of Saskatchewan
Saskatoon, Saskatchewan S7N OWO
Canada

Male sterile lines, genetic (GMS) or cytoplasmic (CMS), of plants have a two-fold value. 1. They are potentially useful in hybrid production, and 2. they are valuable in investigating the gene controlled mechanisms in stamen development. Several studies on different systems have examined various aspects of the histology and biochemistry of normal and male sterile lines (see reviews, e.g., Laser and Lersten 1972; Frankel and Galun 1977; Bhandari 1984). It has been shown that in male sterile lines, the breakdown in microsporogenesis can occur at any stage of development and that one or more than one enzyme system may be affected.

The homozygous recessive, stamenless-2 (sl-2/sl-2) mutant of tomato (<u>Lycopersicon</u> <u>esculentum</u>) is a specially attractive system for investigations into the mechanisms of male sterility, since its expression can be regulated by plant hormones and temperature conditions. For example, gibberellic acid (GA$_3$) and low temperatures restore fertility in the mutant, whereas indole acetic acid (IAA) and high temperatures induce the formation of carpel-like structures in place of stamens (Sawhney and Greyson 1973; Sawhney 1983). Since esterases are implicated to have a role in pollen development (Vithanage and Knox 1976; Abbott et al. 1984) and amylases, through the breakdown of starch, are believed to provide free sugars for developing pollen, a study on the activity and isozymes of these enzymes in the normal and sl-2/sl-2 mutant was conducted. This report deals with some aspects of the histology, and activity of esterase and amylases during the ontogeny of normal and mutant stamens.

MATERIALS AND METHODS

Seed source and the methods of plant cultivation of the normal (+/+) and sl-2/sl-2 mutant were similar to an earlier report (Sawhney 1983). Plants were grown either in a greenhouse with supplemental lighting for 16 hr a day, or in a growth chamber set at one of the following temperature regimes; intermediate (23° day/18°C night, ITR) and low (18° day/15°C night, LTR). Lighting was provided by Gro-Lux wide spectrum tubes at an intensity of 200 μE s^{-1}m^{-2}.

Stamens of eight stages, from the microspore mother cells (MMC) to mature anthers, were processed for light microscopy and esterase and amylase analyses. The methods of light microscopy and esterase activity are reported elsewhere (Bhadula and Sawhney 1987; Sawhney and Bhadula 1988). The amylase assay was according to Jones and Varner (1967). Isozymes of amylase were resolved on 7% polyacrylamide gels and stained following the procedure of Salas and Cardemil (1986).

RESULTS

Histological studies

A detailed study on the histology of normal and mutant anthers at various stages was recently reported (Sawhney and Bhadula 1988). A brief summary of the major distinguishing features of the two types of anthers follows.

In the normal anthers, the development of microspores proceeds from MMC (stage i) to tetrads (stage ii) to separated microspores (stage iii). The tapetal cells enlarged during these stages, were amoeboid, and contained sporopollenin (sp) deposits. In the mutant anthers, the development of microspores was the same as in the normal up to stage iii. However, the tapetal cells were larger than those in the normal and had, at places, divided to form a bilayer.

The major difference in the normal and mutant anthers was at stage iv, when in the former, the tapetal cells had started to degenerate, whereas in the latter they were still intact but had large vacuoles. Also, in the normal, microspores had a deposition of exine, but many of the mutant microspores were devoid of it. The lack of exine resulted in the swelling of mutant microspores which eventually burst and their cytoplasmic contents were dispersed in the locule.

At stage v, the normal microspores were binucleate and progressed from pollen with many small vacuoles (stage vi) to mature pollen containing dense cytoplasm and a few vacuoles (stage viii).

In the mutant, tapetum degeneration was delayed until stage v. At this stage some normal pollen grains were observed, but most had either degenerated or contained large vacuoles. At stages vi and vii there was further degeneration of the tapetum and at stage viii, a mass of a few normal and mostly degenerated microspores was observed in mutant anthers.

Esterase activity

The esterase activity was significantly higher in the normal than mutant stamens at all stages of development (Fig. 1). In the normal, at stages i to iii, the esterase activity was high, it declined at stages iv and v and was further reduced at stages vi to vii (Fig. 1). In the mutant, there was a gradual increase in esterase activity up to stage iv, but then it declined until maturity.

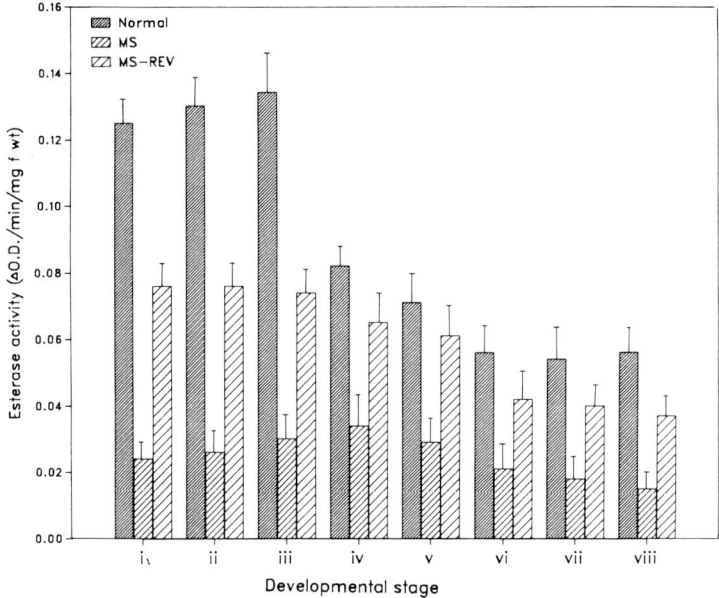

Fig. 1. Esterase activity (ΔO.D./min/mg fr. wt.) of the normal, male sterile stamenless-2 mutant (MS) and low temperature-reverted mutant (MS-REV) stamens at eight (i to viii) developmental stages. Vertical bars represent standard error.

In low temperature (LTR) reverted mutant stamens (MS-REV), the esterase activity was higher than those grown in intermediate temperatures (ITR), or in the greenhouse (Fig. 1). However, the esterase activity of LTR-reverted mutant stamens was less than that of normal stamens at stages i to iii and at anthesis, but was not different at stages iv to vii.

Amylolytic activity

The total amylolytic activity of normal and mutant stamens was
not different at stages i and ii (Fig. 2). But thereafter, the
activity in normal stamens gradually increased through to stage vii,
whereas in the mutant it did not change throughout development, and
was significantly less than the normal. The mutant stamens also
possessed fewer α- and β-amylase isozymes than the normal (data not
presented).

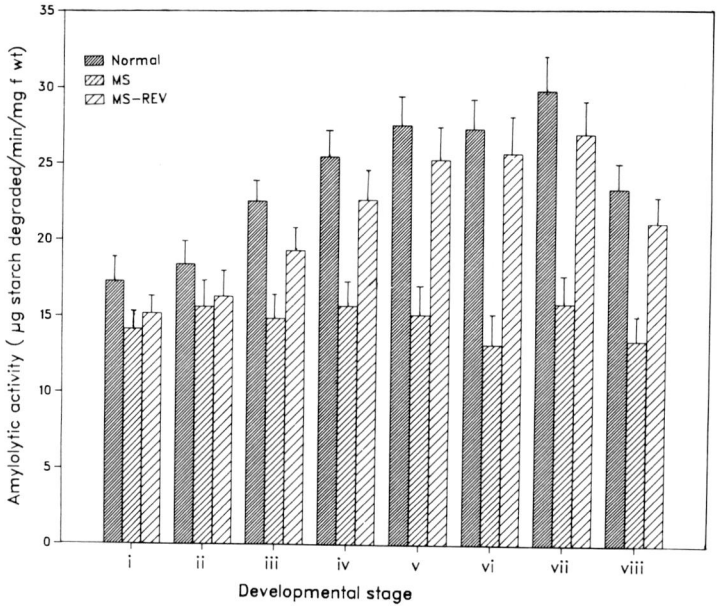

Fig. 2.   Amylolytic activity (µg starch degraded/min/mg fr. wt.) of
the normal, male sterile stamenless-2 mutant (MS) and low
temperature reverted mutant (MS-REV) stamens at eight (i to
viii) developmental stages.   Vertical bars represent
standard error.

The amylolytic activity (Fig. 2) and isozymes of amylases of
LTR-reverted mutant stamens (MS-REV) followed the same pattern as
the normal.

DISCUSSION

This study has shown that there are structural and enzymatic
aberrations during the ontogeny of sl-2/sl-2 mutant anthers. At the
histological level, the tapetal cells of the mutant were abnormally
large, contained large vacuoles, often divided to form a bilayer and
degenerated later than the normal.

The tapetum is a source of a variety of substances required for the normal pollen development (Bhandari 1984, Pacini et al. 1985). A delay in tapetum degeneration, associated presumably with a delay in the release of substances, could conceivably have an adverse effect on pollen development. One of the major effects in the sl-2/sl-2 mutant is evident in the lack of exine deposition in many microspores, leading to their degeneration. The importance of the precise timing of tapetum degeneration in relation to pollen development has been stressed by several workers (e.g. Vasil 1967; Frankel and Galun 1977; Bhandari 1984).

The overall activity of esterase was lower in the mutant than the normal throughout stamen development (Fig. 1). And as shown elsewhere, the mutant stamens also had fewer isozymes and were of lower intensity than the normal (Bhadula and Sawhney 1987). In particular, a major isozyme of esterase was observed in normal stamens at stages i to iii (when the tapetum was intact) and another one at stages v to viii (during pollen maturation). Both these isozymes were of low intensity in the mutant stamens (Bhadula and Sawhney 1987). Esterases have been localized in the tapetum at early stages, and later on the microspores during exine deposition (Vithanage and Knox 1976; Sawhney and Nave 1986). Although esterases are likely involved in a variety of functions, it is possible that some forms of it have a role in the polymerization of sporopollenin in the tapetum and its deposition in the exine (Bhadula and Sawhney 1987). Thus, the lower activity and low intensity of major isozymes of esterases in the sl-2/sl-2 mutant could well be related to the lack of exine deposition and eventual degeneration of many microspores.

The activity of amylases in the mutant was not different from the normal at stages i and ii, but from stage iii onwards, it was significantly greater in the normal stamens (Fig. 2). It should be noted that, in both lines, at stage iii, microspores separate from tetrads and pollen maturation begins.

It is well established that amylases breakdown starch and provide free sugars for the developing organs. The low activity and fewer isozymes of amylases in the mutant would, therefore, cause depletion of sugars, which are essential for pollen development (Stanley and Linskens 1974; Bhandari 1984). We have also analysed the free sugar content in mutant anthers and found it to be lower than the normal (unpublished data). The low level of free sugars could lead to pollen starvation, thus contributing to their abortion.

In some male sterile lines of _Brassica_ also, low levels of sugars were related to pollen starvation (Banga et al. 1984).

The development of pollen grains is a highly complex phenomenon involving, various tissue interactions and several interrelated processes occurring at precise times during stamen ontogeny. Any deviation or abnormality in one, or more than one, process would inevitably lead to pollen abortion. In the sl-2/sl-2 mutant, a delay in tapetum degeneration, large vacuoles in the tapetum, low activity and isozymes of esterase and amylases, and low level of available free sugars, all contribute to pollen degeneration. It should be stressed that these are only some of a large number of events that are involved in pollen development.

REFERENCES

Abbott AG, Ainsworth CC, Flavell RB (1984) Characterization of anther differentiation in cytoplasmic male sterile maize using a specific isozyme system (esterase). Theor Appl Genet 67: 469-473.

Banga SS, Labana KS, Banga SK (1984) Male sterility in Indian mustard (Brassica juncea (L.) Coss.) - a biochemical characterization. Theor Appl Genet 67: 515-519.

Bhadula SK, Sawhney VK (1987) Esterase activity and isozymes during the ontogeny of stamens of male fertile Lycopersicon esculentum Mill., a male sterile stamenless-2 mutant and the low temperature-reverted mutant. Plant Sci 52:187-194.

Bhandari NN (1984) The microsporangium. In: Embryology of Angiosperms. Ed: BM Johri, Springer-Verlag, Berlin, pp. 53-121.

Frankel R, Galun E (1977) Pollination mechanisms, reproduction and plant breeding. Springer-Verlag, Berlin.

Jones RL, Varner JE (1967) The bioassay of gibberellins. Planta 72:155-161.

Laser KD, Lersten NR (1972) Anatomy and cytology of microsporogenesis in cytoplasmic male-sterile angiosperms. Bot Rev 38: 425-454.

Pacini E, Franchi GG, Hesse M (1985) The tapetum: its form, function, and possible phylogeny in Embryophyta. Pl Syst Evol 149: 155-185.

Salas S, Cardemil L (1986) Multiple forms of α-amylase enzyme of Araucaria species of South America: A. araucana (Mol.) Koch and A. angustifolia (Bert.) O Kutz. Plant Physiol 81: 1062-1068.

Sawhney VK (1983) Temperature control of male-sterility in a tomato mutant. J Hered 74:51-54.

Sawhney VK, Bhadula SK (1988) Microsporogenesis in the normal and male sterile stamenless-2 mutant of tomato (Lycopersicon esculentum). Can J Bot (In press).

Sawhney VK, Greyson RI (1973) Morphogenesis of the stamenless-2 mutant in tomato. II. Modifications of sex organs in the mutant and normal flowers by plant hormones. Can J Bot 51:2473-2479.

Sawhney VK, Nave EB (1986) Enzymatic changes in post-meiotic anther development in Petunia hybrida. II. J Plant Physiol 125:467-473.

Stanley RG, Linskens HF (1974) Pollen. Springer-Verlag, Heidelberg.

Vasil IK (1967) Physiology and cytology of anther development. Biol Rev 42:327-373.

Vithanage HIMV, Knox RB (1976) Pollen-wall proteins: quantitative cytochemistry of the origins of intine and exine systems in Brassica oleracea. J Cell Sci 21:423-435.

Pollen Tube and Sperm Cells

The Pollen Tube: Motility and the Cytoskeleton

J. Heslop-Harrison
University College of Wales
Welsh Plant Breeding Station
Plas Gogerddan
Aberystwyth SY23 3EB, UK.

Throughout the 164 years since Giovanni Battisti Amici's first report of the phenomenon, the busy and seemingly purposeful movement of inclusions within pollen tubes has intrigued generations of observers; yet it is only within the last two decades that some understanding of the underlying mechanisms has begun to emerge, mainly in consequence of new evidence concerning the pollen tube cytoskeleton. My purpose in this short talk is to outline some of this recent progress.

Movement within the pollen tube is of course quite comparable with that to be seen in very many plant and animal cells, usually - if misleadingly - referred to as "cytoplasmic streaming". Yet there are certain features that set the pollen tube somewhat aside. Here movement occurs within a cylindrical cell that is continuously growing by extension at the apex, and it involves four more or less distinct activities: (a) circulation along the length of the tube of organelles, lipid bodies and other cytoplasmic inclusions; (b) continuous insertion of numerous polysaccharide-containing precursor particles into the forming wall at the growing tip; (c) forward passage of the vegetative nucleus, sometimes saltatory, which keeps it more or less in register with the growing apex but at some distance behind; and (c) migration along the tube of the generative cell or the gametes produced from it.

Analogy with other intracellular movements, animal and plant, suggests that the motive force is likely to be generated in association with cytoskeletal elements, either microtubules (MTs) or actin microfilaments (MFs). Transmission electron microscopy has been supplemented latterly by immunofluorescence methods using antibodies to tubulin for the localisation of MTs in pollen tubes, and fluorochrome-labelled phallotoxins have come into general use for the identification of actin MFs (Wulf et al. 1979). These new techniques use optical microscopy, and being

applicable to whole pollen tubes, they have facilitated investigation of the spatial organisation of the two components of the
cytoskeleton, in this respect complementing electron microscopy
of thin sections. Table 1 provides a summary of the principal
features as at present understood.

The role of MTs in determining the fusiform shape of
generative cells has been known for some years (Hoefert 1969;
Burgess 1970; Sanger and Jackson 1971a, 1971b). A more recent
survey by Cresti et al. (1984) suggests that a cage- or
basket-like structure made up of peripheral MT bundles is
widespread - probably indeed universal - among angiosperm
generative cells, a view supported also by various
immunofluorescence studies (Derksen et al. 1985; Pierson et al.
1986; Tiezzi et al. 1986). The gametes produced from the
generative cell have a similar MT cytoskeleton (Cass 1974).

Franke et al. (1972) investigated the MT cytoskeleton of the
vegetative cell of the pollen tube by electron microscopy using
standard chemical fixation methods. They reported that MTs
occurred mainly in the peripheral cytoplasm in longitudinal
groups of five or six, those in each group being linked by
cross-bridges 28 nm in length, with the outer linked also to the
plasmalemma. Electron microscopy using physical fixation
techniques likely to preserve cellular structure more faithfully
than chemical fixation has confirmed and extended these earlier
findings (Lancelle et al. 1987), and further clarification of the
3-dimensional organisation of the MT cytoskeleton has come from
immunofluorescence localisation (Derksen et al. 1985; Tiezzi et
al. 1986; Raudaskoski et al. 1987).

The first substantive report of actin MFs in the pollen tube
was by Condeelis (1974), who showed that extended fibrils visible
both in the living tube and in extracted protoplasts were
composed of aggregates of MFs. Actin was identified unequivocally
in the fibrils by the characteristic arrow-head binding pattern
of heavy meromyosin. The availability of fluorochrome-tagged
phallotoxins has provided the means for examining the
distribution of these MF bundles in whole pollen tubes (Perdue
and Parthasarathy 1985; Pierson et al. 1986); and isolated
fibrils, corresponding to those described by Condeelis (1974),
have also been shown to bind phalloidin (Heslop-Harrison et al.
1986). Furthermore, fusiform bodies composed of aggregates of

Table 1. Characteristics of cytoskeletal elements in the
        pollen tube: a summary

MICROTUBULES

(a) Generative cell and gametes. Involved in initial cell definition and shaping; present in longitudinally disposed bundles beneath the wall of the fusiform mature cells, forming a cage
enclosing the nucleus. The complex varies in length during tube
growth, probably through relational movement of the MTs.
Evidence: TEM; immunofluorescence.

(b) Vegetative cell. No organised MTs present in ungerminated
pollen. Longitudinal arrays consisting of bundles of MTs underlying the plasmalemma are formed from the onset of cylindrical
growth from initiation sites near the tube apex. Scattered MTs
occur deeper in the cytoplasm. The peripheral MT cytoskeleton
persists throughout tube extension, implying that there is no
"treadmilling". Evidence: TEM: immunofluorescence; dry-cleaving.

MICROFILAMENTS

(a) Generative cell and gametes. No substantive reports.

(b) Vegetative cell. Present in numerous compact fusiform or
filamentous aggregates throughout the cytoplasm in the ungerminated pollen grain. Longitudinally disposed fibrils composed of
several microfilaments are produced from the beginning of extension growth and extend thereafter throughout the cytoplasm; their
anchorage points vary during tube extension, so that the system
is highly labile. Heavy meromyosin binding indicates that actin
is uniformly polarised in each bundle. Single microfilaments are
associated with and linked to the microtubules adjacent to the
plasmalemma. Evidence: TEM; immunofluorescence; binding of
labelled phalloidin; heavy meromyosin labelling.

---

closely apposed microfilaments have been identified by electron microscopy as an apparently universal - and plentiful - constituent of ungerminated pollen grains (Cresti et al. 1986 and in preparation), and again the presence of actin has been demonstrated by phalloidin binding (Heslop-Harrison et al., 1986).

Franke et al. (1972) and Condeelis (1974) pointed to the likelihood that the microfilament bundles present in pollen tubes were concerned with the movement of organelles and cytoplasmic masses. Condeelis (1974) showed that where the fibrils were well defined in older parts of the tube the movement associated with any one was always unidirectional, predictable from the fact that the actin microfilaments in each are uniformly polarised in the one direction as judged from heavy meromyosin binding. Recently further information about the movement has become available (Heslop-Harrison and Heslop-Harrison 1987; 1988). Individual organelles may travel along single fibrils in the living tube at different velocities, not infrequently passing one another, indicating that the propelling force is generated between components of the fibril and each body individually, without an intervening mass carrier. In a paper presented at the XIVth International Botanical Congress in Berlin in 1987, Shimmen reported that in ATP-containing medium organelles derived from pollen tubes could be caused to move along exposed MF bundles of the giant cells of Chara, behaving in this respect like the organelles of Chara itself in similar experiments. The interpretation to be placed upon this is that the pollen-tube organelles are associated with myosin-like molecules, and that through this agency they are capable of being assembled into a functional actomyosin system with any suitable MF partner. Preliminary experiments aimed at reconstituting an active extracellular system using pollen-tube organelles and the presumed pollen-tube microfilaments themselves have so far met with no success, but in the light of Shimmen's and other results this will no doubt soon be achieved.

Another approach is offered by the use of inhibitors. Franke et al. (1972) and Mascarenhas and Lafountain (1972) showed that cytoplasmic streaming in pollen tubes, assessed by the movement of organelles and other inclusions, was arrested by cytochalasin B (CB), one of a family of fungal metabolites with inhibitory effects on various MF-mediated processes. Franke et al. (1972) found that CB was effective in halting movement in concentrations as low as 0.1 μg/ml. However, when the tubes were transferred to fresh culture medium after this treatment, cyclosis of organelles was re-established within 30 min.

The time course of events in an individual pollen tube following treatment with CB at 5 μg/ml is shown in Table 2. It is known that the cytochalasins readily penetrate cell membranes, and that this must be so with pollen tubes is shown by the rapidity with which the first sign of interference with normal cell activity appeared in this experiment. Table 2 also records structural changes in the cytoplasm following upon the penetration of the drug. The pattern of longitudinal striations seen in the growing tube was progressively lost, and the cytoplasm assumed a beaded or pebbled appearance. In conformity with the findings of Franke et al. (1972) with lily pollen tubes, movement began again after transfer to fresh growth medium. In a further experiment, continuous video records showed that in individual blocked tubes circulatory movement of organelles could be resumed in as short a period as 120 s when preparations were infiltrated with a medium containing 1 mM ATP and 1 mM $MgCl_2$.

Cytochalasins rapidly arrest the apical extension of pollen tubes, and this is associated with the accumulation of the wall precursor bodies (Picton and Steer 1981), which no longer show a net vectorial movement. Concentrations that bring organelle circulation to a stop also arrest the migration of the vegetative nucleus through the tube just as abruptly. The record summarised in Table 2 suggests that the response of the generative cell is more equivocal; in this instance, after a brief acceleration of acropetal movement, the cell began a slow basipetal migration, sustained for a further 60 sec. We have observed continued slow movement of the generative cell after CB treatment in other tubes of this species, but whether such a reaction is generally to be found remains to be established.

While many reports have suggested that CB interferes with the function of the the MF cytoskeleton (Kamiya 1981), the nature of the disturbance it provokes remains obscure. CB in concentrations inhibiting intracellular movement does not depolymerise individual actin MFs (Forer et al. 1972), nor, apparently, does it radically alter the structure of the MF bundles associated

Table 2. Effect of cytochalasin B on the movement of the generative cell (GC) in a 70 min pollen tube of an _Iris_ cv. Cytochalasin at 5 µg/ml in growth medium injected into the culture preparation at zero time.

Time (sec.)	Comment
35	Normal movement of GC and organelles; linear striations apparent in the cytoplasm
45	Acropetal movement of the GC briefly enhanced
60	Organelle movement slowing; GC halts
80	GC begins basipetal movement; cytoplasmic striations disappear
145	GC continues basipetal movement; cytoplasm beaded; organelle movement restricted to irregular local excursions
200	GC continues slow basipetal migration; no appreciable organelle movement.

(unpublished data)

Table 3. Effect of the anti-microtubule agent nocodazole on the rates of movement of vegetative nuclei and generative cells through the pollen tubes of _Endymion nonscriptus_, measured from the time of emergence of 50% of the vegetative nuclei

----

	Mean overall rates	
	Generative cells	Vegetative nuclei
Control	0.156 µm/s	0.185 µm/s
Growth medium with 5 µg/m nocodazole	0.160 µm/s	0.097 µm/s

(unpublished data)

with organelle movement in the characeous algae (Bradley, 1973;
Nothnagel et al. 1981). These bundles are comparable with the
fibrils seen in the cytoplasm of pollen tubes (Heslop-Harrison
and Heslop-Harrison 1988b), although it remains to be established
whether these also survive CB treatment.

Inhibitors have also been used to investigate the possible
involvement of the MT cytoskeleton of the pollen tube in the
movement of the various inclusions. Franke et al. (1972) reported
that colchicine concentrations which led to the complete
disappearance of MTs from lily pollen tubes affected neither
germination nor cyclosis.In experiments with Endymion nonscriptus
we found that colchicine at 1 mM did not radically affect
germination or tube growth, nor prevent the movement of the
generative cell and vegetative nucleus into and through the tube.
Nevertheless, anti-MT agents do have some significant effects.
Notably, the order of precedence of the generative cell and
vegetative nucleus during early growth of the tube of E.
nonscriptus is affected by colchicine at MT-eliminating
concentrations. In one experiment, the generative cell was found
to lead in 20% of treated tubes as compared with some 1% in
controls. Repetition of these experiments with the synthetic
anti-MT agent, nocodazole, in near saturation concentrations
known to eliminate MTs from a wide range of cells without
observable secondary effects (Janssen Pharmaceutica Cell Biology
Note R 17 934), allowed a more detailed analysis. Once again
organelle cyclosis was unaffected. As with colchicine, the order
of migration of generative cells and vegetative nuclei was
modified, but a comparison of the actual rates of movement (Table
3) shows that this was mainly due to a retardation of the
vegetative nuclei, which moved along the tubes at about half of
the rate achieved in controls.

In summary, then, we see that while as yet no specific
function can be attributed to the MT cytoskeleton of the pollen
tube, cytological evidence of various kinds points strongly to
the involvement of the actin cytoskeleton in the circulatory
movement of the organelles and the other lesser inclusions, and
probably also in the migrations of the vegetative nucleus,
generative cell and gametes. The nature of the molecular
interactions that lead to force generation may be surmised, but
most of the critical details have yet to be worked out. It is
obviously important to establish whether myosin or its equivalent

is associated with the moving structures. In the light of the
recent proof by Turkina et al. (1987) that myosin-like molecules
are present in higher-plant vascular tissues this would seem now
to be but a matter of time.

REFERENCES

Bradley MO (1973) Microfilaments and cytoplasmic streaming:
inhibition of streaming with cytochalasin. J Cell Sci 12:327-343

Burgess J (1970) Cell shape and mitotic spindle formation in the
generative cell of _Endymion nonscriptus_. Planta 95:72-85

Cass DD (1973) An ultrastructural and Nomarski study of the
sperms of barley. Can J Bot 51:601-605

Condeelis JS (1974) The identification of F-actin in the pollen
tube and protoplast of _Amaryllis belladonna_. Exp Cell Res
88:435-438

Cresti M, Colni F, Kapil RN (1984) Generative cells of some
angiosperms with particular emphasis on their microtubules. J
Submicrosc Cytol 16:317-326

Cresti M, Hepler PK, Tiezzi A, Ciampolini F (1986) Fibrillar
structures in _Nicotiana_ pollen: changes in ultrastructure during
pollen germination and tube emission. In: Mulcahy DL, Mulcahy GB,
Ottaviano E (eds) Springer, Berlin, Heidelberg, New York, p 283

Derksen J, Pierson ES, Traas JA (1985) Microtubules in vegetative
and generative cells of pollen tubes. Eur J Cell Biol 38:142-148

Forer A, Emmersen J, Behnke O (1972) Cytochalasin B: does it
affect actin-like filaments? Science 175:776

Franke WW, Herth W, Van der Woude WJ, Morre DJ (1972) Tubular and
filamentous structures in pollen tubes: possible involvement as
guide elements in protoplasmic streaming and vectorial migration
of secretory vesicles. Planta 105:317-341

Heslop-Harrison J, Heslop-Harrison Y (1987) An analysis of gamete
and organelle movement in the pollen tube of _Secale cereale_ L.
Plant Sci 5:20-213

Heslop-Harrison J, Heslop-Harrison Y (1988) Organelle movement
and fibrillar elements of the cytoskeleton in the angiosperm
pollen tube. Sex Plant Rep 1:16-24

Heslop-Harrison J, Heslop-Harrison Y, Cresti M, Tiezzi A,
Ciampolini F (1986) Actin during pollen germination. J Cell Sci,
86:1-8

Hoefert LL (1969) Fine structure of sperm cells in pollen grains of Beta. Protoplasma 68:237-240

Kamiya N (1981) Physical and chemical basis of cytoplasmic streaming. Ann Rev Plant Physiol 32:205-36

Lancelle SA, Cresti M, Hepler PK (1987) Ultrastructure of the cytoskeleton in freeze-substituted pollen tubes of Nicotiana alata. Protoplasma 140:141-150

Mascarenhas JP, Lafountain J (1972) Protoplasmic streaming, cytochalasin B and the growth of the pollen tube. Tissue Cell 4:11-14

Nothnagel EA, Barak LS, Sanger JW, Webb WW (1981) Fluorescence studies on the modes of cytochalasin B action on cytoplasmic streaming in Chara. J Cell Biol 88:364-372

Perdue TD, Parthasarthy MV (1985) In situ localisation of F-actin in pollen tubes. Eur J Cell Biol 39:13-20

Pierson ES, Derksen J, Traas JA (1986) Organisation of microfilamnts and microtubules in pollen tubes grown in vitro or in vivo in various angiosperms. Eur J Cell Biol 41:14-18

Picton JM, Steer MW (1981) Determination of secretory vesicle production rates by dictyosomes in pollen tubes of Tradescantia using cytochalasin D. J Cell Sci 49:261-272

Raudaskoski M, Astrom H, Pertyila K, Virtanen I, Louhelainen J (1987) Role of the microtubule cytoskeleton in pollen tubes: an immunological and ultrastructural approach. Biol Cell 61:177-188

Sanger JM, Jackson WT (1971a) Fine structure study of pollen development in Haemanthus katherinae Baker. I. Formation of vegetative and generative cells. J. Cell Sci. 8:280-301

Sanger JM, Jackson WT (1971b) Fine structure study of pollen development in Haemanthus katharinae Baker. II. Microtubules and elongation of the generative cells. J Cell Sci 8:303-315

Tiezzi A, Cresti M, Ciampolini F (1986) Microtubules in Nicotiana pollen tubes: ultrastructural, immunofluorescence and biochemical data. In: Cresti M, Dallai R (eds) Biology of reproduction and cell motility. University of Siena Press, p 87

Turkina MV, Kulikova AL, Sokolov, OI, Bogatyrev VA, Kursanov AL (1987) Actin and myosin filaments from conducting tissues of Heracleum sosnowskyi. Plant Physiol Biochem 25:689-696

Wulf EA, Deboben FA, Bautz H, Faulstick H, Wieland T (1979) Fluorescent phallotoxin, a tool for the visualisation of cellular actin. Proc Natl Acad Sci USA 76:4498-4502

Presence of Kinesin in Tobacco Pollen Tube

MOSCATELLI A., TIEZZI A., VIGNANI R., CAI G., BARTALESI A. and CRESTI M.
Dipartimento di Biologia Ambientale
Università degli Studi di Siena
Via Mattioli n. 4
53100 Siena
ITALY

SUMMARY

The pollen tube is a biological structure in which an intense cytoplasmic movement is obserbed, and cytoskeletal elements appear to be directly involved in this process. Microfilaments are responsible for organelle movement and microtubules, together with other specialized polypeptides could be involved in the transport of vesicles.

Here we report on the presence of kinesin in pollen tubes. Kinesin is a force-generating protein that, as shown for animal cells, cooperates with microtubules in the movement and driving of vesicles. The presence of kinesin in pollen tubes of Nicotiana was demonstrated by immunofluorescence, using a monoclonal antibody to mammalian kinesin. A strong fluorescence reaction was specifically evident at the tube apex.

INTRODUCTION

Organelle and vesicle movement is a complex process hitherto mainly studied in animal cells. Optical and ultrastructural investigations and biochemical evidence indicate that cytoskeletal structures are directly involved in the process and although further investigations are needed to clarify the molecular interactions among vesicles, organelles and cytoskeletal structures, it has been shown that some movements are ATP-dependent and associated with the microtubular system (Sheetz et al. 1986).

The plant cytoskeleton is characterized essentially by two systems, microtubules and microfilaments. It remains to be established what associated or interacting specialized proteins may mediate, or be directly involved in cytoskeletal functions. The plant organelle and vesicle movements need further biochemical investigations to discover whether they are ATP-dependent, and thus similar to observed in animal cells.

Heslop-Harrison and Heslop-Harrison (1988) have recently described of various features organelle movement in the pollen tube of Iris pseudacorus. Their findings show that fibrillar elements corresponding to those described by Condeelis (1974) and shown to be composed of aggregates of actin microfilaments are involved in the process, and that different organelles move with different rates along the tube. At present we know little about the biochemical mechanisms driving the movement, but since it is microfilament-associated it does not seem to be related to the system observed in animal cells.

Some pollen tubes are very rich in microtubules but very little it has been reported about their function. In order to investigate a possible involvement in vesicle movement, we report here some preliminary results concerning the presence of kinesin, a microtubule-based "motor", in Nicotiana pollen tubes.

MATERIALS AND METHODS

Preparation of monoclonal antibodies to calf brain kinesin

Kinesin was prepared following the procedure of Vale et al. (1985) using a concentration of 0.5 mM AMP-PNP as suggested by Amos (1987) and Sepahrose CL 2B for the final step of gel chromatography.

Fractions containing pure kinesin were analyzed by SdS-PAGE using a gel gradient 4-16% following the procedure of Laemmli (1970) Balb/C mice were immunized with purified kinesin. The first injection consisted of 60 ug of antigen. After 21 day each mouse was boosted with 30 ug of antigen. Two weeks later the mice were injected with 15 ug of antigen and boosted again one week later with a further 15 ug. For the first injection kinesin was mixed with complete Freund's adjuvant (1mg/ml). For the others the antigen was mixed with incomplete Freund's adjuvant. Three days after the last injection mice were sacrified and spleen cells were fused with Sp2/0 myeloma cells (Galfrè and Milstein 1981).

Hybridoma supernatants were first tested by a dot blot procedure and further processed by western blot techniques (Towbin et al., 1979). Supernantants were tested by radioimmunoassay in a plastic tube test using a I^{125} goat anti-mouse as secondary antibody.

Immunofluorescence procedures

Pollen tubes of Nicotiana tabacum were grown for four hours in appropriate medium (Brewbaker and Kwack, 1963) and processed for immunofluorescence as previously described (Tiezzi et al. 1986). They were treated with hybridoma supernatants and a polyclonal antibody to tubulin. FITC (fuorescein isothiocyanate) rabbit anti-mouse and TRITC (tetrarhodamineisothiocyanate) goat anti-rabbit were used as secondary antibodies.

RESULTS

SdS PAGE of purified kinesin after the gel filtration process is shown in Fig. 1. The electropherogram clearly shows polypeptides whose molecular weights range between 45 and 116 Kd. Purified kinesin after gel filtration was used for dot-blot for the first screening of hybridoma supernatants. Some positive supernatants were then tested by western-blot procedures to determine polypeptide specificity of the antibodies, but no reactions were detected. Dot blot positive supernatants were than tested by radioimmunoassay. Two supernatants, named K41 and K71, showed high affinity to purified undenatured kinesin.

K71 supernatant was applied to germinated pollen tubes for immunofluorescence investigations. An intense staining was observed in the tube apex and its proximity (Fig. 2). The staining is not diffuse, and seems specifically localized on globular particles present in the sub-apical region and in the tube tip. Double immunofluorescence show different patterns of staining between the polyclonal antibody to tubulin and K71 supernatant (Figs. 3a - 3b). Generative cell microtubules are stained by the polyclonal, whereas no microtubules were detected

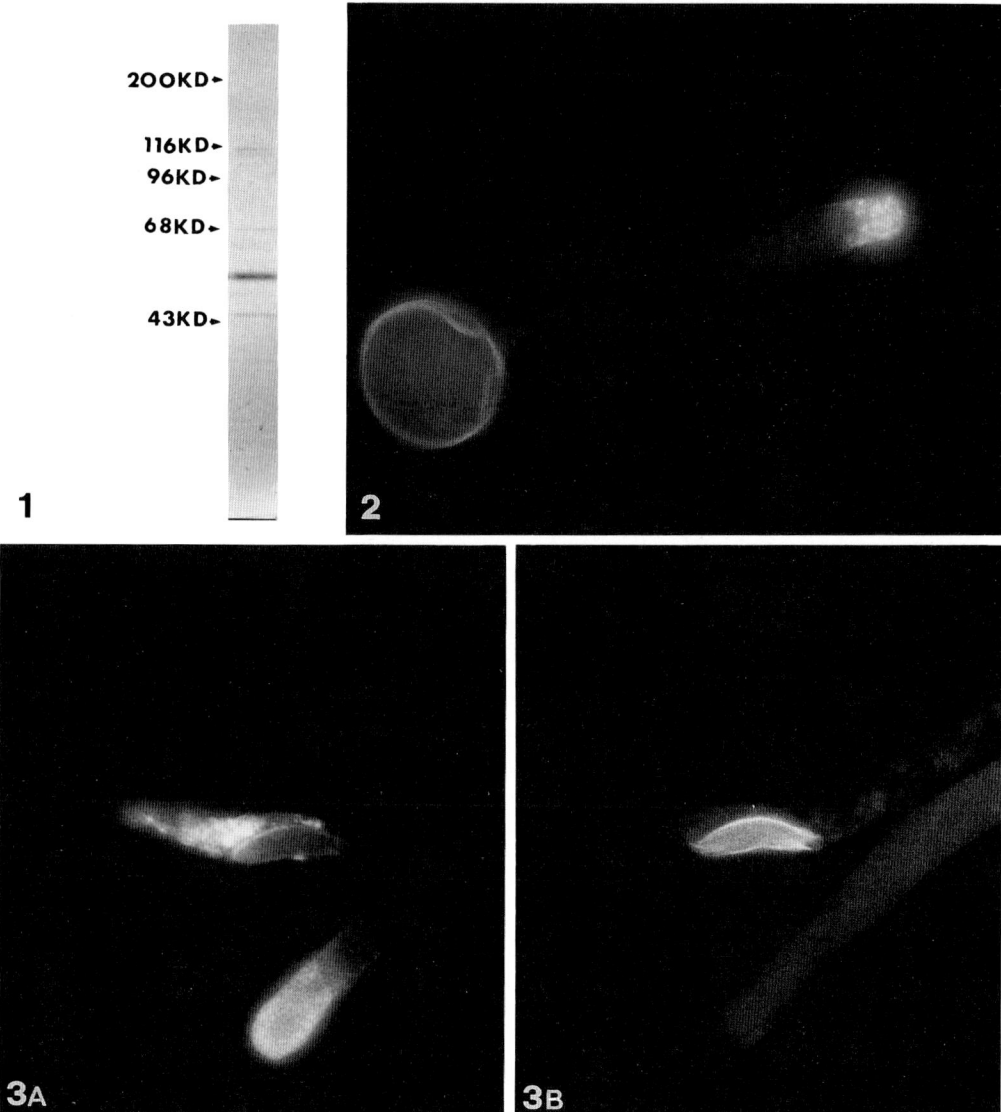

EXPLANATION OF THE FIGURES

Fig. 1 - SdS-PAGE of purified kinesin after gel filtration. Kinesin molecule is
formed by several peptides.

Fig. 2 - K71 supernatant on pollen tube. The staining at the apex is specific for
some globular particles. x 230

Fig. 3 - Double immunofluorescence. a.K71 supernatant stains the tube apex but not
the generative cell. b. Generative cell stained by a polyclonal antibody
to tubulin. x 230

in proximity of the tube apex, and no molecular relationship could be observed between microtubules and the structures stained by the K71 supernatant.

DISCUSSION

Cytoskeletal structures have been described in many pollen tubes from different plants and it has been reported that cytoskeletal proteins are directly involved in the process of cytoplasmic streaming (Lancelle et al., 1987; Pierson et al., 1985). Recently Heslop-Harrison and Heslop-Harrison (1988) have reported on organelle movement in the pollen tube of Iris Pseudacorus where bundles of microfilaments are directly involved in the process. Characteristically, organelles move with different rates, suggesting perhaps different molecular mechanisms responsible in the driving of single organelles. Apparently microtubules have no function in this system (Heslop-Harrison et al., 1988).

In animal cells microtubules are involved together with specialized polypeptides in vesicle movement. In growing pollen tube a lot of vesicles are present, and we are investigating if pollen tube microtubules share a similar role in vesicle movement.

Our K71 monoclonal antibody to calf brain kinesin show an intense staining at the apex of Nicotiana pollen tube where ultrastructural investigations have shown the presence of a large number vesicles. By immunofluorescence we are not able to identify what structures interact with the antibody. We can only say that K71 staining is very selective and interactions with other structures are not present along the tube. Double immunofluorescence does not allow one to relate microtubule distribution to K71 staining structures since microtubules were not observed at the tube tip. Furthermore double immunofluorescence procedures show that generative cell is very intensely stained by the polyclonal antibody to tubulin, but no reaction was detected by the K71 antibody, suggesting that kinesin can not be involved in generative cell behaviour.

Our observations are of course preliminary in nature and further biochemical and ultrastructural investigations are needed.

ACKNOWLEDGMENT

This research was carried out in the framework of contract n° BAP-0204-I of the Biotechnology Action Programe of the Commission of the European Communities.

REFERENCES

AMOS LA (1987), Kinesin from pig brain studied by electron microscopy. J Cell Sci, 87: 105-111

BREWBAKER JL and KWACK BH (1963), The essential role of calcium ion in pollen germination and pollen tube growth. Amer J Bot 50: 859-865

CONDEELIS JS (1974), The identification of F-actin in the pollen tube and protoplasts of Amaryllis belladonna. Exp Cell Res 88: 435-439

GALFRE' G and MILSTEIN C (1981), Preparation of monoclonal antibodies: strategies and procedures. Methods in Enzymol 73: 3-46

LAEMMLI UK (1970) Cleavage of structural proteins during the assembly of the head of bacteriophage T 4. Nature 227: 680-685

LANCELLE SA, CRESTI M and HEPLER PK (1987), Ultrastructure of the cytoskeleton in freeze-substituted pollen tubes of Nicotiana alata. Protoplasma **140**: 141-150

HESLOP-HARRISON J and HESLOP-HARRISON Y (1988), Organelle movement and fibrillar elements of the cytoskeleton in the angiosperm pollen tube. Sex Plant Reprod **1**: 16-24

HESLOP-HARRISON J, HESLOP-HARRISON Y, CRESTI M, TIEZZI A and MOSCATELLI A (1988), Cytoskeletal elements, cell shaping and movement in the angiosperm pollen tube. J Cell Sci (In press).

PIERSON ES, DERKSEN J and TRAAS JA (1985), Organization of microfilaments and microtubules in pollen tubes grown in vitro or in vivo in various angiosperms" Eur J Cell Biol **41**: 14-18

SHEETZ MP, VALE R, SCHNAPP B, SCHROER J and REESE T (1986), Vesicle movements and microtubules-based motors. J Cell Sci **5**: 181-188.

TIEZZI A, CRESTI M and CIAMPOLINI F (1986), Microtubules in Nicotiana pollen tubes: ultrastructural, immunofluorescence and biochemical data. In: Biology of Reproduction and cell motility in plants and animals (Ed. M. Cresti and R. Dallai) pp. 87-94, University of Siena Press

TOWBIN HT, STAEHELIN T and GORDON J (1979), Electrophoretic transfer of proteins from polycrylamide gels to nitrocellulose sheets: procedure and some applications". Proc Natl Acad Sci, USA **76**: 4350-4354

VALE RD, REESE TS and SHEETZ MP (1985), Identification of a novel force-generating protein, kinesin, involved in microtubules-based motility. Cell **42**: 39-50

Analysis of the Structure, Organization and Role of Cytoskeleton During Pollen Germination and Tube Growth in *Pyrus communis* L.

Suresh C. Tiwari and Vito S. Polito
Department of Pomology
University of California
Davis, CA 95616
USA

Behavior of actin microfilaments and microtubules was followed in the vegetative cell of germinating pollen and pollen tubes of Pyrus communis using rhodamine-conjugated phalloidin to visualize actin, indirect immunofluorescence to localize microtubules, and conventional and freeze-substitution electron microscopy. Several previous investigations have focused on the cytoskeleton of pollen tubes at the ultrastructural level (Franke et al., 1972; Miki-Hirosige & Nakamura, 1982; Derksen et al., 1985; Tiezzi et al., 1986 Lancelle et al., 1987) and at the light microscope level (Derksen et al., 1985; Perdue & Parthasarathy, 1985; Pierson et al., 1985; Tiezzi et al., 1986) using fluorescent probes. However, questions remain about the precise organization of the pollen tube cytoskeleton, and there has been scant attention to dynamics of cytoskeletal elements during the periods of pollen hydration and activation that precede germination.

POLLEN ACTIVATION. By using the actin-specific probe, rhodamine-conjugated phalloidin (RP), to study the dynamics of actin microfilaments, we identified several stages of actin organization as dehydrated pollen grains become activated and eventually germinate. In inactivated pollen, actin occurs as course granules around the vegetative nucleus and as circular profiles that are distributed throughout the cytoplasm. With activation, however, the organization becomes clearly fibrillar with intermediate stages evident during early stages (Table 1). Filaments abound in the perinuclear zone, some of them extending well into the cortical areas of the vegetative cells. Parallel arrays of long and thin filaments occur in the cortical areas with two distinct domains of orientation. Those beneath non-apertural areas of the pollen wall are aligned parallel with respect to the polar axis of the pollen grain. Beneath the apertural surfaces, microfilament arrays are oriented at $45°$ to $90°$ angle with respect to the polar axis. With further activation, the microfilaments become organized primarily into large bundles that traverse between pollen wall apertures. Eventually, prior to germination, the microfilaments become confined to a single aperture.

Table 1. Percentage of pollen grains showing different patterns of RP-labelling and percent germination at intervals following incubation in pollen growth medium.

Time (min)	Circular Profiles	Intermediate Patterns	Filamentous Arrays	Perinuclear Label Only	Germi- nation
0	85%	2%	4%	9%	0%
5	40%	36%	18%	5%	0%
10	20%	17%	57%	8%	0%
20	11%	12%	64%	12%	0%
30	7%	11%	73%	7%	17%
40	2%	9%	82%	6%	30%
50	2%	8%	80%	9%	33%
60	0%	9%	82%	9%	40%

Microtubules were localized by using a monoclonal antibody raised against β-subunit of chick-brain tubulin and indirect immunofluorescence with FITC-conjugated secondary antibody. Conventional permeabilization procedures involving incubation with wall-degrading enzymes proved unsuitable for the germinating pollen; however, a freeze-fracture technique was found to produce readily permeable cells. The inactivated pollen is characterized by the presence of few cortical microtubules located at the apertural sites. With activation a system of randomly organized microtubules develops in the cortical areas. Focal sites, which probably represent microtubule organizing centers, are often evident. Gradually, with pollen germination, the microtubules assume an orientation that is predominately perpendicular to the long axis of the developing pollen tube.

GERMINATING POLLEN AND POLLEN TUBES. Organization of the cytoskeleton in germinating pollen and pollen tubes was investigated at the ultrastructural level by conventional (chemical fixation and dehydration) and freeze-substitution electron microscopy. Conventionally fixed cells rarely showed microtubules in the pollen grain; nor could microtubules be seen in the cytoplasm of the pollen tube. Longitudinally aligned cortical microtubules, however, were preserved in the pollen tube. No microfilaments were detected. Freeze-substitution electron microscopy, on the other hand, revealed numerous microtubules in the cortical or cytoplasmic locations in the pollen grain and the pollen tube. Additionally, freeze-substitution demonstrated the presence of cortical and cytoplasmic microfilaments. Cortical microfilaments occur independent of as well as invariably in association with the microtubules. Microfilaments were also found associated with elements of endoplasmic reticulum, cylindrical vacuoles, dictyosomal vesicles and the vegetative nucleus. Short, randomly oriented, single microfilaments occur in the apical region of the pollen tube where they lie amid the vesicles.

Based on the immunofluorescence pattern of microtubules, four zones could be distinguished in germinated pollen: 1) The apical region of the pollen tube, which is devoid of cortical or cytoplasmic microtubules. 2) The subapical region, where principally mitochondria reside, and where microtubules are also absent. 3) The distal part of the pollen tube, where numerous long, closely spaced, longitudinally aligned, cortical microtubules and several cytoplasmic microtubules are found. 4) In the pollen grain, where numerous microtubules occur in the cortex and cytoplasm. These microtubules, especially those in the cytoplasm, appeared oriented in the general direction of the germination aperture. RP-labelling showed the presence of microfilaments in all the locations where freeze-substitution had indicated their presence.

DISCUSSION. It is concluded that germinating pollen possesses a complex cytoskeleton that undergoes spectacular changes during development. In the inactivated pollen (a state which, because of technical reasons, is as close as one could get to the dry pollen) the RP-positive bodies, seen as circular profiles, provide a mechanism for the storage of actin that is, with subsequent activation, drawn upon for the formation of fibrillar arrays. The surface of the vegetative nucleus appears to have a controlling role in the organization of actin arrays. In late stages of activation, a polarized distribution of microfilaments, predicting the direction of future growth, is achieved.

Microtubules have generally been relegated to having an unimportant role during pollen germination. A previous investigation (Derksen and Traas, 1985) indicated that pollen germination proceeded in spite of incubation in colchicine, a microtubule inhibitor. However, in view of the changes observed in the organization of microtubules in P. communis, it seems likely that they do play an important, although as yet undetermined, role.

In the pollen tube, the cortical microfilaments and microtubules, together with the plasma membrane for a structurally integrated cytoskeleton. Cytoskeletal elements also occur in the cytoplasm associated with various organelles and presumably are responsible for their intracellular transport. Furthermore, a comparison of various techniques employed to study cytoskeleton structure in pollen tubes clearly indicates that conventional procedures of electron microscopy are inadequate for the preservation of these elements, but the technique of freeze-substitution preserves them faithfully. Additionally, although a quantitative comparison has not be made, it is our view that the protocol we have followed for the fixation of microfilaments, prior to RP labelling, provides qualitatively identical data to that obtained via the technique of freeze-substitution electron microcopy. Investigations employing cytochalasins, inhibitors of actin, and oryzalin, an anti-microtubular drug, are currently in progress to determine more precisely how these two elements control the development during pollen germination and tube growth.

REFERENCES

Derksen J, Pierson ES, Traas JA (1985) Microtubules in vegetative and generative cells of pollen tubes. Eur. J. Cell Biol. 38: 142

Derksen J, Traas JA (1985) Growth of tobacco pollen tubes in vitro; effects of drugs interfering with the cytoskeleton. In: Willemse MTM, Van Went JL (eds) Proceedings, 8th International Symposium on Sexual Reproduction in Seed Plants, Ferns, and Mosses. Pudoc, Wageningen, p 64

Franke WW, Herth W, Van Der Woude WJ, Morré DJ (1972) Tubular and filamentous structures in pollen tubes: possible involvement as guide elements in protoplasmic streaming and vectorial migration of secretory vesicles. Planta 105: 317-341

Lancelle SA, Cresti M, Hepler PK (1987) Ultrastructure of the cytoskeleton in freeze-substituted pollen tubes of Nicotiana alata. Protoplasma 140: 141-150

Miki-Hirosige H, Nakamura S (1982) Process of metabolism during pollen tube wall formation. J. Electron Microscopy 31: 51-62

Perdue TD, Parthasarathy, MV (1985) In situ localization of F-actin in pollen tubes. Eur. J. Cell Biol. 39: 13-20

Pierson ES, Derksen J, Traas JA (1986) Organization of microfilaments and microtubules in pollen tube grown in vitro, or in vivo in various angiosperms. Eur. J. cell Biol. 41: 14-18.

Tiezzi A, Cresti M, Ciampolini F (1986) Microtubules in Nicotiana pollen tubes: ultrastructural, immunofluorescence and biochemical data. In: Cresti M, Dallai R (eds) Biology of Reproduction and Cell Motility in Plants and Animals. University of Siena, p 87

The Cytoskeletal Apparatus of the Generative Cell in Several Angiosperm Species

TIEZZI A., MOSCATELLI A., CIAMPOLINI F., MILANESI C., MURGIA M. and CRESTI M.

Dipartimento di Biologia Ambientale
Università degli Studi di Siena
Via Mattioli 4
53100 Siena
ITALY

SUMMARY

Generative cell cytoskeletal apparatus from several angiosperms has been investigated both by fluorescence procedures and ultrastructural investigations. In all analyzed plants, microfilaments were not present whereas a basket-like structure formed by microtubules surrounded the generative cell. Microtubules were connected by lateral projections resembling those observed in the axoneme. Microtubules and associated lateral structures could be cooperating together in generative cell reshaping during its movement along the tube.

INTRODUCTION

The cytoskeletal elements of the pollen tube have been identified. Ultrastructural, immunohistochemical and biochemical investigations have shown the presence of two distinct filamentous systems essentially formed by microfilaments and microtubules in the vegetative cell (Cresti et al. 1984; Pierson et al. 1986; Tiezzi et al., 1986; Heslop-Harrison et al. 1987; Lancelle et al. 1987; Heslop-Harrison 1987; Palevitz and Cresti 1988a, 1988b; for a complete review see Heslop-Harrison and Heslop-Harrison 1988). Both systems seem to have a characteristic cytoplasmic distribution and could be involved in different specialized functions. However it remains to be clarified wether any interacting or specialized polypeptides are present and the functional roles that they could possibly play in the cytoskeletal machinary.

The generative cell cytoskeleton is not characterized as well as the vegetative cell cytoskeleton. Ultrastructural and immunohistochemical observations have been done but, because of the small size of the cell, biochemical investigations have not been carried out.

In this paper we present a comparative study, carried out by immunofluorescence and electron microscopical procedures on the generative cell cytoskeletal apparatus of different angiosperms in order to contribute to the characterization of the cytoskeletal structures that are involved in the complex processes of cell shaping and movement.

MATERIALS AND METHODS

Immunofluorescence

Fresh pollen grains from Leucojum verum, Malus domestica and Nicotiana tabacum were grown for 4 h in appropriate medium (Brewbaker and Quack, 1963) and fixed with 3% PFA in PM buffer (0.1 pipes, pH 6.6, 5 mM EGTA, 1 mM MgSO$_4$) for 30' at room temperature. After two rinses in buffer the material was incubated in 2% cellulysin in PM for 5'. After two washes in PM buffer, germinated pollen tubes were divided in two samples and processed for staining of microfilaments and microtubules.

For staining of microfilaments pollen tubes were incubated with Rhodamine-Phalloidine dissolved in PBS (Phosphate Buffered Saline, pH 7.2) for 30' at room temperature. After two washes in PBS, pollen tubes were resuspended in 5% n-propyl-gallate in glycerol, mounted on slides and observed at a Zeiss Axiophot optical microscope.

For staining of microtubules, pollen tubes were treated in absolute methanol at -20°C for 5'. After 2 washes in PM buffer the sample was incubated for 30' at room temperature with a monoclonal antibody to the tubulin submit (Amersham). Pollen tubes were rinsed twice in TBS (TRIS Buffered Saline, pH 7.2) and incubated for 30' with a FluoresceineIsoTioCianate labelled rabbit anti mouse as secondary antibody. After two washes, pollen tubes were resuspended in 5% n-popyl-gallate in glycerol, mounted on slides and observed at a Zeiss Axiophot optical microscope.

Electron microscopy

Pollen grains from Aloe ciliaris, Linaria vulgaris and Nicotiana tabacum were grown for 3h in appropriate medium (Brewbaker and Quack, 1963).

Pollen grains of Aloe ciliaris and Linaria vulgaris were prefixed in 3% glutaraldehyde in 0.066M cacodylate buffer, pH 7.2 at 4°C for 30'. The samples were washed in the same buffer and postfixed in 1% osmium tetraoxide in 0.066M cacodylate buffer, pH 7.2, for 45'. After a dehydration process through increasing concentrations of ethanol, the pollen were embedded in a low viscosity resin (Spurr, 1969). After sectioning by an LKB ultramicrotome apparatus, the samples were stained with uranyl acetate and lead citrate and observed with a Jeol Jem 100BV electron microscope at 80Kv.

Nicotiana tabacum germinated pollens were processed by freeze substitution following the procedure of Lancelle et al. (1987).

RESULTS

The pollen of the Leucojum, Malus and Nicotiana, at maturity have binucleate grain and sperm formation occur during tube growth.

- Immunofluorescence

Our fluorescence observations after Rhodamine-Phalloidine labelling, show that the generative cells from different angiosperms do not contain microfilaments.

By immunofluorescence using a monoclonal antibody to α tubulin subunit we conclude that all analyzed generative cells have a microtubular apparatus. In Leucojum verum the spindle shaped generative cell is surrounded by a basket-like structure of microtubules whose disposition is very regular (Fig. 1). Also in Malus domestica pollen, microtubules form a basket-like structure in the generative cell and they are also present in the tail, an elongated structure present at the distal part of the cell (Fig. 2). In Nicotiana microtubules disposition in the generative cell is similar. Characteristically microtubules seem to converge towards specific points near the proximal part of the cell (Fig. 3).

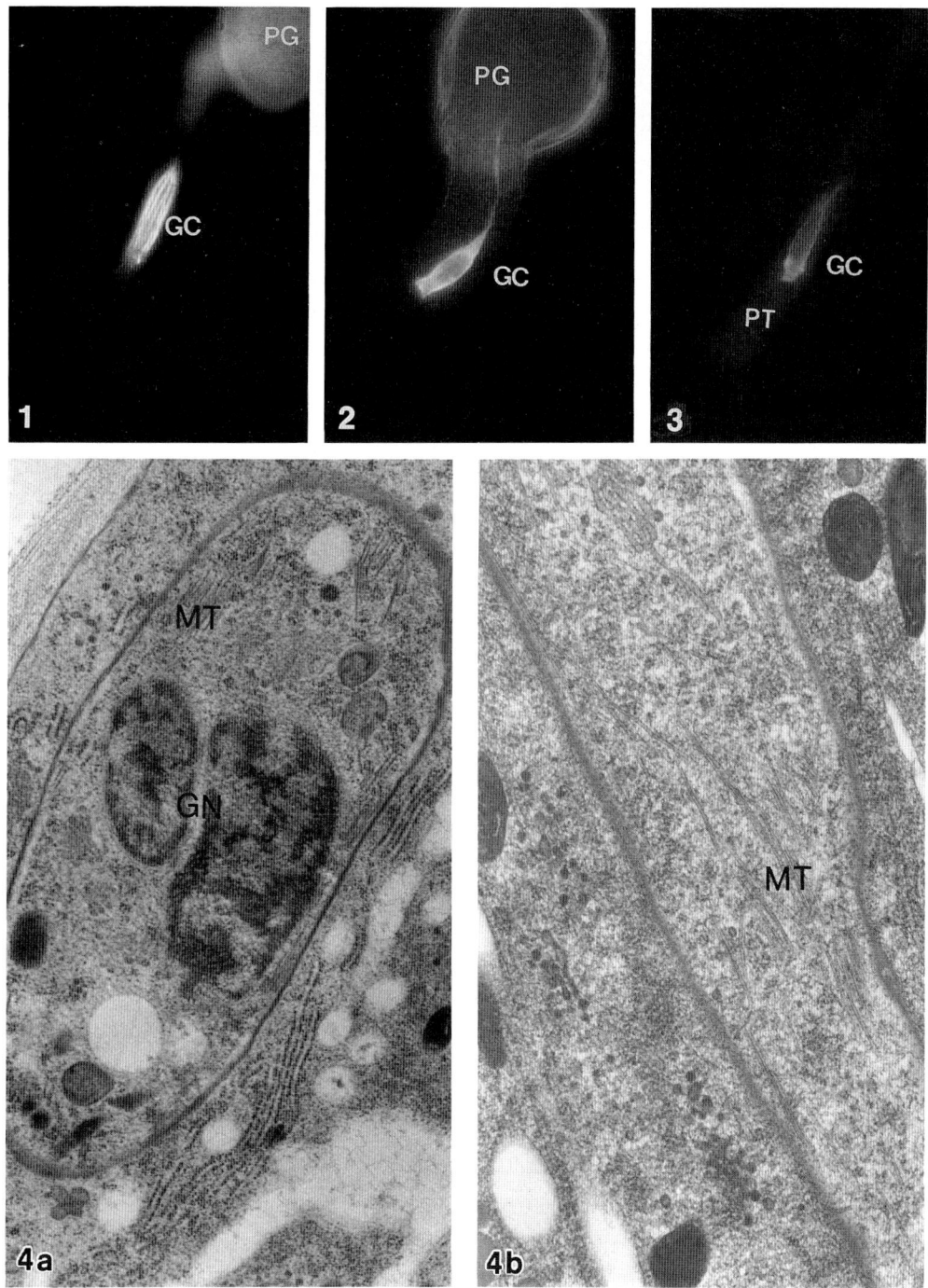

218

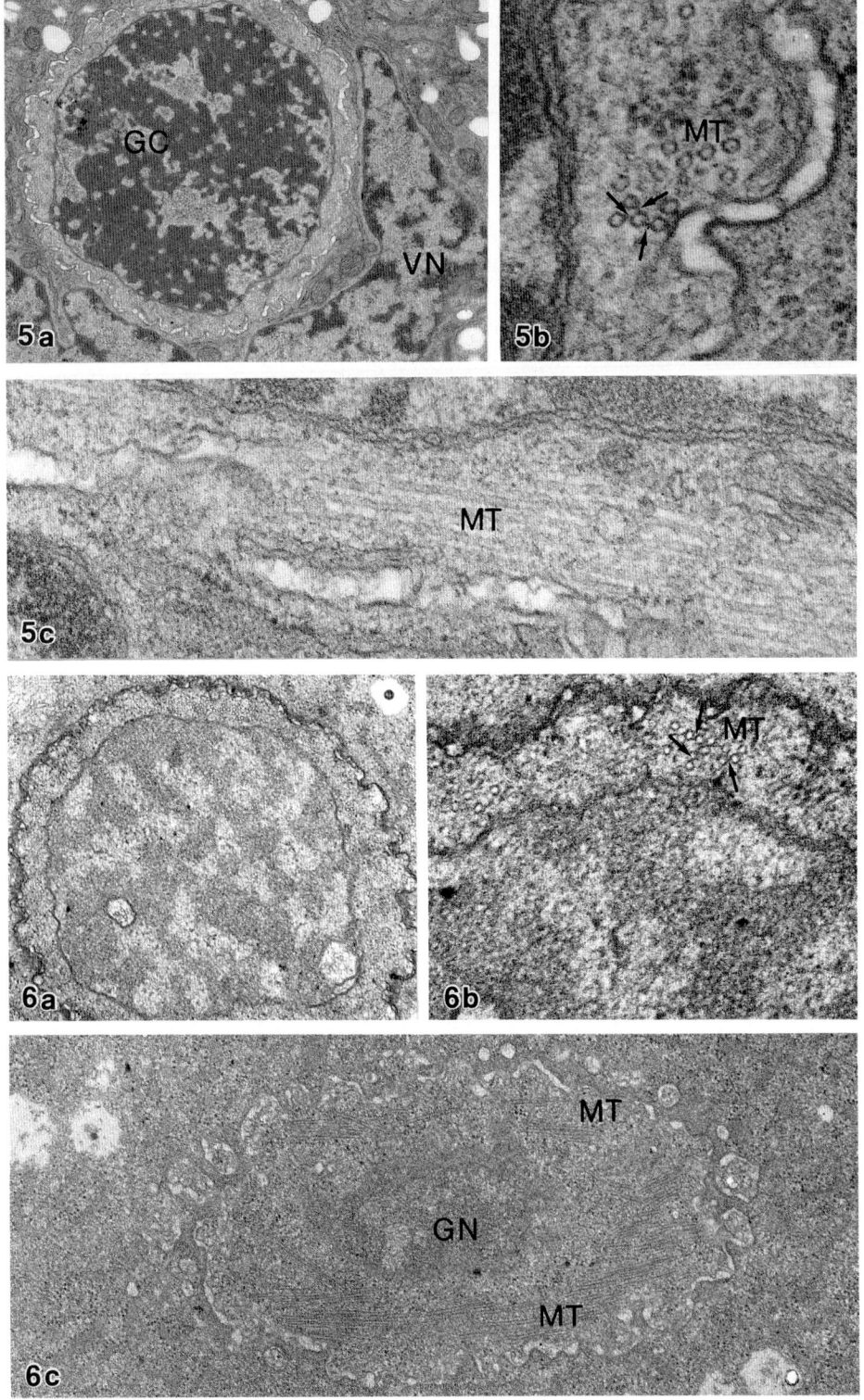

- Ultrastructural investigations

By electron microscopy we have not observed microfilaments in generative cells from Nicotiana tabacum, Aloe ciliaris and Linaria vulgaris.

With respect to their microtubular citoskeleton, generative cells have similar ultrastructural features. After both chemical fixation and freeze substitution, microtubules are organized as bundle and their distribution is principally near the plasma membrane. The microtubules of a bundle appear to be linked together by lateral projections of the microtubular surface (Figs. 4-5-6).

DISCUSSION

Considering all data together, they clearly indicate that the generative cell cytoskeleton is mainly formed by microtubules. The microtubules from bundles which are distributed along the generative cell plasma membrane and extend into the elongated structure defined as tail. Essentially these results confirm what we previously reported (Cresti et al. 1984) and at the same time they are in agreement with the results of other investigators (for complete review see Heslop-Harrison et al., 1988).

Interesting considerations result from the ultrastructural observations of side arms-like structures interconnecting microtubules of the generative cells of the plants investigated. Since analogue structures have not been see at the microtubules of the vegetative cell, it suggests a different molecular organization of the generative cell microtubules, in order to achive and maintain the microtubular bundling. From this point of view side arms-like structures can be of course considered as structural components of the generative cell cytoskeletal apparatus.

The ultrastructural analogy with the axoneme suggests that the microtubule lateral projections shown here, could have a mechanochemical role in microtubules sliding, related to generative cell movement as previously reported by Lancelle et al. (1987). However, since it was recently demonstrated that the vegetative cell microfilament system is responsible for the generative cell movement (Heslop-Harrison and Heslop-Harrison 1988), we believe that a microtubule sliding process could be involved in cell reshaping.

Other cytoskeletal polypeptides could be of course present in the generative cell. In animal cells, for instance, microtubules completely surround the nucleus, while they have tridimensional linkage with all the other cytoskeletal elements. Thus for we don't know whether the generative cell microtubular bundles have molecular linkages with even the nucleus; probably undiscovered cytoskeletal structures could contribute to stabilize the nuclear location during cell reshaping and movement in the pollen tube.

EXPLANATIONS OF THE FIGURES

Fig. 1 - Leucojum verum   - Generative cell (GC) microtubules. A microtubular basket surrounds the cell. (x 210). PG pollen grain.

Fig. 2 - Malus domestica   - GC microtubules. Microtubules are present in the tail-like structure. (x 210).

Fig. 3 - Linaria vulgaris   - GC microtubules. The antibody stains points of the proximal part of the cell. (x 210). PT pollen tube.

Fig. 4 - <u>Nicotiana tabacum</u> - GC after freeze substitution. -a: cross section (x 12.000); -b: longitudinal section (x 16.000) GN generative nucleus, MT microtubules.

Fig. 5 - <u>Aloe ciliaris</u> - GC after chemical fixation. a-: cross section (x 7.000); -b: high magnification. Arrows show side arms-like structures (x 82.000); -c: longitudinal section. Microtubules are organized as bundles (x 59.000). VN vegetative nucleus.

Fig. 6 - <u>Linaria vulgaris</u> - GC after chemical fixation. -a: cross section (x 12.600); -b: high magnification. Arrows show lateral projections on the microtubular surface (x 33.000); -c: longitudinal section. Microtubules are organized as bundles (x 21.000).

Acknowledgement

This research was carried out in the framework of contract n. BAP-02040-1 of the Biotechnology Action Programme of the Commission of the European Communities.

References

Cresti M, Ciampolini F, Kapil RN (1984) Generative cells of some angiosperms with particular emphasis on their microtubules. J Submicroscop Cytol, 16: 317-326

Heslop-Harrison J (1987) Pollen germination and pollen tube growth. Int Rev Cytol 1-78

Heslop-Harrison J, Heslop-Harrison Y, Cresti M, Tiezzi A, Ciampolini F (1987) Actin during pollen germination. J Cell Sci 86: 1-8

Heslop-Harrison J and Heslop-Harrison Y (1988) Organelle movement and fibrillar elements of the cytoskeletal in the angiosperm pollen tube. Sex Plant Repr 16-24

Heslop-Harrison J, Heslop-Harrison Y, Cresti M, Tiezzi A, Moscatelli A (1988) Cytoskeletal elments, cell shaping and movement in the angiosperm pollen tube. J Cell Sci, In Press.

Lancelle SA, Cresti M, Hepler PK (1987) Ultrastructure of the cytoskeleton in freeze-substituted pollen tube of Nicotiana alata. Protoplasma 140: 141-150

Palevitz BA, Cresti M (1988a) Microtubules organization in the sperm of Tradescantia virginiana. Protoplasma (In Press)

Palevitz BA, Cresti M (1988b) Cytoskeletal changes during generative cell division and sperm formation in Tradescantia virginiana. Protoplasma (In Press)

Pierson ES, Bcrksen J, Traas JA (1986) Organization of microfilaments and microtubules in pollen tubes grown in vitro or in vivo in various angiosperms. Eur J Cell Biol 41: 14-18

Spurr AR (1969) A low viscosity epoxy resin embedding medium for electron microscopy. J Ultrastruct Res 26: 31-43

Tiezzi A, Cresti M, Ciampolini F (1986) Microtubules in Nicotiana pollen tubes ultrastructural immunofluorescence and biochemical data. In: "Biology of Reproduction and cell motility in plants and animals" (Cresti M and Dallai R Eds) pp. 87-94.

The Male Germ Unit of *Zea mays:* Quantitative Ultrastructure and Three-Dimensional Analysis

M.L. Rusche and H.L. Mogensen
Department of Biological Sciences
Northern Arizona University
Flagstaff, Arizona 86011-5640
U.S.A.

Introduction

Since the concept of the male germ unit was introduced (Dumas et al 1984) a physical association among the elements of the male germ unit has been observed in mature tricellular pollen grains of dicots, in the pollen tubes of dicots with bicellular pollen and in a monocot with bicellular pollen (see Rusche 1988 for references). In monocots with tricellular pollen, a physical association has not been observed in mature pollen.

Sperm pairs of seven species of angiosperms have been serially ultra-thin sectioned, reconstructed, and analyzed for dimorphism with respect to size, number of organelles, and overall morphology. In the dicots, four species are dimorphic (Russell 1985, 1987, McConchie et al 1985, 1987, Wilms 1986) and one is isomorphic (Wagner and Mogensen 1988). In monocots isomorphism has been reported in the sperm cells of Hordeum (Mogensen and Rusche 1985) and dimorphism in the sperm cells of Zea (McConchie et al 1987, Rusche 1988). As each of the reports on Zea was based on a sample size of one, further examination of this economically important cereal is desirable.

Materials and Methods

Pollen grains were collected from freshly exerted dehiscing anthers of Zea mays Black Mexican and immediately fixed in aqueous 2% $KMnO_4$ for 15 minutes, rinsed in dHOH, dehydrated, and infiltrated with Spurr's resin. Serial ultra-thin sections averaging 70 nm in thickness were cut on an MT-2B ultra-microtome, transferred to formvar coated single slot grids, double stained with uranyl acetate and lead citrate using a LKB automatic stainer, and observed with transmission electron microscopy. Tracings of the sperm cells, their nuclei and mitochondria, and the vegetative nucleus, along with fiducial marks for alignment, were entered into an IBM-AT computer via a digitizing tablet. A series of programs (Kinnamon 1987) was used to 1) reconstruct the male germ unit on a color monitor, 2) rotate the reconstructions, and 3) measure the perimeter and area enclosed by each contour line entered. The measurements were entered into a program (Rusche unpublished) which calculated the surface area and volume of each structure. Surface density of mitochondrial cristae was determined using Weibel's (1979) point/intersect method. Due to the small sample size, ratios were used for comparison of the sperm cells of the present and previous studies (Rusche 1988).

Results

Each sperm is a long, tapering cell with a heterochromatic nucleus and two cytoplasmic extensions (Figs 1, 2). The lengths of the extensions in sperm 1, the larger sperm, are approximately equal (Fig 2, Table 1); in sperm 2, the smaller sperm, one extension is twice the length of the other. Within the thin layer of cytoplasm surrounding the nuclei are the 14 mitochondria of sperm 2 and 3 of the 5 mitochondria of sperm 1 (Figs 1, 3, 4). In sperm 1, two of the mitochondria have a branching filamentous form and extend almost the full length of the nucleus. In sperm 2, three of the four large mitochondria branch. The remaining 13 mitochondria of the two sperm cells are small and oblate in shape. Endoplasmic reticulum associated with the mitochondria continues into the cytoplasmic extensions where parallel ER sheets are observed (Fig 1). Few dictyosomes and no plastids are found in the sperm cells. The dictyosomes are neither as well developed nor as active as those in the vegetative cell cytoplasm. Cytoplasmic inclusions of unknown composition occur in the sperm cell cytoplasm. Additional data are summarized in Tables 1 and 2. No physical association of sperm cells with each other or with the vegetative nucleus was observed (Fig 5). Minimum distances between the two sperm cells, between sperm 1 and the vegetative nucleus, and between sperm 2 and the vegetative nucleus are 0.5 um, 4.3 um, and 3.5 um, respectively.

Discussion

In mature dehiscent pollen grains of the grasses Z. mays (McConchie et al 1987, Rusche 1988, this study) and Hordeum (Mogensen and Rusche 1985), there is no physical association among the elements of the male germ unit. However, in Hordeum (Mogensen and Wagner 1987) after pollination, but before the sperms leave the pollen grain, a physical association is formed. In the pollen tube of Z. mays at least one of the sperm cells is in a close association with the vegetative nucleus (Rusche and Mogensen unpublished). Studies are in progress to determine the extent and nature of the association and when it forms.

In the reconstructed male germ units of Z. mays, grain G (Rusche 1988), grain K (this study), and grain M (McConchie and Knox 1987), there are similar patterns between the two sperm cells of common origin as reflected in similar morphology, and in ratios of sperm cell 2: sperm cell 1 in grains G and K which are close to 1 (Tables 1, 2). There are 3 differences: 1) The nuclear surface area ratio in grain K suggests a difference between the two sperm nuclei as does the face-view surface area of large to small nuclei in sister sperm cells observed by McConchie et al 1987. 2) The relative lengths of the 2 cytoplasmic extensions differ. In grains K and M the larger sperm cell and in grain G the smaller sperm cell have equal extensions. In grain M one of the extensions of the smaller sperm cell is bifid. Evidence of cytoplasmic loss from Z. mays sperm cells (Rusche and Mogensen unpublished) raises the question of the significance of the relative lengths of the cytoplasmic extensions in sister sperm cells. Cytoplasmic loss was indicated in sperm cells of mature pollen of Hordeum (Mogensen and Rusche 1985), in a sperm cell in the embryo sac of Plumbago (Russell 1985), and in Hordeum (Mogensen 1988) at the time of fertilization when the nucleus alone enters the egg cell. The process may be continuous or there may be specific elimination stages and/or taxa specific modes of cytoplasmic loss. 3) The number of mitochondria per sperm cell differs in the two sister sperms of grains K and M, however, their similarity in distribution around the nucleus, and in grain K,ratios which are close to 1 (Table 2), suggest a functional equality. The large and more complex forms of mitochondria in Z. mays persist in the sperm cells in the pollen tube for at least the first few hours post-pollination (Rusche and Mogensen unpublished). The smaller surface area of cristae per mitochondrial volume observed in the sperm cells as compared to the vegetative cell (Table 2) confirms qualitative observations in other species (Jensen 1974) and suggests a lower respiration rate in sperms than in the vegetative cell. Absence of nucleoli, heterochromatic nuclei and less well developed dictyosomes also argue for a reduced activity in the sperm cells.

The data reveal size differences among the male germ units which may reflect different treatments and/or actual variations, possibly correlated with variation in pollen grain size. The larger SA/Volume ratios of the sperm cells and sperm nuclei of grain G compared with grain K correspond to the typically more flattened morphology. Grain M had one flattened and one rounded sperm cell. Although the number of male germ units reconstructed is small, the observations of common differences between sister sperm cells in two varieties of Z. mays and in three different treatments supports the view that these differences are real.

Acknowledgements
This work was supported by a grant from the Pioneer Hi-Bred Seed Company, International. We thank Steve Young, University of California at San Diego for computer assistance, and Carl Ray and Lisa Osborne for technical assistance.

TABLE 1.   ZEA MAYS SPERM CELL QUANTITATIVE DATA

	GRAIN G		GRAIN K		SPERM 2:1*	
	SPERM 1	SPERM 2	SPERM 1	SPERM 2	G	K
Cell Volume	73.4	60.8	234.5	209.7	.83	.90
Cell Surface Area	264.7	232.0	366.2	313.2	.88	.86
SA/Volume*	3.61	3.82	1.56	1.49	1.06	.96
Cell Length	48.4	34.8	67.2	70.6	.72	1.10
Cell Width	6.8	5.7	6.5	6.0	.84	.92
Extension A Length	13.4	13.1	25.7	18.3	–	–
Extension B Length	26.5	13.2	30.5	38.6	–	–
Ratio Extension A:B*	.51	.99	.84	.47	1.94	.56
Ratio Extension B:A*	1.98	1.01	1.19	2.11	.51	1.77
Nuclear Volume	19.0	19.1	85.6	85.8	1.00	1.00
Nuclear Surface Area	54.7	56.6	78.8	109.2	1.03	1.39
SA/Volume*	2.88	2.96	0.92	1.27	1.03	1.38
Nuclear Length	8.5	8.5	10.9	14.2	1.00	1.30
Nuclear Width	4.8	3.2	4.7	4.8	.67	1.02
VN Volume	411.8		1491.9			
VN Surface Area	736.5		1111.1			
SA/Volume*	1.79		.74			

Volume in um³, surface area in um², length and width in um
* = calculated value   VN = vegetative nucleus
Grain G data after Rusche, 1988

TABLE 2.   MITOCHONDRIA IN ZEA MAYS GRAIN K

	SPERM 1	SPERM 2	SPERM 2:1*	GRAIN#
Number of Mitochondria	5	14	2.80	60
Total Volume	5.44	4.86	.91	7.7
Range of Volume/Mito	.04-2.64	.04-1.12	–	.01-.28
Mean Volume/Mito*	1.09	.32	.29	.13
SD Vol/Mito*	1.42	.40	–	.08
Total Surface Area	44.59	45.62	1.02	82.28
Range in SA/Mito	.65-21.5	.51-9.01	–	.75-2.35
Mean SA/Mito*	8.92	3.04	.34	1.37
SD SA/Mito*	11.07	3.27	–	.39
Range of SA/Vol/Mito	7.78-17.79	7.49-15.83	–	7.16-17.5
Mean SA/Vol/Mito*	11.64	10.62	.91	10.62
SD SA/Vol/Mito*	4.23	2.65	–	2.33
% Vol Mito/Sperm*	2.3	2.3	1.00	–
SA Mitochondria/Sperm*	.22	.19	.86	–
SA Cristae/Mito Vol	3.22	3.31	1.03	6.38
SD Cristae/Mito Vol	.6	.7	–	.7

Volume in um³, surface area in um²
* = Calculated value
# = Based on mitochondria within 10 um of the sperm cells, n = 60
SD = Standard deviation of the mean

225

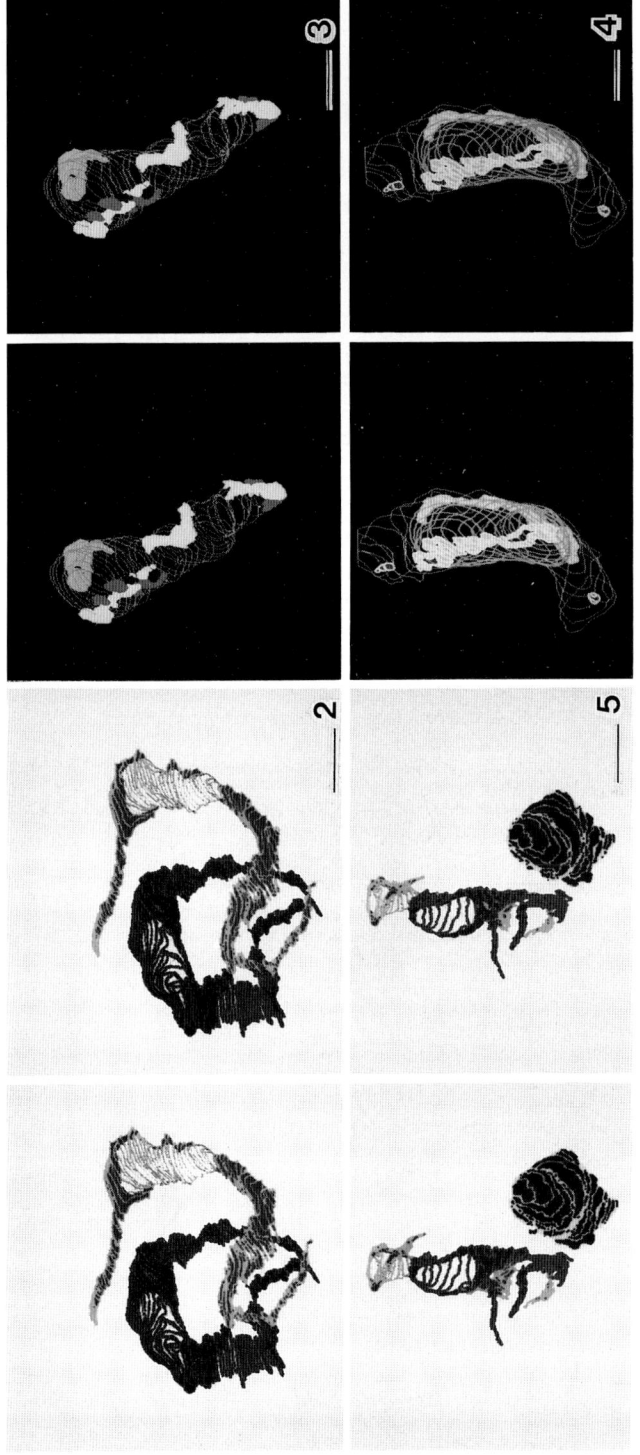

Figs 2-5. Computer reconstructed stereo 3-D images of the sperm cells, their nuclei and the male germ unit. Scale bars are approximate. Fig 2. The two sperm cells. Sperm 1 is on the left, sperm 2 is on the right. Bar = 10 um. Fig 3. Nucleus and mitochondria of sperm cell 2. Note the organization of the mitochondria around the nucleus. Only 12 of the 14 mitochondria are visible. Bar = 2 um. Fig 4. Nucleus and mitochondria of sperm cell 1. On opposite sides of the nucleus are two long mitochondria; one of these branches. Bar = 10 um. Fig 5. The entire male germ unit. The two sperm cells are on the left, the vegetative nucleus is on the right. Bar = 10 um.

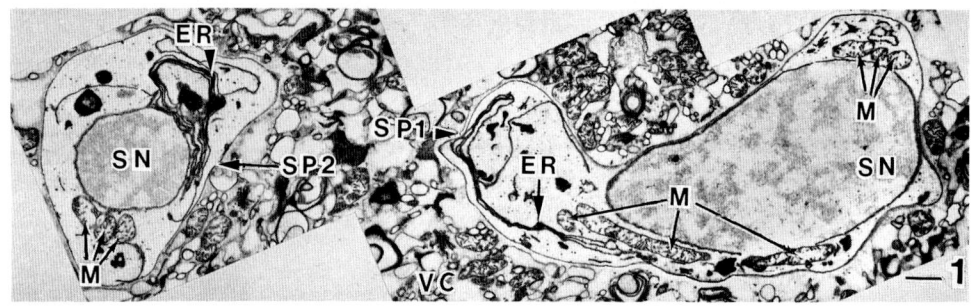

Fig 1.   Electron micrograph of the two sister sperm cells. In sperm 1 the long mitochondrial profiles, arrow, unite to form a single mitochondrion as do the oblate profiles, double arrow.   ER = endoplasmic reticulum, M = mitochondrial profile, SN = sperm nucleus, SP1 = sperm 1, SP2 = sperm 2, VC = vegetative cell cytoplasm.   Bar = 1 um.

References

Dumas C, Knox RB, McConchie CA, Russell SD (1984) Emerging physiological concepts in fertilization.   What's New Plant Physiol 15:17-20

Jensen WM (1974) Reproduction in flowering plants.   In: Robards AW (ed) Dynamic aspects of plant ultrastructure. McGraw-Hill, London, New York, St. Louis, San Francisco, pp.481-500

Kinnamon JC (1987) IBM PC-based three-dimensional reconstruction program. University of Colorado, Boulder

McConchie CA, Hough T, Knox RB (1987) Ultrastructural analysis of the sperm cells of mature pollen of maize, Zea mays. Protoplasma 139:9-19

McConchie CA, Jobson S, Knox RB (1985) Computer-assisted reconstruction of the male germ unit in pollen of Brassica compestris. Protoplasma 127:57-63

Mogensen HL (1988) Exclusion of male mitochondria and plastids during syngamy as a basis for maternal inheritance.   Proc Natl Acad Sci 85:2594-2597

Mogensen HL, Rusche ML (1985) Quantitative ultrastructural analysis of barley sperm: 1.   Occurrence and mechanism of cytoplasm and organelle reduction and the question of sperm dimorphism. Protoplasma 128:1-13

Mogensen HL, Wagner VW (1987) Association among components of the male germ unit following in vivo pollination in barley. Protoplasma 138:161-172

Rusche ML (1988) The male germ unit of Zea mays in the mature pollen grain. In: Wilms H, Kreijzer CJ (eds) Plant sperm cells as emerging tools for crop biotechnology. Wageningen, the Netherlands, In press

Russell SD (1984) Ultrastructure of the sperm of Plumbago zeylanica: 2. Quantitative cytology and three-dimensional reconstruction. Planta 162:385-391

Russell SD (1985) Preferential fertilization in Plumbago: Ultrastructural evidence for gamete-level recognition in an angiosperm. Proc Natl Acad Sci 82:6129-6132

Wagner VW, Mogensen HL (1988) The male germ unit in the pollen and pollen tubes of Petunia hybrida: Ultrastructural, quantitative and three-dimensional features.   Protoplasma 143:101-110

Weibel ER (1979) Stereological Methods Volume 1. Academic Press, London, New York, Toronto, Sydney, San Francisco

Wilms HJ (1986) Dimorphic sperm cells in the pollen grain of Spinacia. In: Cresti M, Dallai R (eds) Biology of reproduction and cell motility in plants and animals. University of Siena, Siena, pp 193-198

Interaction of Vegetative Nucleus and Generative Cell (Then Sperms)

Tang Pei-hua
Institute of Botany
Academia Sinica
Beijing 100044
China

Recently,since appearence of the new conception of male germ unit
in pollen (Dumas et al. 1984), the relationship between vegetative
nucleus (VN) and generative cell (GC, then sperms, S) in development of male gametophyte (MG) becomes more and more interesting
subject for researchers. Dumas et al. (1985), McConchie et al.(1985)
and Kaul et al.(1987) using computer-assisted three-dimensional reconstruction technique and light or fluorescence microscopy proved
that, VN and GC or VN and S are associated structurally together in
a certain developing stage (DS) of MG, while in the meantime Zhou
Chang(1937) succeeded in isolation GC and S (personal communication)
by artificial and enzymatic methods from pollen, but could not see
any associated unit between neither VN and GC, nor VN and S. It is
obvious that is due to the difference between DS of experimental materials which they used. Author has worked on culture of MG of Clivia nobilis and Amaryllis vittata in vitro for many years and observed their whole developing course (from microspore to sperms) by
time-lapse cine-micrography and general cytological observations
(Tang 1973), cytochemical observations on adenosine triphosphatase
(ATPase) distribution (Tang 1979), nucleic acid, protein synthesis
(Tang et al. 1985) and release (Tang et al. 1987a) and the role of
RNA and protein, synthesized in different DS of MG (Tang et al.
1987b), we found that the comportment of VN, GC (then S) in the
pollen grain (PG) and pollen tubes (PT) of above two plants is as
complicated as in Helleborus foetidus (Heslop-Harrison et al. 1986).
Depending on the DS from time to time the VN and GC (then S) conjugated together (or inter-connection, or associated consistantly),
and separated each other, but all the above phenomena occur according to the DS and in a strict order. Author has successfully used
the relative position and comportment of VN, GC and S as the morphological index for division of the developing process of MG into
seven DS. The results of cytochemical observations on ATPase dis-

Table I The correlation of ATPase relative activity to mechanic movement and physiological state of VN and GC in different DS of MG of _Amaryllis_ vittata

DS	The ATPase relative activity (grade)		The mechanic movement and physiological state	
	VN	GC	VN	GC
I The binuclear stage	>2	>3	Displacement and rotation	Displacement and rotation
II The germination of PT	<3	<4	Rapid displacement and rotation	Rapid displacement and rotation
III Entering of VN into PT	>2	3	Movement to the tip of PT at uniform speed	Preparation for entering into PT
IV Entering of GC into PT	>2	<5	Change in form, preparation for connecting with GC	Entering into PT, moving several hundred μ with great speed
V Connecting VN with GC closely to each other	<3	>3	High in hydrous state and nuclear figure	Catching up and connecting with VN, GC and VN connected for 2-3 h, then separated, after that GC continued its division
VI The metaphase of GC	<4	>1	Great change in form and connect with GC	No displacement
VII The formation of two S	>2	<4	Great change in form, moving near S, and associated with S	S formation and movement to the tip of PT

tribution in DS showed that the comportment of VN, GC and S in development is a specific physiological phenomenon and not only GC and S, but also VN present ATPase (Table I, Fig.1).

Further more VN showed itself as a physiologically active organelle, since it has its own ATPase system which appeared very different to the same of GC (Table I, Fig.1).

According to the data in table I and Fig.1, it is no doubt that the change in morphology and relative activity of VN and GC before and after connection are great. We assume that either connection or separation of VN and GC need energy, so the connection may be consi-

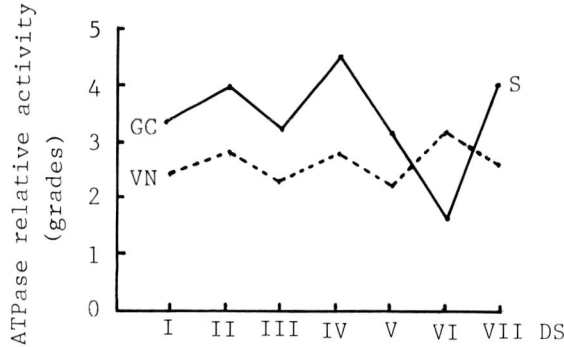

Fig.1 Change of ATPase relative activity in different
DS of VN and GC in MG of <u>Clivia nobilis</u>

dered for a certain physiological function. Based upon the results
of time-lapse cine-micrography and ATPase cytochemical observations,
the behavior of VN, GC (then S), mainly in their interaction, is
conclouded as follows:

Interaction of VN and GC in PG (DS I-II)

Both vegetative and generative nuclei are formed by equtional division of nucleus of microspore (Figs.2,3), but during the succes-
sing development, by unequational distribution of the cytoplasm
(Fig.4), GC finally becomes a cell depending on the nutrition from
the vegetative cell (Fig.5). In mature pollen or towards germination,
VN and GC displaced and rotated first gradually, then rapidly. They
sometimes gathered together, sometimes separated each other, sometimes overlapped (Fig.6), but we have not found structural connection between them.

Interaction of VN and GC during entering into PT (DS III-IV)

5-8 h after germination of pollen VN entered into PT at a speed
much slower than the streaming of the surrounded protoplasm . Under
ordinary conditions VN moves gradually towards the tip of PT. 3-5 h
later, GC entered into PT within a few minutes in general (Figs.7,
8). Occasionally, we observed a GC with a diameter greater than
that of the entrance aperture of the PT had to expend almost half
an hour in changing its shape for entering into PT. After entering
into PT the GC moved at a speed much greater than that of the VN,
toward the tip of PT (Figs.9,10). When it had nearly caught up,
the latter elongated conspicuously, and the two began to connect
with each other (Figs.11,12).

Interaction of VN and GC during connecting (DS V)

After catching up the VN immediately began to actively change its
shape and increase the connecting area. At that time, the relative
activity of ATPase of nuclear membrane of VN near the connected area
(Fig.13, black arrow) and GC (Fig.13, GC) increased to the highest
degree(5th degree). It showed that the active physiological behavior
needs energy. The connection of VN and GC continued about 2-3 h
(Figs.14,15), then slowly separated (Figs.16-18). It took 3-5 h for
them to separate completely (Fig.19). The location of connection was
often the enlarged portion of PT near the tip. The VN at connecting
stage was observed to be very full and transparent (Figs.14,15 VN).
When the VN-GC connecting body was out from the broken tip wall of
PT under pressure and flowing into the culture medium, GC maintained
its original basic shape and size, but VN immediately decreased its
size to about 1/3 of the original and the connecting body separated
(Fig.20). It seems that the connection is rather loose.

Interaction of VN and GC during S formation (DS VI-VII)

After separation the nucleus of GC continued its mitosis, through
metaphase, anaphase, telophase and finally formed two S. In metaphase of GC, VN divided into two parts with a narrow bridge. One
part was located on the PG side of GC, the other part was on the
PT end side of GC (Fig.21). The two parts of VN like two umbrellas
for protection the dividing GC to avoid the effect of the surrounding protoplasmic streaming. The relative ATPase activity of VN reached the highest degree of its life cycle (Fig.1). Thus the shape-
change of VN is an active energy-dependent action. When S were formed, either in PT (Fig.22) or in vitro, isolated artificially, the
association of VN and S was easily to be seen (Figs.23-25). It shows
that the association between VN and S are very consistant.

Conclusion:

The fact may only be explained that, in the PT of two examined
plants exists the structurally associating and physiologically
active S and VN male germ unit.

The author would like to express her thanks to Prof. Wang Fu-hsiung
and Qian Ying-qian for their helpful disccusion.

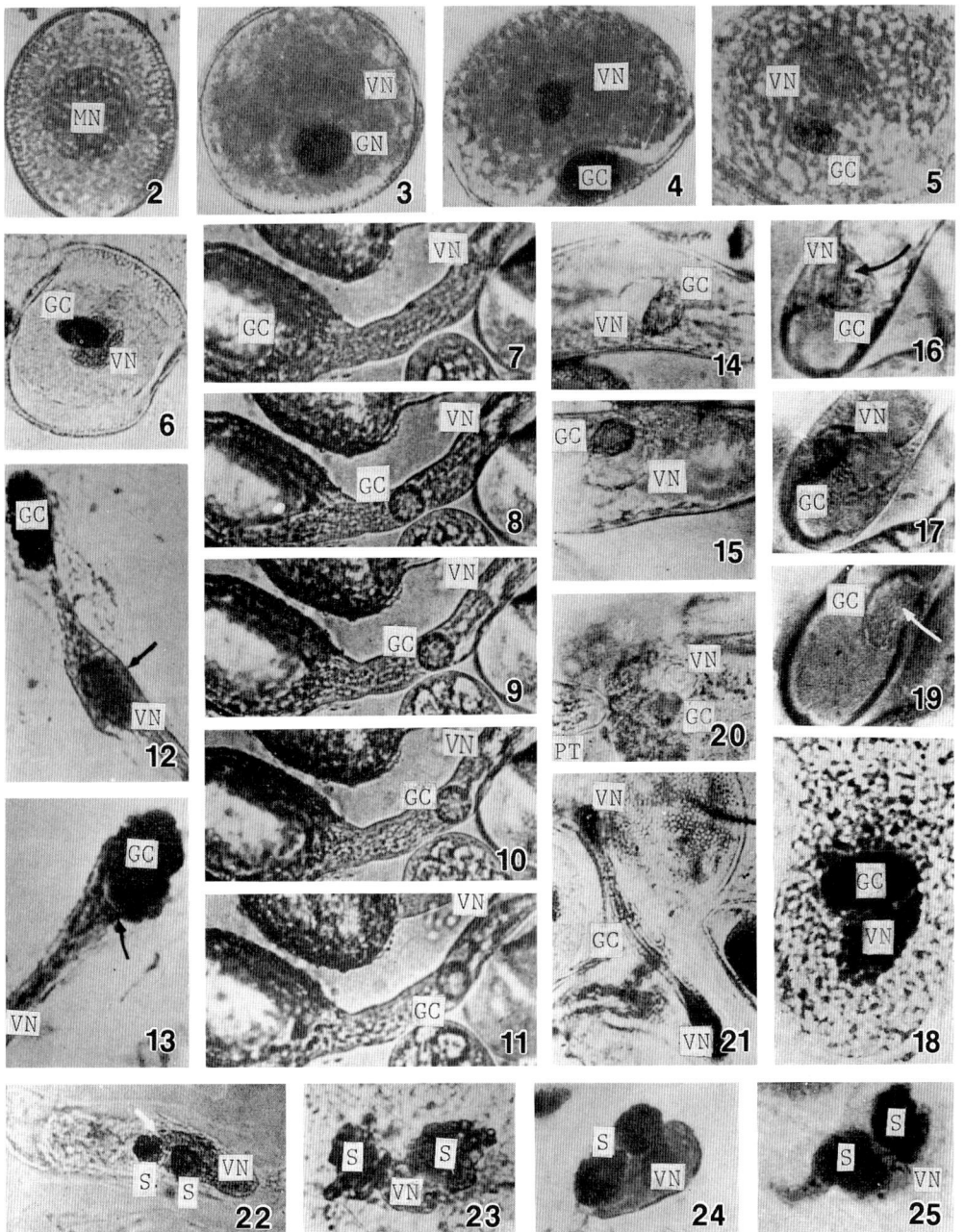

Explanation of plate:
Figs.6,18,22,24 A. vittata; the others C. nobilis. Figs.6,12,13,18,
21-25 ATPase positive. Figs.2-5 MG culture in vitro with 1/2 MS medium; the others are serious photos from time-lapse cine-micrography.

References

Dumas C, Knox RB, McConchie CA, Russell SD (1984) Emerging physiological concepts in fertilization. What's new in plant. Physiology 15:17-20

Dumas c, Knox RB, Gaude T (1985) The spatial association of the sperm cells and vegetative nucleus in the pollen grain of Brassica. Protoplasma 124:168-174

Heslop-Harrison J, Heslop-Harrison JS, Heslop-Harrison Y (1986) The comportment of the vegetative nucleus and generative cell in the pollen and pollen tubes of Helleborus foetidus L. Annals of Botany 58:1-12

Kaul V, Theunis CH, Palser BF, Knox RB, Williams EG (1987) Association of the generative cell and vegetative nucleus in pollen tubes of Rhododendron. Annals of Botany 59:227-235

McConchie CA, Jobson S, Knox RB (1985) Computer-assisted reconstruction of the male germ unit in pollen of Brassica campestris. Protoplasma 127:57-63

Tang Pei-hua (1973) Variation of generative cell and vegetative nucleus in developing male gametophyte of Clivia nobilis etc. cultured in vitro. I. Results of time-lapse cine-micrography and general cytological observation. Acta Botanica Sinica 15:12-21

Tang Pei-hua (1979) -- II. Cytochemical observation on adenosine triphosphatase distribution. Acta Botanica Sinica 21:107-116

Tang Pei-hua, Zhu Ying-min (1985) Studies on nucleic acid and protein synthesis and development of male gametophyte of Clivia nobilis cultured in vitro. Acta Botanica Sinica 27:133-140

Tang Pei-hua, Zhu Ying-min (1987a) Dynamics of newly synthesized RNA and protein diffusion from germinating pollen of Clivia nobilis cultured in vitro. Acta Botanica Sinica 29:9-13

Tang Pei-hua, Liu Yi-min (1987b) The role of RNA and protein synthesized before and after dehiscence of the anther in the development of pollen grains of Clivia nobilis. Scientia Sinica (Series B) 30:1268-1277

Zhou Chang (1987) Cell-biological studies on artificially isolated generative cells from angiosperm. Acta Botanica Sinica 29:117-122

A Technique to Isolate Sperm Cells of Mature Spinach Pollen

C.H. Theunis, J.L. van Went and H.J. Wilms
Agricultural University
Department of Plant Cytology and Morphology
Arboretumlaan 4
6703 BD Wageningen
The Netherlands

Abstract
A technique has been developed to isolate sperm cells of mature
Spinacia oleracea pollen. By using a squashing technique in
combination with centrifugation on a percoll layer is it possible
to isolate the sperm cells in large amounts, and to keep them
viable for several hours.

Introduction
In order to enable modern cell biology studies and manipulation,
many efforts have been made to isolate plant sperm cells. Several
species have been used; Beta (Nielsen & Olesen, pers comm 1988),
Zea (Matthys-Rochon et al. 1987, Dupuis et al. 1987, Roeckel et
al. 1988), Triticum (Matthys-Rochon et al. 1987), Plumbago
(Russell 1986), Gerbera (Southworth 1986), Brassica (Matthys-
Rochon et al. 1987, Hough et al. 1986), Haemanthus (Zhou et al.
1986) and recently generative cells of Lilium (Tanaka 1988). In
Zea, Lilium and Plumbago the cells are isolated in large
quantities, and appear to be vital.
A technique to isolate large numbers of sperm cells of mature
spinach pollen will be described and discussed here.

Material and Methods
Fresh pollen was collected from just opened flowers of Spinacia
oleracea L, cv Peruvianum, grown in the green house.
For each isolation run, approximately 200-300 mg of fresh pollen
was collected and pre hydrated in humid air (100% RH) for 1 hour.
Pollen was suspended in an isolation medium (Brewbaker & Kwack
1963), containing 0.73 M sucrose. A homogeneous cell suspension

was acquired which was used in the isolation experiments. All steps of the isolation procedure where carried out at 4 C.

The vitality of pollen grains and sperm cells was checked with the fluorescein diacetate test (FDA) (Heslop-Harrison et al. 1984). Only pollen samples containing more than 90% viable pollen grains, were used for the isolation procedure.

Results

To isolate sperm cells out of pollen grains, the pollen grain wall has to be broken. A physical breaking is applied in the method described here. A rolling device is made for squashing the pollen grains (fig 1). With this rolling device, the pollen grains are pressed, in the isolation medium, against the underlying glass plate, and will burst. Instead of the complete breakdown of the pollen grain, the cytoplasmic contents including the two sperm cells, will be pushed out of the pollen grain, leaving an empty or partially empty pollen grain behind. Analysis of the material immediately after squashing reveals that the sperm cells directly after release appear still connected to each other, and are spindle shaped (fig 2). After a few minutes they become spherical, and only a few remain connected in pairs. In phase contrast microscopy they are very dense, and in the sucrose concentration used, no cell organelles can be seen (fig 3).

The free floating sperm cells have to be separated from the pollen grain remnants. To achieve this, the mixture of squashed pollen is layered on top of a discontinuous gradient of 50%, 30% and 10% percoll in isolation medium (fig 4A). By centrifugation at 13000 g for 30 min it is possible to separate the cell constituents, empty grains and unbroken grains, according to their densities. This results in four layers of cell parts, one at every boundary of two percoll concentrations, and one at the bottom of the tube. (Fig 4B). At the bottom of the tube the starch grains concentrate, together with the unbroken and the slightly emptied pollen grains. The empty and almost empty pollen grains are collected between the 50% and the 30% percoll. Between the 10% and the 30% percoll, a small number of sperm cells is collected, together with some empty pollen grains and a large amount of very small cytoplasmic organelles. The layer between the original sample of squashed pollen grains and the 10% percoll, is filled with sperm cells. This layer also contains large amounts of small cytoplasmic organelles.

To collect the two bands which contain sperm cells, in one centrifuge step, the complete sample of squashed pollen grains in isolation medium is put on a single layer of 20% percoll in isolation medium (fig 4C). Before doing this, the sample was filtered over a 25 µm filter, to exclude the unbroken pollen grains in order to improve the separation. After centrifugation for 40 min. at 13000 g, this resulted in two layers (fig 4D). At the bottom of the tube the starch grains and pollen walls concentrate. On top of the 20% percoll, the sperm cells gather, together with the small cytoplasmic organelles. This sperm cell band is extracted and examined.

After the isolation and centrifugation procedure, the sperm cells are spherical, and appear intact with an average diameter of 3 µm (fig 5). Testing with FDA reveals that more than 90% of the sperm cells show a bright yellow/green fluorescent staining, indicating their viability (fig 6). If kept on ice, this FDA positive staining persists for several hours. Slowly the percentage of cells with positive FDA staining decreases and after 18 hours, only 50% of the sperm cells are still FDA positive.

With this method approximately 4 million sperm cells per ml. can be isolated. The final yield of sperm cells is 5% to 10%.

Discussion

In _Spinacia_ the mature pollen grains are tricellular. To isolate gametes, the pollen grain has to be broken without breaking the gametes. A physical way of breaking the pollen grain wall, is to be preferred. In this way a suitable osmotic medium can be chosen, to preserve the sperm cells. It also provides an at random breaking of pollen grains. In the developed squashing technique, the pollen grain is gently pressed with a roller device, made of glass, against a slightly rough glass plate. This physical pressure will burst the pollen grain, but will not fractionated it. The cytoplasmic contents, including the sperm cells will be pushed outside. The remaining pollen grain walls will prevent the sperm cells being squashed as well.

Not all pollen grains are broken. A large amount remains intact. The used method is very reproducible, and _Spinacia_ pollen grains are very easy to collect.

Other methods of breaking pollen grains in a physical way (tissue homogenizers and blenders) often damage the sperm cells. They also induce many pollen wall fragments which are difficult to separate in the centrifugation steps. In _Spinacia_ breaking by using osmotic shock as used in _Zea_,(Dupuis et al. 1987; Roeckel et al. 1988) is not effective. Only a small amount of the pollen grains will burst even in pure water. The chemic/enzymatic treatment of pollen grains (Baldi et al. 1986; Loewus et al. 1985; Tanaka 1988) is not usable. The pollen grain wall of _Spinacia_ is rather thick to be digested, and the procedure is time consuming.

As shown in the discontinuous percoll gradient the unbroken pollen grains will pellet down in the tube. A relative high quantity of pollen grains is unbroken. These unbroken pollen grains influence the sperm cell yield in a way that the latter are stuck to the pollen grains and thus lost in the centrifugation step. By filtering the mixture of squashed pollen grains and cytoplasmic material over a 25 µm nylon filter, the unbroken pollen grains are excluded before centrifugation.

The density of a sperm cell can vary. This causes the separation of the sperm cells in two bands in the discontinuous percoll gradient; one on top of the 30% percoll and one on top of the 10% with most of the sperm cells. By using only a 20% percoll layer, the two bands are joined in one layer on top of the 20% percoll. The rest of the sample will be concentrated at the bottom of the tube.

The sperm cells in the pollen grain are spindle shaped, and have long, tail like extensions (Wilms & Van Aelst 1983). Directly after squashing the same spindle shaped sperm cells can be found in the isolation medium. During the isolation and centrifugation procedure the spindle shape disappears and the sperm cells become spherical. The connection between the two sperm cells disappears during the isolation procedure. Similar results are reported for _Zea_ and _Brassica_ (Dupuis et al. 1987, Matthys-Rochon et al. 1987). Microtubules inside the sperm cells of _Spinacia oleracea_ seem to support the cell shape of the sperm cells (Theunis & Wilms, 1988). The change in shape of the sperm cell could be due to the breakdown of the cytoskeleton as also suggested by Tanaka (1988). The FDA test (Heslop-Harrison et al. 1984) is a good measurement for viability. In _Spinacia_ most of the isolated sperm cells are viable. A large percentage looses the viability within 18 hrs. The conditions to keep them viable for a longer period are still to be improved.

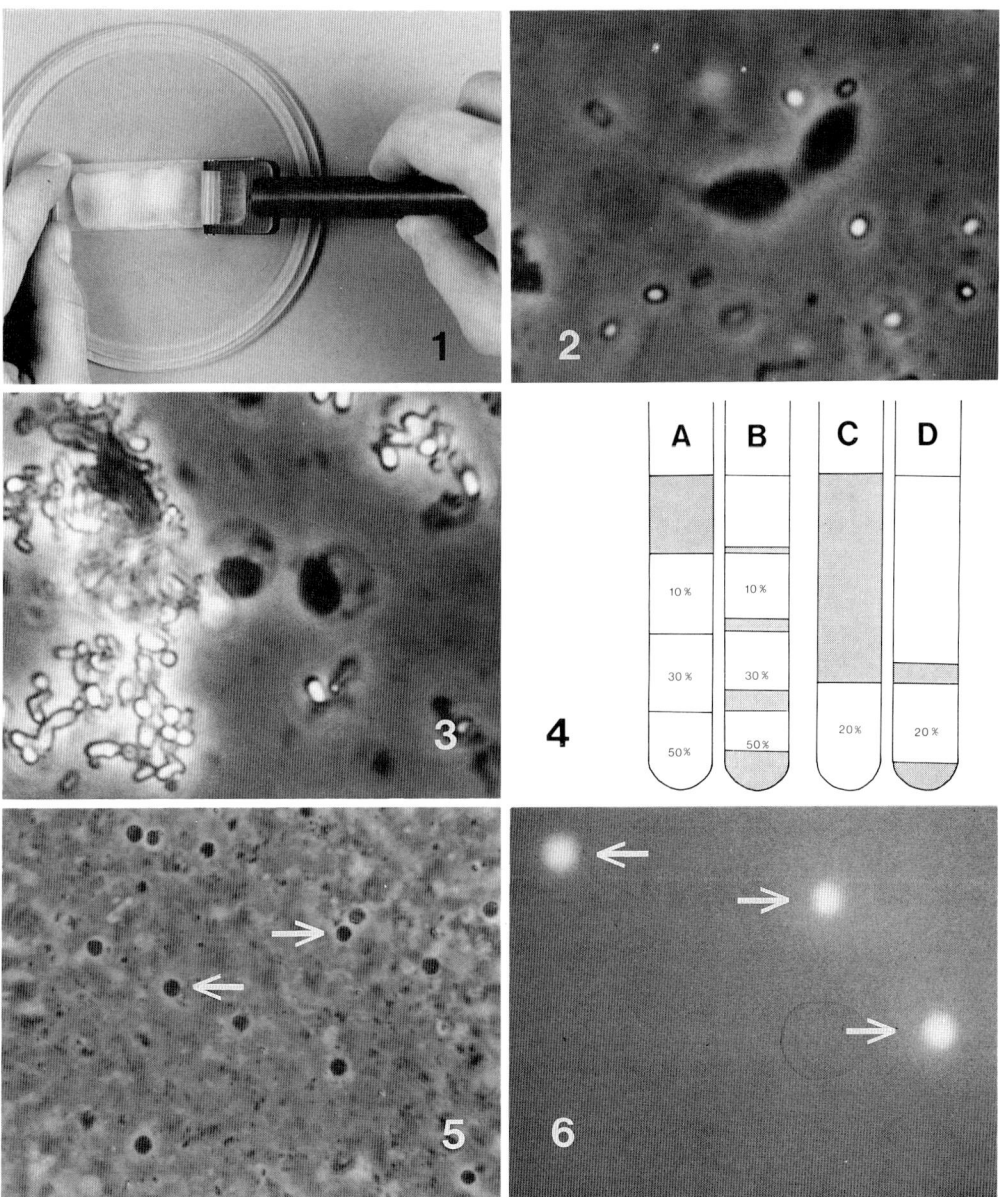

Fig 1. Device for squashing pollen grains. Fig 2. Sperm cells in isolation medium containing 25% sucrose. The two sperm cells are still connected and spindle shaped. Fig 3. Sperm cells in isolation medium containing 5% sucrose. The two sperm cells are separated and round. Fig 4. Schematic drawing of centrifugation. A. Discontinuous percoll gradient before centrifugation. B. Discontinuous percoll gradient after centrifugation. C. 20% percoll layer, before centrifugation. D. 20% percoll layer, after centrifugation. Fig 5. Sperm cell (arrows) after isolation procedure. Fig 6. Positive FDA staining of sperm cells after the isolation procedure (arrows).

Acknowledgement

The authors wish to thank B. van der Swaluw and R. van der Laan for making the squashing device and Dr E. Matthys-Rochon and Dr C.J. Keijzer for helpful discussion. The research was supported by European Community grant Biotechnology Action Programme 0202-NL.

Literature

Baldi BG, Franceschi VR, Loewus FR (1986) Dissolution of pollen intine and release of sporoplasts. In: Mulchay DL, Ottaviano E, (eds) Pollen ecology and biotechnology. Springer Verlag, Berlin Heidelberg New York pp 77-82

Brewbaker JL, Kwack BH (1963) The essential role of calcium ion in pollen germination and pollen tube growth. Amer J Bot 50:859-865

Dupuis I, Roeckel P, Matthys-Rochon E, Dumas C (1987) Procedure to isolate viable sperm cells from corn (Zea mays L.) pollen grains. Plant Physiol 85:876-878

Heslop-Harrison J, Heslop-Harrison Y, Shivanna KR (1984) The evaluation of pollen quality and a further appraisal of the fluorochromatic (FCR) test procedure. Theor Appl Gen 67:367-375

Hough T, Singh MB, Smart IJ, Knox RB (1986) Immunofluorescent screening of monoclonal antibodies to surface antigens of animal and plant cells bound to polycarbonate membranes. J Immuno Meth 92:103-107

Matthys-Rochon E, Verge P, Detchepare S, Dumas C (1987) Male germ unit isolation from three tricellular pollen species: Brassica oleracea, Zea mays, and Triticum aestivum. Plant Physiol 83:464-466

Loewus FA, Baldi BG, Franceschi VR, Meinert LD, McCollum JJ (1985) Pollen sporoplasts: dissolution of pollen walls. Plant Physiol 78:652-654

Roeckel P, Dupuis I, Detchepare S, Matthys-Rochon E, Dumas C (1988) Isolation and viability of sperm cells from corn (Zea mays) and kale (Brassica oleracea) pollen grains. In. Wilms HJ, Keijzer CJ (eds) Plant sperm cells as tools for biotechnology. Pudoc, Wageningen, IN PRESS.

Russell SD (1986) Isolation of sperm cells from the pollen of Plumbago zeylanica. Plant Physiol 81:317-319

Southworth D (1986) Sperm cell structure in Gerbera jamesonii (Asteraceae). In: Knox RB, Williams EG (eds) Pollination '86. Plant Cell Biology Research Centre. Botany School. University of Melbourne, Melbourne, pp 172-177

Theunis CH, Wilms HJ (1988) Immunolabeling of tubulin in mature pollen of spinach (Spinacia oleracea). In. Wilms HJ, Keijzer CJ (eds) Plant sperm cells as tools for biotechnology. Pudoc, Wageningen, IN PRESS.

Tanaka I (1988) Isolation of generative cells and their protoplasts from pollen of Lilium longiflorum. Protoplasma 142:68-73

Wilms HJ, Van Aelst AC (1983) Ultrastructure of spinach sperm cells in mature pollen. In. Erdelska O (ed) Fertilization and embryogenesis in ovulated plants. VEDA, Bratislava, p 105-112

Zhou C, Orndorff K, Allen RD, DeMaggio AE (1986) Direct observations on generative cells isolated from pollen grains of Haemanthus katherinae Baker. Plant Cell Rep 5:306-309

Ultrastructural Observations on Isolated and Cultured Protoplasts of Lily Pollen

H.Miki-Hirosige*, S.Nakamura*, T.Takahashi* & I.Tanaka**
* Biological Laboratory, Kanagawa Dental College, Yokosuka 238,Japan
** Department of Biology, Yokohama City University, Kanazawa-ku, Yokohama 236,Japan

Abstract

Pollen protoplasts were obtained from pollen grains of Lilium longiflorum by dissolving intine layers 2 and 3 with treatment of macerozyme (pectinase) and cellulase. The protoplasts regenerate their cell walls by secretion of polysaccharides from vesicles derived from Golgi bodies. The membranes of these secretive vesicles are reformed in the cytoplasm by the membrane flow system. The ultrastructures of an isolated generative cell and its protoplast are shown.

Introduction

We have reported previously on the isolation of lily pollen protoplasts and the ultrastructural aspects of cell wall regeneration and pollen tube germination of the cultured protoplasts (Tanaka et al. 1987; Miki-Hirosige et al. 1988). The pollen tube normally germinates from the aperture of the pollen grain, but an isolated pollen protoplast lacks an aperture. The location of the tube germination point is therefore not structurally marked, so the question arises of how the point is determined. A possibility is that the cytoplasm and the surface membrane of the protoplast or the regenerated wall of a cultured protoplast possess a polarity which influences or determines the germination point or region. In this report, ultrastructural data on the polarity of the pollen protoplasts is presented. During protoplast cell wall regeneration, coated vesicles and large vesicles were observed in the cytoplasm. The possible significance and roles of the vesicles are discussed. Further, preliminary data on the ultrastructure of a generative cell isolated from a pollen protoplast and a protoplast isolated from the generative cell (Tanaka 1988) will be presented.

Materials and Methods

The experimental material was greenhouse-grown Lilium longiflorum cv. 'Georgia'. The methods for the isolation and culture of protoplasts from the pollen grain (Tanaka et al. 1987) and for the isolation of generative cells and their protoplasts from the pollen protoplasts (Tanaka 1988), were described previously.

Electron microscopy

TEM: The method used was described in Miki-Hirosige et al. (1988). SEM: The materials were prefixed in 1.5% glutaraldehyde in 0.1 M sodium cacodylate buffer, pH7.4, 36 C for 5 min., followed by treatment in 0.5% tannic acid in 1.5% glutaraldehyde in 0.1M sodium cacodylate buffer and then in a 1% tannic acid solution using the same concentration of glutaraldehyde and cacodylate buffer, 10 min. in each solution; washed in 0.1M sodium cacodylate buffer pH 7.4, at 4 C for 1h, and postfixed with 2% osmilum tetroxide in 0.1M sodium cacodylate buffer at 4 C for 2h. The materials were then washed with distilled water, dehydrated in graded dilutions of ethanol and amyl acetate for 3h, dried to the critical point with CO_2, coated with gold, and examined with a JEOL-35C scanning electron microscope at 15 kV.

Histochemistry

(1) The periodic acid-thiocarbohydrazide-osmium tetroxide method (PATCO method: Seligman et al. 1965) was used for polysaccharides, and a description of it was presented in a previous paper (Miki-Hirosige et al. 1988). (2) Cationized ferritin (CF; Sigma) treatment; this method was followed by Tanchak et al. (1984). (3) RCA-HRP conjugate method was used for carbohydrate components which bind RCA, e.g. ß-D-galactpyrenosyl sugar residue (Yokohama et al. 1980); Ricinus communis aggluthinin 120 (RCA_{120}) and horse-radish peroxidase (HRP, Sigma): The protoplasts were fixed by immersion at 4 C in either 2-5 % glutaraldehyde (GLA) in PBS buffer solution for varying periods of time, from 30 min to overnight. After washing the materials in PBS buffer, they were immersed in 10% dimethyl sulfoxide (DMSO) in PBS buffer up to 16 h at 4 C. The protoplasts were treated with the RCA-HRP conjugate at a concentration of 150-200 µg/ml in PBS buffer for varying time periods from 45 min to 5 h at room temperature. Next, the protoplasts were washed thoroughly in PBS buffer, fixed again in cold 2.5 % GLA for 30 min, washed in 0.1 M Tris HCl buffer pH 7.6 and were exposed to 3,3'-diaminobenzidine (DAB, Sigma) for 15 min at room temperature. The materials were osmicated for 30-60 min, dehydrated in ethanol and embedded in low-viscosity resin. The controls were protoplasts (a) preincubated in 0.2 M lactose in PBS and then incubated with the RCA-HRP conjugate and 0.2 M lactose mixed togagher; (b) incubated with HRP at a concentration of 100 µg/ml instead of the conjugate; and (c) treated directly with DAB.

Results

A protoplast which is emerging from the aperture of a lily pollen grain treated with macerozyme R-10 and cellulase Onozuka R-10, shows that intine 1 is attached to the stripped pollen exine, but intine layers 2 and 3 are not evident (Fig. 1a), although the mature pollen grain wall of Lilium longiflorum has exine, nexine and 3 intine layers. The same material, stained by the PATCO method, gives no positive reaction in the area of intines 2 and 3, but some material is visible in this area (Fig. 2a). In a mature pollen grain, this area shows positive reaction with the PATCO method (Fig. 2b, Nakamura, 1979). When a pollen grain is treated with cellulase only, intine 1 and 2 remain beneath the exine layer, but a part of intine 3 is not evident (Fig. 1b). In the treatment with macerozyme only the electron lucent intine 3 is attached to the cell membrane which has coated pits, although intine 2 is not evident (Fig. 1c). In the cytoplasm of the protoplasts (Figs. 1a and 2a), Golgi bodies, mitochondria, amyloplasts, lipid bodies and numerous vesicles of indeterminate shape which appear empty, are observed. During the cell wall regeneration, RCA-HRP positive substances are visible in the regenerated cell wall, Golgi vesicles, and Golgi body (Fig. 3). Although cationized ferritin (CF) is observed only on the cell membrane surface when the protoplast was treated with CF for 2 min, CF is observed in coated vesicles in the cytoplasm when the protoplast was treated with CF for 40 min (Fig. 4).

Figure 5 shows freshly isolated pollen protoplasts, whose surfaces are uniformly smooth. The surfaces of numerous protoplasts were examined with the SEM, but none showed any evidence or indication of polarity or an aperture. Sectioned pollen protoplasts, examined with the TEM, also showed no indication or sign of polarity in the cytoplasm. While protoplasts cultured for 1 week show regenerated cell walls having rough surfaces and some of them germinate pollen tubes (Fig. 6), they still show no sign of polarity, nor is there any evidence in the cytoplasm of such protoplasts when examined with the TEM. Figure 6 also shows pollen tubes developed from pollen protoplasts but their surfaces show no differences from those of the protoplasts from which they have grown.

The generative cell liberated from a disrupted pollen protoplast is spindle-shaped (Figs. 7 and 8), the same shape that it has in an intact, normal pollen grain. Many transparent vesicles, which were changed from lipid bodies (Miki-Hirosige and Nakamura 1977), were observed in the cytoplasm of both the isolated and normal generative cell. These vesicles (Figs. 8 and 9) contain the same PATCO positive substances were seen as in normal pollen grains (Miki-Hirosige and Nakamura 1982; Nakamura and Miki-Hirosige 1985). When a protoplast is isolated from a generative cell, it assumes a spherical shape (Fig. 9).

Discussion

When a pollen protoplast is emerging from the aperture of a lily pollen grain treated with both macerozyme (pectinase) and cellulase, intine layers 2 and 3, which show a positive reaction to PATCO staining in situ (Fig. 2b), are dissolved, while intine 1 remains attached to the stripped pollen exine (Figs. 1a and 2a). Mainly intine 3 is melted with cellulase (Fig. 1b), and intine 2 with macerozyme (Fig. 1c). It appears that intine 3 is composed mainly of cellulose, intine 2 of pectin, and intine 1 is composed of other substances. With regard to pollen tube germination, no evidence of polarity in the cytoplasm or on the surface of the isolated protoplast, was found. Despite the lack of evidence thus far, the determination of the germination point needs to be pursued as a phenomenon of cell differentiation.

During culturing, protoplasts regenerate their cell walls by means of vesicles which contain PATCO positive substances (Miki-Hirosige et al. 1988). In this study, using the RCA-HRP treatment method, Golgi vesicles, a Golgi body, and the regenerated cell wall are shown to contain positively stained substances (Fig. 3). It is assumed that some vesicles containing PATCO and/or RCA-HRP positive substances derived from Golgi body, secrete these substances into the regenerating cell wall. When a protoplast cultured for 2 days is surface-labelled with cationized ferritin for 2 min prior to fixation, CF appears on the surface of the cell membrane. When the CF treatment is extended to 40 min prior to fixation, CF appears in a coated pit and in vesicles (Fig. 4). Tanchak et al. (1984) reported endocytosis of cationized ferritin by coated-membrane system in soybean protoplast. We have observed (1988) numerous coated and smooth vesicles which contain PATCO and/or RCA-HRP positive substances (polysaccharides), and other coated vesicles in the protoplast cytoplasm during the regeneration of the cell wall. It is assumed that some vesicles secrete substances which become part of the cell wall while others play an endocytotic role. After early initiation of the regenerating wall, membrane substances are probably circulated between the cytoplasm and the cell mambrane.

In the isolated generative cell (Fig. 8) and in a mature pollen grain of Lilium longiflorum, many transparent vesicles can be seen. These vesicles derived from lipid bidies contain polysaccharides, increase thier number after pollen germination and fuse to the generative cell wall (Miki-Hirosige and Nakamura 1982; Nakamura and Miki-Hirosige 1985). It is assumed that the vesicles become attached to the generative cell wall and secrete some substances for its enlargement during cytokinesis. The isolated generative cell has a spindle shape (Figs. 7 and 8), but it is not clear from this study that the shape is maintained with the strong forces of the cell wall as reported by Theunis et al. (1985), or if the shape is maintained by the numerous microtubules seen to be scattered alongside the wall.

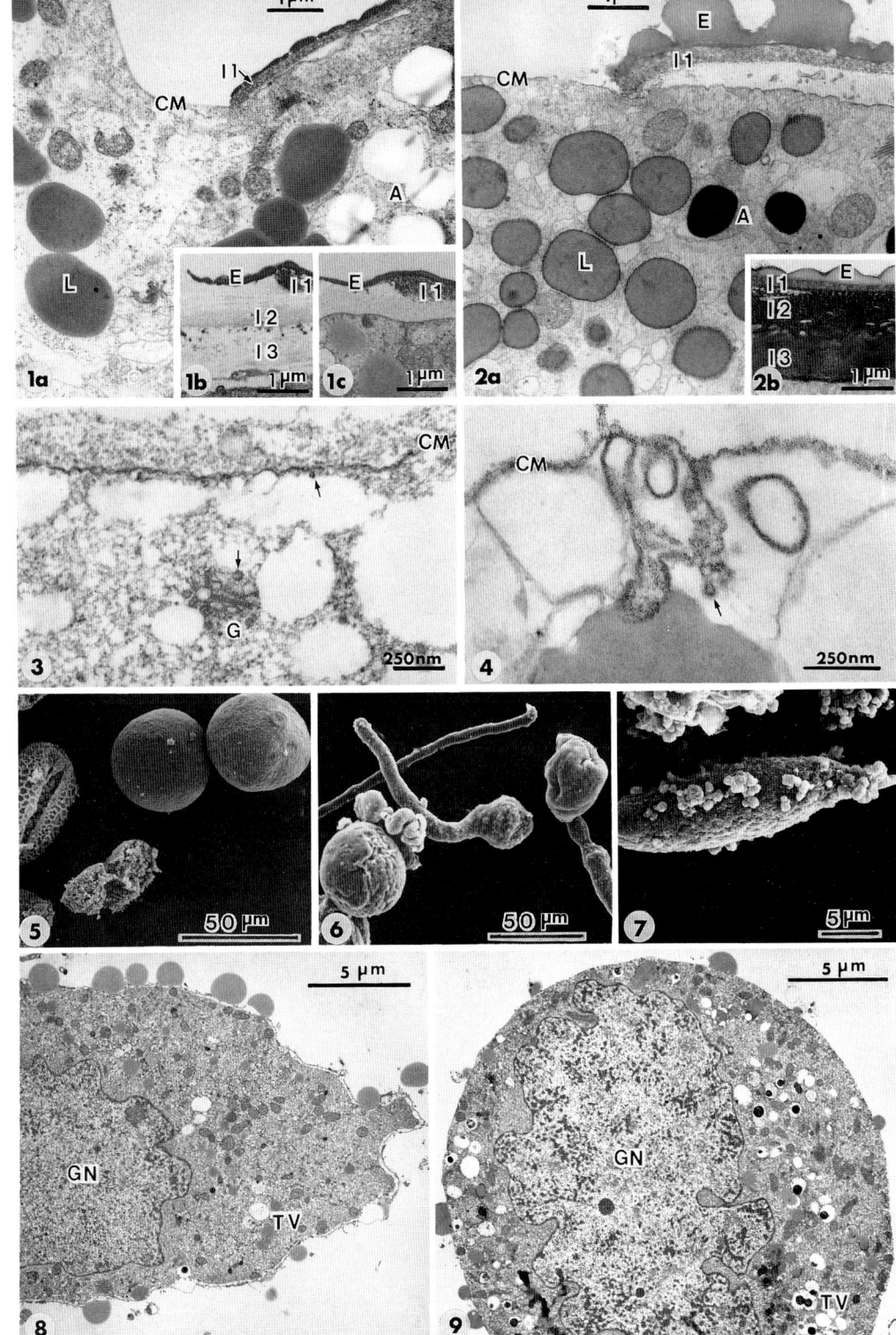

◄ Fig. 1. Protoplast emerging from lily pollen. a. Intine 1 (I1) is attached
to the stripped exine layer (E), but intine 2 (I2) and 3 (I3) are not
evident with macerozyme and cellulase. x10,950 b. Intine 1 and 2
remain beneath the exine layer, but a part of intine 3 is not evident
by treatment with cellulase. x10,000 c. Although intine 2
was melted, electron lucent intine 3 is attached to the cell membrane
(CM) which has coated pits by treatment with macerozyme. x10,000

Fig. 2. a. An emerging protoplast stained with PATCO method. Intine 2 and 3
have not positive reaction for PATCO, but some material is visible
there. Starch grains in amyloplasts (A) are stained with PATCO.
x12,000 b. Intine layers especially intine 2 and 3 of a mature
pollen grain are stained with PATCO. x10,000 Lipid body (L)

Fig. 3. A protoplast cultured 2 days, was stained with RCA-HRP for 45 min.
The regenerated cell wall, Golgi vesicles (arrows), and a Golgi
body (G) are stained positively with RCA-HRP. x40,000

Fig. 4. A protoplast cultured 2 days, was treated with CF for 40 min.
Labelling of coated vesicles by CF (arrow) are shown. x60,000

Fig. 5. Freshly isolated pollen protoplasts and stripped pollen exines.
The protoplasts have smooth surfaces of the cell membrane. x450

Fig. 6. Protopalsts after 8 days in culture, emerging pollen tubes, have
rough surfaces of regenerated cell walls. x360

Fig. 7. An isolated generative cell with a view of SEM. x2,250

Fig. 8. An isolated generative cell with a view of TEM. x4,000
Transparent vesicles (TV) are seen in the cytoplasm.

Fig. 9. A protoplast from the generative cell. x4,000 Generative nucleus (GN)

References

Miki-Hirosige H and Nakamura S (1977) Electron transparent vesicles in the
generative cell of pollen grains. Jap J Palynol No.19 11-19

Miki-Hirosige H and Nakamura S (1982) Growth and differentiation of amyloplasts
during male gamete development in Lilium longiflorum. ed.Mulchay DL &
Ottaviano E "Pollen: Biology and Implications for Plant Breeding"
Elsevier Science Publishing Co. 141-147

Miki-Hirosige H, Nakamura S, Tanaka I (1988) Ultrastructural research on
cell wall regeneration by cultured pollen protoplasts of Lilium longiflorum.
Sex Plant Reprod 1:36-45

Nakamura S (1979) Development of the pollen grain wall in Lilium longiflorum.
J Electron Microsc 28:275-284

Nakamura S, Miki-Hirosige H (1985) Fine-structural study on the formation of the
generative cell wall and intine-3 layer in a growing pollen grain of
Lilium longiflorum. Amer J Bot 72:365-375

Seligman AM, Hanker JS, Wasserkrug H, Dmochowski H, Katzoff L (1965)
Histochemical demonstration of some oxidized macromolecules with thiocarbo-
hydrazide (TCH) or thiosemicarbazide (TSC) and osmium tetroxide.
J Histochem Cytochem 13:629-639

Tanaka I (1988) Isolation of generative cells and their protoplasts
from pollen of Lilium longiflorum. Protoplasma 142:68-73

Tanaka I, Kitazume C, Ito M (1987) The isolation and culture of lily
pollen protoplasts. Plant Sci 50:205-211

Tanchak MA, Griffing LR, Mersey GB, Fowke LC (1984) Endocytosis of cationized
ferritin by coated vesicles of soybean protoplasts. Planta 162:481-486

Theunis CH, McConchie CA, Knox RB (1985) Three-dimensional reconstruction of the
generative cell and its wall connection in mature bicellular pollen
of Rhododendron. Micron and Microscopia Acta 16:225-231

Yokoyama M, Nishiyama F, Kawai N, Hirano H (1980) The staining of Golgi
membranes with Ricinus communis agglutinin-horseradish peroxidase conjugate
in mice tissue cells. Exp Cell Res 125:47-53

Isolation and Characterization of Viable Sperm Cells from Tricellular Pollen Grains

E. Matthys-Rochon, S. Detchepare, V. Wagner, P. Roeckel and C.Dumas
Reconnaissance   Cellulaire et Amelioration des Plantes
Université Claude Bernard LYON I
43, Bd. du 11 Novembre 1918
69622 Villeurbanne Cedex-France

ABSTRACT

A procedure has been developed for *Brassica* sperm cell isolation using pollen breakage followed by a percoll gradient step. Sperm cell viability has been checked by the fluorochromatic reaction. In addition *Zea Mays* sperm cells have been isolated, fixed (Dupuis et al 1987) and characterized for their cell wall. The results show that corn sperm cells   are true protoplasts.

INTRODUCTION

In flowering plants tricellular pollen grains house a vegetative cell with a large nucleus and two sperm cells. When the pollen germinates, these male gametes will go   through the pollen tube, reach and fertilize either the egg or the central cell. The two fusion events are still little known, especially at the recognition processes which seem to occur at the gamete level (Russell 1985). So during the last few years it has become a challenge to isolate and characterize plant gametes.Thus fractions of sperm cells have been purified from different pollen species, *Plumbago* (Russell 1986),*Zea mays* (Dupuis et al.1987), *Gerbera* (Southworth and Knox 1987).Sperm cell characterization is the second step and important topic of this new field of research. The only ultrastructural data has been reported in a previous paper by Dupuis et al. (1986) . Attempts to isolate female gametes have also been recently made in *Torenia* (Mol 1986),*Tobacco* (HU et al. 1985) ,*Zea mays*

(Song et al. 1988). Consequently plant gamete isolation and characterization may lead to a better understanding of plant fertilization. In this paper we present the first results of *Brassica* sperm cell isolation and the cytological demonstration that *Zea mays* sperm cells are true protoplasts.

MATERIALS AND METHODS

Pollen collection and sperm cell isolation

Brassica napus pollen has been collected by hand brushing the dehiscent anthers from plants grown in the fields. Pollen storage was achieved by submersion in liquid nitrogen.Sperm cell release is obtained by breaking pollen with a tissue homogenizer in Brewbaker and Kwack (1963) medium supplemented with 12.5 sucrose (BK12.5).The resulting homogenate is filtered through a 20 μm stainless steel sieve. The filtrate is put on a percoll gradient (15-30-50% percoll in BKS12.5) and sperm cells are then collected between the 15-30 % layers after a centrifugation step (2000g for 40 mn). Sperm cells are washed with BKS12.5 and checked for their viability using the fluorochromatic reaction (Heslop-Harrison et al.1984).To determine the yield, the ratio of the final number of viable sperm cells to the theoretical number of sperm cells contained in the pollen grains is calculated.*Zea Mays* sperm cells have been prepared as described before (Dupuis et al. 1987).

Microscopical studies

Sperm cell observations were conducted with a light microscope (Nikon-labophot type 104) equiped with phase contrast and epifluorescent systems or with a Jeol 1200 EX electron microscope.*Zea mays* sperm cells have been fixed with glutaraldehyde and osmium and stained with uranyl acetate and lead citrate as previously indicated. (Dupuis et al 1987)

RESULTS and DISCUSSION

Sperm cell isolation

Brassica napus sperm cells were released by breaking pollens (fig 1) . The latter are not sensitive to an osmotic shock as it is the case for *Zea mays* pollens (Dupuis et al. 1987). After breakage the homogenate is filtered, but

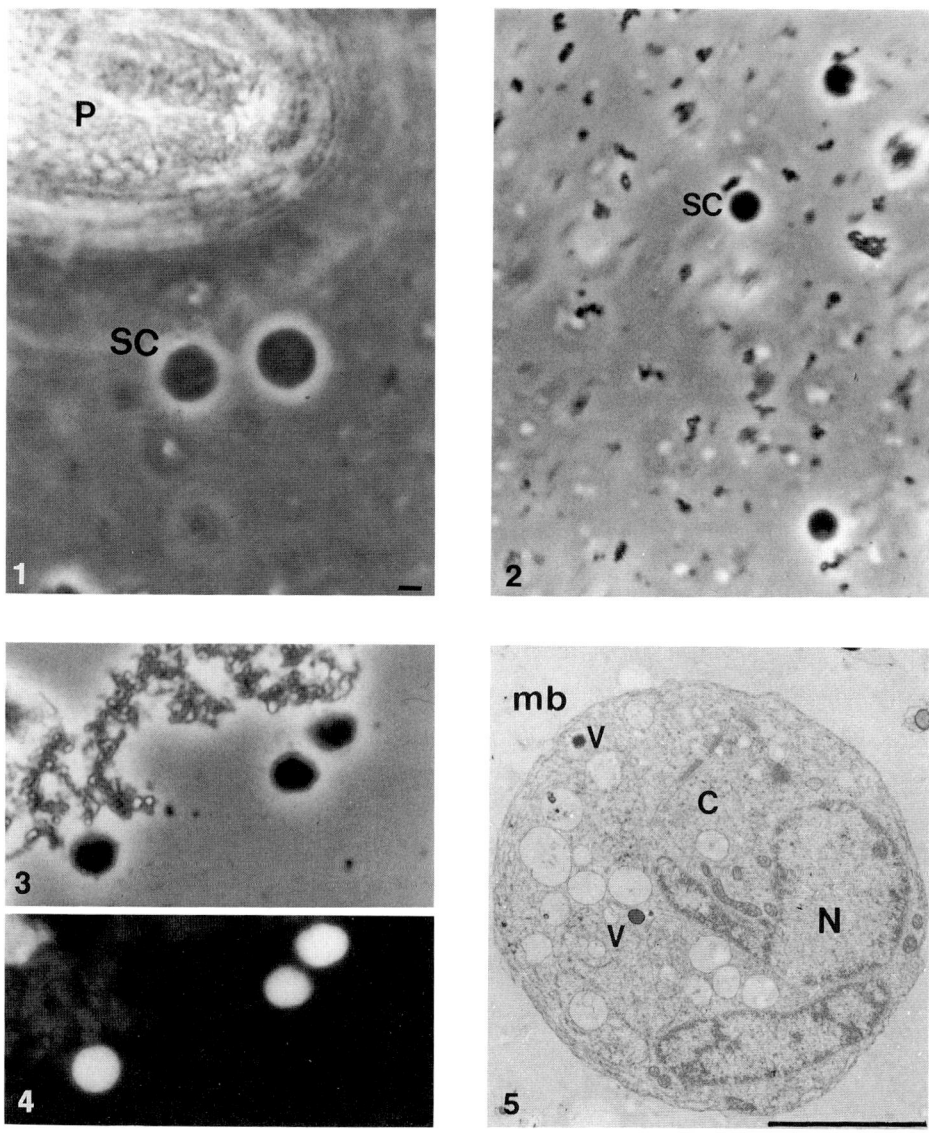

Light microscopical phase contrast observations
1 . *Brassica*   pollen grain (P)   two released spherical sperm cells (SC)
2 . *Brassica*   isolated sperm cells after filtration.Cytoplasmic debris are visible.
3 . *Brassica*   isolated sperm cells
4 . Corresponding epifluorescence observation to number 3 , using   the FCR reaction. The three sperm cells are FCR+
Electron microscopical observation
Maize sperm cell results of Patag test show the presence of polysac charides in cytoplasmic vesicles (v) but no precipitate was observed around the cell. N: nucleus, c: cytoplasm, mb: plasma membrane.bar: 1μ

after filtration the released sperm cells are still mixed with pollen wall debris, organelles and vegetative nuclei.(fig. 2). A percoll gradient allows for the partial purification of the sperm cells. At this stage , some small contaminants remain . Further experiments will be done to increase the purity of the cell fraction. The yield is generally of around 5%. This yield may appear low but the sperm cell isolation is done with frozen pollens and the main problem is to maintain sperm cell survival.

Sperm cell viability

In a previous work (Heslop-Harrison et al. 1984) the FCR test has been described and used to check pollen integrity. We applied this fluorochromatic reaction on sperm cells (Matthys-Rochon et al.1986) and obtained positive results for every sperm cell species checked. In addition, ATP (Adenosine-triphosphate) measurements in sperm cells indicate a direct correlation with the FCR test results (Matthys-Rochon et al.1988). At the end of the isolation procedure the *Brassica* sperm cells appear to be viable in terms of FCR (fig.3-4). Such a data is essential since no future experimentation is possible if sperm cells are not intact.

Polyssaccharide detection

The second aim of our work was to characterize sperm cells. The first electron microscopical views of sperm cells has been given by Dupuis et al.(1987).Following their procedure , *Zea mays* sperm cells have been isolated ,fixed and presence of polysaccharides have been searched. The lack of a cell wall surrounding the sperm cells was proven at both the structural and ultrastructural level.The calcofluor staining and observations using the PATAg test (fig.5) both failed to show the presence of a cell wall surrounding the sperm cells. This result indicates that *Zea mays* sperm cells are true protoplasts.

CONCLUSION

Future fundamental and biotechnological prospects

Enriched fractions of male and also female gametes are of a great interest.In a near future immunological studies may reveal plant gamete specific membrane determinants which seem to have a role in the recognition processes. Russell 1985). Such data may lead to a new dynamic understanding of double fertilization in flowering plants. At the applied level, plant gametes may be cultured as microspores from anthers and give rise to new haploids. At last ,in vitro fertilization may be a way to overcome incompatibility barriers and to obtain new diploids, using wild or transformed sexual protoplasts. Even if all these prospects appear to be speculative, it is clear that isolated plant gametes constitute a promising cellular system for study.

ACKNOWLEDGMENTS

This work was carried out in the framework of contract N° BAP-O2O3-F of the Biotechnology Action Programme of the Commission of the European Communities.

LITERATURE REFERENCES

Brewbaker J.L., Kwack B.H. (1963) . The essential role of calcium ion in pollen germination and pollen tube growth. Am. J. Bot. 50: 859-865

Dumas C., Knox R.B., Gaude T. (1985) The spatial association of the sperm cell and vegetative nucleus in the pollen grain of *Brassica* . Protoplasma 124: 168-174

Dupuis I., Roeckel P.,Matthys-Rochon E., Dumas C. (1987) Procedure to isolate viable sperm cells from maize *(Zea mays)* pollen grains. Plant.Physiol. 85 (4): 876- 879

Heslop-Harrison J., Heslop-Harrison Y., Shivanna K.R. (1984) The evaluation of pollen quality and a further appraisal of the fluorochromatic (FCR) test procedure .Theor. Appl. Genet: 367-375

Hu Shi-Yi , Li Le-Gong and Zhu Cheng. (1985) Isolation of viable Embryo sacs and their protoplasts of *Nicotiana tabacum* Act. Bot. Sin. 27, 4 : 337-344

Matthys-Rochon E., Vergne P., Detchepare S., Dumas C. (1987) Male germ unit isolation from tricellular pollen species: *Brassica oleracea , Zea Mays* and *Triticum aestivum.* Plant Physiol. 83: 464-466

Matthys-Rochon E., Roeckel P., Detchepare S., Wagner V.,Vergne P., Dupuis I., and Dumas C. (1988) Plant sperm cells and potential biotechnological prospects. Proceedings of the International conference on research in plant sciences and its relevance to future.Dehli India

Meadows M.G.(1984) A batch assay calcofluor fluorescence to characterize cell wall regeneration in plant protoplasts. Anal. Bioch. 141: 38-42

Mol R., (1986) Isolation of protoplasts from female gametophytes of *Torenia fournieri* Plant.Cell.Rep. 3: 202-206

Russell S.D. (1985) Preferential fertilization in *Plumbago zeylanica* : Ultrastructural evidence for gamete-level recognition in an Angiosperm Proc. Nat.Acad.Sci. USA 82: 6129-6132

Russell S.D. (1986) Isolation of sperm cells from the pollen of Plumbago zeylanica Plant Physiol. 81: 317-319

Song Y., Wagner V., Matthys-Rochon E., Dumas C. (1988) Observations on isolated embryo sac of corn .Proceedings of the tenth international symposium on sexual reproduction in higher plants. Siena Italy

Southworth D., Knox R.B. (1987) Methods for isolation of sperm cells from pollen,in Plant sperm cells as emerging tools for crop biotechnology. Keijzer C.J. ,Wilms H.J. Pudoc Ed. Wageningen The Netherlands

Thiéry J.P. (1967) Mise en évidence des polysaccharides sur coupes fines en microscopie electronique. J. Microscopie 6: 967-1018

Ultrastructural Aspects of *Epidendrum* Male Gametogenesis

Alfredo E. Cocucci
Laboratorio de Embriología Vegetal IMBIV[1]
Casilla de Correo 495
5000 Córdoba
Argentina

INTRODUCTION-*Epidendrum scutella* Lindl, unlike most flowering plants, is a species with long living pollen tubes; as a matter of fact they remain im the stylar channel for about three months (Cocucci & Jensen 1971) waiting for the female gametophyte development. Becouse of it, most of the cytological events that characterize formation and maturation of sperms are considerably slower. Such feature favours the possibility to catch different stages during the ontogenetic process. Materials were prepared according to the procedure described earlier (*op.cit.*).

RESULTS-Observations are centered concerning the behaviour of the generative cell and the surrounding vegetative cytoplasm involved in the the sperms formation and maturation. For the purpoue of description 4 different stages are recognised: 1.- Germinated pollen; 2.- Sperms formation; 3.-Sperms differentiation.

1.-Germinated pollen: The syphonogenic cell cytoplasm surrounding the generative cell, once in the pollen tube, is characterized by an abundance of cisternoid RER parallel to the generative cell. A great deal of plasmodesmata are then established in order to connect the syphonogenic RER with the generative cell SER at the existing wall gaps. The generative cell in the pollen tube has an elongated nucleus surrounded by an array of parallel microtubules longitudinally oriented (Fig. 1A). The cytoplasm is reduced; the dictyosomic system is active, constantly contributing via exocytosis toward the interphase with the

[1]Instituto Multidisciplinario de Biología Vegetal, Consejo Nacional de Investigaciones Científicas y Técnicas - Facultad de Ciencias Exactas Físicas y Naturales, Universidad Nacional de Córdoba, Argentina.

vegetative cytoplasm.It is also apparent an autophagic process
initiated in early stages of the generative cell, as can be
infered from the occurrence of material made of membrane remnants
in the same interphase.

2.-Sperms formation: After the generative cell mitosis 2
sperm cells result. They are elongated like their parent cell was.
RER and SER together with vesicles derived from the dictyosomic
system are the most abundant membranous component; scattered free
ribosomes can be seen. Remnants of membranous material in the
interphase with the vegetative cell cytoplasm suggest that the
litic process continues. Regarding this matter, has to be remarked
the poorly structured wall with a very irregular contour (Fig
1B).
The nucleous has clumps of chromatic condensations much more
extensive than the previous stages, being an indication of nuclear
sap loss. Plasmodesmata connecting with the vegetative cell are
cut down (Fig.1B, 2B).

An interesting feature of the cytokinetic process during the
sperm formation is that the vesicles involved in it also affects
the thin sheet of vegetative cytoplasm near by, becouse of the
poor organized wall, so the vegetative cytoplasm is cut in 2 parts,
one located toward the pollen grain meanwhile the other is toward
the ovule; the former is then anucleated and provided with a huge
vacuole, and the later is evacuolated but nucleated.

3.-Sperm differentiation: Sperms differentiation is
characterized by the development of a system of intranuclear
microfilamets (fig. 2A).Its double membrane exhibits gaps larger
than the regular nuclear pore which are also present (fig. 2C).
There is a progressive carving of the cytoplasm due to a constant
exocytosis from its vesicles and probably from the ER. Material
from both origins contribute to the interphase with the vegetative
cell, so at this time appears as a fluid matter within which the

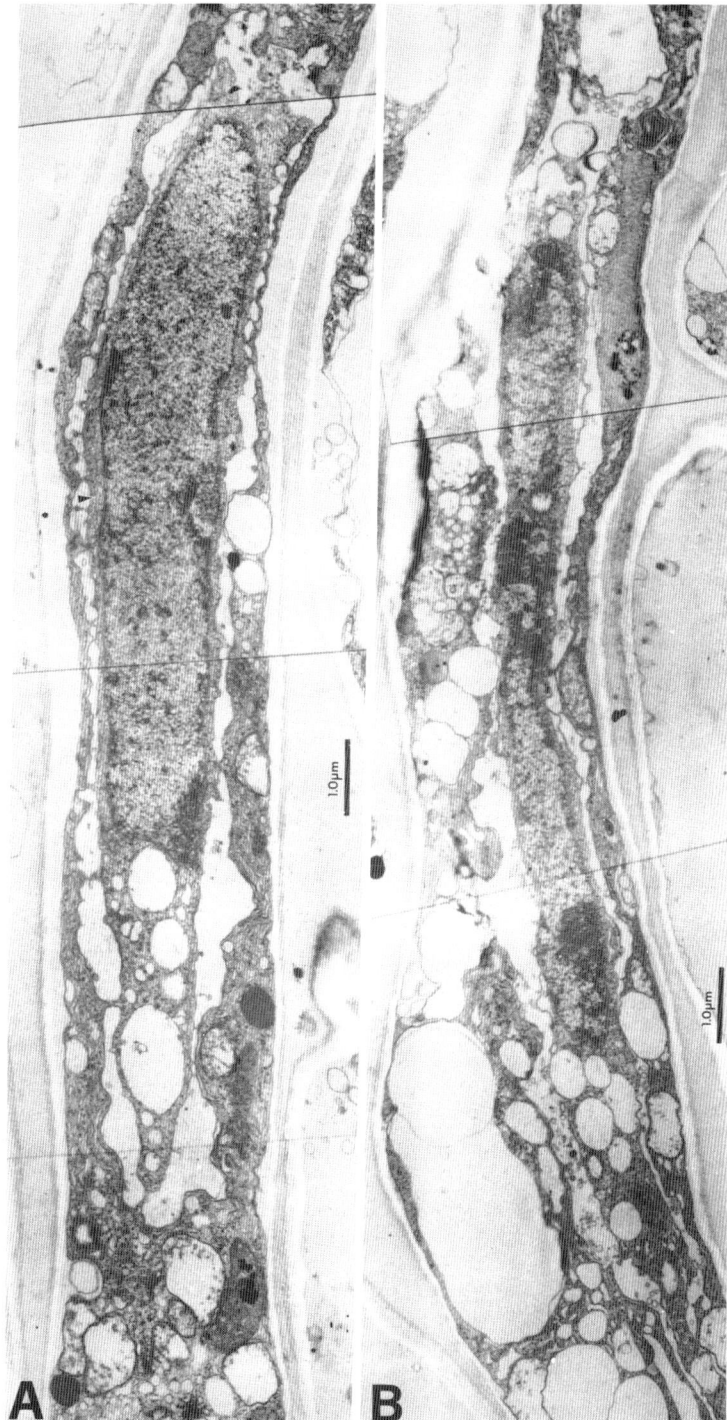

Fig. 1. Gametogenic cell, microtubule (arrowhead), B. newly formed
sperm surrounded by the vegetative cytoplasm.

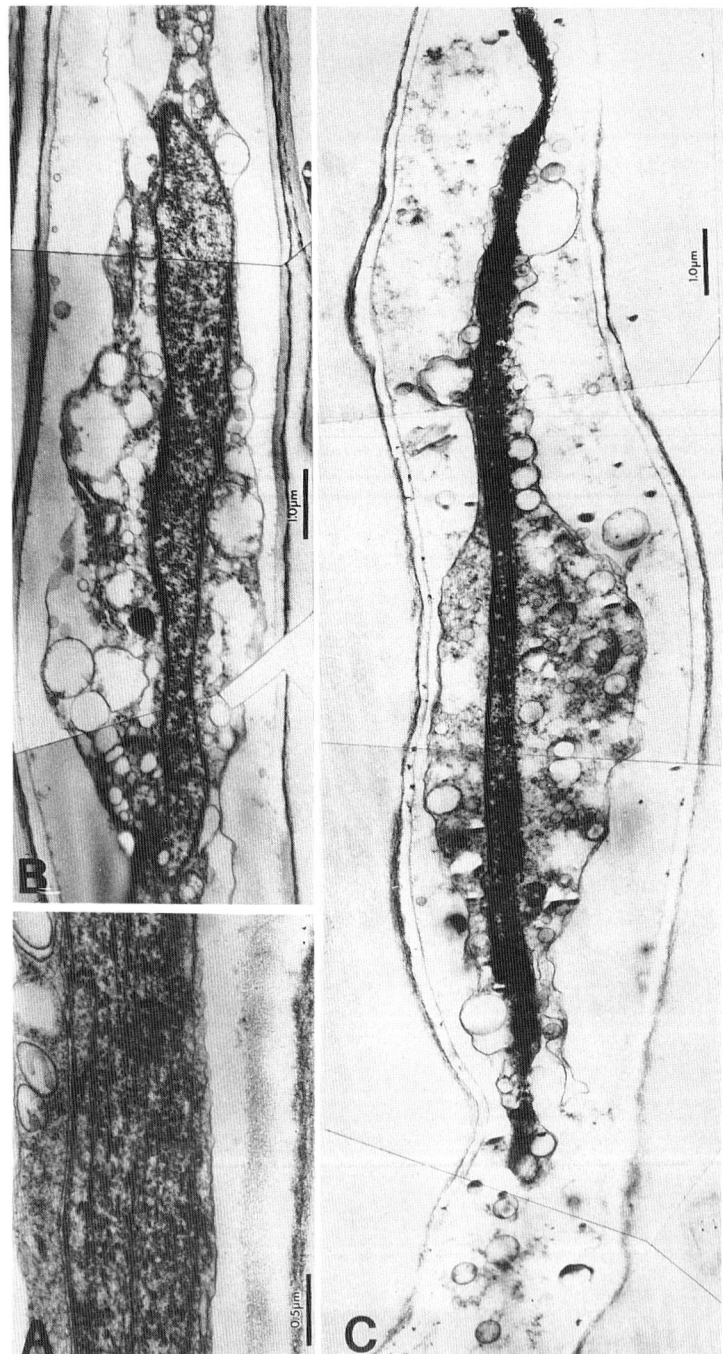

Fig. 2. A. Long. section of a sperm middle part, nuclear pore on top and nuclear fenestra at the bottom, note intranuclear microfilaments, B. sperm in an early stage of differentiation without vegetative cytoplasm around, C. sperm fully differentiated with intranuclear microfilaments.

mature sperms are suspended. If it is considered the origin of such a fluid matter it can be infered that among other components, proteins must be present; they could play a decisive rol in the recognition or in the guiding systems and may also be in facilitating the way of the sperms toward their final destination.

DISCUSSION- The sequence of events in *E. scutella* sperms development is diagramatically sumarized in Fig. 3.

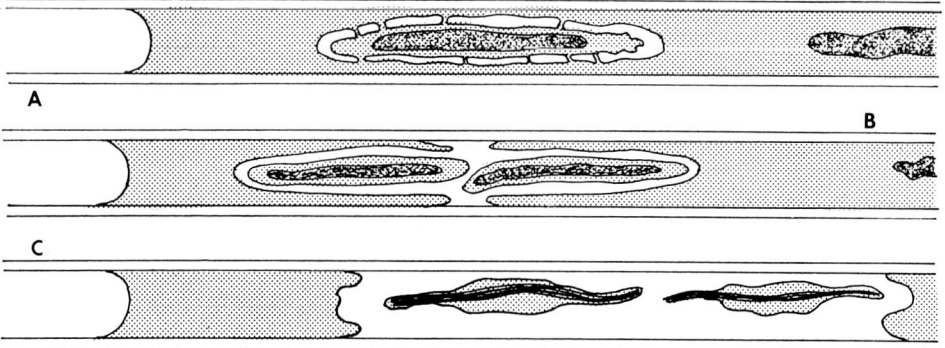

Fig.3. Interpretative diagrams of male gametogenesis. A. generative cell included in the vegetative cytoplasm, at left vacuole, at right pollen tube nucleus, B. newly formed sperms, note the vegetative cytoplasm partition, C. pollen tube with 2 sperms fully differentiated.

A point of interest has to do with the occurrence of a conspicuous wall arround the generative cell since its origin, which is consistent with some previous work (Gorska-Brylass 1967; Heslop-Harrison 1968 Cocucci , Jensen 1969) but markedly different from barley (Cass , Karas 1975), where the generative cell is initially naked and wall formation is a transitory episode initiated during prophase that do not go any further than karyokinesis.

The extreme reduction of the sperm cytoplasm, totally devoided of double membrane organelles stands as an unique feature (Russell , Cass 1981); same thing is to be said of the intranuclear filaments and the fenestration of the nuclear membrane. Both features may be funtionally related. In fact the microfilaments look like actin, on account of their size; if this is so ,they must be engaged in a kind of movement system for what it is very convenient to have an easy way of fuel intake from the cytoplasm. The sigmoid profile of the sperm in Fig.2C suggests that the sperm was frozen by fixation in such an attitude. If such speculation came to be true it would be solved the translation problem from copulative tube tip to the egg an the central cell, through the huge discharge of the tube cell that was forced into the penetrated synergid.

The sperms approaching method of *Epidendrum* appears to be quite different from that of barley and *Beta* (Cass 1973; Cass , Karas 1975; Hoefert 1969),whose sperms provided with clusters of microtubules nearly around the entire periphery of the cell, have no intra nuclear microfilaments and do not display directional movement.

ACKNOWLEDGMENTS- I am pleased to acknowledge finantial support for this work from the Consejo Nacional de Investigaciones de Argentina (PID N.3912805), The John Simon Guggenheim Foundation and for their advice to Dr.W.A. Jensen from the University of California, Berkeley.

Cass D D (1973) An ultrastructural and Nomarski-interference study of the sperms of barley. Can J Bot 51(3):601-605
Cass D D , Karas.I (1975) Development of sperm cells in barley. Can J Bot 53(10):1051-1062
Cocucci A E , Jensen W A (1969) Orchid embryology: the pollen tetrad of Epidendrum scutella in the anther and on the stigma. Planta 84:215-229
-----(1971) Orchid embryology: germinating male gametophyte of Epidendrum scutella.Kurtziana 6:25-39
Gorska-Brylass A (1967) Temporary callose walls in the generative cell of pollen grains Naturwissenchaften 9:230-231
Hoefert L L (1969) Fine structure of sperm cells in pollen grains of Beta..Protoplasma 68:237-240
Heslop-Harrison J (1968) Synchronous pollen mitosis and the formation of the generative cell in massulate orchis. J Cell Sci 3:456-466
Russell S D , Cass D D (1981) Ultrastructure of the sperms of Plumbago zeylanica 1. Cytology and association with the vegetative nucleus. Protoplasma 107: 85-107

Stressed Gametophyte

The Effects of Stresses (Cold or/and Darkness) on Pollen Viability of Two Varieties of Grain-*Sorghum*

S. ALAMI, A. SOUVRE and L. ALBERTINI
Laboratoire de Cytologie et de Pathologie végétales
I.N.P.T.-E.N.S. Agronomique
145 avenue de Muret
31076 Toulouse Cédex, France

Introduction

Unfavorable climatic conditions deleterious to pollen viability are responsible of yield reductions of grain sorghum (*Sorghum bicolor*) observed in the South-West of France these last years (Salgarolo, 1986). Induced pollen sterility in sorghum as well as in other cultures, is the result of low night temperatures (positive temperatures below 13°C) during pollen ontogenesis (Downes and Marshall, 1971 ; Brooking, 1976 ; Corp, 1983 and Salgarolo, 1986).

The aim of the present report is to determine the effects of more and more drastic stresses (darkness, cold, cold + darkness) applied at different stages of pollen ontogenesis. These effects have been analysed in relation to several physiological factors of pollen quality by four cytochemical methods used to test the pollen viability.

Materials and methods

Two varieties of grain sorghum : Esquirol (precocious and with a low content of tannic acids) and Sultan (semi-late and rich in tannic acids) were grown under natural conditions at a day/night temperatures of 25°C/15°C.

Each treatment (see tables 1 and 2) was applied to plants at a stage of pollen ontogenesis according to the development of the panicle sheath (Alami et al., 1988) and followed by moving back to control conditions until flowering, whereas controls were left in natural conditions until flowering.

Four cytochemical tests were applied to evaluate pollen quality : the Alexander test (Alexander, 1969), the Isatine test (Palfi and Köves, 1984), the TTC test (triphenyl tetrazolium chlorid ; Aslam et al., 1964), the FCR test (fluorochromatic test with fluorescein diacetate ; Heslop-Harrison and Heslop-Harrison, 1970 and Heslop-Harrison et al., 1984).

For each test, several groups of pollen viability were defined according to the intensity of the coloured or fluorescent reactions (Table 1) and the importance of these groups was calculated in percentage. In table 2, we reported the groups corresponding to viable pollen only.

Results and Discussion

Test effect. Pollen viability rates fluctuate according to the test used. In controls, the maximal values are obtained with TTC test (Esquirol : 84,3 % ; Sultan : 91,8 %), the values given by Alexander and Isatin are next to each other but of a lower level, and the minimum is given by the FCR test (nearly -20 % of the values given by TTC test).

The determination of viability groups in controls (particularely with TTC and Isatine tests) proves that the responses to stainability tests are more or less important. Our results demonstrate the necessity of using 2 ou 3 tests and analysing accurately the responses to these tests as far as pollen viability studies concerned, in opposition to the majority of the studies concerning cultivated plants. Furthermore, we noted that the rate of aborted pollen differs depending on the sample of spikelets ; it reaches 15 to 18 % in Esquirol and 8 to 16 % in Sultan.

In the case of stressed plants, in spite of the fact that all the results obtained with the four viability tests are consistent, we noted a variability in the reaction level that indicates a deterioration more or less important according to the cytophysiological factor revealed by the test and according to treatments. This is due to the fact that the groups of medium and low viabilities (I^+ and I^* groups of Isatine test by example) are modified as much as the aborted pollen group.

The cytoplasm ripeness (Alexander test) of the Esquirol variety is very sensitive to cold under normal photoperiod conditions (12 h d/12 h n at 5°C). The degradation of proline compounds (Isatine test) in the Sultan variety presents almost the same sensitivity to all the stresses (viability reduction : 10 to 20 %), whereas in Esquirol variety the proline content is most affected by the treatment cited above. The deshydrogenases in the vegetative cell cytoplasm (TTC test) are especially impaired by precocious treatments and particularely by the treatment 12 h d/12 h n, at 5°C. On the other hand, cold during two succes-

Table I - Effects of different treatments on the rate of pollen grains belonging to different viability groups determined by four viability tests in two varieties of *Sorghum bicolor*.

Tests / Treatments		Alexander			Isatine					TTC			FCR		
ESQUIROL		M	S	A	I++	I+	I+-	I-	I^	T+	T	T-	F+	F-	F^
Controls 25°C d/15°C n		77,4	4,4	18,2	63,3	14,3	0	7,2	15,2	71,8	12,5	15,7	66,3	17,4	16,3
12 h n 7°C	VM	78,7	3,7	18,6	63,8	10,7	0	7,7	17,8	-	-	-	52,5	28,5	19,0
12 h n 7°C/12 h d 25°C/12 h n 7°C	VM	68,6	8,8	22,6	54,0	19,4	5,5	5,6	15,5	-	-	-	34,3	42,5	23,2
12 h d 5°C/ 12 h n 5°C	VM	61,5	6,6	31,9	33,8	14,2	23,0	14,2	14,8	26,6	7,7	65,7	48,5	19,8	31,6
	YP	52,5	21,7	25,8	59,9	11,2	0	15,0	13,9	55,0	21,2	23,9	48,5	28,9	22,6
25°C/15°C	VM	64,4	6,9	28,7	60,9	10,5	1,7	5,5	21,4	62,3	4,5	33,2	54,5	18,6	26,9
	YP	78,9	3,4	17,7	70,4	6,9	0	12,1	10,6	71,2	5,9	22,9	58,7	29,6	11,7
24 h n 5°C	VM	60,9	7,7	31,4	57,0	10,9	3,1	6,5	22,5	7,4	50,0	42,6	59,5	17,0	23,9
	YP	51,7	18,5	29,6	64,8	10,6	0	9,7	14,9	0,8	53,7	45,5	56,0	22,4	21,6
SULTAN Controls 25°C d/15°C n		86,9	4,4	8,7	84,8	3,5	0	0	11,7	49,0	42,3	8,7	67,6	15,9	16,5
12 h n 7°C	PM	84,1	1,0	14,9	74,7	3,6	0	4,6	16,7	-	-	-	60,9	24,3	14,8
12 h n 7°C/12 h d 25°C/12 h n 7°C	PM	74,9	9,2	15,9	55,2	13,8	7,7	3,5	19,8	-	-	-	48,0	28,3	23,7
12 h d 5°C/ 12 h n 5°C	PM	79,8	28,0	17,4	39,1	28,7	3,3	2,7	26,2	21,8	28,5	49,7	53,0	14,1	32,9
	M	80,3	5,9	13,8	59,0	14,9	0	8,8	17,3	9,9	50,5	39,6	52,0	22,0	26,0
24 h n 25°C/15°C	PM	73,7	5,9	20,4	51,0	22,0	0	0,8	26,2	41,9	26,9	31,2	50,0	21,2	28,8
	M	83,0	4,4	12,6	70,7	16,9	0	1,5	10,9	26,9	51,5	21,6	57,0	26,1	16,9
24 h n 5°C	PM	79,8	4,7	15,5	67,2	1,4	0	2,7	28,7	29,2	45,0	25,8	35,2	29,5	35,3
	M	74,4	13,8	11,8	69,5	8,2	0	15,8	6,5	57,5	31,3	11,2	36,7	45,5	17,8

Legends - Viability groups of pollen defined in relation to the test reation intensity : Alexander (M = mature pollen ; S = shrinked cytoplasm ; aborted = A), Isatine test (I++ = strong reaction ; I+ = medium reaction ; I+- = low reaction ; I- = no reaction ; I^ = aborted), TTC test (T+ = strong reaction ; T = medium reaction ; T- = no reaction and aborted), FCR (F+ = fluorescent ; F- = no fluorescence ; F^ = aborted).
Stress periods : VM (vacuolised microspore) ; YP (young pollen) ; PM (premeiosis) ; M (meiosis).

sive nights is the one that affects the most esterase content and plasmalemma integrity (FCR test) of the vegetative cell cytoplasm.

A further study (not published) showed that cytochemical treatments induce in stressed sorghum : (i) a swelling of sterile pollen (ii) a sporoderm rupture accompanied by an extrusion of pollen protoplast (= sporoplast) in the pollen made fragile by stresses and unfavorable conditions of culture (room culture or greenhouse).

Table II - Effect of different stresses on pollen viability rates (%) evaluated by four viability tests in two varieties of *Sorghum bicolor*.

Viability Test / Stresses	ESQUIROL								SULTAN							
	Alex	Isat	TTC	FCR	Alex	Isat	TTC	FCR	Alex	Isat	TTC	FCR	Alex	Isat	TTC	FCR
Controls 25°C d/15°C n	77,4	77,6	84,3	66,3	77,4	77,6	84,3	66,3	86,9	88,3	91,3	67,6	86,9	88,3	91,3	67,6
Stress stages	vacuolised microspore				young pollen				Premeiosis				Meiosis			
12 h n 7°C	78,7 (-1,3)	74,5 (3,1)	-	52,5 (13,8)	-	-	-	-	84,1 (2,8)	78,3 (10)	-	60,9 (6,7)	-	-	-	-
12 h n 7°C/ 12 h d 25°C/ 12 h n 7°C	68,6 (8,8)	73,4 (4,2)	-	34,3 (32)	-	-	-	-	74,9 (12)	69,0 (19,3)	-	48,0 (19,6)	-	-	-	-
12 h d 5°C/ 12 h n 5°C	31,9 (45,5)	48,0 (29,6)	34,3 (50)	48,5 (17,8)	52,5 (24,9)	71,1 (6,5)	76,2 (8,1)	48,5 (17,8)	79,8 (7,1)	67,8 (20,5)	50,3 (41)	53,0 (14,6)	80,3 (6,6)	73,9 (14,4)	60,4 (30,9)	52,0 (15,6)
24 h n 25°C/ 15°C	64,4 (13)	71,4 (6,2)	66,8 (17,5)	54,5 (11,8)	78,9 (-1,5)	77,3 (0,3)	77,1 (7,2)	58,7 (7,6)	73,7 (13,2)	73,0 (15,3)	68,8 (22,5)	50,0 (17,6)	83,0 (3,9)	87,6 (0,7)	78,4 (12,9)	57,0 (10,6)
24 h n 5°C	60,9 (16,5)	67,9 (9,7)	57,4 (26,9)	59,5 (6,8)	51,7 (25,7)	75,4 (2,2)	54,5 (29,8)	56,0 (10,3)	79,8 (7,1)	68,6 (19,7)	74,2 (17,1)	64,7 (2,9)	74,4 (12,5)	77,7 (10,6)	88,8 (2,5)	36,7 (30,9)

Legends - Number in brackets = viability rate of controls - viability rate of stressed.

Stress effect. A night (12 h) at 7°C does not modify clearly pollen viability, however we noted a reduction of the percentage of fluorescent pollen grains (F^+) ; the reductions are 13,8 % in the Esquirol variety and 6,7 % in the Sultan variety. When the lowering of temperature occurred two nights sucessively, cold effects are more drastic and pollen viability rates measured by FCR test were reduced of 32 % (Esquirol) or 19,6 % (Sultan). Our results confirm Salgarolo's results (1986) obtained on other varieties with the Alexander test only.

When cold (5°C) is applied continuously during 24 h without modification of photoperiod, it induces the maximal reductions of pollen viability (except for FCR test) and a large increase of the aborted pollen percentage. It seems that night cold is not the most drastic stress for pollen viability of sorghum, in opposition to what was previously believed.

Extended darkness (24 h) alone induces reduction of pollen viability in the two varieties. This result confirms the influence of light modifications of photoperiod on the pollen viability of the Chouan variety observed by Corp (1983). Cold aggravates the negative effects of an extended period of darkness, particularlely it is applied for more than 24 h (not published) ; however, we noted an important variability in the response levels for all the test used.

Stress period effect. According to authors, the most sensitive stage of sorghum pollen ontogenesis to cold is the premeiosis-leptotene (Brooking, 1986), the meiosis (Downes and Marshall, 1971) or the microspore stage (Corp, 1983). Our study proves that negative effects of stresses on pollen viability for each variety are all the more drastic as the stresses occurred at the younger stages of the pollen development in the panicles used : (i) during premeiosis and at the beginning of meiosis for Sultan, this period corresponding to the phases of active syntheses (DNA, RNA, and proteins) and genetic recombinations in meiocytes (Souvré *et al.*, 1987) (ii) at the vacuolated microspore stage in Esquirol variety which corresponds to the phase of DNA duplication before the haploid mitosis.

It seems that among the methods using pollen genetic variability and aiming at a selection of plants with a low temperature resistance, the precocious action of stresses must be considered. This method intends to ensure a selection by gathering at the flowering period a pollen which is supposed to be more resistant to cold.

Variety effect. Among the two varieties studied, Esquirol presents the most altered pollen viability to stresses applied, as it is proved by the results obtained with the four tests used. Its maximal sensitivity is observed after cold action under a normal photoperiod while Sultan is also affected by extended darkness (accompanied or not by low temperature). This kind of differential reaction of grain sorghum genotypes to cold had been previously noted by Salgarolo (1986).

The important sensitivity of Esquirol is perhaps due to its low content in tannic acids ? No certain proof can be given. However we must underline that plants rich in tannic acids are more resistant to biotic alterations (fungus diseases by example) than plants poor in these acids and that Mediterranean varieties of grain sorghum are the one's with the low content of tannic acids (the varieties poor in tannic acids have the most important economic interest).

References

Alami S, Souvré A, Albertini L (1988) Effets de l'obscurité et du froid sur l'ontogénèse et la viabilité du pollen chez deux variétés de sorgho-grain (*Sorghum bicolor* L (Moench)). Incidence sur les correspondances entre le développement végétatif et la formation du pollen. Rev Cytol Biol Végét, Bot 11 : 13-41

Alexander MP (1969) Differential staining of aborted and non aborted pollen. Stain Technol 44 : 117-122

Aslam M, Brown MS, Kohel RJ (1964) Evaluation of seven tetrazolium salts as vital pollen stains in Cotton *Gossipium hirsutum* L. Crop Sci 4 : 508-510

Brooking IR (1976) Male sterility in *Sorghum bicolor* (L) Moench induced by low night temperature. I Timing of the stage of sensitivity. Aust J Plant Physiol 3 : 589-596

Corp D (1983) La microsporogénèse en relation avec la phénologie chez *Sorghum bicolor* (L) Moench. Effet d'un abaissement nocturne de température sur la cytophysiologie (DNA, RNA, protéines basiques) du pollen de sorgho-grain en formation ; froid et stérilité pollinique. Thèse 3e cycle, INP Toulouse

Downes RW, Marshall DR (1971) Low temperature induced male starility in *Sorghum bicolor*. Aust J Ex Agric Anim Hust, 352-356

Heslop-Harrison J, Heslop-Harrison Y (1970) Evaluation of pollen viability by enzymatically induced fluorescence ; intracellular hydrolysis of fluorescein diacetate. Stain Technology 45 : 115-120

Heslop-Harrison J, Heslop-Harrison Y, Shivanna KR (1984) The evaluation of pollen quality and a further appraisal of the fluorochromatic (FCR) test procedure. Theor Appl Genet 67 : 367-375

Palfi G, Koves E (1984) Determination of vitality of pollen on the basis of its amino-acid content. Biochem Physiol Pflanzen 179 : 237-240

Salgarolo P (1986) Etude des stérilités physiologiques chez le sorgho et de l'effet inducteur des abaissements nocturnes de la température. Thèse Doct INP Toulouse

Souvré A, Albertini L, Audran JC (1987) Le grain de pollen des angiospermes. Apports de la biopalynologie et perspectives biotechnologiques. Bull Soc Bot Fr, Actual Bot 134 : 87-112.

Air Pollution Effects on Pollen Germination of Forest Species

L.M. Bellani, E.Paoletti, E.Cenni
Dipartimento di Biologia Vegetale
Laboratorio di Botanica Forestale
Università di Firenze
Piazzale delle Cascine 28
50144 Firenze
Italy

INTRODUCTION

It is now accepted that acid deposition in most areas causes plant deterioration

(Schutt et al., 1983; Evans, 1984; Gellini, 1987). Surfactants also cause plant

injury: they are present in sea-spray and deposited on the foliage of coastal ve-

getation by winds (Gellini et al., 1983,1985). They can be detected very far from

the sea (Cini, pers. comm.). Leaves, being sensitive structures, show necrotic-

yellowish areas, curling, early falling (Clauser and Gellini, 1986). Pollen deve-

lopment and activity are also sensitive botanical indicators of atmospheric pollu-

tion (Feder, 1981). Strong pollen sensitivity is reported to O_3 (Feder et al.,1969;

Davis and Wood, 1972) and to acid precipitation (Smith, 1981; Shidu, 1983; Cox,1984.

Van Ryn et al., 1986). Air pollution can affect pollen viability, germinability and

maximum tube length, can interfere with generative cell formation and division or

its reaching the ovule (Van Ryn et al., 1986). The aim of the present study is to:

1) determine alterations in viability and germinability caused by exposure to air

pollutants for pollen grains of Cedrus atlantica Manetti and Quercus ilex L. col-

lected in different urban and suburban areas; 2) analyse the differing responses of

two Angiosperms (Fagus sylvatica L. and Quercus ilex) and a Gymnosperm (Cedrus a-

tlantica to simulated acid deposition and exposure to surfactants.

MATERIALS AND METHODS

SAMPLE COLLECTION - F. sylvatica and Q. ilex samples were collected in Spring 1987;

C. atlantica in Autumn 1987. Beech branches were taken from a healthy-looking tree at Vallombrosa (Florence, Italy), and kept in a greenhouse at 25°C with natural day-length up to pollen shedding. Holm oak branches were collected from 6 different trees (3 apparently damaged in Florence city, 3 healthy-looking in the suburbs) and kept as beech. Atlas cedar pollen grains were collected from 5 trees growing in different areas of Florence, 3 apparently seriously damaged in the city proper and 2 apparently healthy in the suburbs. The pollen samples were kept individually at -20°C according to Fideghelli (1968). POLLEN VIABILITY AND GERMINATION - Pollen viability was tested by lactophenolcottonblue (0.08%) for the broadleaved species (Darlington and La Cour, 1960) and TTC (0.5%) for the conifer (Cook and Stanley, 1960). Pollen of beech and holm oak was germinated in Petri dishes with the Brewbaker and Kwack medium (1963) at 25°C. Atlas cedar pollen was germinated in a solution of 62 mg/l H_3BO_3 and 147 mg/l $CaCl_2x2H_2O$ at 30°C. Among the sugar concentrations tested the best were: beech 25%, holm oak 20%, atlas cedar 10%. Tube lengths were recorded after 24 h for the first two species and 96 h for Cedrus. SIMULATED ACID DEPOSITION AND SURFACTANT TREATMENTS - Pollen grains were germinated in a solution whose pH was adjusted to values of 2.5, 3.0, 3.5, 4.0, 4.5, 5.0 obtained by 2:1 H_2SO_4:HNO_3, which is the average ratio in precipitation at Vallombrosa (Pantani et al., 1984). The control medium had a pH of 5.6. Pollen was also germinated in the presence of a detergent, dodecyl-benzenesulfonic acid Na salt (ABS) at the following concentrations: 0, 1, 3, 5, 10, 15 mg/l. STATISTICAL ANALYSIS - Each test was replicated 4 times; 300 grains with almost 150 tube lengths were scored for each repetition. A variance analysis (Fisher test) was made except for ABS treatments on beech and holm oak (T-student).

RESULTS

Comparing damaged and undamaged trees of C. atlantica and Q. ilex, a significant decline in germination and tube length for the former was observed. In the latter tube length was not affected, but germination reduced by 40% (Tab.1). Viability was stron-

SPECIES	CATEG.	NR.IND.	%VIAB.	%GERM.	TUBE LENGTH(µ)
ATLAS CEDAR	H	2	80	44	238.50
ATLAS CEDAR	D	3	40	8	163.77
HOLM OAK	H	3	98	43	159.00
HOLM OAK	D	3	83	26	137.80

TABLE NR. 1 - Pollen viability(%) germination (%) and tube length (µ) of atlas cedar (_Cedrus atlantica_ Manetti) holm oak (_Quercus ilex_ L.) trees damaged (D) and not (H) by atmospheric pollution.

SPECIES	5.6	5.0	4.5	4.0	3.5	3.0	2.5
BEECH	94	75	60	0	0	0	0
HOLM OAK	54	49	17	0	0	0	0
ATLAS CEDAR	53	48	44	22	31	3	1

TABLE NR. 2 - Pollen grain germination percentage at different pH in beech (_Fagus sylvatica_ L.), holm oak (_Quercus ilex_ L.) and atlas cedar (_Cedrus atlantica_ Manetti) trees.

SPECIES	CONTROL	1	3	5	10	15
BEECH	94	29	0	0	0	0
HOLM OAK	54	10	0	0	0	0
ATLAS CEDAR	53	48	44	17	8	0

TABLE NR. 3 - Pollen grain germination percentage at different ABS (alchilbenzensulfonate) concentrations (mg/l) in beech (_Fagus sylvatica_ L.), holm oak (_Quercus ilex_ L.) and atlas cedar (_Cedrus atlantica_ Manetti) trees.

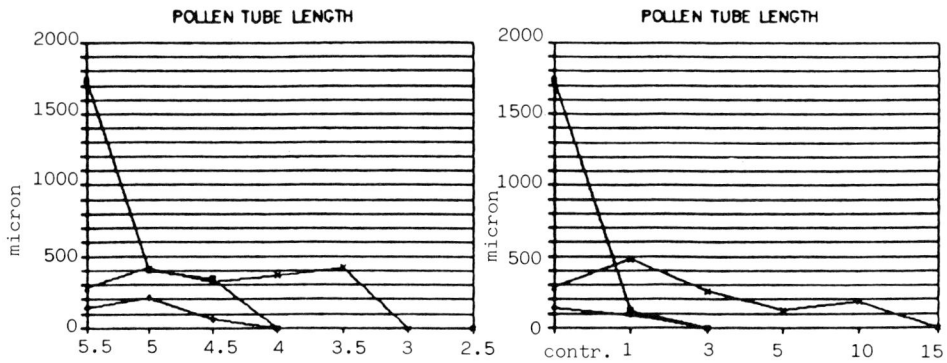

FIGURE NR. 1 - Mean tube length (µ) at different pH in beech (□), holm oak (◊) and atlas cedar (x).

FIGURE NR. 2 - Mean tube length (µ) at different ABS concentrations (mg/l) in beech (□), holm oak (◊), atlas cedar (x).

gly influenced in C. atlantica (80% to 40%), while Q. ilex had a smaller decrease

(98% to 83%). Treatments were effected on samples with the following viability:holm

oak 97%, beech 98%, atlas cedar 86%. In all species, acid pH affected pollen germi-

nation. For holm oak and beech (Tab.2), no germination occurred at pH 4.0 and lower.

When germination began (pH 4.5), the percentage was lower than in the control. Atlas

cedar exibited a decrease in germination across the pH range 5.6 to 2.5. Mean tube

length (Fig.1) was strongly reduced in beech at pH 5.0 and 4.5; holm oak was very

sensitive only at pH 4.5. In atlas cedar a slight increase was detected. No data

were reported for atlas cedar tube length at pH 3.0 and 2.5 because only a few

grains were germinated, but the general trend was a sharp decrease. For all species,

data were statistically significative by Fisher test (P=0.001). ABS influenced both

germination and tube length. 3 mg/l completely inhibited germination of holm oak

and beech; atlas cedar was inhibited   by 15 mg/l (Tab.3). Tube length was sub-

stantially reduced in all species, but atlas cedar showed an increase at 1 mg/l

(Fig.2). For atlas cedar data were statistically significative by Fisher test, for

beech and holm oak by T-student (P=0.001).

DISCUSSION

Our results show that pollen can be an indicator for the "health' of a tree: in fact

pollen viability and germinability are greatly reduced in samples collected from da-

maged trees. The acid and surfactant treatments were intended to simulate a deposi-

tion due to fog or mist more than a proper rain. The latter would only imply a quiçk

pollen washing, the former involves a longer contact with the pollutant. In nature,

during a foggy period, water can collect on the stigma polluting the exudate in which

the pollen grain germinates causing a response similar to the in vitro conditions

(Van Ryn et al., 1986). Broad-leaved foliage shows a greater sensitivity to simula-

ted acid rain than conifer foliage (Evans 1980); the same response was detected in natural conditions (Gellini et al., 1987). Acid treatments inhibit far more stronger germination in the broad-leaved pollens, while the Gymnosperm is less sensitive (see also Cox, 1983). Atlas cedar in fact is capable of germinating up to pH 2.5, although the percentage is low, mean tube length not being influenced by pH. Because tube lengthening is a continuation of germination, it is not surprising that pollen tube elongation is not as sensitive as germination (Cox, 1983). Atlas cedar also demonstrated a greater tolerance to ABS. This general tolerance to treatments would seem to contrast with the greater decrease in viability and germination of atlas cedar damaged plants as compared to healthy. We would conjecture that pollen collected from a damaged tree has been influenced by pollution during its ontogeny. The simulated treatments were on pollen from healthy trees and so with unaffected microsporogenesis. Moreover C. atlantica blooms in late Autumn, when air humidity is higher and pollutant emissions more conspicuous than in Spring (Pantani et al., 1984) when Q. ilex pollen is shed. The area where atlas cedar trees grow is also exposed to more vehicle pollution. pH 5.6 (our control) is the average for unpolluted precipitation (Pantani et al., 1984). However, in industrial regions average rain pH is usually 4.5 with episodes of pH 3.0 or below (Likens and Bormann, 1974; Evans, 1984). Fog shows the highest acidity (Waldman et al., 1982; Waldmann, 1985) with pH even lower than 2.0. Moreover, Gellini et al. (1983,1985) detected a concentration of MBAS (Methylene Blue Active Substances) as high as 18 mg/l in rainwater and 500 mg/kg fresh weight in coastal regions. Our experiments show that pollen germination is sensitive at 3 mg/l ABS for the broad-leaved species and 15 mg/l for the conifer. These results indicate therefore that, as regards human influences on plant reproduction, the present levels of acidity and surfactants in deposition should be a serious cause of concern.

REFERENCES

Brewbaker JL, Kwack BH (1963) The essential role of calcium ion in pollen germination and pollen tube growth. Am J Bot 50:859-865

Clauser F, Gellini R (1986) Moria del bosco: osservazioni sulle latifoglie nel triennio '82-'84 in Toscana. Atti Soc Tosc Sci Nat, Mem, Serie B 93:11-26

Cook SA, Stanley RG (1960) Tetrazolium chloride as an indicator of pine pollen germinability. Silvae Genet 9:134-136

Cox RM (1983) Sensitivity of forest plant reproduction to long range transported air pollutants: in vitro sensitivity of pollen to simulated acid rain. New Phytol 95:269-276

Cox RM (1984) Sensitivity of forest plant reproduction to long range transported air pollutants: in vitro and in vivo sensitivity of Oenothera parviflora L. pollen to simulated acid rain. New Phytol 97:63-70

Darlington CD, La Cour LF (1960) The handling of chromosomes. Allend G, Hunwin Ltd, London UK

Davis DD, Wood FA (1972) The relative susceptibility of eighteen coniferous species to ozone. Phytopathol 62:14-19

Evans LS (1980) Foliar responses that may determine plant injury by simulated acid rain. In: Polluted rain. Plenum press, New York, p 239

Evans LS (1984) Botanical aspects of acid precipitation. Bot Rev 50:449-490

Feder WA (1981) Bioassaying for ozone with pollen systems. Environ Health Perspect 37:117-123

Feder WA, Fox FL, Heck HV, Campbell FJ (1969) Varietal response of petunia to several air pollutants. Plant Dis Reptr 53:506-510

Fideghelli C (1968) Miglioramento genetico delle piante da frutto. Ital Agric:105: 359-373

Gellini R (1987) Inquinamento atmosferico e deperimento delle piante forestali. In: Il bosco e l'ambiente: aspetti economici, giuridici ed estimativi. Atti XVII Incontro CeSET, Firenze, p 347

Gellini R, Clauser F, Rinallo C, Grossoni P, Bussotti F (1987) Danni di nuovo tipo nei boschi italiani. Situazione e probabili cause. Dendrochronologia:in press

Gellini R, Pantani F, Grossoni P, Bussotti F, Barbolani E, Rinallo C (1983) Survey of deterioration of the coastal vegetation in the park of San Rossore in central Italy. Eur J For Path 89:319-332

---, ---, ---, ---, ---, --- (1985) Further investigation on the causes of disorder of the coastal vegetation in the park of San Rossore (central Italy). Eur J For Path 15:145-157

Likens GE, Bormann FH (1974) Acid rain: a regional environmental problem. Science 184:1176-1179

Pantani F, Barbolani E, Del Panta S, Bussotti F (1984) Rilevamento di piogge acide in comprensori della Toscana. Rassegna Chimica 3:135-141

Schutt P, Koch W, Blaschke H, Lang KJ, Schuck HJ, Summerer H (1983) So stirbt der wald. BLV-Verlag, Munchen

Sidhu SS (1983) Effects of simulated acid rain on pollen germination and pollen tube growth of white spruce (Picea glauca). Can J Bot 61:3095-3099

Smith WH (1981) Air pollution and forest. Interactions between air contaminants and forest ecosystems. Springer-Verlag, New York Heidelberg Berlin

Van Ryn DM, Jacobson JS, Lassoie JP (1986) Effects of acidity on in vitro pollen germination and tube elongation in four hardwood species. Can J For Res 16:397-400

Waldman JM, Munger JW, Jacob JD, Flagan RC, Hoffman MR (1982) Chemical composition of acid fog. Science 218:677-680

Waldmann G (1985) Zur anreicherung von sauren in baumkronenbereich. Allg Forst-u J Zeit 156:204-210

Forced Pollen Shedding Effects on Pollen Diameter and Early Seedling Growth in Maize

P.L. Pfahler and R.D. Barnett
Department of Agronomy
University of Florida
Gainesville, Florida 32611 U.S.A.

Introduction

In those crops such as maize, which have been improved for many generations by man, the effectiveness of pollen genotype selection would be decreased each generation because the usual improvement procedure involves pollination with excess quantities of fresh pollen released normally under field conditions. In these crops, variation in this standard pollination scheme may amplify existing pollen transmission differences so that the effectiveness of pollen genotype selection would be enhanced. In maize, such changes as extended pre-pollination mature pollen storage at 2°C and pre-pollination stylar treatments with various chemicals increased pollen transmission differences at both qualitative and quantitative loci (Pfahler 1974b, 1986, Pfahler et al. 1986a,b). Under field conditions, attached maize tassels usually shed pollen daily between 800-1200 h, with all or most of the pollen grains in the tassel released in 7-10 days. Detached tassels with their cut ends submerged in water and exposed to low light intensity high humidity, and 25-30°C, shed pollen continuously, with all or most of the pollen grains in the tassel released in three days.

The purpose of this study was to determine the effect of this forced shedding from detached maize (Zea mays L.) tassels on the diameter of the pollen grains released and the early seedling growth of the resulting progeny.

Materials and Methods

At least 15 tassels from three single cross hybrids (1 = FR9xH55; 2 = B73xOh545; 3 = K64xK55) were cut (about 35 cm below the lowest tassel branch) from field-grown plants at 1600 h (day 0). Each tassel selected was just beginning to exert anthers, and all leaves were removed. The cut ends of the tassels were then submerged in water. The three groups of tassels from each single cross were placed on separate sheets of paper and exposed to low light intensity, high humidity, and 25-30°C conditions. At the end of six collection periods (A = 1600 h, day 0-1000 h, day 1; B = 1000 h, day 1-1600 h, day 1; C = 1600 h, day 1-2200 h, day 1; D = 2200 h, day 1-1000 h, day 2; E = 1000 h, day 2-1600 h, day 2; F = 1600 h, day 2-2200 h, day 2), the papers were replaced. The pollen sample obtained during each collection period was screened to remove the anthers. A small portion of each sample was placed in a killing and fixing solution for diameter measurements. Within 30 minutes after collection, the remaining pollen grains from each single cross were used to pollinate (with an excess amounts of pollen grains), five ears of the same single cross from which the pollen grains were collected to produce three F2 populations.

For the diameter studies, 120 pollen grains from each single cross-collection period combination were measured, using a microprojector at 400x.

For the seedling growth studies, F2 seeds obtained from each single cross-collection period combination were treated with a fungicide and soaked in deionized water for 16 h at 2°C. The seeds were then put in plastic growth pouches (containing 20 mL of deionized water) which were immediately placed in a closed container at 25°C. The embryos were positioned vertically so that the shoot and root would grow normally facilitating more accurate measurement. After 5 days, the shoot and root length from the scutellar node was measured. At least 100 F2 seedlings from each single cross-collection period combination were measured.

Analyses of variance were performed for pollen diameter, shoot length, and root length. Minimum differences for significance (MDS) values were obtained by means of the revised Duncan's ranges using for P only the maximum number of means to be compared (Harter 1960).

Deviations from normality in the F2 distributions in each single cross-collection period combination were determined using skewness ($g1$) and kurtosis ($g2$) values (Snedecor 1956). Positive $g1$ values indicate an excess number of observations smaller than the mean, while negative $g1$ values indicate an excess number of observations larger than the mean. Positive $g2$ values indicate an excess number of observations near the mean with a corresponding deficiency on the flanks of the distribution. In contrast, negative $g2$ values indicate a deficient number of observations near the mean with a corresponding excess number on the flanks of the distribution.

Results

Pollen diameter. The main effects, single cross and collection period, and their interaction were significant at the 0.01 probability level.

The means and g values are presented in Table 1. The mean diameter of the single crosses decreased at the later collection periods but the rate and magnitude of the decrease was greatly influenced by the single cross. A similar relationship between the starch content (after staining with a KI-I$_2$ solution) and collection period was observed. The skewness of the F2 populations was influenced by collection period but the effect depended largely on the single cross and was not consistent with collection period. The same general pattern emerged with the kurtosis of the F2 populations. In general, these results indicated that the effects of successive collection periods on pollen diameter and starch content depended on the single cross but were quite consistent. However, the effects of successive collection periods on the normality of the F2 populations depended on the single cross and were very inconsistent.

Shoot length. The main effects, single cross and collection period, and their interaction were significant at the 0.01 probability level.

The means and g values are shown in Table 1. Single cross greatly influenced the direction, rate, and magnitude of length changes associated with successive collection periods. At the later collection periods, a significant decrease was observed in single cross 2, while, a significant increase was found in single cross 3. The skewness in the F2 populations was generally unaffected by single cross and collection period. The kurtosis of the F2 populations was altered by collection period with successive collection periods reducing the kurtosis in single cross 3. The results indicated that successive collection periods resulted in significant increases or decreases in length depending on the single cross. The effects of successive collection periods on the normality of the F2 populations were influenced greatly by single cross and were generally erratic.

Root length. The main effect, single cross, and the single cross x collection period interaction were significant at the 0.01 probability level.

The means and the g values are presented in Table 1. Single cross greatly influenced the direction, rate, and magnitude of length changes associated with successive collection periods. At the later collection periods, a significant increase was obtained in single cross 1, while, a significant decrease was found in single cross 2. The skewness in the F2 populations was considerably influenced by collection period in all single crosses. The kurtosis in the F2 populations was generally unaffected by either single cross or collection period. The results indicated that successive collection periods resulted in significant increases or decreases in length depending on the single cross. The effects of successive collection periods on the normality of the F2 populations were greatly influenced by single cross and were generally inconsistent.

Table 1. The effect of collection period on the mean, skewness, and kurtosis of pollen diameter, shoot length, and root length in three single crosses

	Single cross	Collection period					
		A	B	C	D	E	F
Pollen diameter							
Mean(μm)[a]	1	92.7	92.1	91.7	90.3	90.3	88.3
	2	96.7	96.6	95.1	94.1	93.9	92.9
	3	92.6	90.6	89.4	87.9	87.5	84.7
Skewness ($g1$)	1	-0.22	0.18	0.00	-0.27	-0.18	0.18
	2	-0.34	-0.08	-0.19	-0.52**	-1.44**	-0.08
	3	-0.47**	-0.27	-0.37	-0.25	-0.22	0.10
Kurtosis ($g2$)	1	0.28	0.57	0.42	0.32	0.43	0.10
	2	0.85	0.01	0.02	0.88*	7.89**	0.03
	3	2.15**	0.56	0.30	1.33**	0.41	0.33
Shoot length							
Mean (mm)[a]	1	83	92	86	82	74	89
	2	81	79	74	82	66	69
	3	78	76	80	88	87	88
Skewness ($g1$)	1	0.23	-0.05	-0.04	-0.21	0.15	0.15
	2	0.30	0.00	0.56**	0.20	-0.01	0.21
	3	0.00	0.20	-0.16	-0.12	-0.38*	-0.32
Kurtosis ($g2$)	1	-0.41	-0.09	-0.30	-0.54	-0.48	-0.18
	2	0.62	-0.40	1.02*	0.35	-0.41	0.07
	3	1.25**	1.65**	0.76*	0.56	0.79*	0.37
Root length							
Mean (mm)[a]	1	106	105	114	111	117	124
	2	105	113	99	108	80	89
	3	97	100	101	104	105	106
Skewness ($g1$)	1	-0.17	-0.56*	0.43*	-0.46*	-0.45*	-0.60**
	2	-0.36*	-0.55**	-0.16	-0.40*	0.08	0.08
	3	-0.19	-0.35	-0.41*	-0.54*	-0.39*	-0.66**
Kurtosis ($g2$)	1	-0.55	-0.53	-0.34	0.10	-0.29	0.56
	2	-0.74*	-0.51	-0.98*	-0.55	-0.90	-0.87*
	3	-0.64	-0.46	0.01	0.14	-0.47	-0.18

*,** g values significant from zero at the 0.05 and 0.01 probability levels, respectively.

[a] MDS values among the means of each character at the 0.05 and 0.01 probability levels, respectively: pollen diameter = 1.1 and 1.5; shoot length = 8 and 10; and root length = 11 and 14.

Discussion

The results of this study indicated that forced shedding reduced the diameter and relative starch content of the pollen grains with single cross, a major influencing factor. In a previous study with maize, forced shedding reduced the diameter, relative starch content, and seedsetting ability of pollen grains from Wxwx, Su1su1, and Sh2sh2 heterozygotes with differences among loci, a major influencing factor (Pfahler et al. 1986a). The specific physiological processes disrupted by this procedure, are unknown. Obviously, the removal of the tassel from the plant eliminates the transfer of various substances and disturbs normal pollen grain maturation. The effect of this disturbance would probably be more pronounced at the later collection periods. In the forced shedding procedure used in this study, no obvious interference with the normal pollen release sequence (lodicule expansion through anther pore opening) was observed, but the cyclical nature of this process was altered so that an almost continuous release of the pollen grains resulted. Since so little is known about this process, no speculation is possible as to why the pronounced single cross effect was obtained.

The results of this study also indicated that forced shedding altered the mean and frequency distribution patterns of the F2 populations resulting from pollination with pollen grains derived from forced shedding. Single cross was a major influencing factor, but no consistent pattern associated with collection period, was evident. For pollen genotype selection to be effective, a relationship among pollen genotype, pollen competitive ability, and progeny traits must exist. Such a relationship in maize has been reported (Mulcahy 1971, Ottaviano et al. 1980). One of the major difficulties in critically examining this relationship is selecting a sporophytic trait which can be accurately measured or classified without genetic complications and/or environmental effects. Qualitative characters such as maize endosperm mutants, fulfill these requirements. Studies have shown that certain pre- and post-pollination treatments alter differential pollen transmission at these loci (Pfahler 1974b, Pfahler et al. 1986a,b). However, when complex quantitative characters in maize whose expression is influenced by inbreeding, environment, etc., are examined, interpretation of the results is difficult and tenuous. The mode of inheritance and environmental influence on early seedling growth in maize are not clear. A positive relationship between early seedling growth, heterozygosity level, and vigor was found in maize when inbred parents were compared to their F1 progeny (Whaley 1952). However, in rye which parallels maize in the degree of inbreeding depression and heterotic response, coleoptile length was not related to vigor expressed later in the life cycle (Pfahler 1974a) and was responsive to selection (Pfahler 1974c).

Thus, heterosis _per se_ probably had limited influence on early seedling growth, the trait measured in this study. Since information on the early seedling growth of the inbred parents of these single crosses is not available, interpretation of the results reported herein is difficult. Apparently, forced shedding produced changes in the mean and distribution patterns of the F2 populations, but the extent of its effect on pollen transmission cannot be accurately determined.

Acknowledgements. Appreciation is expressed to Carla Carter for her skill and diligence in typing and organizing this manuscript and to W.T. Mixon for his outstanding technical assistance in completing this study.

References

Harter HL (1960) Critical values for Duncan's multiple range test. Biometrics 16:671-685
Mulcahy DL (1971) A correlation between gametophytic and sporophytic characters in Zea mays L. Science 171:1155-1156
Ottaviano E, Sari-Gorla M, Mulcahy DL (1980) Pollen tube growth rate in Zea mays: Implications for genetic improvement in crops. Science 210:437-438
Pfahler PL (1974a) Effect of selection for coleoptile length in rye, Secale cereale L. Euphytica 23:515-520
Pfahler PL (1974b) Fertilization ability of maize (Zea mays L.) pollen grains. IV. Influence of storage and the alleles at the shrunken, sugary and waxy loci. In: Linskens HF (ed) Fertilization in higher plants. North-Holland, Amsterdam, p 15
Pfahler PL (1974c) Relationships between coleoptile length and forage production and grain yield in rye, Secale cereale L. Euphytica 23:405-410
Pfahler PL (1986) Pollen storage effects on early seedling growth in maize. In: Mulcahy DL, Bergamini Mulcahy G, Ottaviano E (ed) Biotechnology and ecology of pollen. Springer-Verlag, New York, p 147
Pfahler PL, Mulcahy DL, Barnabas B (1986a) The effect of forced shedding on pollen traits, seedsetting, and transmission at various maize (Zea mays L.) endosperm mutant loci. Acta Bot Neerl 35:195-200
Pfahler PL, Mulcahy DL, Barnabas B (1986b) The effect of pre-pollination stylar treatments on seedset and pollen transmission at various maize (Zea mays L.) endosperm mutant loci. Acta Bot Neerl 35:201-207
Snedecor GW (1956) Statistical methods. 5th edn. Iowa State University Press, Ames, Iowa
Whaley WG (1952) Physiology of gene action in hybrids. In: Gowen JW (ed) Heterosis. Iowa State University Press, Ames, Iowa, p 98

Heat Shock Proteins in Germinating Pollen of *Nicotiana tabacum* Before and After Heat Shock

M.M.A. Van Herpen, W.H. Reijnen, J.A.M. Schrauwen, P.F.M. De Groot &
G.J. Wullems
Department of Experimental Botany
Research Group Molecular Plant Physiology
University of Nijmegen
Toernooiveld
6525 ED Nijmegen
The Netherlands

ABSTRACT

The common response of cells to (heat) stress is an increased synthesis of a specific group of proteins known as HSP's, concomitant with a reduction in the synthesis of the normal proteins. HSP's are believed to play a role in the survival of the organism to (heat) stress. The protein synthesis in pollen of *Nicotiana tabacum,* incubated in vitro, is affected by a heat shock of 30 min applied after germination. The production of HSP's could be established but a concomitant change in the synthesis of the normal proteins was not present. Pollen of tobacco not only synthesizes high molecular weight HSP's but also low molecular weight HSP's. Some HSP's are already present before heat shock, indicating that transcriptionally controlled synthesis of HSP-mRNA's also occurs during normal pollen tube growth. The HSP's in pollen of tobacco are compared with the set of classical HSP's normally found in other cells.

INTRODUCTION

The heat shock induced rapid synthesis of a specific group of proteins (HSP's) is already established for other cells than pollens (Key et al. 1981, Ashburner 1982, Ballinger et al. 1983, Munro et al. 1985, Nover et al. 1984). This characteristic set of LMW-HSP's (M_r

---

Abbreviations: (HMW/LMW)-HSP's, (high molcular weight/low molecular weight)-heat shock proteins; M_r, relative molecular mass; kD, kilodalton(s)

15-30 kD) and HMW-HSP's (M$_r$ 68-100 kD)(Nover et al. 1984) are believed to play a role in the survival of the organism to heat stress (Ashburner 1982, Munro et al. 1985). Pollen of *Tradescantia* responds to a gradual heat treatment by adaptation and finally survive (Mascarenhas et al. 1983, Xiao et al. 1985, Altschuler et al. 1982), but synthesis of the characteristic HSP's could not be demonstrated. In pollen of *Maize* (Cooper et al. 1983,1984), *Lilium* and *Petunia* (Schrauwen et al. 1986) minor differences in synthesized proteins between control and heat shocked samples could be established. However, these proteins seemed to have no resemblance with the HSP's found in other plant cells (Nover et al. 1984). These results seem to indicate that the characteristic set of HSP's (Nover et al. 1984) might not be involved in the thermoprotection of pollen tube growth (Xiao et al. 1985). However, synthesis of HSP's in germinating pollen might be partly if not entirely constitutive (Cooper et al. 1984, Nover et al. 1984). Therefore, we were interested in the presence of HSP-mRNA transcription before and after heat shock.

MATERIALS AND METHODS

Plant Materials. PLants of *Nicotiana tabacum* were grown in the greenhouse under day-light conditions (Van Herpen 1984). Pollen was derived from anthers from flower buds just before anthesis. The anthers were dried at 23°C for 48 h in the dark. Pollen was checked microscopically for the presence of diploid tissue and afterwards stored in small quantities at -20°C in sealed vials. Samples of 3 mg pollen were hydrated in a water saturated atmosphere for 1 h and imbibed for 10 min in 30 µl germination medium (Dickinson 1968). At the end of the imbibition period the concentrated pollen suspension was mixed with 280 µl germination medium. From this moment on the pollen suspensions of tobacco were incubated at 27°C for 1 h.

Inhibition of Transcription. Transcription in pollen was inhibited by adding 50 µg/ml cordycepine in germination medium before imbibition. Cordycepine has no inhibitory effect on the growth of the pollen tubes (Tupý 1983).

Temperature treatment. The temperature during imbibition and incubation of pollen was 27°C. The incubation period of 27°C was followed by a heat shock of 39°C for 30 min, except for the control. These temperature regimes were not lethal to the pollen and caused a maximal change in protein pattern (Van Herpen 1987).

Labeling and Labeling Period. [^{35}S] methionine was applied to the pollen suspension as described previously (Schrauwen et al. 1986). The labeling of proteins in tobacco pollen after incubation at 39°C (for 30 min) was performed during the heat shock period followed by an extra half hour at 27°C (Fig. 1b,c)

Extraction and Analysis of Proteins. Extraction, electrophoretic separation and analysis of pollen proteins were performed as described previously (Schrauwen et al. 1986). Equal amounts of radioactivity were analyzed in the gels.

RESULTS

After heat shock the newly synthesized proteins (proteins indicated with the numbers 3 and 6 in the Figs. 1a,b,c) are LMW-HSP's (M_r 17 and 23 kD respectively). The relative rate of synthesis of proteins 1,2,3,4,6,9,10 and 11 increases and 5,7 and 8 decreases compared to the protein indicated with the asterisk (Fig. 1b versus 1a). The reduction of the synthesis of the normal proteins is restricted to some proteins and not a general decrease as observed in other cells (Nover et al. 1984). In tobacco the pattern of protein synthesis is changed by heat shock (Fig. 1b versus 1a), and for eight proteins (proteins 1,2,3,4,6,9,10 and 11) synthesis is enhanced by heat shock and reduced by cordecypine (Fig. 1c versus 1b) which point to heat shock induced regulation of transcription. Of those

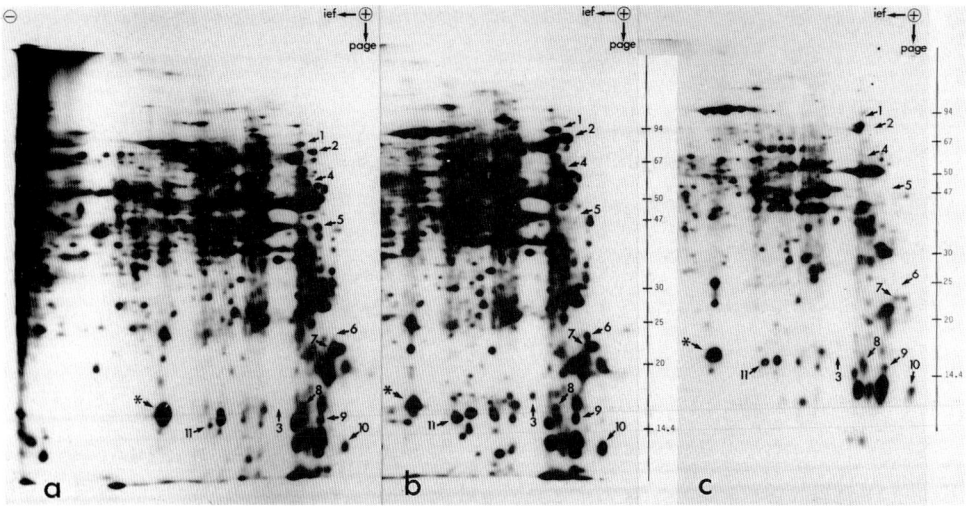

Fig. 1. Fluorograms of proteins synthesized in pollen of tobacco, incubated with 150 kBq [^{35}S] methionine during the last 60 min of an incubation of 2 h at 27°C (a) or an incubation at 27°C for 1.5 h, followed by a 30 min heat shock of 39°C and 30 min 27°C without (b) and with cordycepine as inhibitor of the transcription process (c). Protein standards, 94 kD; Phosphorylase b; 67 kD: BSA; 43 kD: Carbonic anhydrase; 20.1 kD: Soybean trypsin inh.; 14.4 kD: x-Lactalbumin.

proteins some are already present before heat shock (Fig 1a the proteins 1,2,4,9 and 10) and their synthesis at the control temperature of 27°C can be inhibited by cordycepine (2-D pattern equals that of Fig. 1c) which means that the corresponding HSP-mRNA's are already present before heat shock and are transcriptionally controlled during pollen tube growth. In non-heat treated pollen applied with cordycepine (2-D pattern equals that of Fig. 1c) synthesis of several proteins (proteins 1,2,4,5,7,8,9 and 10) is reduced by cordycepine. The HSP's indicated with the numbers 1 (M_r 94 kD), 2 (M_r 85 kD), 3 (M_r 17 kD), 4 (M_r 65 kD), 6 (M_r 23 kD), 9 (M_r 15 kD), 10 (M_r 13 kD) and 11 (M_r 16 kD) in pollen of tobacco (Fig. 1b) correspond with the molecular weight of the HSP's normally observed in plants or plant cell cultures (Nover et al. 1984).

DISCUSSION

In this paper it has been shown that pollen of tobacco, which survives heat shock (Van Herpen 1987) responds to heat shock with the synthesis of HSP's. Untill now survival of pollen after heat treatment was only accomplished for tradescantia (Mascarenhas et al. 1983, Xiao et al. 1985), but only after a gradual increase in temperature, and along with it no changes in protein synthesis could be detected. The results we found in pollen of tobacco, treated with a sudden heat shock, show the opposite and fit in the general finding that plant tissue synthesizes more LMW-HSP's than tissue of other origin (Key et al. 1981, 1985b). Furthermore, the synthesis of the 94, 85 and 65 kD HSP's and some LMW-HSP's (M_r 13-25 kD) in pollen of tobacco (Fig. 1) is in agreement with the HSP synthesis normally observed in plants or plant cell cultures (Nover et al. 1984). The HSP's 1 (M_r 94 kD), 2 (M_r 85 kD), and 4 (M_r 65 kD) in Fig. 1 might be related to the HSP's 95 (M_r 95 kD), 89 (M_r 83-95 kD) and 70 (M_r 68-74 kD) respectively (Nover et al. 1984). The major components of the LMW-HSP's (Nover et al. 1984) are also present in pollen of tobacco i.e. the HSP's indicated with the numbers 11 (M_r 16 kD) and 3 (M_r 17 kD). The HSP's 1,2,4,9 and 10 in Fig. 1 are present before heat shock, but are not similar to the proteins of which Cooper (1984) suggested that they might be constitutive in pollen of maize. The LMW-HSP's, in the relative molecular mass range of 16 kD, and the HMW-HSP indicated with the number 1 (M_r 94 kD) in Fig. 1 might resemble the 16 and 94 kD proteins which Xiao et al. (1985) found present in ungerminated pollen of tradescantia. The conclusions that the great majority if not all of the HSP's are not present in the ungerminated pollen grain of tradescantia (Xiao et al. 1985) and that no HSP's are synthesized in germinating pollen of tradescantia, maize or lily (Xiao et al. 1985 and Cooper et al. 1984) are not valid for pollen of tobacco. Tobacco pollen exhibits both synthesis of HSP's

after heat shock and synthesis of constitutive HSP's during normal
pollen tube growth.

LITERATURE CITED

Altschuler M, Mascarenhas JP (1982) Heat shock proteins and effects
    of heat shock in plants. Plant Mol Biol 1: 103-115
Ashburner M (1982) The effects of heat shock and other stresses on
    gene activity: An introduction. *In* : Schlesinger MJ, Ashburner M,
    Tissieres A (eds) Heat Shock from Bacteria to Man. Cold Spring
    Harbour Laboratory, Cold Spring Harbour, New York, pp 1-9
Ballinger DG, Pardue ML (1983) The control of protein synthesis
    during heat shock in *Drosophila* cells involves altered polypep-
    tide elongation rates. Cell 33: 103-114
Cooper P, Ho T-HD (1983) Heat shock proteins in maize. Plant Physiol
    71: 215-222
Cooper P, Ho T-HD, Hauptmann RM (1984) Tissue specificity of the
    heat-shock response in maize. Plant Physiol 75: 431-441
Dickinson DB (1968) Rapid starch synthesis associated with increased
    respiration in germinating lily pollen. Plant Physiol 43: 1-8
Key JL, Lin CY, Chen YM (1981) Heat shock proteins in higher plants.
    Proc Natl Acad Sci USA 78(6): 3526-3530
Key JL, Kimpel JA, Lin ChY, Nagao RT, Vierling E, Czarnecka E, Gurley
    WB, Roberts JK, Mansfield MA, Edelman L (1985b) The heat shock
    response in soybean *In* : Key JL Kosuge T (eds) Cellular and
    Molecular Biology of Plant Stress. Liss AR, New York, pp 161-179
Mascarenhas JP, Altschuler M (1983) The response of pollen to high
    temperatures and its potential applications. *In* : Mulcahy DL,
    Ottaviano E (eds) Pollen: Biology and Implications for Plant Bree-
    ding. Elsevier, New York, pp 3-8
Munro S, Pelham H (1985) What turns on heat shock genes. Nature 317:
    477-478
Nover L, Scharf K-D (1984) Introduction - The heat shock proteins.
    In : Nover L (ed) Heat Shock Response of Eukaryotic Cells.
    Springer Verlag, Berlin, p 1 -7
Schrauwen JAM, Reijnen WH, De Leeuw HCGM, Van Herpen MMA (1986)
    Response of pollen to heat stress. Acta Bot Neerl 35: 321-327
Tupý J (1983) Transcription activity and the effects of transcrip-
    tion inhibitors in tobacco pollen culture. *In* : Fertilization and
    Embryogenesis in Ovulated Plants. Veda, Bratislava,
    Czechoslovakia, pp 133-136
Van Herpen MMA (1984) Environment and pollen style interaction. PhD
    thesis. University of Nijmegen
Van Herpen MMA (1987) Temperature stress and pollen tubes. *In* : Proc
    XIV Int Bot Congr, Berlin, 24 Jul-1 Aug 1987, p 126
Xiao C-M, Mascarenhas JP (1985) High temperature-induced thermotoler-
    ance in pollen tubes of *Tradescantia* and heat-shock proteins.
    Plant Physiol 78: 887-890

From Pollination to Fertilization

The Relationship Between Potato Pollen and True Seed: Effects of High Temperature and Pollen Size

N. Pallais, D. Mulcahy[*], N. Fong, R. Falcon and P. Schmiediche
International Potato Center (CIP) Apartado 5969 Lima Peru
[*] University of Massachusetts Amherst MA 01003 USA

1. Introduction

In the last ten years, scientists at the International Potato Center (CIP) have been investigating the potential of true potato seed (TPS) as an alternative to traditional potato production from tuber seed in developing countries. The potato is considered a luxury vegetable in many developing countries. Locally produced seed tubers are often of low quality and certified seed is usually imported at high cost from countries with temperate climates. The production of healthy seed tubers in tropical areas is difficult because warmer temperatures are conducive to higher levels of potato pathogens of which viruses are of high importance. Since the most important viruses in potato are not passed on through sexual seeds, TPS may be used to produce potatoes for consumption or for the production of clean "seedling tubers."

The lack of progeny uniformity obtained from tetraploid potato crosses when compared to clonally progagated tubers is a disadvantage of TPS (1) for the farmer during field cultivation and (2) for the consumer due to the segregation of such tuber characteristics as cooking or processing quality. Lack of uniformity in plants and tubers produced from TPS is largely a consequence of the tetrasomic inheritance in progenies derived from the heterozygotic autotetraploid potato, which also displays some segmental allopolyploidy (Larkin et al. 1981). Pollen selection then—if effective—would be a simple biotechnique for reducing the frequency of undesirable genotypes derived from TPS crosses.

In this report, we present the results of two separate studies aimed at investigating the relationship between 1) pollen tolerance to high temperature and 2) pollen size to TPS performance during germination and early seedling emergence. Experiments were conducted to investigate if the TPS produced with heat-treated pollen and pollen of varying sizes was affected in its seed characteristics. The rationale for conducting each of these studies was: (1) Since TPS technology is targeted for application in tropical environments, heat resistant potato pollen may produce seed more adapted to warmer temperatures. Weaver et al.. (1985) reported that a significantly higher percentage of pollen grains from two high temperature tolerant bean siblings survived high temperature treatments which destroyed most pollen grains from the parent and from siblings of a heat sensitive selection.(2) In preliminary studies, we observed that potato pollen in selected male TPS progenitors varied noticeably in size and that the differences in diameter fell into two discreet classes rather than into a continuum (Fig.1). This was confirmed in a subsequent study where pollen grains were photographed at random soon after fluorescence staining, and the diameter of viable grains was later carefully measured at the shortest distance. One example of these results is presented in Fig.2. The possibility of altering segregation ratios in progeny characteristics by producing corn seed with mechanically separated pollen was first reported in 1932 (Mangelsdorf.)

2. Materials and Methods

2.1. Effects of high temperature on pollen

Pollen lots of clones LT-7, LT-5, R128.6 and 104.12LB were collected from field grown plants in Lima, Peru, during 1987. The two LT clones produce TPS progenies adapted to lowland tropical environments. The clone R128.6 is a popular open pollinated selection of a clonal introduction from the breeding program of the University of Cornell, and 104.12LB yields progenies with moderate levels of resistance to Late Blight. The pollen was used for treatments and in vitro evaluations, and for simultaneous field pollinations onto the variety Atzimba. Previously, tagged flowers selected at random in the mother plants were emasculated and pollinated the day after anthesis. The experiment consisted of exposing pollen samples to a temperature of 30°C for 10, 20 and 30 minutes; untreated pollen was used as control. In potatoes, temperatures above 25°C are considered supra-op-

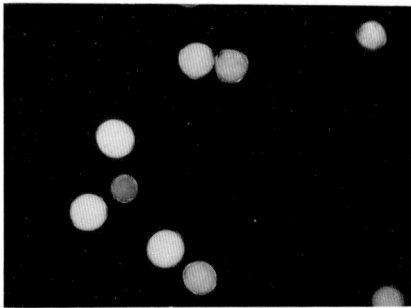

Fig. 1 Pollen sizes in clone LT-7   Fig. 2 Pollen sizes in clone R128.6

timal for plant growth (Burton 1966). The effects of high temperature treatments on pollen viability were evaluated as the percent of in vitro germination in untreated pollen. The number of TPS in each of 10 randomly selected berries produced in the field was used to estimate in vivo pollen fertility. The TPS had been stored dry (3-5% moisture) at approximately 22°C and was tested 4 to 6 months after harvest. The TPS from berries which were not selected for individual measurements was bulked in each treatment and sampled at random with five replications of 100 seeds each for the germination test, and 25 for the seedling emergence test. The germination test consisted of placing untreated TPS evenly over moist filter paper inside covered Petri dishes at 30/25°C under 12 hrs of light and dark periods inside an incubator. The analyses of the results for final percent of germination 11 days after seeding are presented. The seedling emergence test consisted of sowing the TPS in trays containing a 1:1 mix of sterilized peat and sand. The seedlings were grown inside a greenhouse at a mean temperature of 27°C (±6.5°). The analyses of final percent of emergence 17 days after sowing and the dry weight per plant of seedling tops are presented. In the seedling test, the TPS received two different presowing treatments (PST), giberellic acid at 1500 ppm, KNO_3 + K_3PO_4 at -1.0 MPa (priming), in addition to the untreated control as described in a previous paper (Pallais 1988). However, since the PST x pollen treatment interactions were not significant, te data on final emergence and seedling dry weight are the average of 15 replicates. The analyses of percentage data were performed after arcsin √ x transformations.

Table 1. Effects of high temperature on potato pollen

Clones	LT-7					LT-5				
	0'	10'	20'	30'	LSD5%	0'	10'	20'	30'	LSD5%
Pollen % Germ*	100	66	34	24	16.2	100	98	83	62	23.9
TPS per berry**	61	82	61	48	23.1	53	75	51	76	ns
TPS germination (%)	56	61	40	43	15.1	50	45	43	41	ns
TPS emergence (%)	60	74	78	74	ns	68	65	63	63	ns
Dry weight (mg) of plant	3.8	3.8	3.6	3.7	ns	5.7	6.0	5.7	6.8	ns
	104.12LB					**R128.6**				
	0'	10'	20'	30'	LSD5%	0'	10'	20'	30'	LSD5%
Pollen % germ*	100	60	37	35	4.8	100	71	39	34	25.1
TPS per berry**	227	249	240	228	ns	193	186	231	193	ns
TPS germination (%)	81	74	76	46	3.4	63	54	47	54	2.6
TPS emergence (%)	72	73	51	57	11.5	77	82	72	69	ns
Dry weight (mg) of plant	6.7	6.2	6.8	7.4	ns	9.8	9.9	9.2	9.8	ns

Time at 30°C

* pollen germinability as percent of total germination in untreated pollen

** TPS produced on the clone Atzimba

2.2. Effects of pollen size

Two experiments were conducted. In the first experiment, pollen of R128.6 was used unscreened (bulk) and after separation into three pollen sizes: >70μm (large), 70-50 (medium) and <50 (small) for crossings with the clones Atzimba and 65-ZA.5. In the second experiment, pollen of 104.12LB was separated into two pollen sizes: >70μm (large) and <70 (small) and crossed with Atzimba and MEX-750821. The pollen was used for in vitro evaluations and field pollinations as described in the first study.

Table 2

Effects of pollen size on the number (#) of TPS per berry, final percent of TPS germination and seedling emergence, and dry weight (mg) per plant.

	Atzimba x R128.6					65-ZA.5 x R128.6				
	Bulk	Large	Med.	Small	LSD5%	Bulk	Large	Med	Small	LSD5%
TPS #/berry	123	143	190	197	32.3	276	191	279	302	59.9
TPS % germ	66	62	43	52	9.0	61	62	61	57	ns
TPS % emerg	73	80	70	70	3.1	69	71	77	73	ns
dwt/plant	5.2	4.8	4.9	4.9	ns	4.5	5.1	4.6	5.2	0.65

3. Results and Discussion

3.1. Effects of high temperature on pollen

The tolerance to warm growing conditions of clone LT-7 versus R128.6 or 104.12LB was not associated with a higher in vitro germinability of LT-7 pollen after exposure to high temperature (Table 1). However, for LT-5 such association was observed. These results were repeated in our laboratory. Therefore, it appears that adaptation of potato to warmer temperatures may not be necessarily related to the survivability of pollen under high temperature. This is understandable, as in potatoes "adap-

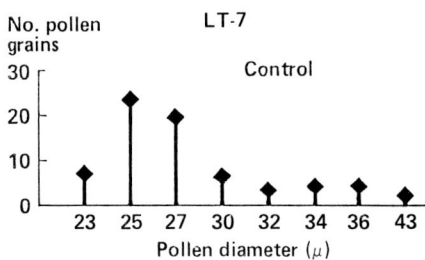

Fig 3. Percent in vitro germination in various pollen sizes

tation" to tropical environments refers to the ability of certain clones to tuberize; clones considered to be "nonadapted" may grow luxuriously under warm temperatures and fail to produce any tubers.

Although is is possible that exposure to high temperatures may have damaged pollen grains as suggested by the reduction on in vitro germinability, their seed setting ability was apparently not affected (Table 1). The effects of high temperature pollen

Table 3				Table 4					
Effects of pollen size of clone 104.12LB on the number of TPS per berry in two crosses				Interactions between pollen and presowing treatments on final percent of emergence and dry weight (mg) per seedling in two crosses with pollen of 104.12LB					
Female Parent	**Bulk**	**Large**	**Small**	**LSD5%**	Female clones		Atzimba		MEX-75821	
Atzimba	215	209	247	20.4		% Emerg			Dwt/plant	
MEX-75821	287	266	310	41.4		Bulk	Large	Small	Bulk	Large	Small
					Priming	89	91	88	4.3	4.4	3.8
					GA1500	60	60	39	2.0	3.1	2.5
					Untreated	68	72	73	2.7	2.6	2.8
					LSD5%		18.11			0.61	

treatments on the performance of TPS were either nonsignificant or detrimental (Table 1). Nevertheless, the description of reduced seed germination or emergence in TPS as a "detrimental" effect may be questioned because seed dormancy may have been enhanced as a consequence of pollen treatment. TPS may retain some dormancy effects up to 18 months after harvest (Simmonds 1964) and these lots were tested 4-6 months after harvest.

To conclude that pollen selection has not occurred as a result of high temperature pollen treatment, then, would be premature. The possiblity still exists that the tolerance to high temperature observed in certain pollen grains could be a genetic factor which may be associated with other sporophytic characters apart from seedling performance. Therefore, the various TPS lots produced in this experiment are currently being evaluated in the field for additional plant growth characteristics, such as flowering and tuberization. This information may be important not only in pollen selection studies, but also in seed production research. Since TPS technology is intended for tropical areas, the seed may have to be produced in warm environments. Nevertheless, our results suggest that it may not be possible to utilize the concept of "genetic overlap" (Mulcahy 1985) on the response in potato pollen viability for selecting high temperature tolerant TPS progenies.

3.2. Effects of pollen size

The results showed that small pollen of R128.6 had a lower percentage of in vitro germination than all the other lots (Fig.3). There were no significant differences in pollen germinability between large and small pollen of 104.12LB (data not shown); in this section only significant effects will be presented and discussed. The data (Table 2) indicates that increasing pollen size of R128.6 resulted in a decrease of seed set per berry in both crosses. However, a defined response to pollen size on the performance of the TPS was not observed. Nevertheless, some significant differences in performance were obtained among seed lots produced with the various pollen sizes. Germination and emergence were higher in TPS of Atzimba x R128.6 when it was produced with large pollen versus medium and small grains (Table 2).

Small R128.6 pollen was less able to germinate in vitro (Fig.3), but still produced more seed in vivo than large grains (Table 2). Therefore, the lower pollen germinability in small grains, as was the case for high temperature treated pollen in the preceeding study, was not a significant factor in seet set. This is explained since it is known that more than 4,000 pollen grains may to be deposited on the stigmas of tetraploid potato clones (Jenssen and Hermsen 1976). Moreover, approximately 700 mature ovules have been counted in potato flowers at anthesis in our laboratory, and only 300 are usually able to set seed. Therefore, the conditions in the style could be the limiting factor to seed set in potatoes; smaller pollen may produce smaller tubes resulting in a higher seed set. Potato pollen size then, appears to play a purely physical role in determining seed set. However, while in the cross of 65-ZA.5 bulk pollen produced as many seeds per berry as medium sized grains, all separated pollen fractions produced more seed than bulk pollen in Atzimba x R128.6; although, the difference with respect to large pollen was not significant (Table 2). Therefore, the female parent (Atzimba) appears to have exerted some type of inhibition in bulk pollen during growth in the style. The lower seed set observed in the Atzimba cross versus 65-ZA.5 (Table 2) supports the suggestion of an inhibiting effect of the mother plant (Atzimba) on pollen tube development in the style.

The effects on seed set per berry of large and small pollen of 104.12LB crossed with either Atzimba or MEX-750821 supported the results obtained with R128.6 pollen; more seed was produced by small pollen (Table 3). Moreover, the Atzimba mother plant also appeared to exert in this case a generally similar, although less strong, inhibiting influence on seed set of bulk pollen. In Atzimba x 104.12LB the significant interaction on final emergence between pollen size and PST shows that when small pollen TPS was treated with GA, emergence was lower than in all the other lots (Table 4). Seedling dry weight in MEX-750821 x 104.12LB of TPS produced with small pollen was also lower than large when the seed had been primed or presowed in GA (Table 4). These results suggest that SP seed was probably less vigorous than large pollen TPS in both crosses.

4. Conclusions

This study showed that the tolerance to high temperature of potato pollen is not necessarily an indication of the ability of these grains to produce TPS progenies adapted to high temperature conditions during germination and early seedling emergence or the ability of the pollen source to produce tubers under tropical conditions. It was also found that decreased in vitro germinability of pollen affected by high temperature was not followed by a reduction in seed set.

The effects of pollen size, however, were found to significantly affect seed set, with small pollen producing more seed than large grains.

Large pollen in a heterozygous species, such as potato, could be the result of a positional advantage or a higher competitive ability of certain grains for the assimilation of the available plant nutrients during pollen development in the anther. For if potato pollen sizes were determined solely by physiological factors, then their size should vary in a continuum; however, pollen size was observed to vary in discreet classes (Fig. 1). Moreover, sufficient significant differences in germination or seedling performance parameters were found among seed lots produced with the various pollen sizes to suggest that in some of these lots the seed was in fact different as a result of the differences in pollen size. Thus, the thesis that pollen selection may be a tool for increasing the uniformity of TPS crosses (Pallais et al. 1984) deserves continued investigation.

5. Bibilography

Burton WG (1966) The potato. Veenman H, Zonen NY, Wageningen, Holland p 382

Jenssen AW, Hermsen, JG (1976) Estimating pollen fertility in Solanum especies and haploid. Euphytica 25:577-586

Larkin PJ, Scowcroft WR (1981) Somaclonal variation—a novel source of variability from cell cultures for plant improvement. Theor Appl Genet 60:197-214

Mangelsdorf PC (1932) Mechanical separation of gametes in maize. J Heredity:289-295

Pallais N, Fong N, Berrios D (1984) Research on the physiology of potato sexual seed production. In: Rep 18 Plann Conf. Innovative methods for propagating potatoes. Int Potato Center (CIP), Lima, Peru, pp 153-158

Pallais, N, Malagamba P, Fong N, Garcia R, Schmiediche (1986) Pollen selection through storage: A tool for improving true potato seed quality. Im: Mulcahy DL, Mulcahy Bergamine G, Ottaviano E (eds) Biotechnology and Ecology of Pollen. Spranger, Berlin, pp. 153-158.

Pallais N. (1988) Osmotic priming of true potato seed: effects of seed age. Potato Res. (In press)

Simmonds NW (1963) Experiments on the germination of potato seeds. I Eur Potato J 6:45-59

Weaver ML, Timm H, Silbernagel MJ, Burke DW (1985) Pollen staining and high- temperature tolerance of bean. J Am Soc Hort Sci 110:797-799

Pollen Maturation and Desiccation Tolerance

Folkert A. Hoekstra, Tineke van Roekel, and Nick ten Pas

Department of Plant Physiology of the Agricultural University

Arboretumlaan 4

6703 BD Wageningen

The Netherlands

Pollen generally is tolerant to severe desiccation (Hoekstra, 1986). The molecular mechanism of this tolerance is far from elucidated, however.

As drought sensitive tissues leak their soluble cellular constituents upon reimbibition, special attention has been paid to the behaviour and composition of membranes during drought stress. Dehydration of PE, particularly, promotes the formation of non-bilayer hexagonal$_{II}$ phase (Crowe et al., 1988a). It has been postulated that hexagonal$_{II}$ phase lipid is responsible for the leakage of rehydrating dry seeds (Simon, 1974). However, no evidence for the occurrence of hexagonal$_{II}$ phase lipid in anhydrobiotic organisms has been provided, thus far (see Crowe et al., 1988a, for details). Membranes of desiccation-stressed wheat seedlings decrease in PE and increase in PC (Huitema et al., 1982). By decreasing the PE content the plant may be able to prevent formation of undesirable hexagonal phase lipid, rendering it more tolerant to desiccation (Vigh et al., 1986).

In microscopic animals and spores of certain fungi and algae, desiccation tolerance has been associated with the presence of high levels of trehalose (Crowe et al., 1984). Rapid drying prevents synthesis of trehalose, and maintains sensitivity to drying, whereas slow drying allows for its accumulation and the acquisition of desiccation tolerance (Madin and Crowe, 1975). The molecular interaction of trehalose with phospholipids during dehydration has been studied in detail (Crowe et al., 1987). In this paper we report on the changes in desiccation tolerance, and in composition of P-lipids and carbohydrates during pollen maturation. We found that desiccation tolerance was associated with high levels of sucrose.

Abbreviations: PC = phosphatidylcholine; PE = phosphatidylethanolamine; PA = phosphatidic acid; FDA = fluorescein diacetate.

MATERIALS & METHODS

Plants of Papaver dubium L. and Nicotiana tabacum cv Mont Calme Brun were grown in the laboratory garden. The various stages of the Papaver flower bud development were recognized according to methods described elsewhere (Hoekstra and van Roekel, 1988). Stage of development of Nicotiana flower buds was selected by length. Anthers were dissected at the appropriate stages of development. When exposed to the ambient air, pollen rapidly dehisced from the anthers. However, dehiscence of Papaver pollen from the -4 stage anthers was inadequate. After sieving, pollen was further dried over silicagel for 24 h prior to storage at -20°C.

Dried pollen was always rehydrated from the vapour phase for at least 1 h prior to incubation in germination medium, to prevent imbibitional damage (Shivanna and Heslop-Harrison, 1981; Hoekstra and van der Wal, 1988). The germination media consisted of the usual Ca, Mg, B, and K-salts, and 0.2 M sucrose in a 2 mM Na-phosphate-citrate buffer, pH 5.9, solidified by the addition of 0.6% agar. Percentage germination was scored on 200 grains. The technique of FDA-induced fluorescence was performed according to Shivanna and Heslop-Harrison (1981). K^+ was analysed by flame photometry. Analyses of soluble sugars and P-lipids were according to standard techniques (see Hoekstra and van Roekel, 1988, for details).

Liposomes (80-100 nm) were prepared and sized by 5 cycles of freezing and thawing of a P-lipid suspension, followed by extrusion 5 times at 500 psi through 2 stacked polycarbonate membranes of 0.1 μm pore diameter. Liposome fusion was recorded by standard techniques of resonance energy transfer, using cholesteryl anthracene-9-carboxylate and N-4-nitrobenzo-2-oxa-1,3-diazole-PE as donor and acceptor probe, respectively (Rudolph and Crowe, 1985). Leakage from liposomes was analysed using 5(6)-carboxyfluorescein as the entrapped hydrophylic dye.

RESULTS

Papaver pollen of the -3 stage of development was unable to germinate in vitro (Table I). Maximal germinative capacity was reached at one day prior to anthesis. Desiccation tolerance was determined in two ways: by the analysis of the percentage of grains that gave fluorescence upon addition of FDA, and by the amount of K^+ in the pollen filtrate as a percentage of the total amount of K^+ present. Both methods indicate that whilst the -3 stage pollen was extremely leaky, the -1 stage pollen was tolerant to drying, the -2 stage having an intermediate tolerance (Table I). At the -4 stage, IKI-staining did not give the typical blue colour specific for the presence of starch, but at the -3 stage starch had filled

Tabel I. Characteristics of Papaver pollen dried to 5% moisture content (on a fresh weight basis) during its development. Pollen was either obtained by drying dissected anthers in air, or by mechanical isolation in a Potter tube in mannitol solution followed by washing and filtration. The stage of development is indicated in days prior to anthesis. In the case of an additional storage in humid air for 30 or 25 h, the drying treatment was given immediately thereafter.

Treatment and Stage of Development	Germination in vitro	FDA-induced Fluorescence	K⁺-Leakage	IKI-stained Grains	Sucrose Content
days	%	%	%	%	% of DW
Anther Drying					
−4	0	0	97	2	9.0
−3	2	3	96	94	7.6
−2	38	55	50	52	12.1
−1	87	88	24	3	17.0
0 (anthesis)	86	90	18	1	16.3
−3 + 30 h humid air	85	86	27	4	15.4
Mechanical Isolation					
−3	0	1	−	91	5.0
−3 + 25 h humid air	59	67	−	15	9.1

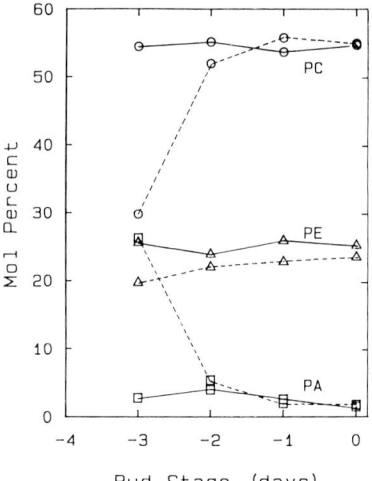

Fig. 1. Changes in composition of the P-lipids, PA, PC, and PE, during maturation of Papaver pollen. Dashed curves: analyses of dried pollen; solid curves: analyses of fresh pollen.

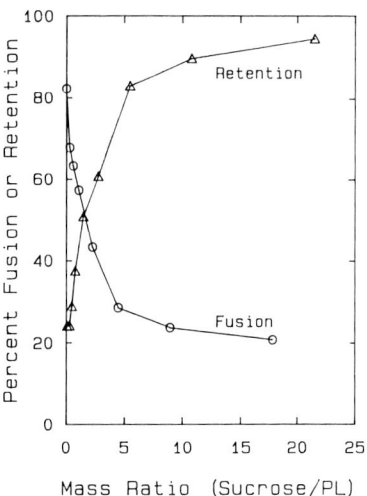

Fig. 2. Effects of a series of concentrations of sucrose on fusion and solute retention of liposomes prepared from Papaver P-lipids (PL), upon a cycle of drying and rehydration.

the grains. Acquisition of desiccation tolerance coincided with degradation of this starch and a considerable increase in sucrose. In Papaver pollen sucrose was the primary soluble sugar, comprising more than 97% of the total (data not shown).

Table I further shows that immature anthers (-3 d) stored for an additional 30 h in moisture-saturated air, shed pollen upon drying that was tolerant to desiccation and had all the other characteristics of a mature grain. The same was true when immature pollen was liberated from the young anthers in a 0.2 M mannitol solution by few strokes in a Potter tube and, after sieving and cleaning, allowed to mature in humid air for 25 h.

An analysis of the P-lipid composition in the dried pollen samples revealed a considerable change in mol percentage of the 3 major P-lipid species during ripening (Fig. 1). As this shift in composition could equally well be the result of desiccation injury rather than a true developmental change, P-lipid analyses have also been done on fresh pollen samples. Non-dried material was obtained by mechanical disruption of anthers in mannitol solution as mentioned above. Fig. 1 shows that changes in P-lipid composition of this fresh material were minimal.

In order to study the effect of sucrose on membranes during drying, liposomes were prepared from purified Papaver pollen P-lipids. Fig. 2 shows that when liposomes were dried in the presence of sucrose, their integrity upon rehydration was preserved best at mass ratios of sucrose to P-lipid of 4 and higher. Less sucrose could not prevent extensive fusion between vesicles and leakage of an entrapped dye from them.

An indication that the changes observed during maturation of Papaver pollen have a more general validity in other species was provided by results of Nicotiana tabacum pollen screened for changes in dehydration tolerance, germinative capacity, starch and sucrose content during ripening (Fig. 3). It is difficult to make an accurate estimation of the developmental stage of this pollen as buds grow slowly during their development. The sucrose content rose from 3% in the immature pollen (bud length 4 cm) to 13% in the mature stage.

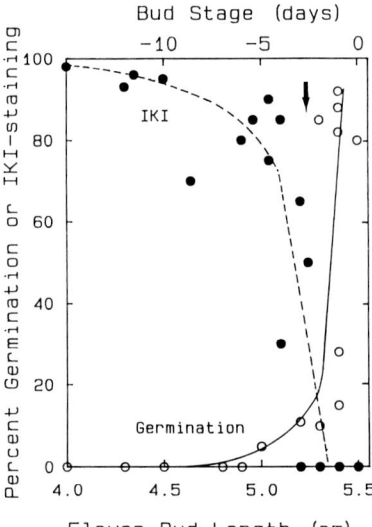

Fig. 3. Some characteristics of fresh Nicotiana pollen during its development. The arrow indicates the transition to desiccation tolerance.

DISCUSSION

Severe drying did not harm mature _Papaver_ and _Nicotiana_ pollen. In contrast, immature pollen had reduced plasma membrane integrity after a cycle of drying and rehydration, as evidenced by the excessive leakage of endogenous K^+ and the inability to retain fluorescein. An attempt to correlate the transition to drought tolerance to possible changes in P-lipid composition failed — we expected a decrease in PE —, as no changes were observed in fresh pollen. On the contrary, the PE content was not exceptionally low in mature _Papaver_ pollen. The high PA and low PC contents in the dried immature pollen can be ascribed to the damaging effect of reduced membrane integrity which may have released phospholipase D from its vesicular confinement.

Interestingly the content of sucrose, the primary sugar present, doubled with the capacity of pollen to survive dehydration. As this may be just a correlation and does not necessarily imply cause and effect, we have conducted the experiments shown in Fig. 2 on effects of sucrose on the stability of liposomes prepared from P-lipids of _Papaver_ pollen. Sucrose prevented fusion between and leakage from the vesicles during a cycle of drying and rehydration. This occurred at a ratio of sucrose to P-lipid not far beyond that encountered in situ. We suggest that sucrose may have a protective function in the intact pollen grain similar to that of trehalose in model membrane systems (Crowe et al., 1984).

Trehalose exerts its effect by interacting with the phosphate head group of the P-lipids, thus preventing dehydration-induced phase separation and formation of gel phase. Recent findings have shown that both trehalose and sucrose also inhibit the transition of PE into the hexagonal$_{II}$ phase during desiccation, thus contributing further to the stability of the dry membrane (Crowe et al., 1988b).

As the rise in sucrose content occurred within the dissected anther, and even in the isolated grain kept at high humidity, it is most likely that the sucrose is derived from the degrading starch. The temporary presence of starch in other pollen species during ripening has been reported (Pacini and Juniper, 1984). We suggest that the breakdown of starch and the rise in sucrose associated with the acquisition of desiccation tolerance and germinative capacity may be phenomena which are common to pollen in general.

ABSTRACT

On the basis of fluorescence and K^+-leakage data, it was concluded that _Papaver dubium_ pollen becomes desiccation tolerant at approximately 2 to 1 d prior to anthesis. At the same time starch degraded, and the level of sucrose, the primary soluble carbohydrate present, increased considerably, but phospholipid composition did not change. Similar processes occurred in <u>Nicotiana tabacum</u> pollen during ripening. During desiccation, sucrose effectively protected the integrity of liposomes prepared from <u>Papaver</u> phospholipids. We suggest that analogous to trehalose in anhydrobiotic microscopic animals, yeasts and algae, sucrose provides desiccation tolerance in pollen.

Immature <u>Papaver</u> pollen isolated at 3 d prior to anthesis, was able to independently become functional and desiccation tolerant when kept at high humidity. During this humid storage starch degraded and sucrose level doubled.

REFERENCES

Crowe JH, Crowe LM, Chapman D (1984) Preservation of membranes in anhydrobiotic organisms: the role of trehalose. Science 223: 701-703

Crowe JH, Crowe LM, Carpenter JF, Aurell Wistrom C (1987) Stabilization of dry phospholipid bilayers and proteins by sugars. Biochem J 242: 1-10

Crowe JH, Crowe LM, Hoekstra FA, Aurell Wistrom C (1988a) Effects of water on the stability of phospholipid bilayers: The problem of imbibition damage in dry organisms. Crop Sci Special Edition (in press)

Crowe JH, Crowe LM, Carpenter JF, Rudolph AS, Aurell Wistrom C, Spargo BJ, Anchordoguy TJ (1988b) Interactions of sugars with membranes. Biochim Biophys Acta (in press)

Hoekstra FA (1986) Water content in relation to stress in pollen. In Leopold AC (ed), Membranes Metabolism and Dry Organisms, pp 102-122 Comstock Publishing Ass Ithaca London

Hoekstra FA, Van der Wal EG (1988) Initial moisture content and temperature of imbibition determine extent of imbibitional injury in pollen. J Plant Physiol (in press)

Hoekstra FA, van Roekel T (1988) Desiccation tolerance of Papaver dubium L. pollen during its development in the anther: Possible role of phospholipid composition and sucrose content. Plant Physiol (in press)

Huitema H, Woltjes J, Vigh L, Van Hasselt P (1982) Drought induced resistance in wheat correlates with changes in phospholipids. In Wintermans JFGM, Kuiper PJC (eds) Biochemistry and metabolism of plant lipids, pp 433-436 Elsevier Biomedical Press Amsterdam New York Oxford

Madin KAC, Crowe JH (1975) Anhydrobiosis in nematodes: carbohydrate and lipid metabolism during dehydration. J Exp Zool 193: 335-342

Pacini E, Juniper B (1984) The ultrastructure of pollen grain development in Lycopersicum peruvianum. Caryologia 37: 21-50

Rudolph AS, Crowe JH (1985) Membrane stabilization during freezing: The role of two natural cryoprotectants, trehalose and proline. Cryobiol 22: 367-377

Shivanna KR, Heslop-Harrison J (1981) Membrane state and pollen viability. Ann Bot 47: 759-770

Simon EW (1974) Phospholipids and plant membrane permeability. New Phytol 73: 377-420

Vigh L, Huitema H, Woltjes J, Van Hasselt PR (1986) Drought stress-induced changes in the composition and physical state of phospholipids in wheat. Physiol Plant 67: 92-96

Pollen Pistil Interaction in the Ovary in Fruit Trees

Herrero, M., Arbeloa, A., Gascon, M.
Unidad de Fruticultura
CENIA DE Aula Dei
Apartado 727
Zaragoza (Spain)

Introduction

Siphonogamy, or the fact that fertilization is mediated through a pollen tube, is emerging as a uniquely adapted system for male-female interaction (Heslop-Harrison, 1983) and for male gametophyte selection (Mulcahy, 1979). However, we still know very little on the physiological mechanisms that control such selection. A better understanding of male-female interaction from pollination to fertilization may enlighten this process. In this respect most of the work has been done regarding pollen germination at the stigma (Heslop-Harrison and Heslop-Harrison, 1985) or concerning pollen tube growth in the style (van Went and Willemsen, 1984). However, there is a paucity of data regarding pollen tube growth in the ovary region (Knox, 1984).

It has been tacitly assumed that, once the pollen tubes enter the ovary, fertilization was an straightforward process. However, working with fruit tree species, we have observed that in the ovary a number of interactive processes are set up from the arrival of the pollen tubes until fertilization occurs. In this paper we discuss these events taking place in the ovary and their possible implications in male-female interaction and in gametophyte selection.

The effect of the ovary on pollen tube growth

In a recent study carried out in peach (Herrero and Arbeloa, 1988), we have observed that a variable lapse of time occurs from the arrival of the pollen tubes at the base of the style until fertilization takes place. This time may be longer than the time required by the pollen tubes to travel down the style.

When the pollen tubes reach the base of the style and enter

the ovary they meet the obturator that is a placental protuberance connecting the style with the ovule micropyle. An histochemical study of this structure (Arbeloa and Herrero, 1987) reveals that when the pollen tubes arrive at the obturator they stop and growth is not resumed until five days later. On the arrival of the pollen tubes the obturator cells are full of starch reserves. Five days later starch fades from these cells and a secretion that stains for carbohydrates and for proteins is produced. Concomitantly with the production of this secretion growth of the pollen tubes is resumed on it. The fact that pollen tubes require this secretion to grow reinforces the idea that pollen tube growth along the pistil is heterotrophic (Herrero and Dickinson, 1979, Herrero and Arbeloa, 1988). However a major difference exists between growth in the style and on the obturator. While in the transmitting tissue starch digestion is triggered by pollination and only occurs in compatible matings (Herrero and Dickinson 1979), in the obturator this process is independent of pollination and appears to be a maturative stage of the pistil for it takes place in a similar way in pollinated than in unpollinated flowers (Arbeloa and Herrero, 1987).

Once the pollen tubes have passed along the obturator, callose starts to accumulate on this structure (Arbeloa and Herrero, 1987). This mechanism confers the obturator a critical role in controlling pollen tube penetration into the ovary since it acts as a leaving bridge either connecting or isolating the ovary to the style. Thus, pollen tube growth is not possible before the secretion phase, neither is it possible later once the obturator degenerates.

It has been observed that pollen tubes grow on the obturator in a number of unrelated species (Tilton and Horner, 1980; Tilton et al., 1984; Hill and Lord, 1987). However, the fact that growth of the pollen tubes on the obturator is controlled by means of a discontinuous secretion has been, so far, only observed in peaches (Arbeloa and Herrero, 1987). It may be worth to evaluate if this process is extensible to other species and if in these species the obturator may also be a physical structure that supports male female interaction and that regulates pollen tube passage to the ovule.

Once the pollen tubes have traversed the obturator a further stop of three days takes place in peach (Herrero and Arbeloa, 1988) until they enter the micropylar exostome and three days later achieve fertilization. Further study is necessary to provide an explanation for these events. However, the fact that pollen tubes may wander before entering the ovule suggest that, like at the obturator, a developmental stage of the ovule may be necessary before the pollen tubes can penetrate and effect fertilization.

It has tacitly been assumed that a chemotropic stimulus may be necessary for a pollen tube directional growth (Welk, Millington and Rosen, 1965) and further work has put forward an effect of the pistil on pollen tube directionality (Mulcahy and Mulcahy, 1987). However directional pollen tube growth along the transmitting tissue and the obturator could be determined just in a trophic way, by following the path of available reserves (Herrero and Arbeloa, 1988). Then the chemotropic effect might be only necessary for the pollen tubes to penetrate the ovule (Lord and Kohorn, 1986)

The effect of the pollen tubes on the ovary

The pollen tubes in turn have an effect on pistil development. Previous work has indicated that pollination induces an activation of the ovary (Linskens, 1973; Deurenberg, 1976). Likewise, working in pear, we have observed an effect of the pollen tubes on the ovule (Herrero and Gascon, 1987). In this study pollen tubes, while not altering embryo sac maturation, they delay embryo sac degeneration. Thus, in a cross-pollinated flower, this degeneration is postponed by about ten days, extending the period over which a successful fertilization can take place. This extension of embryo sac viability is accompanied by an elongation of the embryo sac itself. These two phenomenons are initiated two weeks before fertilization takes place. In peach a similar effect has been observed and degeneration in a proportion of the cross pollinated flowers is delayed for seven days (Arbeloa, 1986).

The interactive effect of the pollen tubes on the embryo sac could be hormonaly regulated since a similar effect is produced by treating unpollinated pear flowers with gibberellic acid (GA_3) (Herrero and Gascon, 1987). It is tempting to putforward that active GA_3 secreted by the pollen tubes may move towards the ovule since pollen is known to be a rich source of GA_3 (Barendse et al., 1970). Alternatively the observed effect on embryo sac life might not be an exclusive effect of GA_3 but rather a consequence of a general activation of the pistil. This activation appears as one of the first manifestation of fruiting, whether it is induced by pollination or parthenocarpy (Martin et al., 1980). Whether extension of embryo sac viability is mediated by GA_3, or is a response to a general activation of the ovary, it does extend the period in which the ovule is receptive to the pollen tube, thus increasing the chances that fertilization will take place.

The significance of male-female interaction in the ovary

The events recorded here point out that, once the pollen tubes enter the ovary, fertilization is not a straightforward process. A lag phase is produced in which a number of mechanisms are set up providing a frame for male-female interaction in the ovary.

Some of these mechanisms, as the elongation of embryo sac viability, appear to be devised to assist fertilization and share this aim with other mechanisms present in the stigma and the style. In some species the stigma releases a post pollination exudate (Sedgley and Scholefield, 1980; Kenrick and Knox, 1981) that promotes germination. In others, pollination stimulates the transmitting tissue to release carbohydrates which are later used to support pollen tube growth (Herrero and Dickinson, 1979). In the ovary the prolongation of embryo sac viability described here, adds to the increasing number of processes known to be stimulated by pollination and that increase the chances that fertilization takes place. It would seem that even those pollen grains destined not to fertilise ovules play an important part in assisting other pollen grains to do so.

On the other hand the meaning of other mechanisms as the lag phase the pollen tubes suffer before entering the ovary or the ovule needs further study. These stops may reflect a lack of synchronism between male and female development. It has been put forward that maturation of the pistil occurs in a basipetal way. This maturation implies a number of secretory processes that take place along the pistilar tract and that play a role in pollen tube nutrition and guidance. Thus, for a succesful fertilization, it would be necessary a synchronism between male gametophyte growth and pistil secretory function (Herrero and Arbeloa, 1988). Failure of synchronism may lead to a lack of fertilisation. Thus male-female synchronism could act playing a role in controlling fertilization.

The evidence discussed here put forward that well controlled mechanisms that regulate male-female interaction are present in the ovary of fruit trees. While some of these mechanisms appear devised to assist fertilization, others appear as barriers or difficulties the gametophytes have to go through, and could play a role in gametophyte selection. A complete knowledge of these mechanisms and their significance will enlighten our understanding of male-female interaction and its implications in the control of fertilization.

Acknowledgments

Thanks are due to the U.S.- Spain Joint Committee for Scientific and Technological Cooperation for financial support.

References

Arbeloa A (1986) Estudio de la biología floral y fructificación en melocotonero (Prunus persica (L.) Batsch). Tesis Doctoral. Univ. Navarra, p 224
Arbeloa A, Herrero M (1987) The significance of the obturator in the control of pollen tube entry into the ovary in peach (Prunus persica). Ann Bot 60: 681-685
Barendse G W M, Rodrigues Pereira A S, Berkers P A, Driessen F M, Van Eyden Emons A, Linskens H F (1970) Growth hormones in pollen, styles and ovaries of Petunia hybrida and Lilium species. Acta Bot Neerl 19: 175-186
Deurenberg J J M (1976) Activation of protein synthesis in ovaries from Petunia hybrida after compatible and incompatible pollination. Acta Bot Neerl 25(3): 221-226
Herrero M, Arbeloa A (1988) The influence of the pistil on pollen tube kinetics in peach (Prunus persica). (In prep.)

302

Herrero M, Dickinson H G (1979) Pollen-pistil incompatibility in *Petunia hybrida*: changes in the pistil following compatible and incompatible intraspecific crosses. J Cell Sci 36: 1-18

Herrero M, Gascon M (1987) Prolongation of embryo sac viability in pear (*Pyrus communis*) following pollination or treatment with gibberellic acid. Ann Bot 60(3): 287-294

Heslop-Harrison J (1983) Self-incompatibility: phenomenology and physiology. Proc R Soc London B 218: 371-395

Heslop-Harrison J, Heslop-Harrison Y (1985) Surfaces and secretions in the pollen-stigma interaction: A brief review. J Cell Sci Suppl 2: 287-300

Hill J P, Lord E M (1987) Dynamics of pollen tube growth in the wild radish *Raphanus raphanistrum* (Brassicaceae). II. Morphology, cytochemistry and ultrastructure of transmitting tissues, and path of pollen tube growth. Amer J Bot 74(7): 988-987

Kenrick J, Knox R B (1981) Post-pollination exudate from stigmas of *Acacia* (Mimosaceae). Ann Bot 48: 103-106

Knox R B (1984) The pollen grain. In: Johri B N (ed) Embryology of Angiosperms. Springer Verlag, Berlin, p 197-261

Linskens H F (1973) Activation of the ovary. Caryologia Suppl 25: 27-41

Lord E M, Kohorn L V (1986) Gynoecial development, pollination and the path of pollen tube growth in the tepary *Phaseolus acutifolius*. Amer J Bot 73: 70-78

Martin G C, Romani R J, Weinbaum S A, Nishijima C, Marshack J (1980) Abscisic Acid and Polysome content at anthesis and shortly after anthesis in pollinated, non-pollinated and non-pollinated 'Winter Nelis' pear flowers treated with gibberellic acid. J Am Soc Hortic Sci 105(3): 318-321

Mulcahy D L (1979) The rise of angiosperm: A genecological factor. Science 206: 20-23

Mulcahy G B, Mulcahy D L (1987) Induced pollen tube directionality. Am J Bot 74(9): 1458-1459

Sedgley M, Scholefield B P (1980) Stigma secretion in the water-melon before and after pollination. Bot Gaz 141: 428-434

Tilton V R, Horner H T (1980) Stigma, style and obturator of *Ornithogalum caudatum* (Liliaceae) and their function in the reproductive process. Amer J Bot 67: 1113-1131

Tilton V R, Wilcox L W, Palmer R G, Albertsen M C (1984) Stigma, style and obturator of soybean *Glycine max* (L.) Herr. (Leguminoseae) and their function in the reproductive process. Amer J Bot 71: 676-686

Welk S M, Millington W F, Rosen W G (1965) Chemotropic activity and the pathway of the pollen tube in lily. Amer J Bot 52: 774-780

Went J V van, Willemsen M T M (1984) Fertilization In: Johri B N (ed) Embryology of Angiosperms. Springer-Verlag, Berlin, p 273-317

In Vitro Pollen-Style Interactions in *Malus Domestica*

Gian Lorenzo Calzoni and Anna Speranza
Department of Biology
University of Bologna
via Irnerio 42
40126 Bologna
Italy

Introduction

Gametophytic self-incompatibility (GSI) is an "ancient" system, perhaps acquired during early stages of the evolution of the Angiosperms (Gibbs, 1986). A very attractive general theory for the GSI mechanism has been advanced by Heslop-Harrison (1983), who proposes that, on the pistil side, S-specific glycoproteins with lectin-like properties, present in surface secretions and/or in the stylar transmitting tracts, are able to bind to specific complementary sugar sequences exhibited by components of the incompatible pollen-tube wall, leading to a disruption of apical growth. Recognition and inhibitory reaction in the Angiosperm self-incompatibility process should thus involve specific adhesion mediated by protein-carbohydrate complementation, as in many examples of cell recognition in various animal and plant systems.

In this context, the success obtained by Sharma and coworkers (Sharma & Shivanna, 1983 and 1986; Sharma et al., 1985) in overcoming self-incompatibility through sugar treatment on pollen (Eruca, Nicotiana, Petunia) before in vivo pollination or in vitro incubation with self-style extract is relevant. The agreement of their results from in vivo and in vitro systems is also comforting. Therefore, despite the difficulties and severe limits of simulating the pistil environment, we attempted to study the response of in vitro germinating apple pollen to self-style proteins, in particular by focusing our attention on what happens in culture following the sugar treatment which, for other species, has been shown to overcome self-incompatibility.

Materials and Methods

Pollen of Malus domestica Borkh. of cvs. Starkrimson (strictly self-incompatible), Golden Delicious and Perleberg 3 (both able to pollinate the Starkrimson), was obtained from plants grown in experimental plots near Ravenna (Italy). It was germinated in mass liquid culture according to Calzoni et al. (1979) for 120 min or less, at 30°C in the dark.

In some specified cases, the rehydrated pollen of Starkrimson was treated, prior to germination, by 10 min soaking in medium supplemented with 100 mM glucose. Pollen was then filtered on Millipore disks (8 μm pore size), re-suspended in fresh medium with or without style extract, and incubated as above.

Styles (including the stigma) were cut at their base from unpollinated flowers of the cv. Starkrimson, immediately frozen in liquid N_2 and stored at -80° C. A diffusate was obtained by suspending styles for 10 min in fresh medium with gentle stirring: when specified, this medium was used for incubating the pollen of the three cultivars. An extract was prepared by crushing styles in a mortar, firstly using liquid N_2, then in an extraction buffer (50 mM Tris-HCl, 10 mM mercaptoethanol, 1 mM Na-EDTA, and 5 mM $MgCl_2$ at pH 6.6). Proteins were precipitated with 80% saturating $(NH_4)_2SO_4$ for 2h at 4° C, pelleted, resuspended and diaconcentrated with an Amicon system, using PM 10 membranes, cut-off 10,000. When specified, this extract was added to the germination medium of Starkrimson.

Protein determination was done according to Lowry et al. (1951) on 10% TCA precipitates of a) pollen culture medium after incubation with or without style diffusate or extract; b) medium supplemented with sugar, which was used for pollen pre-treatment prior to germination.

Results

In the presence of the diffusate from styles of Starkrimson percent emergence significantly decreased in self-pollen whereas it was unaffected in cross-pollen. When medium containing the style diffusate was kept at 100°C for 15 min (prior to being used for suspending pollen) no effect could be observed on germination of Starkrimson (fig.1). Tube length also declined in the presence of self-diffusate, except when the latter was heated before germination. In the case of cross-pollen tubes, there was no significant inhibition (tab.1).

Style extracts also appeared to effectively reduce germination and more strongly elongation of tubes of Starkrimson at increasing protein concentrations (fig.2).

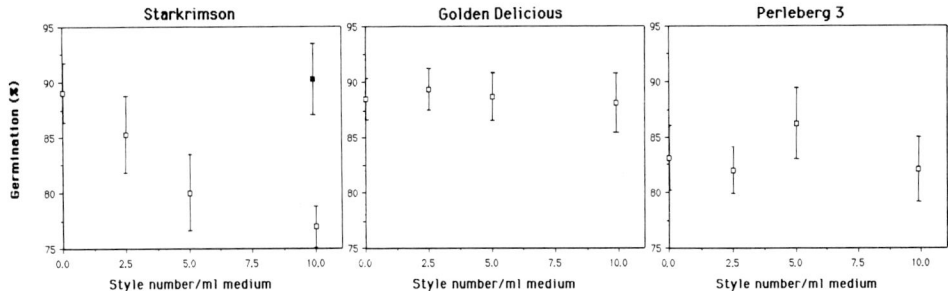

Fig.1. Effect of style diffusate on percent germination at 120 min of self-pollen (Starkrimson) or cross-pollen (Golden Delicious, Perleberg 3). Heating of the diffusate is indicated by (■).

Table 1. Mean tube length (percent of control) of pollen of three apple cvs. germinated in the presence of diffusate of Starkrimson styles. *Diffusate heated at 100°C for 15 min.

Cultivar	Number of styles/ml medium		
	2.5	5.0	10.0
Starkrimson	70+13	58+13	46+13 (95+13)*
Golden Delicious	93+ 8	88+ 6	85+ 5
Perleberg 3	105+17	113+19	105+16

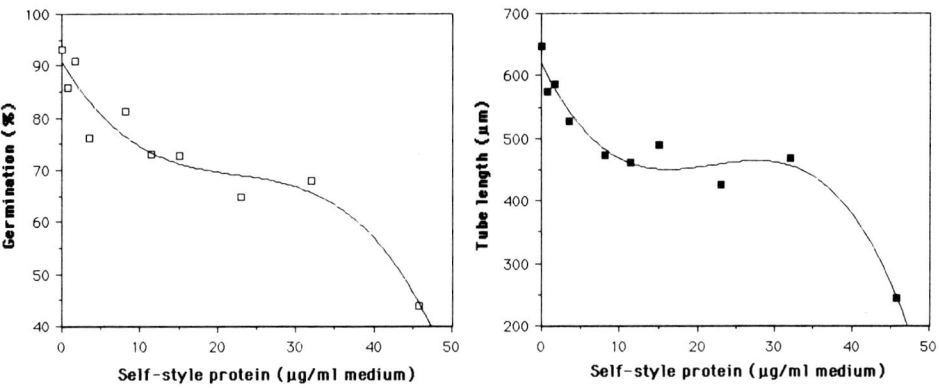

Fig.2. Effect of proteins from self-style on percent germination and tube length at 120 min of pollen of Starkrimson.

Total protein level recovered in the medium at 120 min was studied in the absence or in the presence of increasing amounts of self-extract; in the latter case, the values resulted somewhat lower than expected.

Firstly, we verified that the extract, when alone in the germination medium, maintained a constant level of protein during

incubation at 30°C for 120 min. The time-course of protein content
in the medium plus the extract revealed a dramatic decrease within
30 min of incubation (fig.3, B). This time, in apple pollen,
corresponds to the lag phase of germination, when neither true
pollen tubes nor apical secretion are present. Therefore, the
lowest possible interference exists with protein release by the
pollen within this period. However, a significant amount of
protein originates anyway from pollen because of the diffusion of
cell-wall stored proteins into the medium (see fig.3, A). Among
these, different enzymes are known to be readily leachable from
pollen grains of Angiosperm species (Knox and Heslop-Harrison,
1970). To verify if some pollen protease acting on the stylar
extract could be responsible for the drop in protein levels, pollen
leachate of 30 min (free of pollen grains) was incubated with the
extract: protein level remained unchanged throughout the two hours
of incubation.

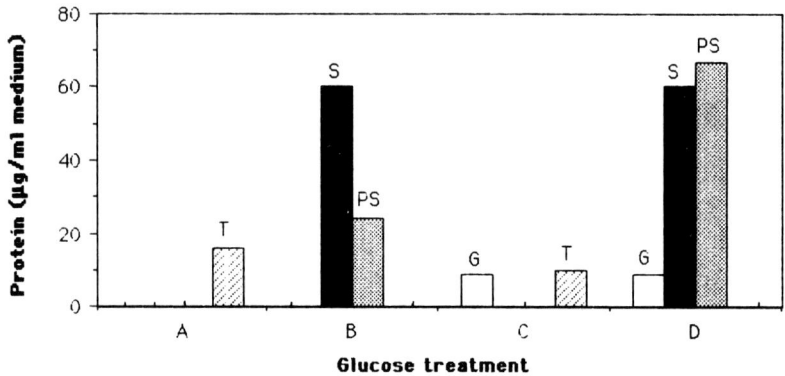

Fig.3. Proteins in the medium without (A,B) or with (C,D) glucose
treatment on pollen of Starkrimson.G: protein release after glucose
treatment ;S: self-style protein added at 0 time ; T: protein release
at 30 min incubation without style extract ; PS: total protein at 30
min incubation in the presence of style extract.

Fig.3 also shows that pre-treatment of pollen with 100 mM glucose
could prevent the drop otherwise occuring at 30 min germination in
the presence of stylar proteins.
It may be observed that the glucose treatment itself induced a
small protein release from pollen (fig.3, column G of C and D);
subsequent release in fresh medium without extract showed normal
value, indicating no apparent damage to the membrane system of the
treated pollen (fig.3, C, column T).
The glucose treatment was also able to reverse the inhibition on

tube elongation observed in the presence of proteins of self-style extract.

Discussion

 The assay we used to study the in vitro expression of self-incompatibility in apple appeared to be selective and sensitive. In fact, in the presence of styles of Starkrimson, the self-pollen was recognized and significantly inhibited (fig.1 and 2, tab.1).
 As also shown in Brassica (Ferrari & Wallace, 1975), substance(s) released by excised pistils (stigma + style) of apple are heat-labile, since complete loss of inhibition on self-pollen was achieved by 100°C treatment (fig.1, tab.1).
Considering that average protein content in the diffusate was 3 μg/style, there is an overall agreement between data obtained with the diffusate or with the extract of self-styles. Since even tube emergence itself was inhibited , it seems that sensitivity of self-pollen to pistil factors should be present, in apple, from the very beginning. In the presence of either diffusate or extract, it was nevertheless tube elongation which underwent the stronger inhibition. In fact, at 30 μg/ml (diffusate of 10 styles/ml, tab.1) or 46 μg/ml (fig.2) of protein, self-tube growth was reduced to about one-half or one-third of the control respectively.
 Measuring protein levels in culture media where pollen (itself a source of proteins) is present together with styles is undoubtedly complicated.
Although the time-course of protein release in culture during "normal" growth of apple pollen is well-known (Speranza & Calzoni, 1986), one can only make assumptions about it in the presence of stylar proteins. Nonetheless, one may consider that the amount of self-style proteins supplied (fig.3, column S) was such as to completely prevent germination, thus eliminating at least the major contribution of active protein secretion from tubes. The dramatic decrease in protein level at 30 min (fig.3, B) should thus be due mainly to a drop in the style proteins. On the contrary, after the glucose treatment, the latter could be entirely recovered after 30 min incubation (fig.3, D).
 These facts could be interpreted on the basis of a sort of adhesion between the apple pollen and the self-style proteins, the latter being sequestered and thus escaping protein determination.

Owing to the glucose treatment (leading somehow to glucose binding to the pollen wall) such an adhesion could not occur, the wall itself not presenting suitable sites for attachment.

In conclusion, it seems that our data on apple pollen, and on the fate of self-style proteins in culture, could be interpreted in terms of the specific carbohydrate-protein complementation which is the "heart" of the self-incompatibility reaction according to Heslop-Harrison's views (1983).

Acknowledgements. Work supported by funds (60%) from Ministero della Pubblica Istruzione, Italy. The authors would like to thank prof. N. Bagni for his critical reading of the manuscript; and Mrs. S. Agostani for technical assistance.

References

Calzoni GL, Speranza A, Bagni N(1979) In vitro germination of apple pollen. Scientia Hortic 10:49-55

Ferrari TE, Wallace DH (1975) Germination of Brassica pollen and expression of incompatibility in vitro. Euphytica 24:757-765

Gibb PE(1986) Do homomorphic and heteromorphic self-incompatibility systems have the same sporophytic mechanism? Pl Syst Evol 154: 285-323

Heslop-Harrison J (1983) Self-incompatibility: phenomenology and physiology. Proc R Soc Lond B 218:371-395

Knox RB, Heslop-Harrison J (1970) Pollen-wall proteins:localization and enzymatic activity. J.Cell Sci 6:1-27

Lowry OH, Rosebrough NJ, Farr AL, Randall RJ (1951) Protein measurement with the Folin fenol reagent. J Biol Chem 193:266-275

Sharma N, Shivanna KR (1983) Lectin-like components of pollen and complementary saccharide moiety of the pistil are involved in self-incompatibility recognition. Curr Sci 52: 913-916

Sharma N, Shivanna KR (1986) Self-incompatibility recognition and inhibition in Nicotiana alata. In: Mulcahy DL, Bergamini G, Ottaviano E (eds) Biotechnology and Ecology of Pollen. Springer, Berlin Heidelberg New York, p 179

Sharma N, Bajaj M, Shivanna KR (1985) Overcoming self-incompatibility through the use of lectins and sugars in Petunia and Erica. Ann Bot 55:139-141

Speranza A, Calzoni GL (1986) Pollen protein during germination in vitro: a dynamics across cytoplasm, tube wall and outer medium. In: Cresti M, Dallai R (eds) Biology of Reproduction and Cell Motility in Plants and Animals. University of Siena, Siena (Italy), p 81

Normal and Abnormal Embryo Development After in Vitro Fertilization in Maize

I. DUPUIS, Z.X. ZHAO and C. DUMAS
Laboratoire de Reconnaissance Cellulaire et Amélioration des Plantes
Université Cl. Bernard - Lyon I
Bât. 741, 5ème étage
43 bd du 11 nov. 1918
F-69622 Villeurbanne
France

INTRODUCTION

The in vitro culturing techniques have allowed for the controlled studies of some processes that could be studied less effectively or not at all in the in vivo condition. The in vitro fertilization techniques have led to the obtention of viable kernels in maize (see Gegenbach, 1977a, b). This method can be applied to analyse the success or failure of fertilization, to study the factors implied in seed set determination, or to overcome incompatibility barriers.

The embryogenesis of the maize embryo grown in vitro has been examined at the ultrastructural level by several authors (Schel and Kieft, 1986; Van Lammeren, 1987)

The objective of this study was to compare the structure of normal and abnormal embryos obtained after in vitro pollination and culture. The importance of the explant age in the subsequent recovery of normal kernels is also discussed.

MATERIALS AND METHODS

Unpollinated ears of the maize inbred line A632 (provided by Association Générale des Producteurs de Maïs, Pau, France) were obtained from field-grown plants. Ear age was estimated by the length of the silks protuding from the husks. For <u>in vitro</u> fertilization, blocks of six spikelets were excised under sterile conditions from the mid-cob region,and then placed in Petri dishes containing a Murashige and Skoog (1962) based medium.The silks were placed outside the Petri dishes and the pollen was applied on their distal part. The pollinated blocks were incubated in the dark at 28°C. After 15 days, developing forms were visually observed and classified in three groups following external appearance (Dupuis and Dumas, in prep.): normal kernels, abnormal kernels and enlarged ovaries. Some of the normal and abnormal kernels were dissected and prepared for optical microscopy. The kernels were fixed in FAA, embedded in parafin, cutted in 8μm sections and stained with the PAS-hematoxylin procedure (Jensen, 1962). Some of the embryos were dissected from abnormal and normal kernels, cultured on appropriate medium and checked for the development of plantlets (Sheridan et al., 1978).

RESULTS AND DISCUSSION

The development observed after <u>in vitro</u> pollination of maize spikelets has been classified into three classes: enlarged ovaries, normal kernels and abnormal kernels (Dupuis and Dumas, in prep.). The enlarged ovary type is probably not a fertilization product but rather an <u>in vitro</u> induced product since it can be recovered from unpollinated spikelets and since no embryo and no endosperm can be detected by dissection (Dupuis and Dumas, in prep.). At the morphological level, normal kernels look like kernels obtained from <u>in vitro</u> conditions (fig 1) whereas abnormal kernels have a shrunken external morphology (fig 1). A wide range of abnormal kernels can be obtained (Gegenbach, 1977a,b; Higgins and Pettolino, 1988; Dupuis and Dumas, in prep.). These kernels contain a variable sized embryo and have different level of endosperm development. Structural observations reveal that embryos from abnormal kernels are reduced in size and do not develop in a normal manner. 15 days after <u>in vitro</u> fertilization the size of abnormal embryo is about four times smaller than the size of normal one (fig 2). The abnormal embryo only begins to develop a scutellar region and the embryo axis. By contrast, the differentiation stage of normal embryos is much more advanced and shows all the main tissues (fig 2A).

In the scutellum tissues of normal embryos, there is an accumulation of polysaccharides reserves; in contrast, no polysaccharides and no nuclei were detected in the scutellum tissues of abnormal embryos. In the normal kernels, the endosperm cells have accumulated starch grains whereas in abnormal endosperm cells the starch grains are much smaller and nuclei were not observed. Furthermore, the volume that the endosperm occupied in the abnormal caryopse is largely reduced.

Schel and Kieft (1986) investigated the _in vitro_ maize kernel development after _in vivo_ pollination in a culture system with a continuous supply of fresh liquid medium. Under these conditions, they showed that several deviations occured during the _in vitro_ culture which were not related to the fertilization process or to the accumulation of toxic metabolites in the culture medium. These authors concluded that defective endosperm development is probably the main reason for a failure of a normal embryo development. As previously stated the scutellum of abnormal embryo contains no detectable polysaccharides. In the normal development of the embryo, glucose derived from the reserve starch in the endosperm is converted to sucrose in the scutellum and then transported to the embryo axis (Esau, 1977). The lack of carbohydrate reserves in the scutellum of abnormal kernel could be a consequence of the low level of starch reserve of the endosperm and this could explain the failure of normal development of the embryo.

The proportion of normal and abnormal kernels obtained has also been studied in relation to the spikelet age. The abnormal kernels never represent more than 10% of the total developmental forms. However, the proportion of normal kernels varies with the spikelet age. Normal kernel set reaches a maximum as soon as the silks emerge from the husks. At this stage, 99% of the developing ovules evolve as normal kernels. When the spikelets become older, the proportion of aberrant development increases (Dupuis and Dumas, in prep.).

The applications of the _in vitro_ fertilization are dependent on the efficient recovery of plants, and thus the conditions which maximize the normal kernel development must be known. The female spikelets must be pollinated during their optimum receptivity period, where a high percentage of normal kernels can be recovered. Abnormal embryos, resulting from a failure of the endosperm development can be rescued and plants can be obtained. The optimization of plantlets recovery after _in vitro_ fertilization is of great importance for incompatible crosses or hybrid seed production.

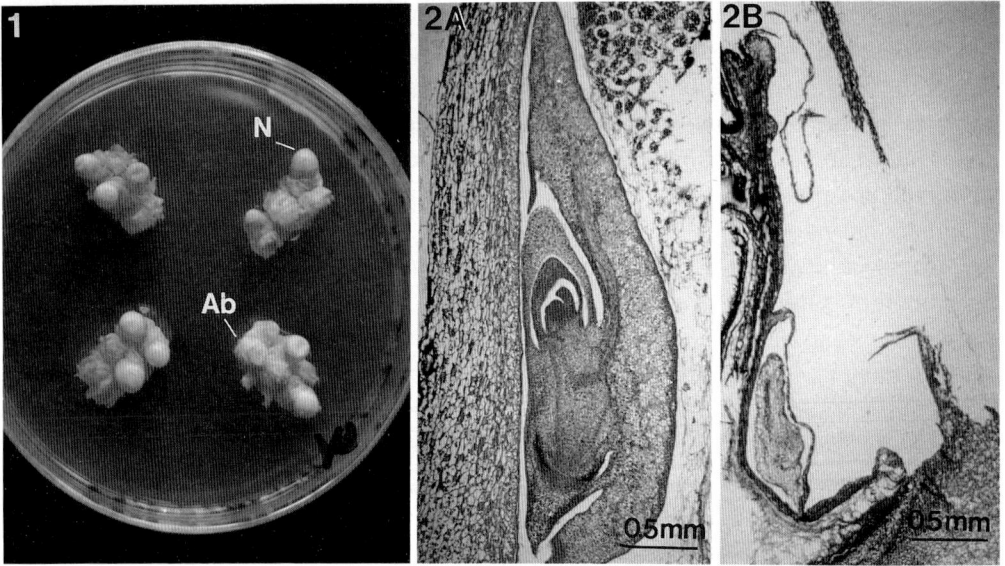

Fig 1: Developing ovules 20 days after <u>in vitro</u> fertilization. Normal kernels (N) and abnormal kernels (Ab) can be distinguished.

Fig 2: Longitudinal section of maize kernel 15 days after <u>in vitro</u> fertilization. Normal kernel (2A) contains an embryo with all its main tissues, and abnormal kernel (2B) contains a much more undifferentiated embryo.

ACKNOWLEDGMENT: This work was partly supported by a grant from Association Générale des Producteurs de Maïs, Pau, France.

REFERENCES

Esau K (1977) Anatomy of seed plants, 2d edition, J Wiley and sons, New York Santa Barbara, London, Sydney, Toronto

Gegenbach BG (1977a) Genotypic influences on in vitro fertilization and kernel development of maize. Crop Sci 17: 489-492

Gegenbach BG (1977b) Development of maize caryopsis resulting from in vitro pollination. Planta 134: 91-93

Higgins RK and Petolino JF (1988) In vitro pollination fertilization of maize: influence of explant factors on kernel development. Plant Cell Tissue Organ Cult 12: 21-30

Jensen WA (1962) Botanical histochemistry, WH Freeman and Company, San Fransisco London

Murashige T and Skoog F (1962) A revised medium for rapid growth and bio-assays with tobacco tissue cultures. Phys Plant 15: 473-497

Schel JHN and Kieft H (1986) An ultrastructural study of embryo and endosperm development during in vitro culture of maize ovaries (Zea mays). Can J Bot 64: 2227-2238

Sheridan WF, Neuffer MG and Bendbow E (1978) Rescue of lethal defective endosperm mutants by culturing immature embryos. Maize Coop News Letter 52: 88-90

Van Lammeren AAM (1987) Embryogenesis in Zea mays L. a structural approach to maize caryopsis development in vivo and in vitro

Pollen-Style Interaction in *Zea mays* L.

P. Landi and E. Frascaroli

Institute of Agronomy

University of Bologna

Via Filippo Re 6

40126 Bologna

Italy

INTRODUCTION

In higher plants pollen grain fertilization ability is a rather complex trait which depends on the rate of germination (i.e. germinability and-or germination time) and on the rate of tube growth. The expression of such components is in turn largely affected by the genotype of the stylar tissue, even in regard to compatible pollinations (for a review, see Heslop-Harrison, 1987).

The role played by the stylar tissue in the fertilization processes has proved to be of particular importance in maize, as evidenced by the poor tube growth achieved in vitro even when a rather complex medium is used (Sari Gorla et al., 1975).

To obtain information on the pollen-style interaction in maize, the pollen fertilization ability of two inbred lines and their cross was studied by the pollen mixture technique, using as pistillate parents two different single crosses. A second objective was to evaluate the effect on pollen-style interaction of the varying dose of the investigated pollen genotypes in the mixture.

MATERIALS AND METHODS

The pollen sources were the two inbred lines Iabo78 and B73 and their F_1. Their pollen fertilization ability was studied by means of the pollen mixture technique (Ottaviano et al. 1982) in relation to that of a standard line (W22) carrying an aleurone dominant marker. For each source (P1, P2 and F_1) seven mixtures were prepared by varying the pollen percentage (as weight) from 20 to 80% at a regular step of 10%. Each mixture was then split into two equal parts which were used to pollinate the two single crosses Iabo78 x B73 and A632 x WF9 (i.e. related and unrelated female parents, respectively).

For each entry 10 well fertilized ears were transversely divided into five segments of equal size and scored from 1 (apex) to 5 (base). The percentage of uncoloured kernels per segment was regressed on the corresponding score, according to the procedure described by Armitage (1955). Two parameters were calculated: the apex intercept ($\hat{Y}_1$) and the regression coefficient (b). According to Ottaviano et al. (1982), the former depends on differences in proportion and-or germination rate of the two competing pollen genotypes in the mixture, whereas the latter is mainly due to differences in pollen tube growth rate. The regression analysis was also made for each pollen source over the seven doses: in this case the average proportion of the investigated pollen is expected to be 50%.

RESULTS AND DISCUSSION

For each pollen source, the apex ordinate ($\hat{Y}_1$) showed an upward trend at increasing pollen dose in the mixture, with a slope not significantly different from 10% (Table 1). As this value was the interval between pollen doses, we can presume that the formulation of the pollen mixtures on a weight basis was not biased by large errors.

Table 1 - Regression coefficients and corresponding standard errors of the apex ordinate ($\hat{Y}_1$ x 100) on the pollen dose.

	Female parent	
Pollen source	Iabo78 x B73	A632 x WF9
	(%)	(%)
Iabo78	9.0 ± 0.5	9.6 ± 0.6
B73	9.3 ± 0.6	11.4 ± 0.6
Iabo78 x B73	8.0 ± 1.1	9.3 ± 0.9

With respect to the regression coefficient (b), a curvilinear type of response was noted, with the intermediate pollen doses showing the lower values. The magnitude of the responses varied, however, depending on pollen source and female parent genotypes (Fig. 1). The line Iabo78 evidenced quite moderate changes with the related female parent and slightly more pronounced with the unrelated one. The tendency to a greater variation with the unrelated female was even more

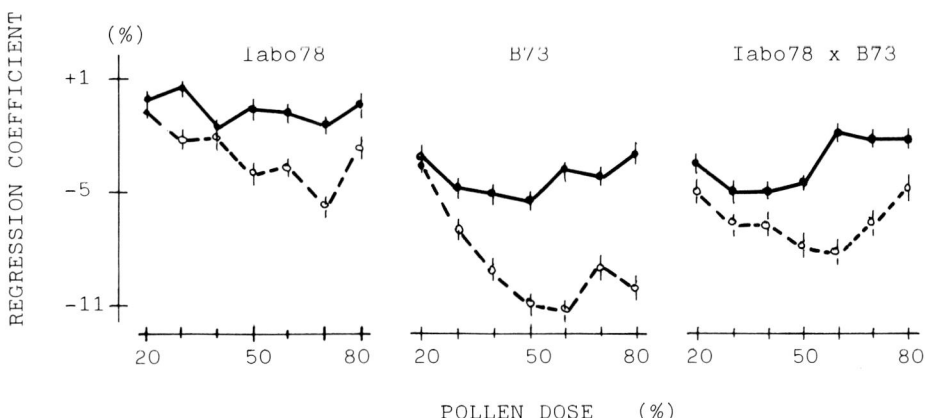

Fig. 1 - Regression coefficients of the pollen sources as affected by their pollen dose (continuous line: related female parent; dashed line: unrelated female parent). Vertical bars indicate standard errors.

evident with the line B73; in contrast, moderate changes with both female parents were shown by the F_1. These results indicate that the relative growth rate of a pollen genotype can be affected by its relative amount in the mixture, and that the magnitude of such an influence can be modulated by the female parent.

A poor knowledge of the physiological aspects regarding the pollen - style interaction does not allow us to formulate an explanation about such findings. However, a curvilinear response could be interpreted by hypothesizing two contrasting trends controlled by different gene systems.

When pollen performance was analysed over the seven doses, the F_1 exhibited an apex ordinate largely exceeding that of the parents in both female hybrids (Table 2). Differences in apex ordinate among the three pollen sources can arise from variations in pollen grain size or germination rate. As Johnson et al. (1976) showed that F_1 pollen grains are not smaller than those of parental lines, the higher F_1 apex ordinate could be mainly accounted for by a higher germination rate. The heterotic pattern found for this trait shows that it is prevailingly under the control of the diploid sporophyte as evidenced by Ottaviano et al. (1982).

Table 2 - Apex ordinates ($\hat{Y}_1$ x 100) and corresponding standard errors over the pollen doses.

	Female parent	
Pollen source	Iabo78 x B73	A632 x WF9
	(%)	(%)
Iabo78	41.5 ± 0.4	39.5 ± 0.4
B73	55.6 ± 0.4	53.4 ± 0.4
Parental mean	48.2 ± 0.3	46.2 ± 0.3
Iabo78 x B73	66.9 ± 0.4	64.9 ± 0.4

Table 3 - Regression coefficients (b x 100) and corresponding standard errors over the pollen doses.

Pollen source	Female parent	
	Iabo78 x B73 (%)	A632 x WF9 (%)
Iabo78	-0.6 ± 0.2	-3.0 ± 0.2
B73	-4.5 ± 0.2	-8.6 ± 0.2
Parental mean	-2.5 ± 0.1	-5.7 ± 0.1
Iabo78 x B73	-3.5 ± 0.2	-6.3 ± 0.2

Each of the three sources exhibited a slightly higher apex ordinate with the related hybrid, indicating that pollen grain germination was slightly favoured in the related stylar tissue.

In contrast with what was observed with the apex ordinate, the best performance for the regression coefficient was exhibited by the line Iabo78, while that of the F_1 was even lower than the parental mean in both female hybrids (Table 3). These results indicate that pollen tube growth rate is not dependent on the heterotic vigor of the pollen mother plant, thus supporting the hypothesis that this trait is mainly under the control of the gametophytic haploid genotype (Ottaviano et al., 1982). This would imply a prevalence of additive gene actions, even though non allelic interactions (epistasis) could be at work as well. Such non allelic interactions, evidenced by Sari Gorla et al. (1975) in a study carried out in vitro, could explain why the F_1 pollen tube growth rate was lower than parental mean.

It should also be noted that the regression coefficient of the three pollen genotypes was higher with the related hybrid, indicating that pollen tubes grew faster in the related stylar tissue. These results are consistent with the findings of Jones (1928) and Johnson and Mulcahy (1978), who studied inbred lines, and of Ottaviano et al. (1983), who studied both inbred lines and hybrids.

ACKNOWLEDGEMENTS: Research work supported by C.N.R., Italy. Special grant I.P.R.A. - Subproject 1.
The authors are grateful to Prof. M. Sari Gorla and to Prof. S. Conti for their useful suggestions.

REFERENCES

Armitage P (1955) Test for linear trends in proportions and frequencies. Biometrics 11: 375-386

Heslop-Harrison J (1987) Pollen germination and pollen-tube growth. International review of cytology 107: 1-78

Johnson CM, Mulcahy DL, Galinat WC (1976) Male gametophyte in maize: influences of the gametophytic genotype. Theor Appl Genet 48: 299-303

Johnson CM, Mulcahy DL (1978) Male gametophyte in maize: II. Pollen vigor in inbred plants. Theor Appl Genet 51: 211-215

Jones DF (1928) Selective fertilization. University Press Chicago, p 163

Ottaviano E, Sari Gorla M, Pe E (1982) Male gametophytic selection in maize. Theor Appl Genet 63: 249-254

Ottaviano E, Sari Gorla M, Arenari I (1983) Male gametophytic competitive ability. Selection and implications with regard to the breeding system. In: Mulcahy DL, Ottaviano E (eds) Pollen: biology and implications for plant breeding. Elsevier, New York, 367-373

Sari Gorla M, Ottaviano E, Faini D (1975) Genetic variability of gametophyte growth rate in maize. Theor Appl Genet 46: 289-294

Pistil Treatments for Improved Fertility in Hybridization of *Eucalyptus gunnii* (Hook)

B. CAUVIN
ASSOCIATION FORET CELLULOSE
AFOCEL REGION SUD
98, Route de Tournefeuille
31270 CUGNAUX
FRANCE

1 INTRODUCTION

Pistil treatments, especially on the stigma, have become more and more used since a few years in breeding programs. The main purpose is to avoid some barriers at different levels of incompatibility : this will occur, for example, in a mating system with high level of autogamy or in inter generic crosses with large genetic distances between the two parents.

Dumas and al (1983), Hervé, Gaudé (1984) think that incompatibility is regulated for a large part mainly at a sporophytic level by a complexe stigma x pollen recognition system. That interface stigma x pollen and its recognition is based on a strong interaction between the pollinic molecules and the stigmatic receptors operating as a reading system.
Those works are more important about herbaceous plants than on forests trees.
The most used applications are done :
- at the pollen level, by application of irradiated pollen or pollen mentor
- at the pistil level, by some chemical treatments on the stigma e.g. exogene solvents (Ether, KDA, AIA, ANA)
- at both levels, by pollination at bud stage (Shivanna D.R. ; Heslop-Harrison 1978) or in moist or CO_2 enriched environment.

The first works on Eucalyptus genus have been done by Pryor and Willing (1974) to remove incompatibility between species belonging to different sections of Eucalyptus, in the same subgenus. They were inspired by works made with Populus, using organic solvents and pollen mentor.

We focused on artificial pollination of E. gunnii (mother) crossed with E. ovata (pollen) just after cutting or mutilation of stigma Those treatments have been applied at anthesis stage.
The results have been measured by counting the number of progenies seedlings and using a phenotype marker at the leaves color level.
The different applications and its possible effects in breeding programs might be important enough to justify our interest in the realisation of that research.

2 ARCHITECTURE AND PHENOLOGY OF THE FLOWER

The morphology of E. gunnii flower is shown in figure n'1.

322

Fig.1   Eucalyptus gunnii : cutting split of a flower

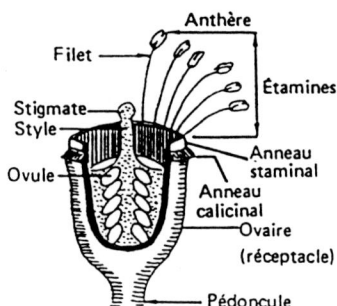

The flower is hermaphrodite, and requires an emasculation of staminal ring before artificial pollination. Anthesis is characterised by a yellow color of the second opercule which falls very quickly (figure n°5).

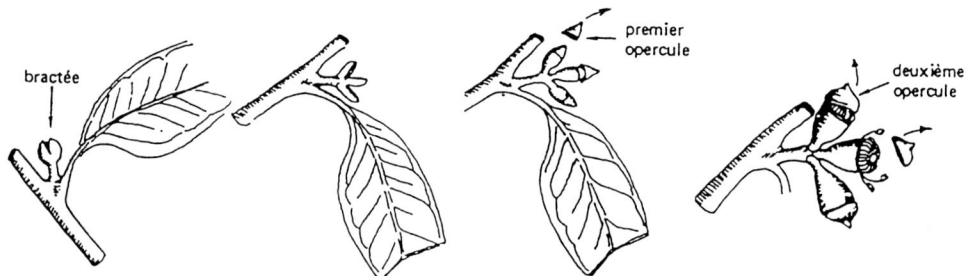

Fig.2 E.gunnii :   Fig.3 E.gunnii :   Fig. 4 E.gunnii :   Fig. 5 E.gunnii :
bracts including   appearance of buds   dehiscence of the   dehiscence of the
floral buds   first opercule   second opercule
                                                                 (anthesis)

In several species of Eucalyptus, there is a time-lag between the maturity of pollen and receptivity of stigma. In E. gunnii, the pollen is mature four to six days before the complete receptivity of the stigma.

3 DESCRIPTION AND CHARACTERISTICS OF THE TREATMENTS

Choice of parents - Reminder of glaucous character heredity

We used E. gunnii Hook as mother and Eucalyptus ovata Labill as father ; the choice of both parents has been made because F1 progenies inherit the green character of the E. ovata leaves.
The glaucoussness character of E. gunnii leaves is recessive (CAUVIN, POTTS B., POTTS W., 1987).
We always observed in our artifical pollination works that F1 progenies and even

F2 back-cross towards dominant genotype (E.ovata) in the E.gunnii x
E.ovata crosses have green colors of the leaves. The disjonction
observed in the F2 progenies (self) is related to a normal
segregation like achieved with a monogenic character in Mendelien
sense.
Only the F2 back-cross towards recessive parent E. gunnii showed
green leaves deficit in the ratio. The difference between expected
and observed ratios seem to be linked with the presence of two genes.
The ambiguity at that level still exists. These dominant characters
allow us to estimate and control in the progenies the part of the
pollinic pollution by counting the percentage of glaucous phenotypes.
The experimentation has been applied on thirty years old E. gunnii,
growing at Coussergues (Hérault-France). Its neighbouring trees were
only Eucalyptus belonging to the same species.
The treatments applied on the styles have been as follow :
- control (no mutilation)
- cutting style between stigma and ovary
- lengthwise split styles
- stigma abrased
all treatments are illustrated in figures 6 to 9.

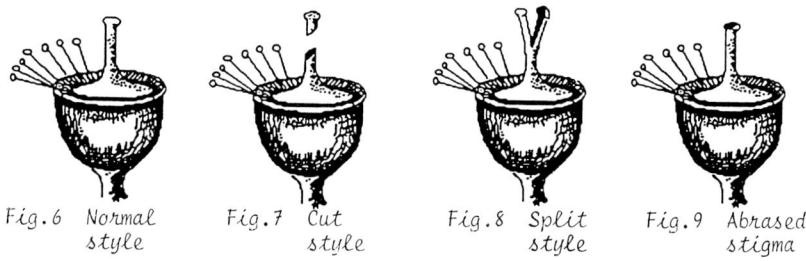

Fig.6 Normal style   Fig.7 Cut style   Fig.8 Split style   Fig.9 Abrased stigma

Pollination was applied once with a brush, with E. ovata pollen about
25 months old, stored in frozen temperature (-18°C). We brought
pollen immediately after treatment on the style and before the
complete oxydation of its surface. The flowers were not protected by
cellulosic bags.

Table 1   Number of pollinated flowers for each mother and
treatments on pistil (*treatments no realized).

♀ \ ♂	E.ovata (n°67)			
	normal style	abrased stigma	split style	cut style
E.gunnii n°309	9	*	*	*
E.gunnii n°310	16	15	9	13
E.gunnii n°311	11	*	18	16

4 RESULTATS

The results are shown in table 2. The higher number indicates the quantity of seedlings achieved and the lowest, the ratio between the number of seedlings and the number of pollinated flowers.

Table 2   Number of seedlings (high) and relationship between number of seedling/number of pollinated flowers (low) for each mother and treatment.

♀ \ ♂	E.ovata (n°67)			
	normal style	abrased stigma	split style	cut style
E.gunnii n°309	0 0	* *	* *	* *
E.gunnii n°310	0 0	5 0,33	0 0	8 0,61
E.gunnii n°311	2 0,18	* *	12 0,67	16 1

The ratio of seedlings pollinated flowers is the best way to estimate the results of each treatment considering in view of the difference in the pollinated flowers number. It clearly appears that pollen germination and growth and fertilisation are still possible when the stigma and part of the style have been removed.
The result obtained with the control (normal pollination) is the poorest of all treatments. We can explain easily, by the time-lag existing between maturity of pollen and receptivity of stigma (pollen has been applied once at the anthesis stage, in fact when the pistil was not yet full receptive).
A phenotypic study of the progenies from all treatments has been undertaken. All the seedlings for each treatment have shown green leaves phenotypes (excepted only one which was glaucous in cutting style treatment). It proves that it is really E. ovata pollen which has fertilized the ovules of E. gunnii.

5 POSSIBLE APPLICATIONS

Those treatments mainly cutting style, introduced in a breeding program, could have good applications as for example :
- easiness and reduction of handling ; as the emasculation of the style and the staminal ring can be done together at the anthesis stage
- decrease from 3 to 1 the number of pollen application on the stigma
- protection bags are not necessary. The oxydation of the stigma surface is fast enough to avoid the germination of pollen coming from abroad. Everybody knows the lot of time spent by the staff in this kind of handling.

6 CONCLUSION

All the treatments applied either on the style or the stigma of E. gunnii flowers, show that cutting style have proved successfull. A lot of green leaves phenotypes seedlings were obtained in the F1 progenies. In an improvment program using sexual way by artificial pollination, very important saving time could be done either by reducing the number of pollinations hand, or by avoiding preparation of cellulosic protection bags.

Some attempts being aimed to reduce the incompatibility effects at intra or interspecific levels could be undertaken by removing all the recognition pollen-stigma system. It could be also an other mean to cross some species for which the length of the style do not correspond to the lenght of pollinic tubes.
However, neither interspecific or intraspecific pollinations have been successfull using this approach to cut and reduce the length of the style of E. globulus and the mechanism of cross incompatibility in E. globulus requires further investigations (P.Volker and W. Tibbitts, pers.com.).
Nevertheless, that technic could be tried with other species of Eucalyptus.

REFERENCES

Dumas C, Gaude T (1981) Stigma-pollen recognition : a new look. Acta Soc. Bot. Pol. 50: 235-247.

Gaget M, Teissier du Cros E (1984) Création d'un hybride par minipulations génétiques à l'aide de techniques physico-chimiques chez Populus In Hervé Y and Dumas C (Eds) Incompatibilité pollinique et Amélioration des Plantes. ENSA Rennes Publi., 40-50.

Gaude T, Fumex B, Dumas C (1983) Are lectin-like compounds involved in the stigma-pollen adhesion and/or recongnition in Populus and Brassica ? In "Pollen : Biology and Implications for Plant Breeding" Mulcahy DL and Ottaviano E (Eds) Elsevier Biomedical, New York, Amsterdam, 265-272.

Hervé Y, Gaude T, Dumas C (1984) L'incompatibilité pollinique et son rôle en amélioration des plantes In Hervé Y and Dumas C (Eds) Incompatibilité pollinique et Amélioration des Plantes. ENSA Rennes Publi., p 9.

Heslop-Harrison J (1978) Genetics and physiology of Angiosperm incompatibility systems. Proc. R. Soc. Lond. B. 202: 73-92.

Heslop-Harrison J (1983) Self-incompatibility : phenomenology and physiology. Proc. R. Soc. Lond. B. 218: 371-395.

Knox RB (1984) Pollen-Pistil interactions. Encyclopedia of Plant Physiol. 17: 508-556.

Mulcahy DL, Mulcahy GB (1983) Gametophytic self-incompatibility reexamined. Science 220: 1247-1251.

Palloix A, Hervé Y (1984) Levée de l'autoincompatibilité chez Brassica Oleracea application au chou-fleur d'automne de CO_2 et d'humidité relative contrôlée In Hervé Y and Dumas C (Eds) Incompatibilité pollinique et Amélioration des Plantes. ENSA Rennes Publi., p 95.

Shivanna KR (1979) Recognition and rejection phenomena during pollen-pistil interaction. Proc. Indian.Acad. 88: 115-141.

Shivanna KR, Heslop-Harrison Y, Heslop-Harrison J (1978) The pollen stigma interaction : bud pollination in the Cruciferae. Acta bot. Neerl. 27: 107-119.

Teissier du Cros E (1984) Hybridations interspecifiques et incompatibilités chez les arbres forestiers In Hervé Y and Dumas C (Eds) Incompatibilité pollinique et Amélioration des Plantes. ENSA Rennes Publi., p 175.

Induced Polarity as an Index of Pollination-Triggered Stylar Activation

Gabriella Bergamini-Mulcahy and David L. Mulcahy
Department of Botany
University of Massachusetts
Amherst, Massachusetts 01003
U.S.A.

Summary

Pollen inserted in mid-style of Nicotiana alata (cv. sensation) will grow equally in both directions (i.e. toward the stigma or toward the ovary); but if the mid-style pollination is immediately preceded by a stigmatic one, the pollen tubes will preponderantly grow toward the ovary. The present research is aimed at investigating the causes of this observed induced polarity.

Introduction

The activation of the pistil, as a consequence of pollination, is a well known fact. Several aspects of it have been studied through the years by innumerable researchers. I would like, here, to consider another aspect of such activation, namely the pollination induced directionality on pollen tubes originating from an intrastylar pollination. We chose the term directionality, rather than trophism, since "trophism", with its specific meaning of "attraction" carries a need for a prerequisite: the attractant. In the present experiments, while we observe the induction of a significant directional pattern we do not come close to know what the specific cause, or the attractant, could be.

Buchholz (1932) and Iwanami (1959), in the past, had noticed that pollen inserted into the style grows more or less equally in both directions. Working with Nicotiana alata cv. sensation, we could confirm the above finding, that is, the lack of prevalent directionality in pollen tubes from an intrastylar pollination. We also observed that, if such intrastylar pollination was immediately preceded by a stigmatic one, a downwards pattern of directionality was clearly established (Bergamini-Mulcahy and Mulcahy, 1987).

Methods

Pistils of <u>Nicotiana</u> <u>alata</u> cv. sensation were collected and kept in a saturated atmosphere at 20°C 24 hours before using them for pollinations - to avoid the effect of a surge of ethylene due to wounding. Using Brown Jeweler forceps (Fisher catalog #08953E) we made a 2 mm incision in the mid-style, precisely at 2.4 cm from the stigma. This distance was calculated to give enough space for overnight tube growth, avoiding overlapping of tubes growing down from the stigma and up from the intrastylar incision. While, with the microforceps, we kept the mid-style cut open, with the help of another microforceps, we inserted the pollen in the wound. In this manner we could insert enough pollen to be comparable to a stigmatic pollination. When not indicated otherwise, pollinations were done with compatible pollen just collected from open flowers, growing in greenhouse, 18-24°C 14 hour days. Prewashing of pollen was done with 15 minutes suspension in Brewbaker media (Brewbaker and Kwack, 1963) followed by 5 minutes centrifugation at 1000 rpm and removal of supernatant. Pollen tubes in style were measured with a stage micrometer after clearing and staining with decolorated aniline blue and visualized with epifluorescence (Bergamini-Mulcahy and Mulcahy, 1987).

Results and Discussion

Table 1 includes data from 6 separate experiments, with a total of 46 replicas for each treatment. Both treatments were simultaneously carried on with flowers from the same individual plant. For different experiments, two groups of sib plants were used: one group with an average style length of 3.5 cm and another of 4.5 cm.

Table 1: Effect of stigmatic pollination on intrastylar pollen tube directionality (46 replications each treatment; units = mm)

	pollen tube length upward	pollen tube length downward	down/up
pollinated stigma	X̄=5.52 SE=0.3	X̄=8.85 SE=0.4	X̄=1.9 SE=0.24
nonpollinated stigma	X̄=6.84 SE=0.45	X̄=6.99 SE=0.5	X̄=1.1 SE=0.08

The stigmatic pollination obviously influenced the direction of the pollen tubes originating from mid-style pollination, reducing the upward growth and, even more, stimulating the downward one. This finding, per se, interesting, is particularly attractive to us as another means to detect pistil activation in response to pollination. Both Deurenberg (1977) and Linskens (1974), studying different parameters, had come to the conclusion that an incompatible pollination evokes a different (at least quantitatively different) response of the various parts of the flower when compared to a compatible pollination. Such a difference is noticeable also in the induction of pollen tube directionality. Table 2 shows how an incompatible stigmatic pollination failed to induce the directionality of the intrastylar compatible pollen tubes, to the extent exhibited by a compatible stigmatic pollination (10 replicas for each treatment). This can be considered as more evidence of impaired crosstalk between incompatible pollen and pistil.

Table 2: Effect of compatible vs. incompatible stigmatic pollination on compatible intrastylar pollen tube directionality (10 replications each treatment; units = mm)

on stigma	pollen tube length upward	pollen tube length downward	down/up
compatible pollination	$\bar{X}=5.3$ SE=0.4	$\bar{X}=6.8$ SE=0.8	$\bar{X}=1.34$ SE=0.2
incompatible pollination	$\bar{X}=5.7$ SE=0.7	$\bar{X}=5.2$ SE=0.7	$\bar{X}=0.96$ SE=0.1

Gilissen (1976 and 1977) defines stigma and style as sensor organs, collectors of stimuli (i.e. the occurrence of pollination) to be transformed and communicated rapidly to the rest of the female apparatus. Nichols (1983) advances the hypothesis that ACC (1-aminocyclopropane-1-carboxylic acid) could act as a hormone, translocated from the stigma to the rest of the flower, (an hypothesis not confirmed by the observations of Gilissen and Hoekstra (1984) in Petunia). Tobacco pollen (Hill, 1987) contains very high levels of ACC; and Hill suggests that the pollen ACC could be responsible for the pollination induced ethylene production in this species. Ethylene, in turn, is the recognized cause of floral wilting after pollination. Always according to Hill, 20-30 minutes of pollen suspension in germination media will elute 60% of the pollen ACC, and, ethylene induction, from isolated styles, pollinated with washed pollen was significantly lower than that from those pollinated with unwashed pollen (but still more than the ethylene production from unpollinated styles).

When we washed pollen in germination media (the pollen to be used for stigmatic pollination) the consequent induced directionality on the intrastylar pollination was somewhat reduced as shown in Table 3. This finding, while suggestive, does not necessarily prove that ACC is directly implicated in inducing the directionality of the intrastylar pollination, since not only ACC but amino acids, proteins and carbohydrates are also rapidly released from the pollen wall by suspension (Stanley and Linskens, 1965; Linskens and Schrauwen J, 1969; Heslop-Harrison J, Heslop-Harrison Y, Knox RB, and Howlett B, 1973).

Table 3: Effect of stigmatic pollination with fresh vs. washed pollen on intrastylar pollen tube directionality (6 replications each treatment; units = mm)

on stigma	pollen tube length upward	pollen tube length downward	down/up
fresh pollen	$\overline{X}=7.5$ SE=0.5	$\overline{X}=11.9$ SE=1.05	$\overline{X}=1.60$ SE=0.12
washed pollen	$\overline{X}=6.6$ SE=1.2	$\overline{X}=6.88$ SE=0.8	$\overline{X}=1.16$ SE=0.2

Nichols (1983) found, in _Dianthus_, that while pollination caused a rapid increase in ethylene in stigma and style, only much later a similar reaction occurred in the ovary. When we observed the effect on intrastylar pollination directionality, as determined by a stigmatic pollination on pistils without ovary, it was clear (Table 4) that, while the presence of the ovary had a bearing on the overall tube length, it did not have influence on the induced directionality.

Wishing to find out how much the stigma, _per se_, was implicated in inducing the directionality when pollinated (Nichols, 1983, believed in a qualitative difference between the enzymes in the stigma converting ACC to ethylene and those in other parts of the flower) we repeated the intrastylar pollinations, but this time, in one-half of the pistils, the stigmatic surface was removed 24 hours before pollination. As Table 5 shows, the removal of the stigmatic surface

did not impair the ability of pollination, in this case on the decapitated style, to induce the downward directionality in the intrastylar pollination.

Table 4: Ovary influence on intrastylar pollen tube directionality as response to stigmatic pollination (18 replications
each treatment; units = mm)

pollinated stigma	pollen tube length upward	pollen tube length downward	down/ up	down + up
ovary present	$\overline{X}=5.67$ SE=0.5	$\overline{X}=8.42$ SE=0.5	$\overline{X}=1.85$ SE=0.4	$\overline{X}=14.09$ SE=0.9
ovary removed	$\overline{X}=3.06$ SE=0.5	$\overline{X}=4.49$ SE=0.8	$\overline{X}=1.745$ SE=0.24	$\overline{X}=7.56$ SE=1.3

Table 5: Effect on intrastylar pollen tube directionality, of
stigmatic pollination on intact stigma vs. pollination on
the stump after stigma removal (18 replications each treatment; units = mm)

pollionation on:	pollen tube length upward	pollen tube length downward	down/up
stigma intact	$\overline{X}=7.23$ SE=0.4	$\overline{X}=10.694$ SE=0.5	$\overline{X}=1.53$ SE=0.08
stump, after stigma removal	$\overline{X}=6.53$ SE=0.37	$\overline{X}=9.873$ SE=0.47	$\overline{X}=1.57$ SE=0.1

While the removal of the stigma surface did not seem to have
any effect on the directionality of the intrastylar pollination, it
has an effect on seed set. We made two sets of crosses performed on
a single plant (seven crosses on each of two different branches).
On one branch, the stigmatic surfaces were first pollinated and
then, after 10 minutes, removed and the stump repollinated. On the
other branch, the stigmatic surfaces were removed without previous
pollination, and only the stump was pollinated. In the first set,
100% of the crosses produced capsules with a total seed number =
4862, in the second set only 56% of the crosses produced capsules,
and the total seed number = 1697. Pollen tubes in the ovary, as
seen with epifluorescence, after pollination only after stigma
removal showed an unusual winding and a general lack of direction

toward micropyles. A very specific, as yet unknown, interaction must occur between pollen and stigmatic surface, and in a similarly unknown manner, the occurrence of such an event must be, very rapidly, communicated to all parts of the gynoecium.

References

Brewbaker JL, Kwack BH (1963) The essential role of calcium in pollen germination and pollen tube growth. Am J Bot 50:859-865

Buchholz JT, Doak CC, Blakeslee AF (1932) Control gametophytic selection in Datura through shortening and splicing of styles. Bull Torrey Bot Club 59:109-118

Deurenberg JJM (1977) Differentiated protein synthesis with polisomes from Petunia ovaries before fertilization. Planta 133:201-206

Gilissen LJW (1976) The role of the style as a sense organ in relation to wilting of the flower. Planta 131:201-202

Gilissen LJW (1977) Style-controlled wilting of the flower. Planta 133:275-280

Gilissen LJW, Hoekstra FA (1984) Pollination induced corolla wilting in Petunia hybrida. Rapid transfer through the style of a wilting-inducing substance. Plant Physiol 75:496-498

Hall IV, Forsyth FR (1967) Production of ethylene by flowers following pollination and treatments with water and auxin. Can J Bot 45:1163-1166

Heslop-Harrison J, Heslop-Harrison Y, Knox RB, Howlett B (1973) Pollen wall proteins: gametophytic and sporophytic fractions in the pollen walls of the Malvaceae. Ann Bot 37:403-412

Hill SE, Stead AD, Nichols R (1986) Pollination induced ethylene production of 1-aminocyclopropane-1-carboxylic acid by pollen of Nicotiana tabacum cv. white barley. Jour of Plant Growth Regulation 6:1-13

Iwanami Y (1959) Physiological studies of pollen. Yokohama Municipal Univ 116 (C-34, Biol 13)

Linskens HF, Schrauwen J (1969) The release of free amino acids from germinating pollen. Acta Bot Neerl 18:605-614

Linskens HF (1974) Translocation phenomena in the Petunia flower after cross- and self-pollination. In: Linskens HF (ed) Fertilization in Higher Plants. Elsevier pp 285-292

Mulcahy-Bergamini, G, Mulcahy DL (1987) Induced pollen tube directionality. Am J Bot 74:1458-1459

Nichols R, Bufler G, Moz Y, Fujino DW, Reid MS (1983) Changes in ethylene production and 1-aminocyclopropane-1-carboxylic acid content of pollinated carnation flowers. Jour of Plant Growth Regulation 2:1-8

Schrauwen J, Linskens HF (1967) Mass culture of pollen tubes. Acta Botanica Neerl 16:177-179

Stanley RG, Linskens HF (1965) Protein diffusion from germinating pollen. Physiol Plant 18:47-53

Stead AD, Moore KG (1979) Studies in flower longevity in Digitalis. Planta 146:409-414

Evidence for Non-Random Fertilization in the Common Zucchini, *Cucurbita pepo*

A. G. Stephenson
J. A. Winsor
C. D. Schlichting
Department of Biology
The Pennsylvania State University
University Park, PA 16802 USA

Over the last 15 years, several studies have shown that the progeny produced under conditions of intense pollen tube competition are more vigorous than progeny produced under conditions of little or no pollen tube competition (see review by Mulcahy and Mulcahy 1987). Recent studies from our lab have shown that zucchini (*Cucurbita pepo* L., Cucurbitaceae) seeds resulting from large pollen loads (intense pollen competition) germinate more rapidly, exhibit greater growth as seedlings and have greater reproductive output as adult plants than seeds resulting from low pollen loads (Stephenson et al. 1986; Davis et al. 1987; Winsor et al. 1987).

Mulcahy (1979) hypothesized that such differences in progeny vigor are due to differences in the vigor (growth rate) of the microgametophytes that sired the seeds. When pollen competition is intense, only the fastest growing pollen tubes achieve fertilization whereas both fast and slow growing pollen tubes achieve fertilization in the absence of pollen tube competition. The correlation in vigor between the microgametophytes and the resulting sporophytes is presumably due to substantial overlap in gene expression between the two stages of the life cycle (Tanksley et al. 1981; Willing and Mascarenhas 1984). Pollen competition studies, however, do not directly demonstrate a relationship between pollen tube vigor and progeny vigor and are, therefore, subject to alternative interpretations. In this chapter, we report that the first pollen tubes into the ovary of zucchini fertilize ovules in a specific region of the ovary and that the seeds produced by this region are more vigorous than the seeds produced in the last region of the ovary that is fertilized.

BACKGROUND, MATERIALS AND METHODS

Black Beauty Bush cv. Zucchini is a short-internode vine with indeterminate reproduction. The flowers are born individually and each plant produces 8-12 pistillate flowers (4 to 6 mature fruits) and 20 to 30 staminate flowers over the course of the growing season. Each flower lasts for only one day. Our previous studies showed that mature fruits resulting from very large pollen loads (either hand-pollinations or natural pollinations by bees) contain approximately 300 seeds. However, when flowers receive small pollen loads (240 ± 36 ($\bar{x}$ ± SD) pollen grains), the mature fruits contain approximately 40 seeds (Winsor et al. 1987). Moreover, when we divided the mature fruit into four regions, each containing 25% of the ovules (A = stylar end and D = peduncular end), we found that greater than 80% of the seeds in fruits resulting from low pollen loads were located in region B. The remaining seeds were located in regions A and C.

In order to determine the sequence of fertilization when larger pollen loads are deposited onto a stigma of zucchini, in July 1987 we made 36 controlled pollinations on 15 plants growing at The Pennsylvania State University Agriculture Experiment Station in Centre County, PA. Each pistillate flower in the experiment was covered with a cheesecloth bag throughout anthesis to exclude bees and other visitors. The following protocol was followed: each flower received four pollen loads which were distributed over the tips of the three stigmatic lobes. Each pollen load consists of a thin, uniform layer of pollen (462 ± 152 grains, N = 10) on the end of a 2.5 mm in diameter stainless steel rod. The pollen was collected from 10 staminate flowers, one from each of 10 plants. The pistillate flowers were harvested at 12, 24 and 48 hours after pollination and preserved in FAA. Later, the exocarps were removed from the ovaries. The stigmas, styles and ovaries were then softened in 8N NaOH for 24-48 hours at 65°C, stained with 3% aniline blue, and pollen tube growth was examined under a microscope with epifluorescent U.V. light (see Martin 1959). In April and May 1988, pistillate flowers from plants growing in a greenhouse were pollinated as above. After 24 hours, the style was excised just above the ovary and 21 days later, the fruits were harvested, sliced into four regions and the developing seeds were counted.

In a second study (1986), each pistillate flower on 20 plants at the field site received two pollen loads, (462 ± 152 grains each). Pollen was gathered and applied as above. In September, the mature fruits were harvested and cut into four regions each containing about 25% of the ovules. The seeds in each region of each fruit were counted and weighed. We then selected five fruits, one from each of five plants. All five fruits (a) contained at least 15 seeds in each region, (b) were produced during the last three weeks of July, (c) contained a similar number of mature seeds (133 to 152) and (d) had similar mean seed weights. Nine seeds, each weighing .21-.23 g, from each section of each fruit were selected and randomly assigned to four liter pots, one seed per pot, and grown in a greenhouse.

We recorded the number of days to emergence, the number of days from emergence to the first true leaf, the number of leaves at 20 d and the dry weight at 28 d after emergence.

RESULTS

Table 1 summarizes the results from the time series study of pollen tube growth. At 12 hours after pollination, pollen tubes were clearly observed but only in the upper portion of the style (stigma end). By 24 hours, maximum pollen tube growth ranged from the lower portion of the style to the top (stylar end) portion of region C. Pollen tubes were always observed in the ovary at 48 hours after pollination. Typically, these tubes were restricted to regions A, B and C. Unfortunately, we were unable to identify fertilizations by ovary region in this study. Only preliminary data are available from the greenhouse study in which styles were excised after 24 hours. By mid-May 1988, only two fruits have been examined. However, in each fruit, the developed seeds (less than 40 in each) were located exclusively in region B.

Table 1. Percentage of the pistillate flowers in which pollen tubes or fertilizations were observed in the gynoecia. Each ovarian region contains approximately 25% of the ovules. A = stylar end, D = peduncular end.

Time After Pollination	Flower Number	Pollen Tubes				
		Style	A	B	C	D
12 hours	12	100%	0	0	0	0
24 hours	12	100%	58%	42%	25%	0
48 hours	12	100%	100%	92%	50%	8%

There were 85 mature fruits produced from the field portion of the second study. On average, 17% of the 120 ± 58 seeds in these fruits were from region A, 34% from B, 36% from C and 13% from D. Mean seed weight ranged from .182 g in region B to .167 g in region D (A = .181 g, C = .178 g). The range of seed weights (.21-.23 g) that we used for the greenhouse portion of this study was above the average for all four regions of the ovary (32% above average for region D).

The greenhouse performance of the progeny produced from the four regions of the ovaries are summarized in Table 2. A contrast of the performance of the progeny from regions B and D reveals that the seeds from B emerge significantly earlier from the soil and have significantly more dry weight at 28 days after emergence. When all four measures together were subjected to a multivariate analysis of variance (MANOVA), which accommodates any interdependence of separate measures, the effect of the two regions that produced the seeds was highly significant (Wilks' L = .93, $F_{4,153}$ = 2.70; p = 0.033). A standard analysis of variance using the entire data set (all four regions) also reveals that the maternal plant/fruit had a significant effect on all four measures of seedling vigor and the interaction of the maternal plant by region had a significant effect on three of the four measures of vigor (dry weight being the exception).

Table 2. Performance of progeny from the four regions of the ovary (A = stylar end, D = peduncular end). Means (S.D.).

Ovarian Region	Days to Emergence***	Days to First Leaf*	Leaves At 20 Days	Dry Weight (g)**
A	3.99 (.80)	8.87 (.80)	5.04 (.47)	2.86 (.90)
B	3.77 (.81)	8.87 (.96)	5.06 (.77)	3.12 (1.05)
C	3.97 (.64)	8.82 (.66)	4.93 (.59)	2.94 (1.0)
D	4.24 (.87)	9.13 (.61)	4.86 (.75)	2.70 (.95)

Contrast B vs. D: *0.10 > p > 0.05; **p < 0.05; ***p < 0.005

DISCUSSION

Our previous study (Winsor et al. 1987) showed that small pollen loads (240 ± 36 grains) produced mature fruits containing about 40 seeds. Although the exact fates of the pollen grains and pollen tubes that failed to achieve fertilization are unknown, it can be noted that similar (or higher) ratios of pollen grains to seeds have been reported for a wide variety of species (e.g., Snow 1986). A closer examination of the fruits produced in this study also revealed that the seeds were concentrated in region B of the ovary and in the adjacent portions of regions A and C, indicating that the first fertilizations, at least with low pollen loads, occur within or near region B.

The time series study of pollen tube growth reveals that by 24 hours the fastest pollen tubes have entered the ovary in 58% of the flowers. The excised style protion of this study was designed to identify the ovules that are fertilized by these fastest tubes by preventing the slower tubes from entering the ovary. The preliminary data we obtained are consistent with the observations from the previous study using small pollen loads. The first tubes into the ovary fertilize ovules in region B. Hill and Lord (1986) in a study of the dynamics of pollen tube growth in *Raphanus raphanistrum* also concluded that "prezygotic mechanisms of gamete selection operate to sort pollen tubes nonrandomly to different ovule positions in the ovary."

In the second study reported in this chapter, pollen deposition on the stigmas (2 x 462 ± 152) was insufficient to produce a full complement of seeds in the mature fruit. That is, seeds were produced in the absence of pollen competition. In these fruits, it is likely that the seeds were fertilized by microgametophytes representing a large portion of the available variance in pollen tube growth rates. Moreover, seeds in regions B and D of the ovaries represent the extremes in this variance. Our greenhouse studies of progeny vigor show that the seeds from these two regions of the ovary differ significantly in traits related to germination and seedling growth to 28 days after germination. These data are consistent with Mulcahy's (1979) hypothesis that a correlation exists between the growth rates of pollen tubes and the growth rates of the resulting progeny.

While these data are consistent with the hypothesis, it is not possible to completely rule out the possibility that these differences in progeny vigor are due to some maternal effect. For example, the seeds in region B may enjoy some spatial or temporal nutritional advantage which, in turn, may lead to increased seedling vigor. Such advantages have been reported for many species and are typically manifested by differences in seed size (e.g. Stanton 1984). In this study, seeds from region D were also significantly smaller (on average) than seeds from region B. However, our study of germination and seedling growth controlled for differences in seed size by only using seeds within a narrow size range that was well above the mean seed size for region D. Our study is, therefore, a conservative test of the hypothesis. This study does not, however, identify the sources of variation in pollen tube growth rates which could include both variation among pollen genotypes and pollen-pistil interactions.

ACKNOWLEDGMENTS. We thank Bob Oberheim, his staff, and the Department of Horticulture for use of the Agricultural Experiment Station. This project was supported by funds from The Pennsylvania State University Agricultural Experiment Station, Project 268? and by grant BSR-8315612 A01 from the National Science Foundation.

REFERENCES

Davis LE, Stephenson AG, Winsor JA (1987) Pollen competition improves performance and reproductive output of the common zucchini squash under field conditions. J Amer Soc Hort Sci 112:712-716

Hill JP, Lord EM (1986) Dynamics of pollen tube growth in the wild radish, *Raphanus raphanistrum* (Brassicaceae). I. Order of fertilization. Evolution 40:1328-1333

Martin FW (1959) Staining and observing pollen tubes in the style by means of fluorescence. Stain Technol 34:125-128

Mulcahy DL (1979) The rise of the angiosperms: a genecological factor. Science 206:20-23

Mulcahy DL, Mulcahy GB (1987) The effects of pollen competition. Amer Scientist 75:44-50

SAS Institute Inc (1982) SAS User's Guide: Statistics. Cary, NC

Snow AA (1986) Pollination dynamics in *Epilobium canum* (Onagraceae): consequences for gametophytic selection. Amer J Bot 73:139-151

Stanton ML (1984) Development and genetic sources of seed weight variation in *Raphanus raphanistrum* L (Brassicaceae). Amer J Bot 71:1090-1098

Stephenson AG, Winson JA, Davis LE (1986) Effects of pollen load size on fruit maturation and sporophyte quality in zucchini. In: Mulcahy DL, Mulcahy GB Ottaviano, E (eds.). Biotechnnology and Ecology of Pollen. Springer, Berlin Heidelberg New York

Tanksley SD, Zamir D, Rick CM (1981) Evidence for extensive overlap of sporophytic and gametophytic gene expression in *Lycopersicon esculentum*. Science 213:453-455

Willing RP, Mascarenhas JP (1984) Analysis of the complexity and diversity of mRNAs from pollen and shoots of *Tradescantia*. Plant Physiol 75:865-868

Winsor JA, Davis LE, Stephenson AG (1987) The relationship between pollen load on offspring vigor in *Cucurbita pepo*. Amer Nat 129:643-656

An Ultrastructural Study of Fertilization in Douglas Fir [*Pseudotsuga menziesii* (Mirb.) Franco]

John N Owens and Sheila J Morris
Department of Biology
University of Victoria
Victoria, British Columbia
Canada, V8W 2Y2

INTRODUCTION

There has been renewed interest in fertilization in conifers since the early ultrastructural studies of archegonial development (Camefort 1962, 1967; Chesnoy and Thomas 1969) and fertilization (Camefort 1969; Chesnoy 1973; Singh 1978). This results from recent molecular studies showing that the inheritance of chloropast DNA (cpDNA) in conifers is strictly paternal (Neale et al, 1986) which is unlike most angiosperms. Preliminary studies of mitochondrial DNA (mtDNA) inheritance suggest that inheritance is maternal (Neale, personal communication) as in other higher plants. Ultrastructural studies of fertilization in angiosperms have recently demonstrated the structural mechanisms by which cpDNA and mtDNA inheritance occur (Russel 1983; Connett 1987). In conifers these mechanisms need further investigation in light of results from molecular studies.

OBSERVATIONS

Pollination occurred in April and was followed by 6 weeks when pollen elongated within the micropylar canal. A pollen tube then formed, penetrated the nucellus and fertilization occurred by mid-June (Fig. 1). Embryos developed over the next 2 months.

The engulfed pollen swelled, ruptured the exine and elongated. The nucellus tip formed a minute secretion stimulating pollen-tube formation. A narrow pollen tube penetrated between the loose outer nucellar cells. Deeper in the nucellus, cells which came in contact with the tip collapsed. Two to four pollen tubes commonly penetrated each nucellus. During pollen-tube growth the tube nucleus remained near the tip of the pollen tube followed by the large body cell and small stalk cell. Two degenerated prothathial

340

cells remained enclosed by a thickened intine attached to the inner pollen wall. The highly vacuolate tube-cell cytoplasm (Figs. 2, 3) contained abundant starch, electron transparent and electron dense vesicles, rough ER, dictyosomes, and mitochrondria. Many plastids appeared to degenerate. In contrast, the large body cell contained little starch, few vesicles, and abundant ribosomes, tubular ER, mitochondria and plastids with no evidence of organelle degeneration. The body cell was bounded only by a plasma membrane which became invaginated with sheets of tube cell cytoplasm penetrating deep within the body cell (Fig. 3). Small projections extended from the body cell membrane and attached to the tube cell membrane.

Pollen tubes grew to the distal end of the megagametophyte, penetrated the megaspore wall then grew into one of the archegonial chambers found above each group of neck cells (Fig. 2). The body cell settled into this pocket and divided to form the two male gametes.

Four to six archegonia per megagametophyte developed. Plastids within the central cell enlarged by inclusion of cytoplasm and condensation of the electron dense stroma. Numerous large and

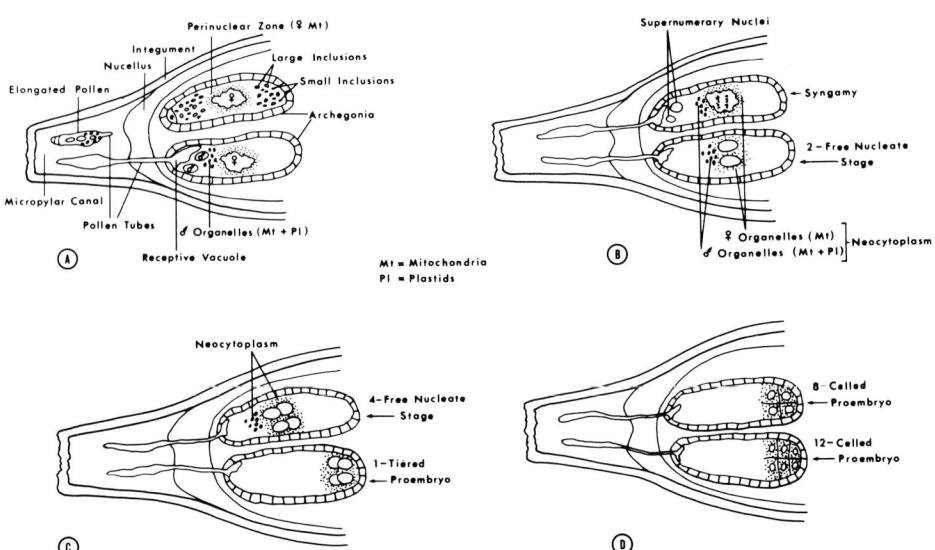

Figure 1. Diagram of fertilization in Douglas fir

small cytoplasmic regions combined with the electron dense plastid
stroma to produce complex double membrane-bound large inclusions.
In early June, the central cell divided unequally to form the small
ventral canal cell, which contained little cyoplasm (Fig. 2) and
the large egg cell. The egg nucleus migrated to the centre of the
egg and all of the mitochondria from the egg aggregated in the
perinuclear zone (Fig. 1). These mitochondria lacked distinct
cristae and became electron dense and irregular in shape. The egg
nucleus had many embayments. Four to six small neck cells formed a
uniseriate layer at the base of each archegonial chamber (Fig. 2).
These showed many features of secretory cells.

A large receptive vacuole formed where the contents of the pollen
tube entered the egg. The two male gametes and paternal
organelles, derived from the body cell, migrated towards the egg
nucleus (Figs. 1A, 4). Microtubules were visible in this region.
During migration the paternal organelles formed membrane bound
groups (Fig. 6) and many of these groups formed the cluster visible
in light micrographs (Figs. 1, 5). One male gamete fused with the
egg nucleus and the cluster of paternal organelles remained visible
in the neocytoplasm (Figs. 1B, 5). The zygote nucleus formed four
free nuclei. These, with the neocytoplasm migrated to the chalazal
end of the archeogonium (Fig. 1C). During this migration the
cluster of paternal organelles intermingled with maternal
organelles which originated from the perinuclear zone. The four
free nuclei settled as a single tier in the neocytoplasm divided
and cell walls formed a 2-tiered, 8-celled proembryo (Fig. 1D).
At this time paternal and maternal organelles may be morphologically distinguishable (Fig. 7). Cells in the lower tier divided
transversely interposing a middle suspensor tier between the apical
and open tiers. This formed the 12-celled proembryo characteristic
of certain conifers (Fig. 1D).

DISCUSSION

Several new observations were made which add to our understanding
of reproduction and the mechanism of cpDNA and mtDNA inheritance in
conifers. The secretory cells of the nucellar tip resemble those
of nectaries (Fahn 1979) in other conifers where they are responsible for pollination-drop secretion (Owens et al. 1987). However,

in Douglas fir no pollination drop occurs and the minute secretion functions in pollen tube induction. The deeply lobed body-cell cytoplasm has not been reported (Singh 1978) and may be the mechanism by which the large body cell is pulled down the narrow pollen tube by the tube cytoplasm. The archegonial chamber, allows the large body cell to settle over the neck cells, where it divides to form the male gametes. Pollen-tube penetration of the megaspore cell wall, some distance from the neck cells has not been reported in other conifers (Singh 1978). Growth of pollen tubes towards the neck cells, and the secretory appearance of the latter, suggest that substances are secreted which attract pollen tubes.

In Douglas fir and perhaps other conifers, the mechanism by which cpDNA, thus plastid-associated characters, are paternally inherited, is by compartmentalisation of plastids in the body cell and exclusion and destruction of maternal plastids during egg development. Our ultrastructural observations of Douglas fir and earlier reports for Pinus (Camefort 1962), Larix (Camefort 1967), Douglas fir (Chesnoy and Thomas 1969) and Chamaecyparis (Chesnoy 1973) agree with the pattern of strictly paternal cpDNA inheritance reported by Neale et al. (1986) using restriction fragment length polymorphism (RFLP) techniques on Douglas fir and other members of the Pinaceae, Cupressaceae and Taxodiaceae. Preliminary studies using RFLP techniques in conifers (Neale, personal communication) indicate that mtDNA inheritance is maternal. The mechanism by which mtDNA, thus mitochondrial characters, are maternally inherited is by the aggregation of mitochondria in the perinuclear zone of the egg. However, there is some paternal contribution resulting from the compartmentalization of paternal organelles which migrate as a cluster with the neocytoplasm. Ours and earlier ultrastructural studies suggest that mitochondria are primarily of maternal origin with some contribution from the cluster of paternal organelles. These observations suggest further studies of mitochondrial inheritance are needed using RFLP techniques.

Figure 2. Electron micrograph of a pollen tube in the archegonial chamber beneath the megaspore cell wall (mw). X 640. Figure 3. Electron micrograph of a portion of the lobed body cell (bc) cytoplasm, body nucleus (bn), tube cell (tc) and cell membranes (arrows) separating them. X 7,500. Figure 4. Light micrograph of a portion of an archegonium showing the receptive vacuole (rv)

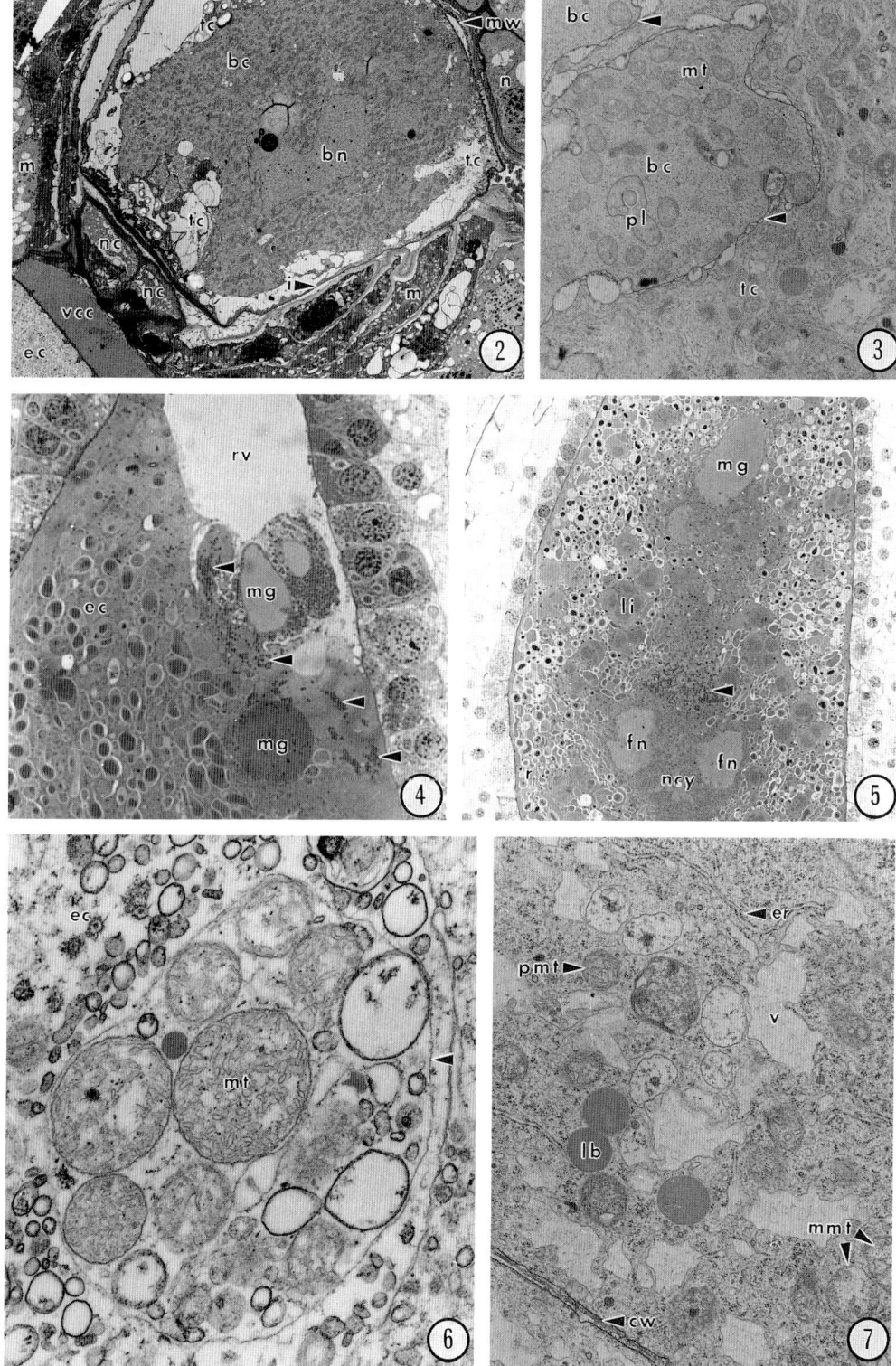

formed by the release of the contents of the pollen tube including two male gametes (mg) and many paternal organelles (arrows). X 375. Figure 5. Light micrograph of a portion of an archegonium showing two free nuclei (fn) of the proembryo and a cluster of paternal organelles (arrow) in the neocytoplasm (ncy). X 200. Figure 6. Electron micrograph of a membrane bound (arrow) group of paternal organelles comprising the cluster. X 30,000. Figure 7. Electron micrograph of a portion of the cytoplasm and cell wall (cw) between the distal tier of cells in an 8-celled proembryo. Maternal (mmt) and paternal (pmt) mitochondria appear to be distinguishable. X 1,100. ec, egg cytoplasm; er, endoplasmic reticulum; i, intine; lb, lipid body; m, megagametophyte; mt, mitochondria; n, nucellus; nc, neck cell, ncy, neocytoplasm; pl, plastid; rv, receptive vacuole; v, vesicle; vcc, ventral canal cell.

References

Camefort H (1962) L'organisation du cytoplasme dans l'oosphere et la cellule centrale du Pinus laricio poir. (var. austriaca). Ann Sci Nat Bot Biol Veg **12**: 269-291.

Camefort H (1967) Observations sur les mitochondries et les plastes de la cellule centrali et de l'oosphere du Larix decidua Mill (Larix europea DC). CR Acad Sci Paris **265**: 1293-1296.

Camefort H (1969) Fecondation et proembryogenese chez les Abietacees (notion de necytoplasme). Rev Cytol Biol Veg **32**: 253-271.

Chesnoy, L (1973) Sur l'origine paternelles des organites de proembryon du Chamaecyparis lawsonia A. Murr (Cupressacees). Carylogia 25 Suppl: 223-232.

Chesnoy L and M J Thomas (1969) Sur la presence de mitochondries Feulgen positives dans la zone perinucleaire du gamete femelle, du Pseudotsuga menziesii (Mirb.) Franco. Etude cytochimique et ultrastructurale. CR Acad Sci Paris **268**: 55-58.

Connett M B (1987) Mechanisms of maternal inheritance of plastids and mitochondria: developmental and ultrastructural evidence. Pl Mol Biol Rep **4**: 193-205.

Fahn A (1979) Secretory tissues in plants. Academic Press, New York.

Neale, D B, Wheeler, N C, Allard R W (1986) Paternal inheritance of chloroplast DNA in Douglas-fir. Can J For Res **16**: 1152-1154.

Owens, J N, Simpson, S J, Caron, G (1987) The pollination mechanism of Engelmann spruce (Picea engelmannii Parry). Can J Bot **65**: 1439-1450.

Russel S D (1983) Fertilization in Plumbago zeylanica: gametic fusion and fate of the male cytoplasm. Amer J Bot **70**: 416-434.

Singh H (1978) Embryology of gymnosperms. Gebruder Bontraeger, Berlin.

Secretion and Composition of the Pollination Drop in the *Cephalotaxus drupacea* (Gymnosperm, Cephalotaxeae)

R. Seridi-Benkaddour and L. Chesnoy
Laboratoire des Membranes Biologiques
Université Paris VII
2 place Jussieu
75251 Paris Cedex 05
France

INTRODUCTION

The pollination drop exudation on the micropylar orifice of the ovule ensures in numerous Gymnosperms the collecting of pollen grains and the first stages of their germination. The drop carries out a sorting of the wind-borne pollens and the contact of the male gametophyte with a female sporophyte product. This first contact often takes place before that of the pollen with the ovule tissues if the drop is completely external to the female organ. However little is known about the mechanism of the nucellar secretion and the composition of the pollination fluid.

Drops successively produced by different ovules of each strobilus always represent a very reduced volume (about 0,06 µl, 0,25 µl in case of the isolated ovules in Taxus baccata). Drops are secreted only a few days a year. In order to proceed to different analyses, it is necessary to have enough volume of nucellar fluid. The difficulties to obtain sufficient quantities each year explain the scanty informations available up to now, particularly concerning the sugar content of the drop in Pinus nigra (Mc William 1958), the carbohydrates, amino acids and inorganic phosphates in Taxus baccata and Ephedra campylopoda (Ziegler 1959).

Previous ultrastructural studies of the micropylar area of the nucellus in Cephalotaxus drupacea have shown the secretory nature of the cells bordering the pollen chamber which is well delimited in this species (Seridi and Chesnoy 1988).

RESULTS

The drops of <u>Cephalotaxus</u> <u>drupacea</u> Siev and Zucc. were taken
with 5 µl micropipettes from ovules of strobili on freshly collected branches brought to the laboratory and kept in water for several days. For comparison, drops of <u>Thuya</u> (= <u>Biota</u>) <u>orientalis</u>
L. and <u>Taxus</u> <u>baccata</u> L. were collected. Male strobili of the mono-
ecious <u>T</u>. <u>orientalis</u> were removed in order to avoid the pollen
grains presence in the fluid. After collection, drops were lyophi-
lized and preserved until further studies.

In the pollination drop of <u>C</u>. <u>drupacea</u>, oligosaccharides, an
unidentified polymer which we tried to analyse, uronic acids and
free amino acids were detected. The results were tentatively compared to those obtained with secretory products of <u>T</u>. <u>orientalis</u>
and <u>T baccata</u>.

SUGARS

They were searched for in <u>C</u>. <u>drupacea</u> drops as well as in the
two other species. Thin layer chromatography of lyophilised drops
suspended in water shows the presence of a polymer which does
not migrate in a butanol-acetic acid-water saturated atmosphere
(20/10/10 V/V/V). The polymer is therefore separated from the
migrating sugars.

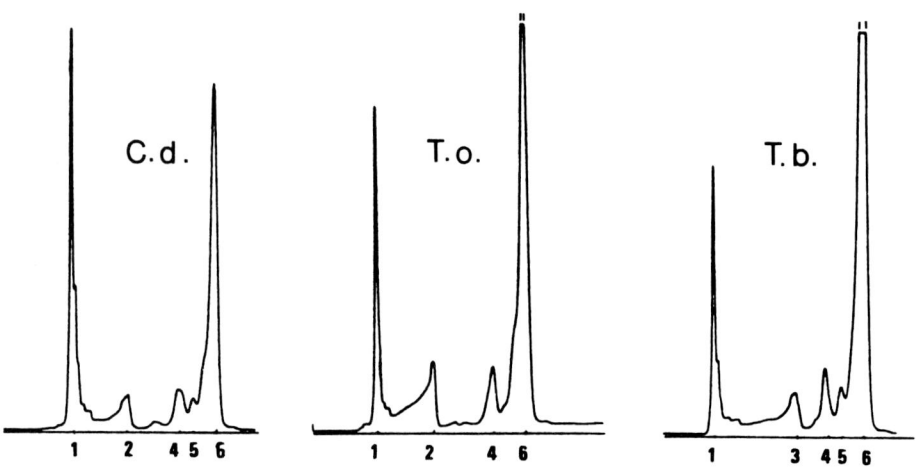

Figure 1- H.P.L.C. separation of the carbohydrates of <u>Cephalotaxus</u>
<u>drupacea</u> (Cd), <u>Thuya</u> <u>orientalis</u> (To) and <u>Taxus</u> <u>baccata</u> (Tb).
1: polymer, 2 and 4: non identified sugars; 4: sucrose; 5: glu-
cose; 6: fructose.

A high performance liquid chromatography (HPLC) analysis indicates the existence of a sugar which was found in abundance in the three extracts. By comparison with reference sugars it was identified as fructose. It represents 77,31% of the total sugar content in C. drupacea. The other components were incompletely identified with the exception of glucose which is present in the C. drupacea fluid (2,39%) and in the T. baccata drop. Sucrose was not detected in the extracts except in that of T. baccata (Figure 1).

C. drupacea polymer which amounts for 15,29% of the total sugars was analysed by gas chromatography (PGC) after conversion of the neutral sugars into alditol acetates according to the Sloneker technique (1972). The polymer yielded two fractions after they were hydrolysed with HCl, 2N, 2h. It can be assumed that the glycanic fraction represents the following composition: galactose (57%), arabinose (18%), glucose (13%), rhamnose (8%) and mannose (4%). The most common aglycone fraction appears to be a phenolic compound, but its mass spectrum still remains unknown.

AMINO ACIDS

Free amino acids analysed by the Hubac et al technique, are different in C. drupacea and T. orientalis. There are 5 main amino acids in the first species, whereas only 4 were found in the second (Table 1). Several other amino acids were detected but only as traces (Table 2). T. baccata drop was not analysed for amino acids. However data are available from Ziegler (1959).

Table 1: MAIN AMINO ACIDS
(Concentrations are given in μmol. ml^{-1} and in %)

	C.d		T.o		T.b(Ziegler 1959)
Serine	0,52	4,7%	4,59	50%	
Asparagine	3,60	32,7%	-	-	+ +
Glutamic ac.	2,34	21,3%	0,68	7,4%	+ + + +
Proline	3,79	34,5%	-	-	+ + + +
Glycine	-	-	2,39	26%	
Alanine	0,73	6,6%	1,52	16,5%	+ + + +

Table 2: AMINO ACIDS PRESENT ONLY AS TRACES
C. drupacea: Leucine, isoleucine, threonine, glutamine, aspartic acid.
T. orientalis: Phenylalanine, tyrosine, leucine, isoleucine, threonine, aspartic acid.
T. baccata (Ziegler 1959): Glutamine (+++), lysine, tryptophane, valine. In addition, this author had hydrolysed the peptids.

URONIC ACIDS

Research for uronic acids was carried out in the three species drops by the Blumenkrantz and Asboe-Hansen method (1973).

Galacturonic acids were identified in all samples: C. drupacea: 22,60 µg. ml^{-1}; T. orientalis: 13,88 µg. ml^{-1}; T. baccata: 44,53 µg. ml^{-1}.

DISCUSSION

Analysis of the nucellar fluids shows that fructose is the most common sugar in the three species studied: C. drupacea, T. orientalis and T. baccata, although belonging to different families. Fructose also appears to be present in large amount in Pinus nigra fluid reaching up to 40 mM, whereas glucose and sucrose reach respectively 33 mM and 2,5 mM in this species (Mc William 1958). Nygaard (1977) has tested the uptake and the utilization of various carbohydrates in suspension cultures of Pinus mugo pollens and concluded that exogenous fructose is more rapidly taken up than glucose and is preferentially used for starch synthesis by germinating pollen. In addition he showed that sucrose must be hydrolyzed before the fructose uptake. Thus the composition of Pinaceae pollination drop seems to be well adapted to the immediate needs of pollen germination. The male gametophytes of Cephalotaxaceae, Cupressaceae and Taxaceae possibly require the same carbohydrates, especially fructose, although this has not been demonstrated by in vitro cultures.

Our results are in agreement with those of Ziegler (1959) concerning the sugar content of the ovular fluid of T. baccata. This author has indeed noted the presence of hexoses (80 mg. ml^{-1}) and sucrose (26 mg. ml^{-1}). But our analysis indicate that the fructose is the most abundant hexose into the pollination drop of this species. The sucrose is less abundant as in the case also in P. nigra (Mc William 1958). On the other hand the presence of sucrose is doubtful in the pollination drop of C. drupacea and T. orientalis. However high concentration of sucrose has been identified in the drop of an Ephedra (25% Ziegler 1959). In Ephedra pollination is entomophilic.

In our studies evidence are given for the presence of galacturonic acids and a partly identified polymer. The glycanic fraction of the latter yields a neutral monosaccharide distribution characteristic of the pectic wall polymers. This suggests a parietal origin of the two constitutive materials, probably resulting from the break up of the middle lamellae of the nucellar superficial cells. These constituents might also be provided by the secretory activities of the nucellar cells which produce simultaneously parietal fibrills and mucilage-like substances (Seridi and Chesnoy 1988).

A certain similarity exists between the free amino acids composition of the drops in C. drupacea and T. baccata which, because of their putative affinities, were classified in the Taxales.

Duhoux and Pham Thi (1980) have shown that an amino acid complementarity occurs between the pollen tube and the ovular tissues and that a supply with ovular amino acids improves the male gametophyte development in suspension pollen cultures.

Our investigations show the complexity of the composition of the fluid secreted by the nucellus during pollination. This complexity results from an intense secretory activity associated with a tissue desorganization of the micropylar apical region of the nucellus. However the nature and the concentration of carbohydrates need to be completed. Others organic constituents as well as ions should also be searched for.

It has been pointed out that the origin of the drop components proceeds from cellular breakdown. In fact at the pollination time, the nucellar superficial cells are unaltered. Some isolated cells are poor in cytoplasm as if it has been progressively used up by their secretory activity. Later, after the pollination, cellular degeneration is important at the apex of the nucellus.

The production of the micropylar liquid has been interpreted as a phenomenon comparable to nectar secretion or guttation. Nectaries and hydathodes enclose conducting cells, phloem in the nectaries (Fahn 1979), xylem in the hydathodes (Perrin 1972). The unvascularized gymnosperm ovule cannot be attributed to such structures (Seridi and Chesnoy 1986). In addition the composition of the drop is comparable neither to that of the nectars nor to the guttation fluids. The pollination drop represents an original product, highly adapted to its functions: pollen collecting and germination.

Acknowledgements. Our thanks to Mrs Hubac, Mrs Laval-Martin, Mr Morvan and Mr Kharamanos for their help and advices.

References

Blumenkrantz N, Asboc-Hansen G (1973) New method for quantitative determination of uronic acids. Ana Biochem 54:484-489

Duhoux E, PhamThi A (1980) Influence de quelques acides aminés libres de l'ovule sur la croissance et le développement cellulaire in vitro du tube pollinique chez Juniperus communis (Cupressacées) Physiol Plant 50:6-10

Fahn A (1979) Secretory tissues in plants. Acad Press, London New-York

Hubac C, Guerrier D, Ferran J (1969) Resistance à la sécheresse du Carex pachystylis (G Gay), plante du désert du Negev. Oecol Plant IV:325-346

Mc William JR (1958) The role of micropyle in the pollination of Pinus. Bot Gaz 120:109-117

Nygaard P (1977) Utilization of exogenous carbohydrates for tube growth and starch synthesis in Pine pollen suspension cultures. Physiol Plant 39: 206-210

Perrin A (1972) Contribution à l'étude de l'organisation et du fonctionnement des hydathodes: Recherches anatomiques, ultrastructurales et physiologiques. Thèse Université de Lyon

Seridi R, Chesnoy L (1986) Ultrastructure et cytochimie des cellules superficielles du nucelle de Thuya orientalis L (=Biota orientalis (L) Endl) au moment de l'émission de la goutte micropylaire. Bull Soc bot Fr Lettres bot 135:111-124

Seridi R, Chesnoy L (1988) Cytologie du nucelle du Cephalotaxus drupacea Siev et Zucc, lors de la sécrétion de la goutte de pollinisation. Ann Sc Univ Reims Champ-Arden Arers, 23:64-66

Sloneker JH (1972) Gas-liquid chromotography of alditol acetats. In: Wistler RL, Miller JN (eds) Methods in carbohydrate chemistry. Acad Press, New-York London, p 20-24

Ziegler H (1959) Über die Zusammensetzung des Bestäubungstropfen und den Mechanismus seiner Sekretion. Planta 52:587-599

Pollination and Reproduction in Dianthus Silvester Wulf

Andreas Erhardt
Botanisches Institut der Universität Basel
Schönbeinstrasse 6
CH-4056 BASEL
Switzerland

INTRODUCTION

Dianthus silvester Wulf. is a widely distributed but not too common plant species which occurs in Central and South European mountain ranges (Hegi 1979).
The morphology of the flower has been described already by Müller (1881). However, only scanty observations on pollinators have been reported so far. Müller (1881) observed the sphingid Macroglossum stellatarum L. but stated, based on the morphology of the flower, that Dianthus silvester is pollinated by long tongued butterflies, e.g. Papilio machaon L. Knuth (1898) also reports one observation of Macroglossum stellatarum. Although no observations of pollinating butterflies have been reported so far, the statement of Müller (1881) has been accepted for almost a full century (Hegi 1979). The aim of the present study was to observe pollinators of Dianthus silvester in the field, to analyze the chemical composition of the floral nectar and to investigate the breeding system of Dianthus silvester.
Observations at night disclosed that the noctuid moth Hadena compta Schiff. is most likely the prime pollinator of Dianthus silvester. This is of particular interest since this moth also uses Dianthus silvester as a larval host.

MATERIALS AND METHODS

The study was carried out in the Swiss Alps (Tavetsch valley, 1550 m, 46°40'N, 8°44E) where large and easily accessible populations of Dianthus silvester occur. Observations of the plant and its breeding system were made in the field. Nectar samples were taken from open flowers, using carefully drawn out micropipettes flamed at the tip to avoid scratching the floral tissue. The nectar samples in the micropipettes were also inspected with a strong hand-lens (28x) to check for contaminating pollen grains. Pollen contaminated nectar samples were rejected for nectar analysis.
Techniques used for nectar analysis are described in Baker and Baker (1977, 1980). Sugar concentrations were read with a pocket refractometer (Brand: Krüss HRN-32). Nectar volumes were calculated from the spot sizes of the nectar samples spotted onto filter paper (Whatman #1) (Baker 1979) or were measured with a gastight hamilton syringe (#1701).
To test for self-compatibility in Dianthus silvester, flowers were bagged when they were still in bud to prevent contamination with foreign pollen by pollinators. Plants were also tested for apomixis. Using a scalpel, a slit was carefully cut into the calyx of flowers still in bud, and the two stigma lobes were removed from the ovary. Anthesis of the treated flowers was normal as was the development of the ovary.

Observations of pollinators were made on 15 nights during the peak flowering period of Dianthus silvester (July) in 1983 and 1984. Pollinators were observed with a strong torch. Voucher specimens of pollinators were collected with a regular butterfly net (diameter 50 cm). The collected pollinators were inspected under a dissecting microscope for adhering pollen grains. Pollen grains collected from pollinators were inspected microscopically, compared with pollen samples of Dianthus silvester and counted. Observations on ovipositing females of Hadena compta were also made during the observation nights.

RESULTS AND DISCUSSION

Breeding system of Dianthus silvester

Dianthus silvester is protandrous, gynodioecious and gynomonoecious (Müller 1881, Knuth 1898). Since flowers did not set seed when the two stigma lobes were removed from flower buds (n=37) they are most likely not apomictic although it cannot be excluded that they are pseudogamous. However, 82% of bagged flowers were found to set seed, if they were not destroyed by herbivores or fungi (n=51). This strongly suggests that Dianthus silvester is self-compatible. Self-compatibility seems to be a general trait in Caryophyllaceae (H.G. Baker, pers. comm.).

Besides hermaphroditic and female flowers, cleistogamous flowers were also observed in Dianthus silvester. These flowers do not unfold their petals. Only the two stigma lobes stretch and can be seen protruding from the calyx. The stigma lobes could still receive some foreign pollen, however, this seems unlikely since no corolla is present in such flowers to attract pollinators. When first observing these flowers I judged them to be not fully developed and infertile. It was therefore surprising to find that such flowers do indeed set seed. A careful investigation showed that in these flowers the anthers open and shed their pollen within the calyx. Since the plant is most likely self-compatible, these flowers are probably selfing.

Scent and Nectar

It was initially observed that in the investigated population of Dianthus silvester the light sweetish fragrance of the flowers slightly increases in the evening.

Nectar is produced in fairly high quantities (fresh flowers: $\bar{x} = 7.8 \,\mu l$; 4.2 - 14.3 μl, n=6). The "standing crop" may nevertheless be low ($\bar{x} = 0.5 \,\mu l$; 0.2 - 1.5 μl, n=34). The sugar concentration is moderate ($\bar{x} = 21.1\%$; 18.5 - 24.3%, n=5). The dominant sugar is sucrose. (Sugar composition, proportions: Sucrose 0.74, Glucose 0.05, Fructose 0.19, ratio: $\frac{Sucrose}{Glu+Fru} = 3.24$.) The amino acid concentration is also relatively low ($\bar{x} = 48.5 \,\mu g/ml$, n=2, ca. 3.5 on histidine scale), but many different amino acids were detected in the nectar (alanine 2, arginine + (Sakaguchi test), asparagine 3, aspartic 1, glutamic 1, glutamine 3, glycine 2, isoleucine 3, leucine 2, lysine 1, methionine 2, phenylalanine 3, proline 2, serine 3, tyrosine 2, valine 3, γ-amino but. 2, unknown 1i scores from 1 - 5 indicate increasing levels of concentration with 1 for the least concentrated amino acid in the sample, Baker and Baker 1977). A high number of these amino acids are essential (arginine, isoleucine, leucine, lysine, methionine, phenylalanine, valine) or quasi-essential (glycine, proline, serine) for insect nutrition (Haydak 1970, Dadd 1973). Tests for lipids (osmic acid), phenolics (folin), alkaloids (Dragendorff), protein (bromphenol) and organic acids (2,6-dichlorophenol-indophenol) were all negative.

Pollinators

In table 1 the observed visitors of Dianthus silvester are listed. The table clearly shows that Hadena compta was the most frequent visitor observed. Although only a low number of pollen grains was found on the moths, this does not necessarily mean that these noctuids are not efficient pollinators. The length of their proboscis corresponds strikingly well to the calyx length of Dianthus silvester and forces them to touch stigma and anthers not only whith their proboscis but also with with their labial palps or other parts of their head and possibly also with their thorax and legs. Since pollen of Dianthus silvester forms a loose powder it might have been easily lost during the collecting procedure of the voucher specimens. Brantjes (1976) showed that Hadena bicruris Hufn., a closely related species of Hadena compta, efficiently pollinates the female flowers of Silene alba. However, Brantjes (1976) did not investigate the moths for adhering pollen grains but observed the number of pollen grains deposited on the stigmas of female flowers after the moths had visited male flowers. The observation, that one individual of Hadena compta carried spores of Ustilago violacea on its proboscis indicates that lepidopteran pollinators are also the vectors of this parasitic fungus as reported by Jennersten (1983). It is noteworthy that Hadena compta moths had a distinct preference for the flowers of Dianthus silvester. I never observed them visiting other flowers although other plant species such as Knautia arvensis (L.) Coulter em. Duby or Centaurea scabiosa L. were also in bloom at the study site and were frequently visited by other nocturnal noctuid moths (e.g. Mythimna conigera Schiff.). These moths in turn regularly skipped the abundant flowers of Dianthus silvester, most likely because their proboscis is too short to reach the nectar which in Dianthus silvester is secreted at the base of the filaments and is deeply hidden within the calyx (Müller 1881).

The high number of pollen grains found on the proboscis of Macroglossum stellatarum indicates that hawkmoths may be more efficient pollinators than Hadena compta due to their fast and numerous flower visits. This may be a selecting force towards a sphingophilous syndrome as is observed in Dianthus superbus L. (Erhardt, in prep). However, the observed individuals of Hadena compta clearly outnumber the observed hawkmoths. The identification of Herse convolvuli is tentative, since I failed to catch a voucher specimen.

Only one butterfly species, Thymelieus lineola O., visited the flowers of Dianthus silvester. However, the proboscis of this species is too short to reach the nectar (table 1). Of all the butterflies occurring in the study area, only Papilio machaon L. would have a proboscis long enough to reach the nectar of Dianthus silvester flowers. However, individuals of Papilio machaon, although present at the study site, were never observed to visit the flowers of Dianthus silvester, but regularly flow past them to visit flowers with more easily accessible nectar such as the purple capitula of Knautia arvensis, Scabiosa columbaria L. and Centaurea scabiosa.

Flies were frequent visitors of Dianthus silvester flowers during late afternoon and early evening. They collected pollen from the anthers of hermaphroditic flowers but hardly ever touched the stigma and thus can be disregarded as effective pollinators.

The presented results clearly show that Dianthus silvester is not butterfly pollinated as stated by Müller (1881) but is adapted to nocturnal noctuid moths. The pale pink colour of the petals, the long calyx, the slight crepuscular increase of the scent as well as the nectar composition are all features which agree well with the syndrome of moth pollinated flowers (Faegri and van der Pijl 1980, Baker and Baker 1980, 1986). However, Dianthus silvester is not especially adapted to hawkmoths as is Dianthus superbus L. (Erhardt, in prep.).

Table 1: Visitors of Dianthus silvester

Species	Length of proboscis (in mm)	Number of observed individuals	Number of observed flower visits	Number of investigated individuals	Number of Pollengrains found on visitors		
					Dianthus silvester	Others (Indet.)	Part of Body
REGULAR POLLINATORS NOCTUIDAE							
Hadena compta Schiff.	19-26 $\bar{x}$=23.2 (n=12)	16+4?	28	5	0-20 $\bar{x}$=6.6	0-3* $\bar{x}$=0.6	proboscis
Others	15-24	5	5	3	1-2 $\bar{x}$=1.3	0-4 $\bar{x}$=2	proboscis
SPHINGIDAE							
Herse convolvuli L.(?)	65-80**	3	9-11	–			
Macroglossum stellatarum L.	24-25(n=2) 25-28**	2	10	1	391 7	18 0	proboscis labial palps
Sphingids (indet.)	?	2	6	–			
COINCIDENTAL POLLINATORS GEOMETRIDAE							
Eupithecia sp.	ca. 5	1	1 (trial)	–			
HESPERIIDAE							
Thymelicus lineola O.	13-14(n=2)	2	2 (trials)	–			
BOMBYLIIDAE							
indet.species	4	1	2 (trials)	1	0	1	thorax
MUSCIDAE							
indet.species	ca. 2	***	***	2	1-8 $\bar{x}$=4.5		thorax, legs

* one individual with numerous spores of Ustilago violacea (Pers.) Roussel / ** after H. Müller (1881) /
*** visit flowers often in late afternoon for collecting pollen from the anthers. Stigma mostly not touched
Note: Calyx length of Dianthus silvester: $\bar{x}$ = 20.1 mm (18 - 24, n=26)

Life history of Hadena compta

At the study site, Dianthus silvester is also the larval host of Hadena compta. On the observation nights, females were repeatedly observed to oviposit. A female ready to oviposit flies slowly and closely above the herbaceous vegetation, eventually landing on a flower stem of Dianthus silvester. It then climbs up the stem, inspecting it carefully. Eggs are deposited only on flowerbuds, never on open flowers. The females oviposit into the small pocket formed by the bracts at the base of the calyx and the calyx itself. Only one egg is deposited per flowerbud. The eggs are not glued onto the floral tissue, but are deposited loosely and fall off when the bracts are turned down. Since I observed single infestations in

93 % of all observed parasitized flowers (eggs and young larvae, n=59), and since infestations were evenly distributed among the whole plant population, it seems likely that superinfestations are prevented by the ovipositing females, for example by leaving an odour mark.

The freshly hatched larvae eat a hole through the tender tissue at the base of the calyx and enter the ovary where they feed on the young ovules. By about the third instar, they have consumed most of the ovary and the basal floral tissue and move to other flowers, where they again feed on developing ovaries and seeds. Older larvae also feed on young fruits, eating a characteristic round hole through the calyx and the ovary and feeding mainly on developing seeds.

Symbiontic and Parasitic Aspects of the Relationship

The combination of a lepidopteran pollinator and seed predator as it is observed in Hadena compta and Dianthus silvester, is paralleled only by Hadena bicruris Hufn. and Silene alba (Miller) E.M.L. Krause (Brantjes 1976), by Diaphone eumela Cramer and Thuranthos macrathum (Bak.) C.H.Wr. (Stirton 1976) and by the famous Yucca and the yucca moths (Tegeticula spp., Parategeticula sspp., Baker 1986). Although the investigated Hadena compta is clearly an important seed predator of Dianthus silvester, this relationship also includes strong symbiontic aspects:

(1) Hadena compta is an important, if not the main pollinator of Dianthus silvester. (2) Seed set of Dianthus silvester is, at least in some years, clearly pollinator limited (Erhardt, unpubl). (3) Moths of Hadena compta do not oviposit on the same flowers they visit for feeding. (4) Moths of Hadena compta have a preference for feeding on flowers of Dianthus silvester. (5) Hermaphroditic flowers of Dianthus silvester are probably in part functionally male flowers (Erhardt, unpubl.). Therefore the plant is not severely damaged if sterile ovaries are consumed by caterpillars after anthesis of these flowers. (6) Unlike Silene alba, Dianthus silvester is not dioecious. Therefore, not only half of the flower visits of the moths result in seed set as is the case with Hadena bicruris and Silene alba.

Consequently, seed set of Dianthus silvester is clearly less severely reduced by Hadena compta than is Silene alba by Hadena bicruris, if the caterpillars of Hadena compta consume about the same amount of ovaries as caterpillars of Hadena bicruris, and if the moths of both species visit about the same number of flowers.

Clearly the population size of Hadena compta will also affect its relationship to Dianthus silvester. An outbreak of Hadena compta could result in saturated pollination of Dianthus silvester, but could severely damage its seed set. If on the other hand the population of Hadena compta is low and pollinator limitation is severe (which was the case in 1987), the moths would still ensure a minimal amount of cross fertilization while the damage by the caterpillars would not be too serious.

The presence of cleistogamous flowers could also indicate a response to pollinator limitation.

Evolutionary Aspects

The observation that some hermaphroditic flowers are probably functionally male flowers indicates a trend to dioecy in Dianthus silvester. Hadena compta could affect the breeding system of Dianthus silvester in being a selective agent for maintaining hermaphroditic, but functionally male flowers as larval food for its caterpillars, but this is speculative.

It is also of interest that species of Silene and Lychnis are reported to be additional larval hosts of Hadena compta (Forster and Wohlfahrt 1971). Since bracts at the base of the calyx are lacking in these genera, females of Hadena compta have to oviposit at different sites on the flowers of these larval hosts. Thus, different ecotypes or even ecospecies differring in their oviposition sites may well occur in different populations of Hadena compta. Since oviposition habits of females incaptivity in are less specific and differ strongly from the habits observed in the field, only field observations would be valid in revealing such differences.

ACKNOWLEDGEMENTS

I am grateful to Irene and Herbert Baker for helping me with the nectar analysis and for stimulating discussions. I also thank St. Whitebread and Th. Boller for valuable comments and for reading the manuscript. I especially thank S. Engeli for typing the manuscript.

This paper is dedicated to Kaspar Wolfensberger for his personal help and encouragement during this work.

REFERENCES

Baker HG, Baker I (1977) Intraspecific constancy of floral nectar amino acid complements. Bot Gaz 138: 183 - 191

Baker HG , Baker I (1980) Studies of nectar-constitution and pollinator-plant coevolution in: Coevolution of Animals and Plants, 2nd edition (Ed. LE Gilbert, PH Raven). Texas Press Austin London: 100 - 140

Baker HG., Baker I (1986) The occurrence and significance of amino acids in floral nectar. Pl Syst Evol 151: 175 - 186

Baker HG (1986) Yuccas and the Yucca moths - a historical commentary. Ann Missouri Bot Gard 73: 556 - 564

Baker I (1979) Methods for the determination of volumes an sugar concentrations from nectar spots on paper. Phytochemical Bulletin 12: 40 - 42

Brantjes NBM (1976) Riddles around the pollination of Melandrium album (Mill.) Garcke (Caryophyllaceae) during the oviposition by Hadena bicruris Hufn. (Noctuidae, Lepidoptera), I, II. Proceedings K Nederlandse Akademie van Wetenschappen Ser C 79: 1 - 12 (I), 127 - 147 (II)

Brantjes NBM (1978) Sensory responses to flowers in nightflying moths. In: The pollination of flowers by insects (Ed. AJ Richards). Linn Soc Symp Ser Nr 6, Academic Press London: 13 - 19

Dadd RH (1973) Insect nutrition: current development and metabolic implications. Ann Rev Entomol: 18, 381 - 420

Faegri K and van der Pijl L (1980) The principles of pollination ecology, third revised edition. Pergamon Press Oxford New York Toronto Sydney Paris Frankfurt

Forster W and Wohlfahrt ThA (1971) Die Schmetterlinge Mitteleuropas Bd. IV Eulen (Noctiudae). Franck'sche Verlagshandlung Stuttgart

Haydak MH (1970) Honey bee nutrition. Ann Rev of Entomol 15: 143 - 156

Hegi G (1979) Illustrierte Flora von Mitteleuropa. Bd. III, Teil 2, 2. Aufl. (Ed KH Rechinger). Paul Parey Berlin Hamburg

Jennersten O (1983) Butterfly visitors as vectors of Ustilago violacea spores between caryophylla ceous plants. Oikos 40: 125 - 130

Knuth P (1898) Handbuch der Blütenbiologie Bd. II, 1. Teil. Verlag von Wilhelm Engelmann Leipzig

Müller H (1881) Alpenblumen, ihre Befruchtung durch Insekten und ihre Anpassungen an dieselben. Verlag von Wilhelm Engelmann Leipzig

Stirton CH (1976) Thuranthos: notes on generic status morphology, phenology and pollination biology. Bothalia 12: 161 - 165

Ovular Development and Pollen Tube Growth in the Ovary of *Gasteria verrucosa* (Mill.) H. Duval as Condition for Fertilization

M.T.M. Willemse and M.A.W. Franssen-Verheijen

Department of Plant Cytology and Morphology

Agricultural University Wageningen

Arboretumlaan 4

6703 BD Wageningen

The Netherlands

SUMMARY

A heteromorphic pollen tube of Gasteria grows in the locular fluid produced by placental papillar cells at the base of the ovules. The sugar and protein concentration of this fluid decreases from the moment of stylar elongation. From ovary pollination experiments it becomes clear that the appearance of the stigmatic exudate coincides with the permission to penetrate the micropyle marks the period of receptivity.

In the ovule the zones with glucose containing cells do not change during the penetration of the micropyle. Also the callose present in the filiform apparatus does not inhibit pollen tube penetration. A nutritive pathway leads the pollen tube towards the micropyle. The ovule influences the penetration of the pollen tube in the micropyle.

INTRODUCTION

In Gasteria verrucosa the structure and function of stylar development and pollen tube growth are studied (Willemse and Franssen-Verheijen 1986a,b, 1988). The results are comparable with Ornithogalum caudatum (Tilton and Horner 1980, 1983).

In this study some conditions for the pollen tube growth in the loculus of the ovary are reported and related to the conditions for fertilization.

MATERIAL AND METHODS

The material and methods applied for ultrastructural observations, 15% SDS PAGE and thin layer chromatography, were described earlier (Willemse and Franssen-Verheijen 1986a,b, 1988).

For ovary pollination one inflorescence was used and firstly from some flowers the style, stamens and a upper part of the perianth were cut off and the ovary was cross pollinated. For ovular pollination one loculus was opened and the ovary was stored in a wet petri disc. For detection of pollen tubes a callose staining with water blue and UV light was used. To detect free glucose in the cell the method of Okamoto et al. (Gabe 1976) was used. The osmolarity of three samples of 50 µl diluted liquid collected from different developmental stages of the ovary was measure with a Knauer osmometer.

RESULTS

Ovary and pollen tube growth

Three carpels form the ovary. Each locule contains two rows of 10-15 hemitrope ovules; the placenta is axillar. Between the ovules placental papillar cells develop from the epidermal layer of the placenta. These papillar cells produce a fluid which fills up the placental part of the locule, covering the ovules just above their micropyles, placed at the lateral part of the locule.

The pollen tubes enter the locules from the base of the style through a triangular canal, which widens in the ovary. In the first traject no secretory papillar cells are present, but yet a thin film of locular fluid covers the top of the locular cavity at the time of receptivity.

Pollen tubes run in the fluid on the lateral side of the locule or between the two rows of the ovules. These last pollen tubes may change their direction and grow between two ovules to the lateral side of the locule. When the thin pollen tubes arrive in the ovary a callose wall lacks, but gradually more callose plugs can be observed and a subsequently callose wall reappears (fig. 1). Pollen tubes penetrating the micropyle are thick and possess a thick callosic wall. After cross pollination one pollen tube penetrates in nearly every micropyle.

Cross ovary pollination from one inflorescence composed of different flowers, shows that the bud of receptive flower, bearing a drop of exudate, produces seed. The ovaries of flowers near this flower show pollen tubes in the locular fluid, but no penetration of the micropyle. So, in non-receptive ovaries there is no opportunity for penetration. On younger ovaries, pollen tubes germinate very scarce.

Ovular pollination results in germination of the pollen in the locular fluid in all developmental stages of the flower. In young flowers pollen tubes are short and twisting. Near the receptive stage the pollen tubes germinate normal and grow up and downwards following the normal pathway along and between the rows of ovules, but they pass the micropyles. Only in the receptive stage a number of the pollen tubes penetrate in the micropyles.

A barrier exists which prevents the penetration of the micropyle in young and old ovaries while the condition when penetration may take place coincides with the appearance of the exudate on the stigma.

The locular fluid and the condition of the ovule

The composition of the locular fluid is analyzed by thin layer chromatography for carobohydrates and by 15% SDS PAGE for proteins. Figure 2 represents scematically the carbohydrate composition of the locular fluid during flower development. Fructose, glucose and sucrose appear to be the main components. Trehalose and four other components are present in minor quantitiy in the open flower with a short style. At receptivity, the quantity of glucose and the minor components is decreasing in the locular fluid. The carbohydrate pattern of one ovule during its developmental stages remains similar to that of the locular fluid of a young flower.

2. Scheme of thin layer chromatography of the locular fluid and ovule of different developmental stages and the stylar liquid and stigmatic exudate of a receptive flower. For each spot the same volume of about 0,5 µl is used. S = short style; EL = elongated style; EX = with exudate; W = wilted. Gu/Fru = glucose/fructose; Su = sucrose; Tre = trehalose; f = front; Cz = concentration zone; O = origin.

3. Scheme representation of the 15% SDS PAGE gel of 20 µl locular fluid during different stages of development. A pattern of 10 µl stylar liquid is also represented. CF = closed flower; M = marker. For the abbreviations see figure 2.

Figure 3 represents scematically the protein patterns obtained by 15% SDS PAGE of the locular fluid in relation to the development of the flower. The concentration of the proteins diminishes when the period of receptivity is reached. The protein composition of a small quantity of stylar liquid is very similar to the locular fluid.

The osmolarity of the locular fluid taken from ovaries of different developmental stages expressed in milliosmol are as follows: in the open flower with short style: 657 ± 65; with elongated style: 858 ± 166; with stigmatic exudate 693 ± 190 and a wilted flower: 529 ± 72.

These data show an increase of osmolarity up to the receptive stage, followed by a decrease during the receptive stage and wilting of the flower.

During ripening of the ovary the concentration of carbohydrates and proteins decreases together with the value of the osmolarity. The composition of the locular fluid differs from the stylar liquid in the lower quantity of glucose/fructose.

On ultrastructural level no clear changes are observed in the embryo sac before and during pollination, except for the synergids in which the filliform apparatus is developing. Both synergids remain in good condition and no distinct

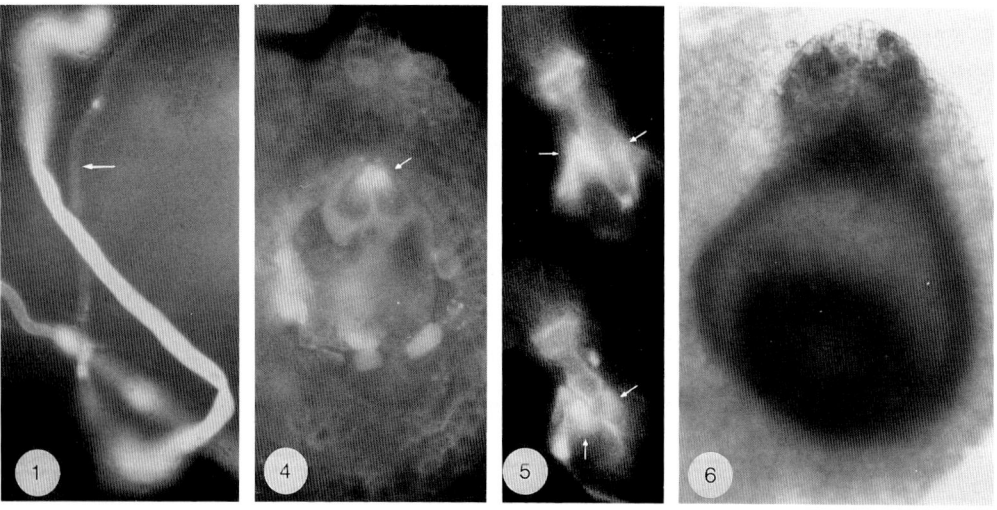

1. Pollentube in the locule with partly a tube with callose plugs without callose wall and a callosic wall. x100.

4. Filliform apparatus partly reacting positive on callose, arrow. x120.

5. After penetration of the pollen tube the filliform apparatus still shows the callose. x480, arrow, focussed on two levels. x300.

6. Reaction of the ovule after the glucose staining, note the zones around the hypostase, embryosac and nucellar top. x30.

previous degeneration is observed. The top of the filliform apparatus reacts positive on callose staining (fig. 4). Also after penetration of a pollen tube this reaction persists (fig. 5).

The nucellar cells bordering the top of the embryosac show smaller vacuoles then the more remote nucellar cells. In the nucellus and first integument, zones develop in which the cells possess a higher glucose content. At the receptive stage in the nucellus around the hypostase and basal as well as apical around the embryo sac, rings of glucose containing cells are present. The whole boundary of the nucellus reacts weak. The top of the nucellus at the basal part of the micropyle has a strong local reaction. The first integument reacts positive in its micropylar region (fig. 6). After penetration of the pollen tube, this pattern remains and the glucose increases after fertilization.

On structural level and glucose localisation no distinct changes can be observed in relation to pollen tube penetration.

CONCLUSIONS

In Gasteria the receptive stage marked by the exudate droplet on the stigma is operative for the whole pathway. The ovary pollination succeeds only in the receptive stage where the penetration of the micropyle is permitted. The production of the exudate droplet coincides with the permission to penetrate the micropyle.

At the time of receptivity, the osmolarity of the locular fluid decreases. This general preparation of a good condition for pollen tube growth, can be the result of a delution of the stylar liquid and locular fluid by uptake of water.

The increasing volume also results in the appearence of the exudate. A retarded and twisted growth of the pollen tubes after ovular pollination in early developmental stages, can be the result of a hypertony.

The condition of the pollen tube pathway through the stigmatic droplet, stylar liquid and locular fluid, is not the same. The coarse analyses shows that the stylar liquid contains more glucose/fructose than the locular fluid. The pollen tube growth reacts to it and becomes heteromorfic. The reappearence of the callosic wall of the pollen tube may be a result of the present papillar cells and a gas fase as on the stigma, but also of the change in composition of the fluid and its decreasing osmolarity. The reappearance of the callosic wall around the pollen tube offers it more strength to penetrate the micropyle.

At the moment of receptivity only changes in the synergids, especially the filliform apparatus are observed. The presence of callose in the filliform apparatus means that a molecular sieve exists at the tip of the embryo sac. This has consequences for uptake or excretion. The callose seems not to be a barrier for the pollen tube penetration. Remarkable is the absence of a degenerated synergid.

The localized glucose in the ovule seems to be related to the ovular development and can function as storage for the development of the embryo sac after fertilization. A role as nutrient for the pollen tube cannot be excluded because of the presence of glucose in the micropylar part together with a decreasing level of glucose in the locular fluid.

From exudate to the micropyle, the pathway of the pollen tube is nutritionned and directed by stylar liquid and locular fluid. Only the penetration of the micropyle, the target, needs a special condition to catch one pollen tube. An attractans, a growth substance or a source of nutrition, probably all originating from the ovule, can be supposed. In the area of the micropyle no remarkable change at the period of receptivity can be observed on structural level and after some histochemical tests. So, the micropylar area needs more study.

ACKNOWLEDGEMENT

The authors thanks Dr. R.W. den Outer for critical reading of the manuscript, Mrs J. Cobben-Molenaar for typing the manuscript and Mr S. Massalt for the photographs.

LITERATURE

Gabe M (1976) Histological technics. Springer-Verlag Paris Berlin.
Tilton VR, Horner Jr HT (1980) Stigma, style, and obturator of Ornithogalum caudatum (Liliaceae) and their function in the reproductive process. Amer J Bot 67(7):1113-1131
Tilton VR, Horner HT (1983) Carpel development, anatomy, and function in the reproductive process in Ornithogalum caudatum (Liliaceae). Flora 173:1-31
Willemse MTM, Franssen-Verheijen MAW (1986a) Stylar development in the open flower of Gasteria verrucosa (Mill.) H. Duval. Acta Bot Neerl 35(3):297-309
Willemse MTM, Franssen-Verheijen MAW (1986b) Pollination in Gasteria verrucosa (Mill.) H. Duval. In: Biology of reproduction and cell motility in plants and animals. Cresti M, Dallai R, eds. University of Siena
Willemse MTM, Franssen-Verheijen MAW (1988) Pollen tube growth and its pathway in Gasteria verrucosa (Mill.) H. Duval. Annales Scientifiques de L'Universite de Reims Champagne-Ardenne et de L'A.R.E.R.S. 23:119-123

Cytochemical Study of Adenylate Cyclase in Pollen-Pistil Interactions and its Relation to Incompatibility

M. Rougier, N. Jnoud and C. Dumas

Reconnaissance Cellulaire et Amélioration des Plantes

Université Lyon I

43 Bd du 11 Novembre 1918

69622 Villeurbanne Cedex

France

SUMMARY

The cellular and molecular events governing interspecific incompatibility in poplars are poorly understood. We have utilized this cell to cell recognition model to test the hypothetical mechanism evoked by Clarke et al (1985) by which signal molecules secreted by an animal or a plant cell bind to a target cell to produce cyclic AMP as a second messenger by activating adenylate cyclase. For the cytochemical localization of adenylate cyclase in our material, we have employed an original procedure using strontium as the capture ion and adenylylimidodiphosphate as the specific substrate. The specificity of the reaction has been checked by several controls. Our results suggest that in *Populus* , adenylate cyclase activity is correlated to pollen adhesion and germination at the stigma surface and that the abolition of this enzyme activity could be one of the cellular events governing incompatibility in the cross between *P. deltoïdes* and *P. alba*.

INTRODUCTION

The interaction of pollen and pistil represents one of the plant systems known to involve the recognition of self and non-self (Heslop-Harrison and Linskens 1984) and the implication of messenger molecules and specific receptors (Clarke et al. 1979, Dumas et al. 1984, Clarke et al. 1985). Among components involved in the pollen-pistil recognition are proteins, enzymes and lectins. According to Clarke et al (1985),the activation of adenylate cyclase by signal molecules to produce cyclic AMP as a second messenger represents one hypothetical mechanism that has emerged from studies of cell-cell recognition in animals and might apply to plant cell recognition.

To obtain information supporting the hypothesis of the involvement of adenylate cyclase activity in the pollen-pistil interaction, we have detected cytochemically this enzyme in an interspecific incompatibility system, i.e. *Populus* . Except the study of Gaude and Dumas (1986) devoted to the cytochemistry of the *Brassica* stigmas, there has been no cytochemical evidence for adenylate cyclase activity in reproductive organs of flowering plants. Nevertheless, in support of a physiological role of cyclic AMP in pollen germination are the data collected from Malik et al. (1976) and Katsumata et al. (1978). Malik et al (1976) have demonstrated the action of exogeneously supplied cyclic AMP on pollen tube growth in *Tradescantia paludosa.* On another hand Katsumata et al. (1978) have detected changes in cyclic AMP level and adenylate cyclase activity at the germination of pine pollens.

In the present study, pistils of *Populus deltoïdes* belonging to the section Algeiros have been preferentially used since Leuce pollens i.e. *P. alba* are able to adhere, hydrate and germinate at their stigma surface, the site of arrest of pollen tubes being further in the style (Gaget et al. 1984). Adenylate cyclase localization has been studied on self-, cross- and non-pollinated styles of *Populus deltoïdes* . This has enabled us to correlate the enzyme activity to pollen adhesion and germination at the stigma surface and to the penetration of pollen tubes following compatible pollination.

MATERIAL AND METHODS

Material of *Populus deltoïdes* and *Populus alba* was obtained from the INRA Station des Arbres Forestiers, Orléans (France) and grown in a greenhouse at the University Lyon I, Villeurbanne (France). Both types of crosses were examined: *P. deltoïdes* X *P. deltoïdes* (compatible intraspecific cross) and *P. deltoïdes* X *P. alba* (incompatible interspecific cross). Pollinations were carried out either with dry pollen freshly collected from mature anthers or with pollen that had been previously stored at -20°C. Compatible and incompatible pollinations were performed in separate rooms, for 2 h or more at 25°C.

For enzyme cytochemical studies with transmission electron microscopy (TEM), pistil and pollen samples were briefly fixed in 1% glutaraldehyde in 0.1M sodium cacodylate buffer (pH 7.4) for 15 min at room temperature and washed for 1 h with several changes in the above mentioned buffer. Their incubation was performed with a reaction mixture for the detection of adenylate cyclase adapted from Poeggel et al. (1982) and modified by the introduction of strontium by Zajic and Schacht (1983) instead of lead as the capture ion. The composition of the incubation medium and the conditions used are described elsewhere (Rougier et al.1988)

After incubation and washings, postfixation was done with 1% OsO_4 in cacodylate buffer for 1 h at room temperature. Following dehydration in ethanol, the samples were embedded in Spurr medium (Spurr, 1969). Several controls for the nonspecific reactivity were also performed

:incubation in presence of 10 mmol/l alloxan hydrate as an adenylate cyclase inhibitor, incubation without substrate, incubation in a medium lacking strontium chloride

Ultrathin sections were cut on a Reichert OMU2 ultra-microtome and examined unstained with a Hitachi HU 12A microscope at CMEABG, University of Lyon.

RESULTS

No reaction deposits indicating adenylate cyclase activity were detected either at the surface or within the cytoplasm of unpollinated *P. deltoïdes* papillae. Non-hydrated pollens of both species used for compatible and incompatible pollinations showed also very few reaction products lacking evident specificity.

When adenylate cyclase activity was detected on self-pollinated *P. deltoïdes* stigmas, a strong reaction was observed following the germination of pollen at the stigma surface (Fig.1). Deposits of electron opaque material were detected on the stigmatic pellicle and on the underlying cell wall surrounding the germinating pollen tube. On the contrary, the internal cell wall layer close to the plasmalemma limiting the cytoplasm of the reactive stigma papilla remained unreactive at this stage. The penetration of the pollen tube through the stigma surface results in pronounced deposit of electron opaque material on the radial cell walls of papillae in the area surrounding the tip of the growing pollen tube (Fig. 2). Reaction products are also deposited on the plasmalemma of the stigmatic papillae at the sites of pollen tube penetration. The tip of the pollen tube itself appears enriched in numerous electron-transparent vesicles and is surrounded by a strongly reactive plasmalemma and cell wall.

In the incompatible cross between *P. deltoïdes* and *P. alba*, reaction products indicating adenylate cyclase activity were still encountered at the sites of penetration of pollen tubes on the stigma surface. This situation is illustrated in figure 3. By contrast, figure 4 illustrates the penetration of *P. alba* pollen tube between adjacent papillae of *P. deltoïdes* and the concomitant decrease in the intensity of adenylate cyclase activity within the stigma tissues. Reaction products are no longer detected within the stigma in the area where the pollen tube has grown. The pollen tube itself seems to swell at its extremity and contains dense lipid-like droplets which have not been observed after compatible pollinations.

Controls have been performed to check the specificity of the cytochemical detection of adenylate cyclase With the addition of alloxan in the incubation medium, the intensity of the reaction was reduced in most samples but not abolished When the substrate was omitted from the incubation medium, no deposits were found at the stigma surface following the pollen adhesion and growth of the pollen tube on the stigma surface

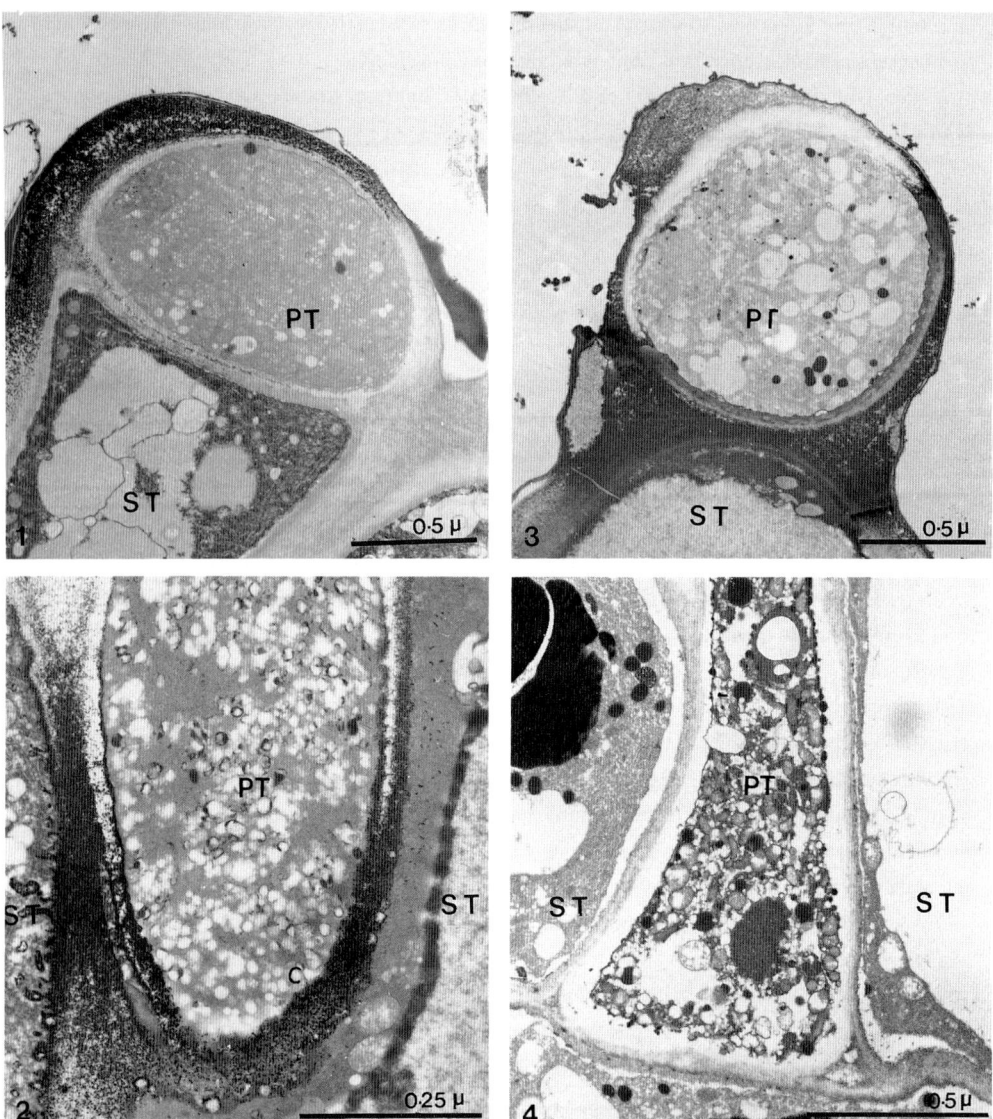

Figures 1 -2 : Cytochemical reaction for adenylate cyclase at sites of pollen-stigma interactions 2 h after compatible pollination between *P. deltoïdes* and *P. deltoïdes*. Reactions products are observed around the pollen tube at the stigma surface or after its penetration betweeen receptive papillae.
Figures 3-4 : Cytochemical reaction for adenylate cyclase at sites of pollen -stigma interacions 2 h after incompatible pollination between *P. deltoïdes* and *P. alba.*. Reaction products remain at sites of pollen tube adhesion at the stigma surface but are no more visible at sites of pollen tube penetration within the stigma.

CONCLUSION

The present data devoted to the localization of adenylate cyclase activity in *Populus* indicate that on both pollen and stigma sides, adenylate cyclase activity was found to be associated to membrane structures and cell wall components. In compatible cross between *P. deltoides* X *P. deltoides* we have demonstrated that the reaction products are concentrated at the sites of the pollen tube - stigma surface interaction or at the pollen tube tip after its growth between the receptive papillae. Thus the cytochemical localization of adenylate cyclase in *Populus* pollens germinated *in vivo* confirms and extends the data previously mentioned and related to the presence of cyclic AMP, adenylate cyclase and protein kinases in pollen germinated *in vitro* (Katsumata et al. 1978, Polya et al. 1986). Following compatible pollinations in *Populus* we have also demonstrated an adenylate cyclase activity on the stigma surface in association with the pellicle which is known to offer similarities with a biological membrane (Gaude and Dumas 1986). . A weak adenylate cyclase activity (Gaude and Dumas 1986) has previously been detected cytochemically in the *Brassica* pellicle .

Our study brings new information on the factors involved in the pollen tube inhibition in the style by showing that the arrest of pollen tubes may be dependent on an abolition of adenylate cyclase activity. At present we have no indications on the mechanisms involved in the disappearance of adenylate cyclase activity in cross-pollinated stigmas. Further investigations will have to be performed to discriminate between inhibition or degradation of the enzyme. Another cytochemical study performed on the gametophytic self-incompatibility system of *Petunia* (Carraro et al. 1986) has reported the involvement of wall peroxidases present in the outer portion of the transmitting tissue in incompatibility reactions. As ours, these results suggest that enzymes are some of the components involved in incompatibility reactions.

We may therefore conclude that our present data collected from *Populus*, which represents an example of interspecific incompatibility, add strong experimental evidence to the mechanism suggested by Clarke et al (1985) to govern the cellular recognition events during pollen-stigma interactions. Investigations in progress in our laboratory are correlated with the possible involvement of adenylate cyclases during the whole process of fertilization and during the germination of *Populus* pollen *in vitro* . The extension of our cytochemical investigations to other systems of incompatibility in flowering plants represents another future promising project.

REFERENCES

Carraro L, Lombardo G, Gerola FM (1986) Stylar peroxidase and incompatibility reactions in *Petunia hybrida*.. J Cell Sci 82 : 1-1O

Clarke AE, Gleeson P, Harrison S, Knox RB (1979) Pollen-stigma interactions: identification and characterization of surface components with recognition potential. Proc Nat Acad Sci U.S.A. 76: 3358-3362

Clarke AE, Anderson MA, Bacic T, Harris PJ, Mau SL (1985) Molecular basis of cell recognition during fertilization in higher plants. J Cell Sci suppl 2: 261-285

Dumas C, Knox RB, Gaude T (1984) Pollen-pistil recognition: New concepts from electron microscopy and cytochemistry. Int Rev Cytol 9O: 239-272

Gaget M, Said C, Dumas C, Knox RB (1984) Pollen-pistil interactions in interspecific crosses of *Populus* (sections Algeiros and Leuce) : pollen adhesion, hydration and callose responses. J Cell Sci 72: 173-184

Gaude T, Dumas C (1986) Organization of stigma surface components in *Brassica* Cytochemical study. J Cell Sci 82: 203-216

Heslop-Harrison J, Linskens HF (1984) Cellular interaction: a brief conspectus. In: Linskens HF, Heslop-Harrison J (eds) Cellular interactions (Encyclopedia of Plant Physiology New Series Vol. 17) Springer-Verlag, Berlin Heidelberg New-York Tokyo, pp 2-17

Katsumata T, Takahashi N, Ejiri SI (1978) Changes of cyclic AMP level and adenylate cyclase activity during germination of pine pollen. Agric Biol Chem 42: 2161-2162

Malik CP, Chhabra N, Vermani S (1976) Cyclic AMP-induced elongation of pollen tubes in *Tradescantia paludosa* . B P P 169: 311-315

Poeggel G, Luppa H, Weiss J (1982) Multiple localizations of adenylate cyclase in rat hippocampus. Histochemistry 74: 139-147

Polya GM, Micucci V, Rae AL, Harris PJ, Clarke AE (1986) Ca^{++} dependent protein phosphorylation in germinated pollen of *Nicotiana alata* , an ornemental tobacco. Physiol Plant 67: 151-157

Rougier M., Jnoud N., Dumas C. (1988) Localization of adenylate cyclase activity in *Populus* : its relation to pollen-pistil recognition and incompatibility (in press)

Spurr A.R. (1969) A low viscosity epoxy resin embedding medium for electron microscopy. J Ultrastruct Res 26: 31-43.

Zajic G, Schacht J (1983) Cytochemical demonstration of adenylate cyclase with strontium chloride in the rat pancreas. J Histochem Cytochem 31: 25-28

The Influence of Double Pollination and Pollen Load on Seed Set and Seedling Vigour of Apple and Pear

T. VISSER & J.J. VERHAEGH

Institute for Horticultural Plant Breeding

Mansholtlaan 16

6708 PA Wageningen

The Netherlands

INTRODUCTION

In fruit breeding, pollination is a simple operation: viz. with a small brush the pollen is dusted quickly on the stigmas. Although in apple, pear and rose such a single pollination generally appears to produce a fair seed set, this can be considerably improved by repeating the pollination at an interval of 1-2 days (VISSER & VERHAEGH, 1980b; VISSER, 1986; DE VRIES & DUBOIS, 1983). While it is obvious that the quantity of pollen affects, up to a limit, the quantity of seeds, the breeder is less aware of the fact that the pollen quantity may influence seed quality as well, e.g. as was shown for the vigour of Zucchini seedlings by STEPHENSON et al. (1986). Therefore, after having investigated the influence of pollen load on pollen tube growth in apple, pear and rose styles (MARCUCCI & VISSER, 1988; SNIEZKO et al., 1989; VISSER et al., 1988), the effect of pollen load on the seed quality of apple and pear was studied subsequently.

METHODS AND MATERIALS

Pollen and flowers were handled as described previously (VISSER & VERHAEGH, 1980a). In 1986 two pollination trials were carried out on the pear 'Doyenne du Comice' (at Elst and SAP) and two on the apple cvs 'Golden Delicious' (at Elst) and 'Melrose' (at SAP). In three trials treatments were: pollination once (1x) and twice (2x) at a one day interval with diluted (1 part viable: 7 parts dead pollen) and with pure (viable only) pollen; in the Melrose trial they were: normal pollination with diluted and pure pollen and abundant pollination using ample pure pollen twice in quick succession (1+1x). Per treatment 7-10 replicates were used, totalling 150-250 flowers. The pollen used for dilution was killed in methanol followed by drying. When pollinating, care was taken that the brush was visibly yellow with pollen.

As before, the results of the pollination trials were measured by the seed set (seeds/fruit) and pollination index or PI (seeds/pollinated flower) (VISSER & VERHAEGH, 1980a). Mean seed weight per treatment was determined for the 'good' seeds only. For the seed germination tests it was seen to that the replicates (35 seeds each) of the treatments in the same block had seed weights which differed no more than 5-10%. In the pear trials the treatments were represented by 4-6, in the apple trials by 6-7 replicates. The seeds, sown in seed boxes, were after-ripened in a cold store and then placed in a warm glasshouse to germinate. The seedlings were subsequently potted, hardened off and transferred to the nursery in May 1987. The same statistical outlay as used for the germination tests was maintained, but in the pear trials some replicates had to be excluded later due to adverse soil conditions. The length of the seedlings was measured early autumn.

RESULTS

The general trend (Table 1) was that pure pollen produced a higher seed set than diluted pollen and pollinating twice a higher one than pollinating once, though differences were significant only in the pear trials. In terms of pollination efficiency, the PI shows that pollinating twice with diluted pollen was always as effective as pollinating once with pure pollen, but using diluted pollen once was always less efficient than using pure pollen once, and even less so when compared to pure pollen applied twice. The pollination treatments had neither a significant effect on the mean weight nor on the germination of the seeds (Table 2). The origin of the seeds as related to the pollination treatment (Table 3), did not influence the performance of the pear seedlings, possibly because poor soil conditions interfered with the results. Pollinating Golden Delicious once or twice with pure pollen (1:0) compared with diluted (1:7) pollen increased seedling length and reduced the proportion of weak growing seedlings, as also resulted from abundant pollination in the case of the (inbred) Melrose population.

Table 1. Mean seed set (seeds/fruit) and pollination index PI (seeds/pollinated flower) after pollinating once or twice with diluted (1:7) or pure (1:0) pollen.

Pollination treatment	Seed set				Pollination index			
	D.d.Comice		Golden Delicious	Mel-rose	D.d.Comice		Golden Delicous	Mel-rose
	Elst	SAP			Elst	SAP		
1:7 once	3.4a	3.6a	8.0a	6.2a	0.44a	0.78a	1.68a	1.26a
1:0 "	5.8b	5.2bc	8.1a	7.2a	0.94b	1.15a	2.55b	1.59ab
1:7 twice	6.2bc	4.5ab	8.4a	-	1.26b	1.06a	2.33b	-
1:0 "	7.2c	5.9c	9.5a	7.6a[*]	1.26b	1.96b	2.20b	1.89b[*]

N.B. Different letters indicate a significant difference;
*)abundant pollination of Melrose.

Table 2. Mean weight per 35 seeds (g) and mean seed germination (%) after pollinating once or twice with diluted (1:7) or pure (1:0) pollen

Pollination treatment	Doyenné du Comice Elst SAP				Golden Delicous		Melrose	
	g	%	g	%	g	%	g	%
1:7 once	1.343	98	1.177	95	1.262	85	1.109	94
1:0 "	1.268	93	1.276	91	1.329	85	1.057	89
1:7 twice	1.271	91	1.165	87	1.301	77	-	-
1:0 "	1.249	95	1.203	98	1.312	77	1.070	91[*]

N.B. Differences not significant; [*]abundant pollination of Melorse.

Table 3. Mean seedling length (cm) and the percentage small seedlings (<55 cm) after one growing season in the nursery in relation to pollination treatment;

Pollination treatment	Seedling length				% Small seedlings			
	D.d.Comice Elst	SAP	Golden Delicious	Mel- rose	D.d.Comice Elst	SAP	Golden Delicous	Mel- rose
1:7 once	75a	79a	104a	95	9.3a	10.9a	13.8a	20.3a
1:0 "	75a	77a	120b	96	10.7a	7.1a	7.8b	13.3ab
1:7 twice	75a	76a	106a	-	10.8a	11.8a	11.8ab	-
1:0 "	74a	75a	119b	105	11.2a	12.4a	6.7b	9.0b[*]

N.B. Different letters indicate a significant difference; [*]abundant pollination of Melrose.

Although the seed weight per treatment differed little on average, differences between the replicates were much greater, amounting to 35-50% between the lightest and heaviest sample. Calculating the correlation between seed weight (g per 35 seeds) and seedling length (cm) of all the replicates in a trial, showed that the two parameters were highly significantly related in the Comice and Golden Delicious trial at Elst and just not significantly in the Comice and Melrose trial at SAP: respectively $r=0.654$ (n=22) & 0.696 (n=18) and $r=0.483$ (n=14) & 0.360 (n=26). In view of this relationship, the distributions shown in Fig.1 are based on only those replicates (in the same block) which had similar seed weights (varying less than 5%). It is seen that there was still no difference between the distribution of the pear seedlings originating from the heaviest (1:0, 2x) and lightest (1:7, 1x) pollen load. However, the distributions for Melrose and Golden Delicious illustrate that with the higher pollen load a shift occurs from fewer weak to more vigorous seedlings.

DISCUSSION AND CONCLUSIONS

In view of the only one or two ovules per style of cherry, plum, apple, pear and rose flowers, the pollen load usually far exceeds the theoretical need of only one or two viable pollen grains for fertilization (see VISSER et al., 1988). Such an excess also appears from the fact that diluting viable, compatible, pollen 7-9

372

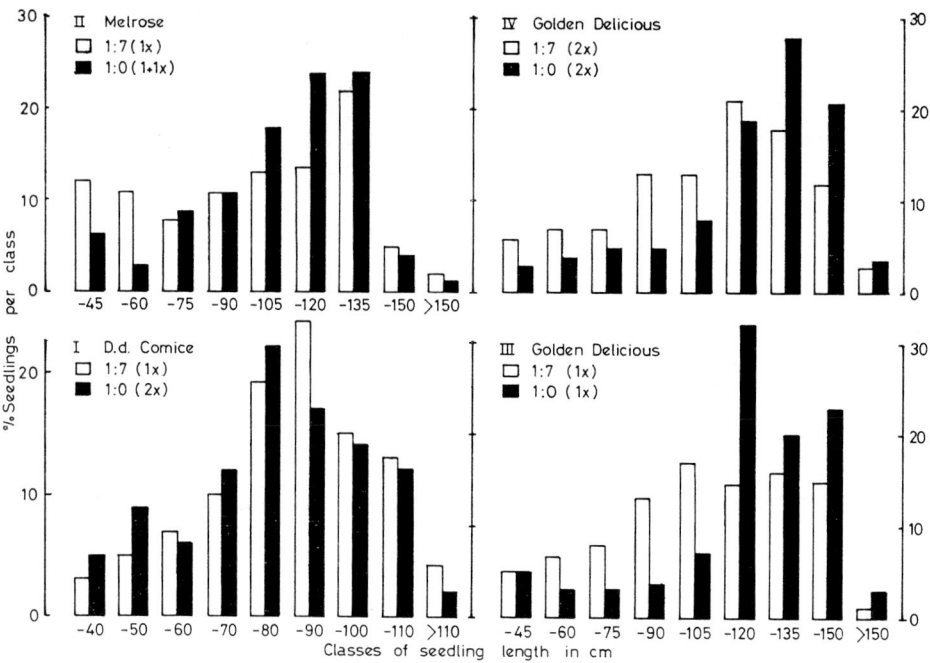

Fig.1. Distribution of seedling length in relation to pollination treatment;
80-100 seedlings/treatment/experiment.

times with dead or incompatible pollen decreased the PI with apple and pear on
average by one third only (VISSER, 1955; VISSER & MARCUCCI, 1984; this paper). On
the other hand, optimal tube growth in the style required for rose more than 20
and for apple and pear 40-50 pollen grains on the stigma (MARCUCCI & VISSER, 1988;
SNIEZKO et al., 1989). Presumably, in a single pollination the excess pollen
promotes as 'mentor pollen', the other simultaneously present pollen (VISSER &
MARCUCCI, 1983). In a double pollination, the first pollen applied in advance,
the 'pioneer pollen', 'paves the way at its own cost', the second pollen to the
extent that this may produce twice as many seeds as the first (VISSER & VERHAEGH,
1980b; VISSER & MARCUCCI, 1983). The pioneer pollen situation scores over the
mentor pollen one (see also VISSER & OOST, 1982), because in a double pollination
the first pollen had time to form tubes in the style apex, a requirement for
optimal stimulation of the second pollen (VISSER & MARCUCCI, 1983), while in a
single pollination the time lapse between the faster, or 'stronger', and the
slower-germinating, or 'weaker', pollen is relatively brief and the interaction
accordingly less. Translated in terms of pollen competition, it means that in a
single pollination the weaker pollen, being promoted by the stronger, is in a
favoured position to reach the ovules, though chances decrease with increasing
pollen load.

This concept of the stronger pollen, 'altruistically' promoting the weaker, agrees well with the data of JANSE (1987) on rye, showing that aneuploid pollen - with probably slower tube growth than euploid pollen (BUCHHOLZ & BLAKESLEE, 1930) - contributed much more to seed set with restricted than with abundant pollination (20 versus 4%). It also agrees with the results of STEPHENSON et al. (1986) with Zucchini, indicating a significant relation between pollen load and seedling vigour and with our results with apple showing a distribution shifting to fewer weak and more vigorous seedlings with a heavier pollen load. As to double pollination, in our trials this did not produce significantly stronger seedlings than single pollination, but it did significantly decrease the percentage abnormal pollen tubes penetrating rose ovaries (SNIEZKO et al., 1988). This fits our concept of pollen competition in which the stronger pollen profits more from double than from single pollination, because the pollen first applied, acting as pioneer pollen, would promote in the second dose both the stronger and the weaker pollen to the greater advantage of the former. In this sense, repeated pollination, which in nature is normal for many plants, is - apart from the pollen load per se - an instrument promoting the survival of the fittest. A final conclusion is that in the practice of plant breeding one should be aware of the possible advantages of abundant pollination.

REFERENCES

BUCHHOLZ JT, BLAKESLEE AF (1930) Pollen tube growth and control of gametophytic selection in cocklebur, a 25-chromosome Datura. Bot.Gaz. 90: 366-383

JANSE JOKE (1987) Certation between euploid and aneuploid pollen grains from a testiary trisomic of rey Secale cereale L. Genome 29: 353-366

MARCUCCI CLARA M, VISSER T (1988) Pollen tube growth in apple and pear styles in relation to self-incompatibility, incongruity and pollen load. (In press.)

SNIEZKO RENATA, PIJNACKER HORDIJK JP, DUBOIS LIDWIEN AM, VISSER T (1989) Pollen tube growth and seed set as affected by double pollination and pollen load of the hybrid tea-rose cv. 'Sonia'. (In press.)

STEPHENSON AC, WINSOR J, DAVIS L (1986) Effects of pollen load size on seed number, fruit maturation and seed quality in Zucchini. Biotechnology and Ecology of Pollen (eds.: D.L. Mulcahy & E. Ottoviani). Springer-Verlag N.Y.: 429-434

VISSER T (1955) Handling of pollen of deciduous fruits. (Dutch with English summary). Meded. Dir. Tuinbouw 18: 856-865

VISSER T (1981) Pollen and pollination experiments. IV 'Mentor' pollen and 'pioneer' pollen techniques regarding incompatibility and incongruity in · apple and pear. Euphytica 36: 363-369

VISSER T (1986) The interaction between compatible and compatible or self-incompatible pollen of apple and pear as influenced by pollination interval and orchard temperature. Biotechnology and Ecology of Pollen (eds.: D.L. Mulcahi, G.B. Mulcahi & E. Ottoviani). Springer-Verlag: 167-172

VISSER T, MARCUCCI CLARA M (1983) Pollen and pollination experiments IX. The pioneer pollen effect in apple and pear related to the interval between pollinations and temperature. Euphytica 32: 703-709

VISSER T, MARCUCCI CLARA M (1984) The interaction between compatible and self-incompatible pollen of apple and pear as influenced by their ratio in the pollen cloud. Euphytica 33: 699-704

VISSER T, OOST EH (1982) Pollen and pollination experiments V. An empirical basis for a mentor pollen effect observed on the growth of incompatible pollen tubes in pear. Euphytica 31: 305-312

VISSER T, VERHAEGH JJ (1980a) Pollen and pollination experiments. I. The contribution of stray pollen to the seed set of depetalled, hand-pollinated flowers of apple. Euphytica, 29: 379-383

VISSER T, VERHAEGH JJ (1980b) Pollen and pollination experiments. II. The influence of the first pollination on the effectiveness of the second one in apple. Euphytica 29: 385-390

VISSER T, SNIEZKO RENATA, MARCUCCI CLARA M (1988) The effect of pollen load on pollen tube performance in apple, pear and rose styles. 'Sexual Reproduction of Plants', Siena 1988, Springer Verlag:

VRIES DP DE, DUBOIS LIDWIEN AM (1983) Pollen and pollination experiments. X. The effect of repeated pollination on fruit and seed set in crosses between the hybrid tea-rose cvs. 'Sonia' and 'Ilona'. Euphytica 2: 685-689

From Ovule to Seed

Cytology of Seed Development in Perennial Ryegrass, *Lolium perenne* L.

A. Elgersma and R. Sniezko[*]

Foundation for Agricultural Plant Breeding (SVP)

P.O. Box 117, 6700 AC Wageningen, The Netherlands

ABSTRACT

In seed crops of perennial ryegrass, <u>Lolium perenne</u> L., many florets do not produce a seed and yields are low. This study was conducted to determine if seed abortion was a major cause of this low seed yield. Spikes were fixed at weekly intervals one to five weeks after first anthesis. Whole spikelets were microtome sectioned. The cytology of normal seed development, ovule degeneration and seed abortion is presented. The histology of the pro-embryo is presented in relation to the development of the endosperm.

In 90 % of the unproductive florets, embryo sac degeneration occurred within a few days after flowering. A few florets were sterile or showed embryo sac degeneration before flowering and in about 5 % of the unproductive florets seeds aborted in later stages, sometimes as a result from insect attack.

Abortion of developing seeds was not important to low seed yield in perennial ryegrass. Degeneration of embryo sacs shortly after flowering was the major cause of unproductive florets in all genotypes. In field crops, however, also other factors limit high seed yield.

INTRODUCTION

Perennial ryegrass, <u>Lolium perenne</u> L., is an important temperate grass species used for both forage and turf. Seed crops of perennial ryegrass have a high yield potential, but realized yields are low (Griffiths et al., 1973). New cultivars can only be commercialized, however, if the seed production is satisfactory.

Detailed observations are necessary for a better understanding of the problems associated with low seed production in perennial ryegrass (Marshall, 1985).

Developmental morphology of the embryo is described in cereals (Suetsugu, 1951; 1953), but in forage grasses little is known about seed development and abortion (Elgersma, 1985). Reusch (1959) observed the early growth rates of

---

[*] Present address of R. Sniezko is Zaklad Anatomii i Cytologii Roslin, Instytut Biologii UMCS, ul. Akademicka 19, 20-033 Lublin, Poland.

the endosperm and the embryo of perennial ryegrass.

In this study seed development, degeneration of ovules, and seed abortion were studied cytologically in four genotypes of perennial ryegrass. The fruit, consisting of the single seed fused to the pericarp, is botanically a caryopsis but will be referred to as a seed in this paper.

MATERIALS AND METHODS

We used four unrelated diploid genotypes of perennial ryegrass, Lolium perenne L. with contrasting seed production. In 1985, spikes were collected from spaced plants growing on sandy soil in an experimental field in Wolfheze, the Netherlands. As perennial ryegrass is a wind-pollinating, self-incompatible species, the plants were allowed to open-pollinate with compatible pollen. Starting one week after the beginning of anthesis, whole spikes were collected and fixed in FAA (900 ml 15 M ethanol, 50 ml acetic acid, 50 ml formaldehyde 1^{-1}) at one-week intervals until five weeks after anthesis. The material was stored at 4 ^{o}C in 15 M ethanol.

The spikes were dissected and spikelets were observed with a binocular microscope. Observations were made on the number of spikelets and the number of florets per spikelet. A distinction was made between normally developed florets and sterile, rudimentary florets at the tip of each spikelet, which were omitted in determinations. Florets were checked for the presence of a seed, and the size of the seed. The developmental stage was assessed per floret, and a wide range of stages (from immature embryo sacs to ripe seeds) was found. Sixty-five spikelets containing 587 florets of different developmental stages were chosen for cytological and histological investigation. They were dehydrated in ethanol and embedded in paraffin. Whole spikelets were sectioned at a thickness of 12 μm. The sections were stained with safranin and fast green and examined with a light microscope.

RESULTS AND DISCUSSION

Flowering sequence within a spike

Along the rachis of a spike of perennial ryegrass, 20 to 30 spikelets are present. Each spikelet contains 5 to 13 normally developed florets and some rudimentary florets at the top. A normal floret consists of a lemma and a palea enclosing two lodicules, one ovary with two stigmas and three anthers.

In perennial ryegrass, flowering generally starts in the central spikelets. It continues along the ear rather rapidly to the top and more slowly to the base. Within a spikelet the basal floret always flowers first, then the second and so on. Under favorable weather conditions, more than one floret within a spikelet may flower during the same day. At anthesis, the lodicules enlarge and the glumes open. Stamens and stigmas emerge simultaneously from the glumes. Pollen is released within a few minutes after anther emergence. The time between pollination and fertilization depends upon both genetical and environmental factors. In perennial ryegrass, pollen germinated within 0.5 hour, and 2 to 5 hours after pollination pollen tubes had reached the micropyle at 14 to 26 ^{o}C (Elgersma and Stephenson, unpublished results). Reusch (1959) observed divisions in endosperm nuclei 12 hours after pollination.

Normal seed development

The grass ovule is anatropous and bitegmic. The mature embryo sac is of the Polygonum type and consists of an egg cell and two small synergids located at the micropylar end, a central cell, and three prominent antipodals at the chalazal end.

In the observed spikes, the developmental stages ranged from immature florets to ripe seeds. Prior to flowering the ovary was round and had an average diameter of 0.5 mm. In florets with recently fertilized embryo sacs the ovary increased in length to 0.75 to 1 mm. The ovule enlarged as well. Within the embryo sac, the zygote was located at the micropylar end and synergids were no longer visible. We were not able to observe synergid degeneration after pollination. Shortly after pollination, the primary endosperm nucleus divided repeatedly and free-nuclear endosperm was formed (Fig. 1). Nuclei were scattered mainly near the zygote, around the antipodals, and in a thin layer of cytoplasm lining the periphery of the embryo sac. The antipodals enlarged. Due to the growth of the endosperm, the antipodals became located in a lateral position towards the adaxial side.

Some days after pollination, the ovary elongated to 1.5 mm. The zygote divided, by forming a transverse wall, into an apical cell at the chalazal end and a basal cell at the micropylar end. This pro-embryo was surrounded by a thick layer of free-nuclear endosperm. Then the endosperm cellularized. Cell formation started around the embryo and proceeded to the chalazal end of the ovary. The antipodals persisted and enlarged further (Fig. 2).

As cellularization of the endosperm continued, the first layer of endosperm cells at the chalazal end formed along the periphery of the embryo sac. The apical cell of the pro-embryo divided perpendicularly to the plane of the first division. After the three-celled stage, division took place in the two upper cells in a plane parallel to the first division, resulting in a five-celled pro-embryo. After this stage, the pro-embryo developed further into a multi-cellular, pear-shaped embryo.

The endosperm continued cellularizing and the antipodals declined. When the cellularization of the endosperm was completed and the central part of the embryo sac was filled with cells, the embryo was globular (Fig. 3). This stage was observed approximately one week after pollination, in basal florets of central spikelets.

Two weeks after pollination, embryo differentiation had begun and starch was found in the endosperm cells. The outer layer of the endosperm consisted of one layer of smaller cells without starch, which became the aleurone layer.

In ripening seed, differentiation of the provascular and meristematic tissue could be observed (Fig. 4). In a fully differentiated embryo, a scutellum, plumule, and radicle were visible.

Our results demonstrate the big lag in early embryonic versus endospermatic division. The early growth rate of the embryo is much slower than that of the endosperm which agrees with the observations of Reusch (1959).

The pattern of cell divisions in the pro-embryo of perennial ryegrass has not been described before, as far as we know, and is similar to the pattern observed in barley and rye (Suetsugu, 1951).

The rate of seed development depends on the position of the floret within the ear. Apical spikelets ripen faster than intermediate and basal spikelets, although the intermediate spikelets flower first. Within each spikelet, top florets are the last to flower but first to ripen. Basal florets of central spikelets have the longest period between flowering and ripening. Not surprisingly, Anslow (1964) found that seeds from basal florets are larger than seeds from apical florets within each spikelet.

Abnormalities

Among the normally developing florets, many florets did not set seed or had arrested seed development. We found various types of degeneration.

Non-apical rudimentary florets occurred in very low frequencies. In a low percent of the sectioned florets the embryo sac collapsed prior to anthesis (Fig. 5). In about 90 % of the unproductive florets, degeneration started within a few days after flowering, affecting the nucellar tissue. Some nucellar cells were swollen and stained more intensely with safranine than normal cells (Fig. 6). In more advanced stages of degeneration, such ovules withered completely. The contents of the cells disappeared (Fig. 7). The ovary walls then withered and collapsed in the same way. It is not clear whether

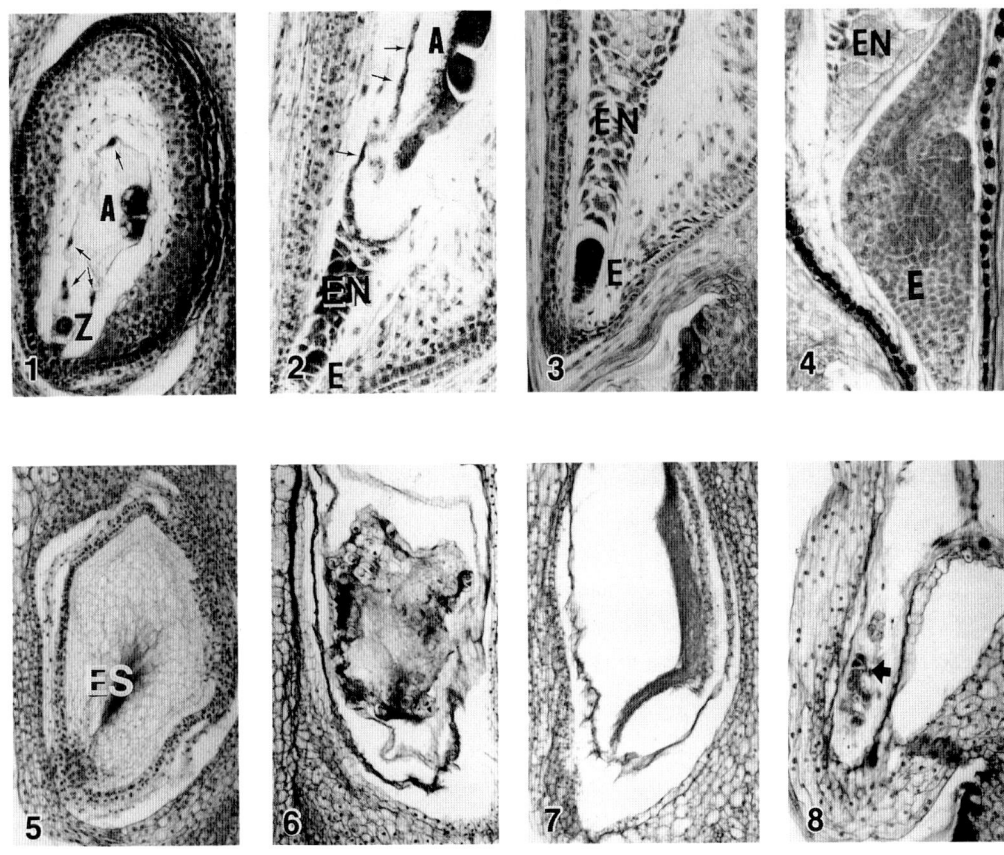

Fig. 1. Ovule with zygote (Z), free-nuclear endosperm (arrows), and two
antipodals (A) visible (x 90).

Fig. 2. Ovule with proembryo (E), as the cellularization of endosperm (EN)
starts. Free-nuclear endosperm is indicated by arrows. Two of the three
antipodals (A) are prominently visible (x 90).

Fig. 3. Part of ovule with globular embryo (E) and cellular endosperm (EN)
(x 90).

Fig. 4. Differentiated embryo (E), scutellum and plumule are visible. EN is
endosperm (x 90).

Fig. 5. Ovule with collapsed embryo sac (ES) in pre-flowering stage (x 90).

Fig. 6. Ovule degeneration shortly after pollination (x 90).

Fig. 7. Ovule degeneration as in Fig. 6, but in a more advanced stage of
degeneration (x 90).

Fig. 8. Seed abortion showing the degenerating embryo (arrow) and deteriorated
endosperm (x 90).

this was caused by lack of fertilization or of cell divisions of the primary endosperm nucleus or the zygote. Also, irregular chromosome behaviour may contribute to sterility. However, we did not check meiosis.

In a few of the observed florets, seed abortion occurred in a more advanced developmental stage (Fig. 8). The endosperm in these seeds was in the early cellular stage and the embryo was globular. In aborting seeds, the embryo shrank and stained very darkly. In the endosperm, the content of cells was hydrolyzed and whole cells deteriorated. In some cases, late abortion may have been caused by insects, which were found on top of the seeds, inside the lemma and palea. Insect damage has also been mentioned in other grass seed crops, e.g., in cocksfoot, where florets with thrips infestation were found, larvae destroying the developing seed (Johnston, 1960).

We conclude, that ovule degeneration occurring shortly after anthesis was the major cause of unproductive florets in all genotypes and contributes to low seed yield. However, in seed crops also other factors such as seed shattering, uneven ripening, diseases and harvest losses will affect yield. Seed abortion was not an important contributer to low seed yield in perennial ryegrass.

A more complete description of this work will be published elsewhere shortly.

REFERENCES

Anslow RC (1964) Seed formation in perennial ryegrass. II. Maturation of seed. J Br Grassl Soc 19:349-357.

Elgersma A (1985) Floret site utilization in grasses: definitions, breeding perspectives and methodology. J Appl Seed Prod 3:50-54.

Griffiths DJ, HM Roberts and J Lewis (1973) The seed yield potential of grasses. Welsh Plant Breeding Station, Annual Report, 1973. pp. 117-123.

Johnston MEH (1960) Investigations into seed setting in cocksfoot seed crops in New Zealand. NZJ Agric Res 3:345-357.

Marshall C (1985) Developmental and physiological aspects of seed production in herbage grasses. J Appl Seed Prod 3:43-49.

Reusch JDH (1959) Embryological studies on seed development in reciprocal crosses between Lolium perenne and Festuca pratensis. S Afr J Agric Sci 2:429-445.

Suetsugu I (1951) Developmental morphology of the embryo in barley, rye and oats. Bull Nat Inst Agr Sci, Ser D 1:83-93.

Suetsugu I (1953) Developmental morphology of the embryo in rice varieties. Bull Nat Inst Agr Sci, Ser D 4:23-52.

Embryology of Barley II: Synergids and Egg Cell, Zygote and Embryo Development

Kirsten Engell
Botanical Laboratory
University of Copenhagen
Gothersgade 140
DK-1123 Copenhagen K.

Abstract.

In the embryo sac of barley Hordeum vulgare cv. Bomi one synergid is degenerated before pollination. A statistically survey of the mutual position of the cells in the
egg apparatus shows that the egg cell in 87.3% of the investigated 158 cases is pla -
ced distally for the placental region. In a vast majority of cases the synergids are
placed proximally to the placental side. The pollen tube reaches the degenerated sy -
nergid 42-43 minutes after pollination (m.a.p.) and releases sperms and organelles e -
specially plastids with rod-shaped starch grains but also mitochondria, dictyosomes
and lipid bodies in the synergid, and 45 m.a.p. one of the sperm nucleus is fused with
the nucleus of the egg cell. The first division in the zygote occurs 22-24 hours after
pollination (h.a.p.). The embryo consists of 3 cells 34 h.a.p., 4 cells 38 h.a.p. and
7 cells 50-52 h.a.p.. Protoderm formation takes place 72-80 h.a.p..

Introduction.

The present report which is a part of a comprehensive investigation of the reproduc -
tive biology of barley, will cover mutual position of the cells in the egg apparatus,
entrance of pollen tube in embryo sac cells, fertilization and especially the
earliest embryo formation.
Earlier as well as the previous investigators have analysed the position of the cells
in the egg apparatus (Mogensen 1984), the fertilization (Cass & Jensen 1970, Bennett
1975, Cass 1981) and parts of the embryo formation (Norstog 1972) but no one have in
details described fertilization and embryo formation in relation to the time course
for these events, based on serial sections. (see Engell 1988).

Materials and methods.

Plants of Hordeum vulgare cv. Bomi were grown under controlled environmental conditions in growthrooms at Research centre, Risø, Roskilde, Denmark.
The fixations were carried out at precisely determined time intervals after emasculation and pollination (see Engell 1988). The pistils were excised, the two style bran -
ches cut off and the ovary fixed in glutaraldehyde (3% in 0.025 M Phosphate buffer at
pH 7.0). For light microscopy (LM) preparation see Engell 1988. For transmission electron microscopy (TEM) the material was postfixed in 1% OsO_4 in the same buffer type.
The whole procedure including dehydration in acetone took place in rotating vials at

384

room temperature. The material was embedded in Epoxy resin sectioned with a dia-
mond knife on a LKB Ultratome III and double-stained in uranyl acetate and lead citrate. The sections were observed in a JEOL-100 CX electron microscope at 80 KV.

Observations.

The egg apparatus.

The final maturing process of the embryo sac in Hordeum vulgare cv. Bomi is independant of the pollination. Approximately 20-24 hours before stigma is receptive for
pollen grains one of the synergids begins to degenerate.
An investigation of the mutual position of the cells in the egg apparatus will show
that both synergids are fastened against part of the embryo sac wall just inside the
micropyle, whereas the egg cell itself is fastened a little more chalazally and
there-fore always in connection with the embryo sac wall a little beyond the
micropylar region (fig.1). A statement of 158 ovules with healthy egg cell/zygote
before and after fertilization shows that the egg cell in 138 cases (87.3%) is
fastened distally i.e. away from the placental region and forward in the locule of the
fruit, only in 20 cases the egg cell is fastened against the placental region
(proximally). An enumeration of 111 ovules with all three cells of the egg apparatus
intact will show that it is extremely infrequent that all three cells are placed to
the same side (1.8%). (Table I). This statement shows that the egg cell is clearly
preferably placed distally (83.7% of 111 cases) and of these 93 cases the degenerated
synergid will be placed near to the placental region (proximally) in about 60%. Some
variation in the position of the persistent synergid can occur but the most common is
opposite to the egg cell - 61% of 93 embryo sacs with distally placed egg cells have a
proximally placed persistent synergid.

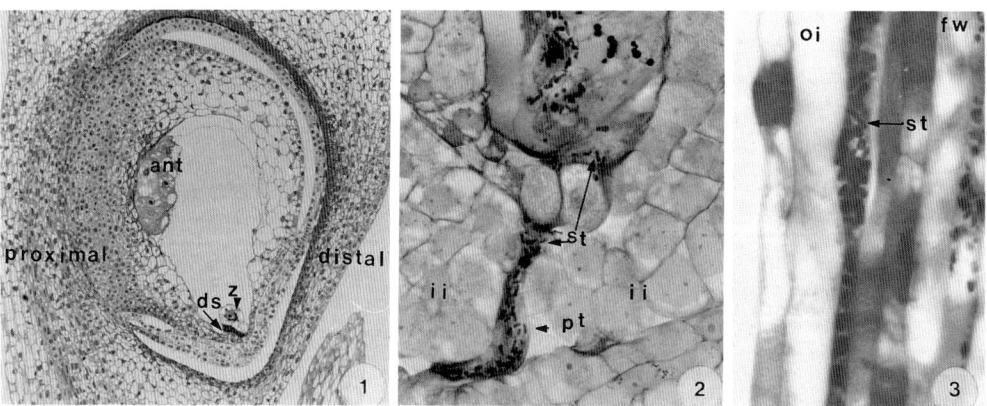

Figs 1-3. Hordeum vulgare cv. Bomi - LM - Fig.1. An anatropous, bitegmic ovule.
The egg cell is placed distally, the degenerated synergid proximally. X 90. Fig.2.
The pollen tube entrance through the micropyle. Rod-shaped starch grains inside the
tube and in the egg apparatus. Fig.3. Pollen tube with rod-shaped starch grains
between the fruit wall and the outer integument. Figs 2 & 3 X 1 350

Table I. Percentage distribution of the position of the cells in the egg apparatus.

Egg cell dist.	prox.	D.synergid dist.	prox.	P.synergid dist.	prox.	Numbers	Percentage
	x		x		x	1	0.9
	x		x	x		7	6.3
	x	x			x	8	7.2
	x	x		x		2	1.8
x			x		x	20	18.0
x			x	x		35	31.5
x		x			x	37	33.3
x		x		x		1	0.9

TABLE I

Both synergids are in about 22% of the cases (table I) placed to the same side for the micropyle and in an overwhelming majority of these cases they are proximally placed.

The pollen tube.

The pollen tube grows through the micropyle (fig.2) between the nucellar cells and advances into the degenerated synergid in the first 40 m.a.p.. Here the pollen tube contents will be disharged, and the material is fan-shaped spread in the degenerated cytoplasm (fig.4). The disharge is easily recognizable owing to the rod-shaped starch grains and they are splendid indicators for the pollen tube on its whole way through the tissue of the style, in the ovary, between the carpel and the outer integument (fig.3) and further on into the ovule.

The pollen disharge consists of numerous plastids with starch grains (figs 4,6,7), some mitochondria, dictyosomes with only few dilated cisternae, lipid bodies present as numerous of small bodies, and lots of small vesicles.

The cytoplasm of the degenerated synergid, which is forced to the cell periphery, has much more amount of large and small lipid bodies. The other organelles are difficult to recognize because of the degree of degeneration and density of the degenerated cytoplasm but a great deal of visicles with ribosomes on the membranes are recognizable.

The fertilization.

The nucleus of one of the sperm cells will penetrate the egg cell 42-43 m.a.p. - and 44 m.a.p. a small sperm nucleus could be seen just outside the large nucleus of the egg cell, but there are no contact between the outer membranes of the nuclei (fig.5). During the next minute the fusion seems to take place and 45 m.a.p. the two nuclei are in connection. 50 m.a.p. only one nucleus with two distinct compartments is observed, and this persists in the following hours. As late as $4\frac{1}{2}$ h.a.p. material from the sperm nucleus can still be recognized, but later the zygote will have an uniform nucleus with two nucleoli - a large one from the egg nucleus, a small one from the sperm nucleus. Triple fusion - the endosperm nucleus - is beyond the subject of this paper.

386

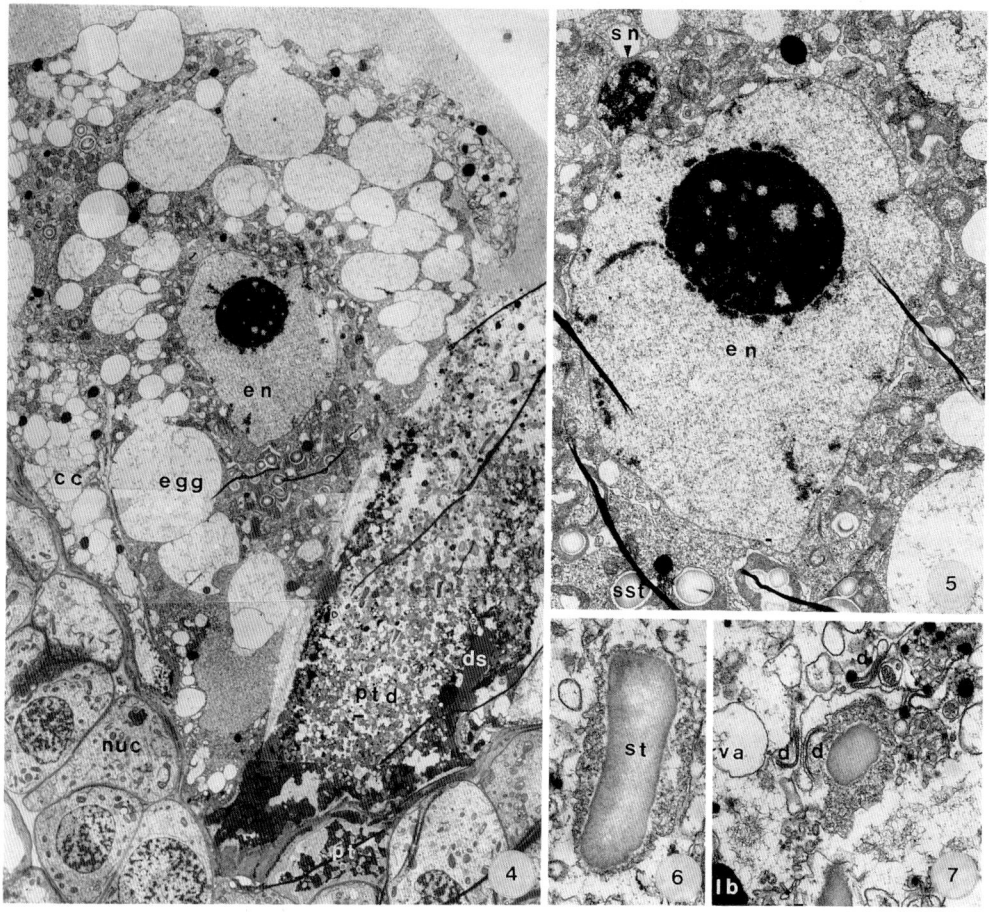

Figs 4-7. <u>Hordeum vulgare cv. Bomi.</u> - TEM - Fig.4. The egg cell and the degenerated synergid with pollen tube. The cytoplasm of the degenerated synergid is forced against the wall. 40 m.a.p. X 2 300. Fig.5. The nuclei of the egg- and the sperm cell just before contact. 44 m.a.p. Sphaerical starch grains in the egg cell. X 5 600. Fig.6. A plastid with rod-shaped starch grain from the pollen tube disharge. Fig.7. Organelles from the pollen tube disharge. Dictyosomes with dilated cisternae, lipid bodies, vacuoles and a few mitochondria. Figs 6 & 7 X 20 000.

The embryo.

The first division in the zygotic nucleus occurs about 22-24 h.a.p. (fig. 8) and af-ter this cell division the 2-celled embryo consists of a large, pyriformed cell - the hypobasal cell - against the micropyle and a short, wide cell - the epibasal cell - chalazally placed (fig.10). The next division occurs in the hypobasal cell (fig.11) 34 h.a.p. and the result will be a micropylarly placed initial cell (ci) for the suspensor, a middle placed cell (m) which is Initial for the radicle and the lower part of the hypocotyl, and most chalazally the still undivided epibasal cell. The third division happens 38 h.a.p. in the epibasal cell, which divides vertically forming two juxtaposed cells (fig.12).

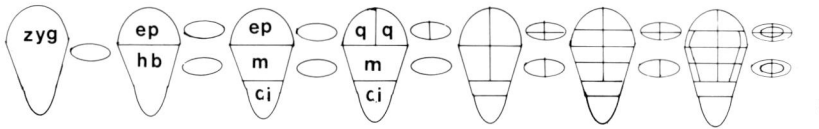

Fig.8. Schematically drawings of embryo development in Hordeum vulgare cv. Bomi.

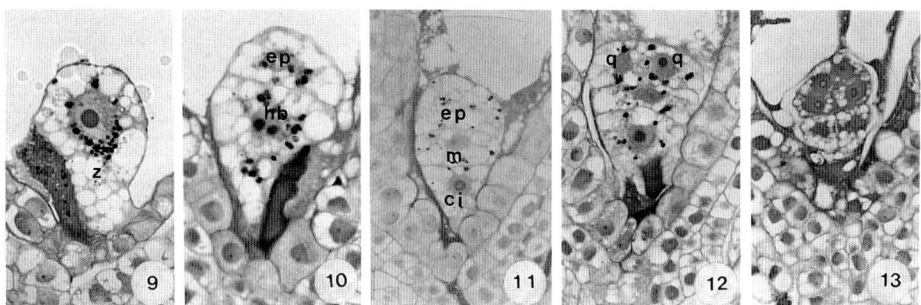

Figs 9-13. Hordeum vulgare cv. Bomi. The embryo development. Fig.9. The zygote. $4\frac{1}{2}$ h.a.p. Fig.10. 2-celled embryo. 24 h..a.p. Fig.11. 3-celled embryo. 34 h.a.p. Fig.12. 4-celled embryo. 38 h.a.p. Fig.13. 13-14 celled embryo.$80\frac{1}{2}$p.a.h. All X 360.

As soon as 4 h. later the epibasal cell divides vertically once more resulting in four juxtaposed quadrant cells (q). Now the embryo has 6 cells. About 50-52 h.a.p. the "m" cell too divides vertically and some times the suspensor initial cell (ci) divides resulting in two more or less inclined cells. The embryo consists now of 7-8 cells. Before 70 h.a.p. the four "q" cells will divide into two planes of four cells each (fig.13) and now, after a horisontal division in the two "m" cells, all the derivates from "q" and "m" divide periclinally forming the protoderm initials and the initials for the inner tissues in the embryo. 72-80 h.a.p. the embryo thus consists of about 26 cells. In the further embryo formation the numbers and divisions of the cells are more difficult to follow and therefore the events later than about 80 h.a.p. has not been followed.

Discussion.

In Hordeum vulgare there are different opinions about the time for degeneration of the synergids. This investigation shows that degeneration of one synergid in cv. Bomi is started before pollination. Cass & Jensen (1970) reported in cv. "Atsel" that one of the synergids has degenerated just after pollination but Mogensen (1984) reported in cv. Bonus that both synergids degenerated just after pollination in about 50% of the investigated cases. The results are very different on that point, and investiga-tions on synergids will therefore be continued.

The investigators, which have made studies for determining the length of the periods from pollination to fertilization or development of the embryo have cultivated different varieties of <u>Hordeum vulgare</u> and particularly under different conditions. In the older investigations (references see Engell 1988) the temperatures were often constant and the plants were cultivated under continous light. Results from such investigations are difficult to compare because the conditions are too far from natural conditions. The later investigations are more comparable (see Engell 1988) but the cultivated materials are still of different varieties and this could be the reasons for the differences in the results. Mogensen (1982) reported not the exact time for fusion between sperm- and egg nuclei but his figures show that it occurs before 2 h.45 m.a.p.. Bennett (1975) reported contact between sperm- and egg nuclei 40 m.a.p.. Cass & Jensen (1970) say nothing about the time interval between pollination and fertilization. Pope (1937) found 45 m.a.p..

Many investigators concluded that the first division of the zygote happens about 24 h.a.p. but other information about the time course are very sparsely. Norstog (1972) gives very little indication of time course.

Thus – the time course for all the events in embryo development in <u>Hordeum vulgare</u> under in vivo conditions were missed until these investigations.

Acknowledgements.

The author wish to thank Dr. C. John Jensen, Research centre, Risø, Roskilde, Denmark for his advice during this investigation, Lise Girsel for her excellent technical assistance and never-failing help, and H. Elsted Jensen for photographic work.
The research was supported by the Danish Agricultural & Veterinary Research Council (grant no.5.17.4.3.04).

Literature.

Bennett MD, Smith JB and Barclay I (1975) Early seed development in the Triticeae. -Phil. Trans. Royal Soc. London, Ser.B.272. p. 199-227.
Cass DD (1981) Structural relationships among central cell and egg apparatus cells of barley as related to transmission of male gametes.-Acta Soc. Bot. Pol.50 (1-2). p 177-180.
Cass DD and Jensen WA (1970) Fertilization in Barley. - Am.J.Bot. 57(1). p.62-70.
Engell K (1988) Embryology of barley I: Time course and analysis of controlled fertilization and early embryo formation based on serial sections. - (in prep.).
Mogensen HL (1982) Double fertilization in Barley and the cytological explanation for haploid embryo formation, embryoless caryopses, and ovule abortion. - Carlsberg Res. Commun. vol. 47. p.313-354.
Mogensen HL (1984) Quantitative observations on the pattern of synergid degeneration in barley. - Am.J.Bot. 71 p.1448-1451
Norstog K (1972) Early development of the barley embryo: Fine structure. - Am.J.Bot. 59(2). p.123-132.
Pope MN (1937) The time factor in pollen tube growth and fertilization in barley.- Journ. Agricult. Res. 54(7). p.525-529.

ABBREVIATIONS.

ant: antipodals, ci: micropylar embryo cell, d: dictyosome, ds: degenerated synergid, e: egg cell, en: egg nucleus, ep: epibasal cell, fw: fruit wall, hb: hypobasal cell, ii: inner integument, lb: lipid body, m: middle cell, nuc: nucellus, oi: outer integument, pt: pollen tube, ptd: pollen tube disharge, q: quadrant cell, sn: sperm nucleus, sst: sphaerical starch grain, st: starch grain, va: vacuole, z: zygote.

Quantitative Data on *Petunia* Embryogenesis: Mitotic Activity and Characteristics of the Cell Cycles

Jean VALLADE

Biologie du Développement et de la Différenciation chez les Végétaux.

Université de Bourgogne, Faculté des Sciences de la Vie, Bât. Mirande. 21000 DIJON. France.

The studies on the embryonal development of Angiosperms are essentialy descriptive and some quantitative data are only available at the present time. Within Dicotyledons, one must mentioned the works of Rietsema and al. (1955) on the growth of ovules and embryos of Datura stramonium, the interesting comparison between the growth of Capsella and Gossypium embryos effected by Pollock and Jensen (1964), the studies of Simoncioli (1974) and Tykarska (1980) on two Brassicaceae, respectively Diplotaxis erucoides and Brassica napus, and the work of Gray and al. (1984) on the embryo development of Daucus carota.

On the Petunia, first quantitative data on the mitotic activity and cell cycles of early embryogenesis have been given previously (Vallade, 1980). In this paper, quantitative analysis of the late proembryogenesis and organogenesis is presented particularly with regard to cell number, embryo length and differential participation of different embryonic tiers to the construction of the embryo. The cell doubling time and the main characteristics of the cell cycles are defined at the different stages of the embryonic development.

MATERIALS AND METHODS.

The embryos were extracted from ovules and young seeds, fixed in formaline (4%) for 24 hr, stained with Schiff's reagent after hydrolysis in HCl 5N at the room temperature. They were then transferred to a pectinase solution for 30 minutes which allowed a good dissociation of cells after squashing on the slide. From these dissociated embryos, quiescent and dividing nuclei were counted and the relative amount of DNA per nucleus was measured. These last measurements were realized with MPV2 Leitz cytophotometer using the two-wavelength method according to Patau (1952). The detail of the method was described previously (Vallade, 1976).

Plants of TL-h1, S-h2 and S-KR4 lines of Petunia hybrida were grown under fluorescent light with an 16 hr photoperiod in rooms at 19°C (+ 1°C) for the TL-h1 x S-h2 cross and at 21°C (+1°C) for the TL-h1 x S-kR4 cross. In all cases, the pollinations were effected either by S-h2 or by S-kR4 pollen. Before squashing, camera-lucida drawings were made to determine the length of each embryo.

RESULTS.

1- <u>Relation between the embryo length and the corresponding cell number.</u>

The length of an embryo (excluding the suspensor) and the corresponding cell number were established on thirty embryos resulting from TL-h1 x S-h2 cross and on thirty five embryos resulting from TL-h1 x S-kR4 cross. The sixty five pairs of measurements are plotted on the figure 1. The two series of measurements show positive correlations. The correlation coefficients (c.c.) are respectively 0.992 and 0.991. The slopes of the two regression lines are not very different and the regression line from the 65 pairs of measurements can be plotted (c.c. : 0.989 and regression equation : y = 0.075x + 30.8 [1]).

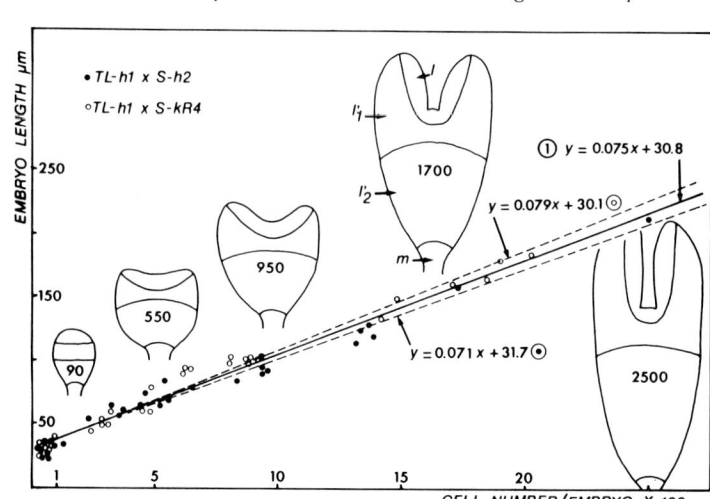

FIG. 1. The relationship between embryo length and cell number from the globular stage to the "torpedo" stage.

2- <u>Evaluation of the mean cell doubling time (MCDT) during the embryogenesis.</u>

a/ <u>Proembryogenesis.</u>

Flowers of the TL-h1 line pollinated by S-h2 line have been cut off at 24 hr intervals between the 8th and the 14th days after pollination. For each sample, the mean cell number per embryo was estimated after observation of histological sections. The growth of the proembryo is represented in figure 2 (dashed lines) with logarithms of cell number plotted against the age in days after pollination. The increase of cell number seems to be nearly linear (c.c. : 0.995). The regression equation is y = 3.55 logx + 6.46 [2]. The cell number in proembryos increases exponentially with time. The slope of the line represents the relative growth rate ; the computation shows that in our experimental conditions, the mean cell doubling time is about 26 hr during this early period of the embryogenesis.

b/ <u>Organogenesis.</u>

In the TL-h1 x S-h2 cross, the growth of embryos is related in figure 2 with the logarithm of the length plotted against the age in days after pollination. The pairs of measurements show a linear correlation (c.c. : 0.986) and the regression equation is $y = 7.3 \log x + 1.62$ [3]. From [3] and [1] equations, the MCDT was calculated (table 1A) : it increases from 23 hr (between the 14th and 15th days) to 56 hr (between the 19th and 20th days after pollination).

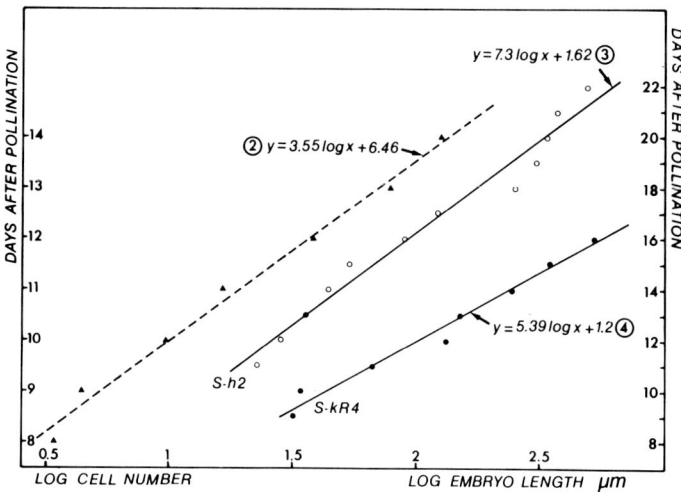

FIG. 2. The relationship between log cell number (proembryogenesis) or log embryo length (organogenesis) and days after pollination.

TABLE 1. Growth of embryos : evaluations of length, cell number and MCDT in TL-h1 x S-h2 (A) and Tl-h1 x Sk-R4 (B) crosses.

A				B			
days after pol.	embryo length (μm)	cell number	MCDT (hr)	days after pol.	embryo length (μm)	cell number	MCDT (hr)
14	49.6	252		10	42.9	162	
15	68.1	513	23.2	11	65.8	467	13.4
16	93.3	868	34.7	12	100.9	896	24.4
17	127.9	1355	42.8	13	154.6	1576	31.6
18	175.3	2023	48.7	14	237	2619	36.3
19	240.3	2938	53.1	15	363.3	4217	39.3
20	329.5	4194	56.1	16	557		

In TL-h1 x S-kR4 cross, the growth is indicated in figure 2. The length increases exponentially with time also (c.c. : 0.990) and the regression equation is $y = 5.31 \log x + 1.2$ [4]. In this experience the embryo growth is faster than the previous one. From [4] and [1] equations, the MCDT was calculated (table 1B) : it increases progressively from 13 hr (between the 10th and 11th days) to more than 39hr (between the 14th and 15th days).

3- Nuclear DNA levels in the embyo cells at different stages of the embryo-genesis (TL-h1 x S-h2 cross).

The measurements were taken on the one hand on 10 globular proembryos with 20 to 50 cells (sample 1), and on the other hand on 7 embryos (samples 2 to 8) at different stages of the embryonic organogenesis. The cytophotometric measurements of relative DNA content of half-telophases and prophases or anaphases enable one to determine respectively the 2C and 4C reference levels. Moreover, the relative DNA content of quiescent nuclei show the values 2C and 4C are generally distributed in two distinct populations. For each sample, the dispersions of 2C and 4C values are estimated with a probability of 95% according to the method of Landré (1973). It is still possible to evaluate the respective percentages of nuclei wih 2C and 4C DNA levels and, by deduction, the percentage of nuclei with 2C-4C level. The results are indicated in table 2.

TABLE 2. Percentages of nuclei with 2C, 2C-4C and 4C DNA levels at different stages of development ; mitotic indices corresponding.

embryos (samples)	cell number	number of nuclei	2C	2C-4C	4C	mitotic index
1	(10 embryos from 20 to 50 cells)	127	58.3	14.2	27.6	7.7
2	99	57	63.2	21.0	15.8	6.1
3	235	103	70.1	7.8	21.4	4.3
4	436	101	61.4	18.8	19.8	3.2
5	924	226	65.0	11.1	23.9	5.0
6	1373	220	64.5	14.1	21.4	3.1
7	1725	219	68.9	8.2	22.8	3.5
8	2504	176	73.3	13.6	13.1	3.9

4- Respective participation of the different proembryonic tiers at the building of the embryo.

TABLE 3. Proportions of cell number per tier per embryo at different stages of embryogenesis.

total cell number (T)	cell number/ longitudinal section (t)	T/t	percentage of cell number per tier			
			l	l'1	l'2	m
90	21	4.3	28.6	28.6	28.6	14.2
230	36	6.4	28.2	32.4	31.0	8.4
600	90	6.7	20.2	36.5	33.2	10.4
1150	175	6.6	17.7	32.3	40.8	9.2
1800	264	6.8	15.2	31.1	46.6	7.2

The previous results exhibited means calculated for the whole embryos. But, an unequal distribution of mitotic activity, according to the different embryonic tiers, occurred early on the embryogenesis. Later divisions are noticeably few in the mid-central portion of the flattened globular to young heart stage (Vallade, 1976). A quantitative evaluation of the differential growth is proposed here. The exact cell lines being known, it is possible to calculate the relative proportion of cells originating from tiers l, l'1, l'2 and m, on each axial section (table 3).

DISCUSSION AND CONCLUSIONS.

1- The number of cells per embryo increased lineary as a function of increase in embryo length, as shown in figure 1. The same results were obtained with TL-h1 x S-h2 and TL-h1 x S-kR4 crosses. This relation presents a practical interest : it is enough to measure the length of an embryo to determine its cell number.

2- During the proembryogenesis of Petunia, like the cotton (Pollock and Jensen, 1964), the cell number increases exponentially with time. During the organogenesis, there is a linear relation between the logarithm of the embryo length and the time. This constatation is in agreement with the indications given by Simoncioli (1974) with regard to the embryo growth of Diplotaxis erucoides.

3- The mean cell doubling time (MCDT) corresponds with the mean duration of the cell cycle. The MCDT is evaluated at about 26 hr in the proembryogenesis period, which agrees with the results of Pollock and Jensen (1964) in cotton (MCDT = 20-22 hr). During the organogenesis period this duration increases progressively from 23-26 hr to 56 hr. When pollen provided from another line (S-kR4 instead S-h2) and when the temperature is slightly higher (about 2°C), the rate of the growth increases : the MCDT is only about 13 hr in a globular embryo and rises to 39 hr for a "torpedo" stage. These results are close to the works of Forster and Dale (1983) on Seccale cereale embryos (15,7 hr to 22,7 hr) and Gray and al. data (1984) on Daucus carota embryos (42 hr). The common major fact in our two experiences, is the progressive extension of the duration of the mean cell cycle during the second period of the embryogenesis.

4- The measurements of relative DNA contents in the embryos of different sizes, show a slight tendency to an increase of percentage of 2C nuclei balanced by a decrease of the 4C nuclei (table 3). The pointed differences are especially perceptibly between the young globular embryos and the 2500 cells embryo (respectively 58 and 73% of 2C nuclei and 28 and 13% of 4C nuclei). For the other embryos, the percentage change little and the means obtained are about 66% of 2C nuclei and 21% of 4C nuclei. The percentage of nuclei in S-phase seems relatively constant (13%) for all embryos. The mitotic index (table 3) is higher during proembryogenesis (6 to 8%) than during organogenesis (3 to 5%) which is in agreement with the evolution of the duration of cell cycles in terms of embryonal development.

From these data, taking the durations of cell cycles evaluated precedently into account, it can be concluded that the extension of the cell cycle results especially from the increase of the G1 phase (table 4).

TABLE 4. Evaluation (in hours) of durations of different phases of the mean cell cycle during the embryogenesis (T : Total cell cycle duration).

embryos (cell number)	T	G1	S	G2	M
20-50	23	12	3	6	2
800-1300	40	25	5	8	2
2500	56	40	7	7	2

5- The above data illustrate the evolution of the mean behaviour of the embryonic cells but ignore the differential growth which takes place early inside the embryo (Vallade, 1980). In the leptembryate species such as Petunia, it is possible to determine, on histological sections, the exact cell lines from the first formed cells ; thus, the rate of participation of each proembryonic tier can be defined : the largest part of the embryo results in the proliferative activity of tiers l, l'1 and l'2 ; the cotyledons and the precaulinar zone originating from the two former tiers and the radicle from the latter (fig. 1). The tiers l'1 and l'2, issuing from l', constitute the main morphogen areas.

These data agree with our concept of the existence of different kinds of growth processes during the embryogenesis (self acting areas in l'1 and l'2 ; areas that are only induced to be active, for example in l and protoderm cells) as it was developed elsewhere (Vallade, 1976, 1980).

References.

Forster BP, Dale JE (1983) A comparative study of early seed development in genotypes of barley and rye. Ann Bot 52 : 603-612

Gray D, Ward JA, Steckel JRA (1984) Endosperm and embryo development in Daucus carota L. J Expt Bot 35 : 459-465

Landré P (1973) Etude cytophotométrique des types cellulaires d'un épiderme adulte II. Pluralité des valeurs en DNA nucléaire des cellules épidermiques et des cellules stomatiques de Solanum nigrum. C R Acad Sci : 2673-2676

Patau K (1952) Absorption microphoyometry of irregular-shaped objects. Chromosoma 5 : 341-362

Pollock EG, Jensen WA (1964) Cell development during early embryogenesis in Capsella and Gossypium. Amer J Bot 51 : 951-921

Rietsema J, Blondel B, Satina S, Blakeslee AF (1955) Studies on ovule and embryo growth in Datura I. A growyh analysis. Amer J Bot 42 : 449-455

Simoncioli C (1974) Ultrastructural characteristics of Diplotaxis erucoides (L.) DC suspensor. Giorn Bot Ital 108 : 175-189

Tykarska T (1980) Rape embryogenesis III. Embryo development in time. Acta Soc Bot Poloniae 49 : 369-385

Vallade J (1976) Contribution à la connaissance des problèmes fondamentaux de l'embryogenèse chez les Angiospermes. Recherches sur le Petunia hybrida Hort. et quelques autres espèces. Thèse Dijon

Vallade J (1980) Données cytologiques sur la proembryogenèse du Pétunia ; intérêt pour une interprétation morphologique du développement. Bull Soc bot Fr 127 : 19-37.

Chalaza-Micropyle Element Concentration Gradients in the Endosperm Tissue During Embryogenesis

M.Ryczkowski and W.Reczyński
Institute of Molecular Biology
Jagiellonian University
Al. Mickiewicza 3
31-120 Cracow
Poland

Abstract

Ovules of Haemanthus Katharinae Bak. were used as experimental material. Quantitative determinations of concentration of K,Na,Ca,Mg,Fe, Mn,Zn and Cu in the micropylar and chalazal parts of the endosperm tissue during exponential phase of embryo growth were made using the atomic absorption spectrophotometer (Perkin-Elmer, Model 503). It has been established that: a. there are distinct chalaza-micropyle concentration (μg/g fr wt) gradients of elements (K,Ca,Mg,Fe,Mn,Zn,Cu) in the endosperm tissue, b. the elongation of the embryo proceeds from the micropylar to the chalazal end of the ovule i.e. in the direction opposite to that of the chalaza-micropyle element concentration gradients in the endosperm tissue, c. generally concentrations of all determined elements increase in the endosperm tissue during embryo growth, d. it is suggested that during proembryo stage there are also some chalaza-micropyle element concentration gradients in the endosperm tissue.

Key words: Haemanthus Katharinae Bak., element concentration gradients, endosperm tissue, embryogenesis.

Introduction

This paper contains results of investigations which are a continuation of two problems previously started, namely: 1. on concentration of elements in the developing ovule (central vacuole sap, embryo; Ryczkowski et al.,1986a, Ryczkowski and Reczyński,1986b), and 2. on physico--biochemical and physiological gradients in the developing ovule (coat, endosperm tissue, embryo; Ryczkowski,1967, 1980) during embryogenesis. It should be stressed that the problem of physiological and biochemical gradients in animal zygote and their influence on embryogenesis is well elaborated (Brachet 1964). The determined elements - K,Ca,Na,Mg,Fe,Mn,Zn and Cu in the endosperm tissue play a decisive function in maintenance of the ultrastructure and physiological activity of the cell cytoplasm (Wyn Jones and Pollard,1983; Sandmann and Boger,1983). This paper contains the results of element determination in the micropylar (Mpen) and chalazal (Chpen) part (hal-

ves) of the endosperm tissue during the exponential phase of the embryo growth. These determinations were carried out with the aim to establish wheather there exist or not chalaza-micropyle element concentration gradients in this tissue during embryogenesis.

Material and method

Ovules of Haemanthus Katharinae Bak. cultivated in green house collected between 7 and 8 a.m. were used as experimental material. The
number of days counted from the day the perianth wilted to the day of
sampling, and dimensions of embryos (Ryczkowski,1962) were adopted
criteria of age of ovules and endosperm tissue. The technique of endosperm preparation its division into the chalazal (Chpen) and micropylar (Mpen) part (halves) was given in paper Ryczkowski,1967. Both
parts were weight (separately; fresh weight - fr wt), and dried to a
constant weight in a vacuum dryer at $100^{\circ}C \pm 2^{\circ}C$. Then the tissue
was powdered in an agate mortar. The procedure of the powdered sample digestion was given in previous papers (Ryczkowski and Reczyn-
ski, 1986b). A quantitative analysis of element concentration was
made by atomic absorption spectrometry, using the Perkin Elmer spectrophotometer Model 503. Concentration of K,Ca,Na,Mg and Zn was
determined in airacetylene flame in standard conditions, while of Fe,
Mn and Cu in an electrothermic graphite atomizer HGA-74.

Results

Element concentration (μg/g fr wt) was determined in the micropylar
(Mpen) and chalazal (Chpen) part of the endosperm tissue during the
exponential phase of embryo growth (embryo dimensions - 1,88x1,04 -
8,91x2,05 mm; Table 1,2). Concentration of K in the Mpen was within
the limits 1978-3030 μg/g fr wt, and was lower than this value found
in the Chpen - 2456-3267 μg/g fr wt (Table 1). Ca concentration in
the Mpen was within the range 26,2-61,7 μg/g fr wt, and in the Chpen
this value was 16,2-81,1 μg/g fr wt. Concentration of Ca was higher
in the Chpen than in the Mpen with the exception of two first determinations (embryo dimensions - 1,88x1,04 - 3,64x1,50 mm; Table 1).
Concentration of Na in the Mpen and Chpen was within the range 20,8
-51,7 μg/g fr wt. In this case a chalaza-micropyle concentration gradient was not found. Mg concentration in the Mpen was within the
limits 80,5-121,6 μg/g fr wt, and was markedly lower than this value

Table 1. <u>Haemanthus Katharinae</u>. Elements concentration ($\mu g/g$ of fresh weight) in the micropylar (Mpen) and chalazal (Chpen) parts of the endosperm tissue during exponential phase of embryo growth. Age - days counted from the day the perianth wilted to the day of sampling.

Age in days	Dimensions of embryos mm	K		Ca		Na		Mg	
		Mpen	Chpen	Mpen	Chpen	Mpen	Chpen	Mpen	Chpen
60	1.88x1.04	1978 <	2456	26.2 >	21.0	32.2 >	20.8	80.5 <	160.0
67	3.64x1.50	2439 <	2623	29.3 >	16.2	25.8 >	22.2	85.3 <	165.9
74	4.98x1.59	2424 <	2609	41.7 <	50.6	25.1 <	29.0	80.6 <	190.4
82	6.12x1.76	3030 <	3267	50.3 <	60.9	29.2 <	33.6	90.2 <	229.4
88	8.29x2.05	2837 <	3105	49.4 <	58.1	33.3 <	51.7	107.0 <	243.6
96	7.84x1.98	2739 <	3026	57.5 <	64.0	27.5 >	26.6	114.5 <	257.2
105	8.19x2.00	3006 <	3140	59.3 <	63.1	32.2 >	23.6	119.5 <	247.2
114	8.91x2.05	2956 <	3116	61.7 <	89.1	31.7 <	35.7	121.6 <	301.6
Sr %		1.1		2.9		8.5		5.0	

Table 2. Haemanthus Katharinae. Elements concentration (μg/g of fresh weight) in the micro-pylar (Mpen) and chalazal (Chpen) parts of the endosperm tissue during exponential phase of embryo growth. Age – days counted from the day the perianth wilted to the day of sampling.

Age in days	Dimensions of embryos mm	Fe Mpen	Fe Chpen	Mn Mpen	Mn Chpen	Zn Mpen	Zn Chpen	Cu Mpen	Cu Chpen
60	1.88x1.04	2.86	3.63	0.23	0.51	1.76	4.54	0.95	1.25
67	3.64x1.50	2.60	4.33	0.20	0.49	2.06	4.74	0.89	1.30
74	4.98x1.59	3.24	4.43	0.24	0.70	1.96	6.22	0.99	1.51
82	6.12x1.76	2.85	4.82	0.25	0.92	2.12	5.54	1.16	1.22
88	8.29x2.05	3.11	4.99	0.26	1.15	2.26	5.39	1.07	1.17
96	7.84x1.98	3.45	5.97	0.28	0.95	3.03	6.45	2.09	1.81
105	8.19x2.30	3.83	4.79	0.26	1.03	3.15	6.74	1.36	1.91
114	8.91x2.05	4.21	5.59	0.29	1.06	3.16	8.03	1.62	2.12
Sr %		14.3		4.8		8.1		6.4	

established in the Chpen - 160,0-301,6 μg/g fr wt (Table 1). Concentration of Fe in the Mpen was within the range 2,60-4,21 μg/g fr wt, and was lower than this value found in Chpen - 3,65-5,59 μg/g fr wt (Table 2). Mn concentration in the Mpen was within the range 0,20-0,29 μg/g fr wt, and was lower than this value found in Chpen 0,49-1,15 μg/g fr wt (Table 2). Zn concentration in the Mpen was within the limits 1,76-3,16 μg/g fr wt, and was evidently lower than this value determined in the Chpen - 4,54-8,03 μg/g fr wt (Table 2). Concentration of Cu in the Mpen was within the range 0,89-2,09 μg/g fr wt, and was generally lower than this value found in the Chpen - 1,17-2,12 μg/g fr wt (Table 2).

Discussion

Basing upon the performed determinations of particular elements in the micropylar (Mpen) and chalazal (Chpen) part of the endosperm tissue during the exponential phase of embryo growth the following facts were established: a. evident chalaza-micropyle element concentration gradients of the following elements: K,Ca,Mg,Fe,Mn,Zn and Cu, b. the elongation of the embryo proceeded from the micropylar to chalazal end of the ovule i.e. in the direction opposite to the chalaza--micropyle element concentration gradients in the endosperm tissue, c. generally concentration of elements determined in the endosperm tissue increased during the exponential phase of embryo growth. Results of the chalaza-micropyle element concentration gradients in the endosperm tissue are in good agreement with Wardlaw´s (1955) hypothesis based on morphological-anatomical observations that there is a con - centration gradient of nutrient compounds in the developing ovule particularly during the embryo intensive growth. The obtained results are also in good agreement with the earlier author´s observations (Ryczkowski,1960) on anathomy and structure of Haemanthus Katharinae ovule. It has been established that the tissue of nucellus and nuclear (and cellular) endosperm tissue are better developped in the chalazal than in the micropylar part of the ovule i.e. in the region which is better supplied with nutrient compounds (Ryczkowski,1960). The results on chalaza-micropyle element concentration gradients are consistent with quantitative data on chalaza-micropyle osmotic value, non reducing sugars and respiration rate gradients (Ryczkowski,1967,1980) in this tissue. The occurence of distinct chalaza-micropyle element concentration gradients in the endosperm tissue at the begining of embryo proper differentation (embryo dimensions – 1,88x1,04 mm; Table 1,2) suggests that these gradients exist in the endosperm tissue also during the proembryo stage. The present results and previous author´s data (Ryczkowski,1967,1980) on the different kinds of chalaza micropyle gradients in the developing ovule (coat, embryo, endosperm tissue) during embryogenesis show some similarities with the results concerning physiological and biochemical gradients found in the developing animal zygote (Brachet,1964). It seems that these similarities suggest a possibility of a common basic mechanism controlling embryogenesis in animal and plant kingdoms.

References

Brachet J (1964) The Biochemistry of development (in Polish). PWN, Warszawa

Ryczkowski M (1960) Changes of the osmotic value during the development of the ovule. Planta 55: 343–356

Ryczkowski M (1962) Changes in the osmotic value of the sap from embryos, the central vacuole and the cellular endosperm during development of the ovules. Bull Acad Polon Sci Ser sci biol 10: 375–380

Ryczkowski M (1967) Osmotic gradients in the developing ovule and embryo. Acta Soc Bot Pol 36: 627–638

Ryczkowski M (1980) Physico-biochemical and physiological gradients in the ovule during embryogenesis. Bull Soc bot Fr 127: (3/4) 51–58

Ryczkowski M, Kowalska A, Reczyński W (1986a) Element concentrations in the sap surrounding the developing embryo of Aesculus hybrida. J Plant Physiol 122: 467–472

Ryczkowski M, Reczyński W (1986b) Concentration of elements in the central vacuole sap and in the developing embryo of Aesculus glabra Willd. Acta Bot Neerl 35: 217–222

Sandmann G, Boger P (1983) The enzymological function of heavy metals and their oole in electron transfer processes of plants. In: Lauchli A, Bieleski R L (ed) Encyclopedia of plant physiology, new series. Springer-Verlag, Berlin Heidelberg New York Toronto, vol XV, p 563–596

Wardlaw C W (1955) Embryogenesis in plants. London New York

Wyn Jones RG, Pollard A (1983) Proteins, enzymes and inorganic ions. In: Lauchli A, Bieleski R L (ed) Encyclopedia of plant physiology, new series. Springer-Verlag, Berlin Heidelberg New York Toronto, vol XV, p 528–562

Immunocytochemical Visualization of the Microtubular Cytoskeleton in Developing Kernels of Wheat *(Triticum aestivum)*

A.A.M. van Lammeren

Department of Plant Cytology and Morphology

Agricultural University Wageningen

Arboretumlaan 4

6703 BD Wageningen

The Netherlands

INTRODUCTION

Morphogenesis is based upon processes such as cell differentiation and tissue interaction. In a series of studies concerning the function of the microtubular (MT) cytoskeleton during plant cell development various functions of microtubules (MTs) were established in developing pollen (Van Lammeren et al. 1985), megaspores and embryo sacs of <u>Gasteria</u> (Willemse and Van Lammeren 1988), in developing somatic embryos of <u>Daucus</u> (Van Lammeren et al. 1987) and in developing endosperm of <u>Zea</u> (Van Lammeren & Kieft 1987). Because of the impermeability of cell membranes for antibodies, used for the immunological detection of MTs, tissues were first embedded, then sectioned and after the removal of the embedding medium processed for immuno labelling.

Caryopsis development in wheat offers the opportunity to investigate the MT cytoskeleton in coenocytic and cellular conditions. It was asked what roles MTs would play during tissue differentiation. Semithin sectioning and the labelling of sections from which the embedding medium was removed proved a successfull method.

MATERIALS AND METHODS

Wheat plants (<u>Triticum</u> <u>aestivum</u>) were grown under greenhouse conditions at a temperature of 16-22°C. Ovaries were excised at various stages of development up to 22 days after pollination. The average caryopsis length was determined for each stage. The fixation in 4% paraformaldehyde, the dehydration and embedment in either London Resin White or polyethylene glycol, the sectioning and staining or indirect immuno labelling for tubulin with fluoresceine isothiocyanate (FITC) were as described previously (Van Lammeren 1988).

402

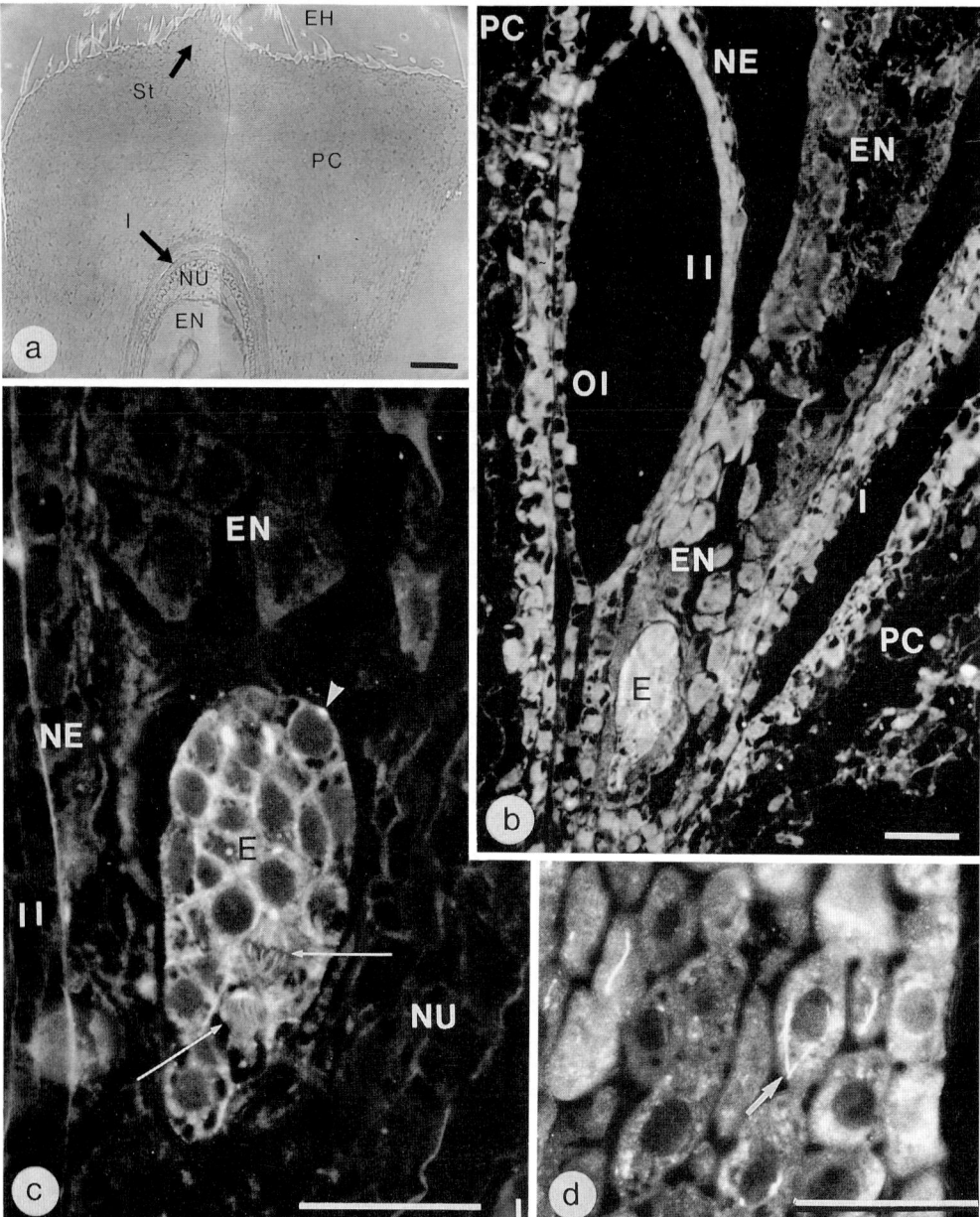

Fig. 1. a. Phase contrast micrograph of the upper side of the developing ovary and ovule with the endosperm at coenocytic stage. Silk attachment is at the upper side. b–d. Epifluorescence micrographs of FITC labelled microtubular cytoskeletons in sections of wheat embryos At 7–8 (a,b) and 20 (c) days after pollination. b. Low magnification view of embryo and cellular endosperm. c. Higher magnification of embryo surrounded by endosperm. d. Detail of embryo suspensor with thick bundles of microtubules (arrow) in the central cytoplasm. Note the presence of microtubules in spindles, (c, arrows), preprophase bands (c, arrows head), the central cytoplasm (d, arrow) and in the cortical cytoplasm. The bars represent 50 µm.

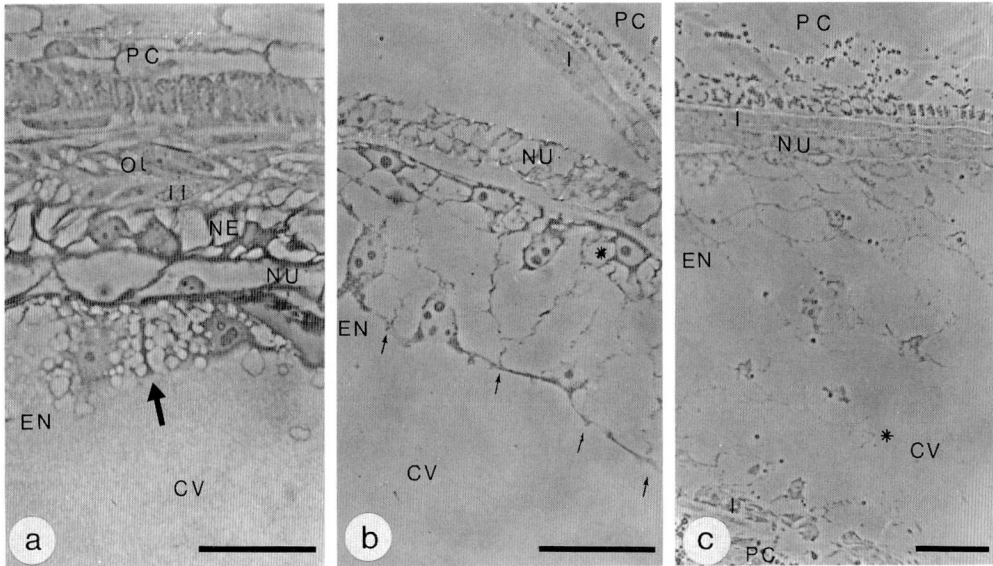

Fig. 2. Phase contrast micrographs of wheat endosperm surrounded by tissues of the ovule and ovary. a. Coenocytic endosperm surrounded by nucellus, integuments and pericarp. First sign of phragmoplast formation is indicated by an arrow. b. Cellularization of the endosperm by the formation of cell cylinders (arrows) and a first periphere layer of mononucleate endosperm cells. c. Progressing cellularization of the endosperm. The central area is still coenocytic (*). The bar represents 50 μm.

RESULTS

Initially the caryopsis of wheat consists of a well developed pericarp which surrounds the ovule. Figure 1a shows a caryopsis containing an ovule with coenocytic endosperm. Epidermal hairs characterize the distal side of the ovary where the styles are attached. The ovule consists of two integuments, the nucellus, the endosperm and the embryo which are shown in more detail in Fig. 1b. Here the PEG-embedded sections were labelled for MTs and the fluorescence indicates the presence of MT configurations in the endocarpium, the integuments, the endosperm and especially in the embryo.

At higher magnification MTs or bundles of MTs were observed in spindles (Fig.1c, arrows) preprophase bands (Fig. 1c arrow head) and in the cortical cytoplasm. At late stages of embryo development bundles of MTs running through the cytoplasm were seen in the suspensor and coleorrhizal regions of the em-

Abbreviations. AL = aleurone layer; CE = central endosperm; CU = cuticle; CV = central vacuole; E = embryo; EH = epidermal hairs; EN = endosperm; I = integument; IE = inner epidermis; II = inner integument; NE = nucellus epidermis; NU = nucellus; OE = outer epidermis; OI = outer integument; PC = pericarp; St = style.

404

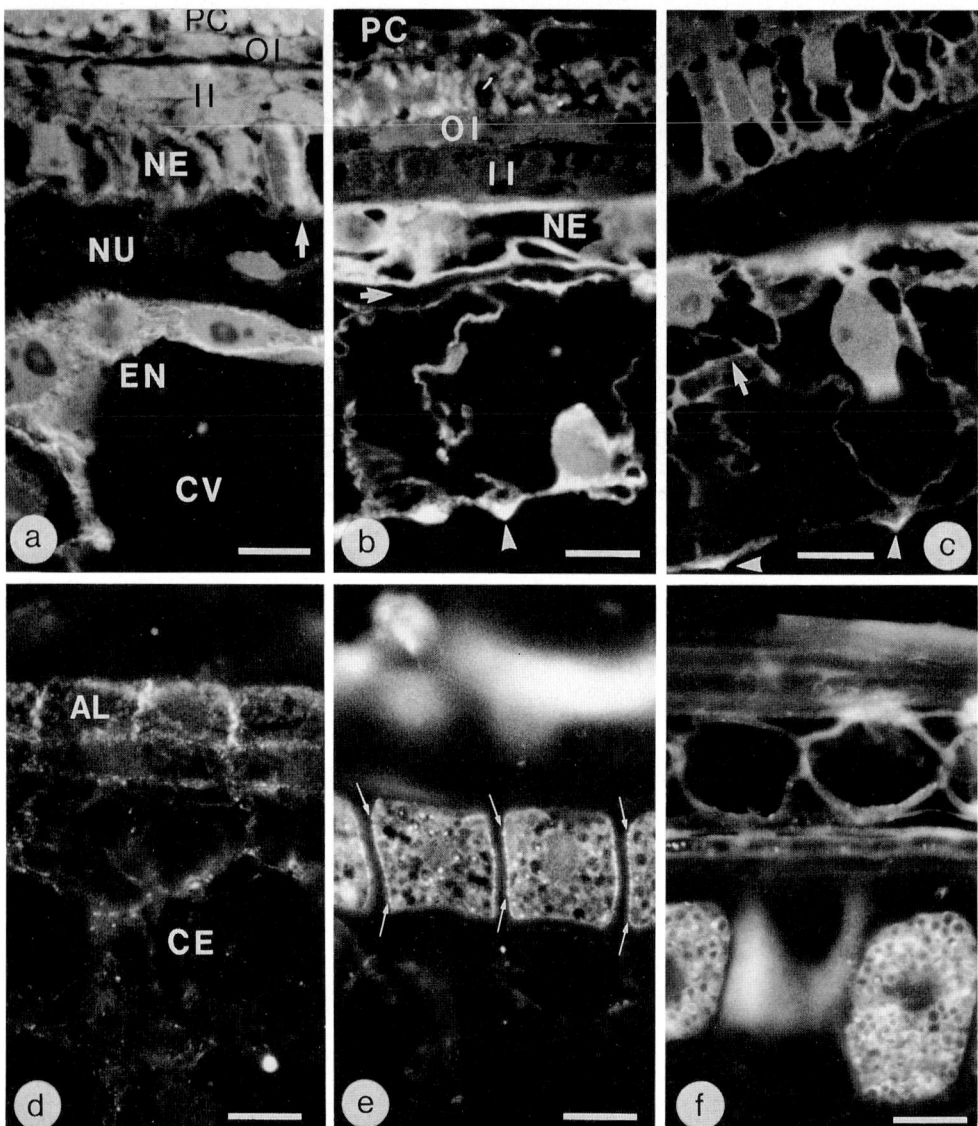

Fig. 3. Epifluorescence micrographs of FITC labeled microtubules in sections of wheat ovaries. a. Endosperm at the coenocytic stage. Note the microtubules radiating from the endosperm nuclei. The nucellus epidermis shows MT arrays in a phragmoplast (arrow). The integuments and the endocarp epidermis exhibit intensive fluorescence. b. Formation of cell cylinders in the still coenocytic endosperm. Note that only one layer of subepidermal nucellus cells remained (arrow). The growing cell walls of the cylinders are bordered by bundles of MTs which are arranged in a phragmoplast like fashion (arrow heads). c. First stage of cellular endosperm. Note the cell wall between the periphere cell and the central coenocyte (arrow). The cell cylinders of the coenocyte still exhibit the microtubular arrays at the growing ends of the cell walls (arrow heads). d. Cellular endosperm with the formation of an epidermal cell layer which will differentiate into the aleurone layer. Especially the outer endosperm cells exhibit cortical microtubules along their periclinal cell walls. e. Further differentiation of the aleurone layer. Note the bundles of MTs along the anticlinal cell walls of

bryos. Sometimes these bundles twisted around each other, sometimes they divided into two individual bundles running from one side of the cell towards the opposite side (Fig. 1d).

The development of the endosperm, its enlargement and cellularization coincide with the degeneration of the nucellus cells. At the nuclear stage of endosperm development nucellus cells are most numerous at the side of style attachment (Fig. 1a). At the lateral sides of the ovule only 2-3 cell layers of nucellus cells were found. The nucellus epidermis persisted during the early phases of endosperm development whereas the inner nucellus cells disappeared after the coenocytic stage of the endosperm. A phase-contrast view of endosperm cellularization and the tissues surrounding the endosperm is given in Fig. 2.

The cellularization of the endosperm and the arrangement of the MTs during that process are shown in Figs. 2 and 3 respectively. At the coenocytic stage the nuclei align the cell membrane of the former central cell. Microtubules radiate from the nuclear envelopes and often meet MT arrays from the adjacent nuclei (Fig. 3a). As soon as small vacuoles are formed in the layer of cytoplasm it was observed that many MTs ran just alongside the tonoplast of the central vacuole. First signs of cellularization are shown in Fig. 2a where phragmoplasts developed after karyokinesis. A cell plate is formed which meets the original cell wall but which grows continuously towards the centre of the central vacuole (Fig. 2b). The growing ends are indicated by arrows (Fig. 2b) and contain numerous MTs which are arranged in a phragmoplast fashion (Fig. 3b, arrow head). Thus cell cylinders were formed containing one nucleus each. The nuclei still exhibited radiating MT arrays by which they were kept near to the tonoplast of the central vacuole (Fig. 3b). Mononucleate cells appeared at about 7 days after pollination when the nuclei of the cell cylinders divided and periclinal cell plates were formed (Fig. 3c). Repeated division resulted in rows of cells (Fig. 3d) of which the outer ones differentiated into aleurone cells. These cells were characterized by MT arrays near to the anticlinal cell walls. Especially during further aleurone development the bundles of cortical MTs were arranged in a hoop-like fashion along the cell membranes and in periclinal planes (Fig. 3e). The central endosperm exhibited less cortical MTs. In full grown, yellowing wheat grains diffuse fluorescence was still observed in living cells of the pericarp and in aleurone cells (Fig. 3f) but MTs were not seen anymore.

---

the aleurone cells (arrows). f. Structure of pericarp, seed coat and aleurone at 22 days after pollination. There is still diffuse fluorescence in some pericarp and aleurone cells, but microtubules were not found anymore. The bars represent 50 μm.

DISCUSSION

Microtubules act upon cytomorphogenesis, organelle movement, karyokinesis and cytokinesis. In the present paper the function of MTs is related to tissue differentiation in embryo, endosperm and surrounding sporophytic tissues. It was found that especially in the endosperm the MT cytoskeleton functioned at various ways. First the MTs acted upon in the positioning of the endosperm nuclei in the coenocytic stage, a phenomenon which was also mentioned earlier for coenocytic dyads and tetrads during pollen development (Van Lammeren et al. 1985). Then they formed phragmoplasts at the cellularization stage in a way already mentioned by Bajer and Molè-Bajer (1986). After cellularization phase the MTs in the cortical cytoplasm influence the direction of cell enlargement. This phenomenon was studied in various somatic plant tissues (see a.o. Lang Selker and Green 1984) but also during endosperm development (Van Lammeren and Kieft(1987).

Microtubular arrays in nucellus epidermis, integuments and endocarp were often found in the cortical cytoplasm in an orientation perpendicular to the direction of net cell enlargement. Especially these MTs functioned in cytomorphogenesis. Microtubular arrays such as preprophase bands, in spindles and phragmoplasts enabled cell division. All these phenomena were observed in developing embryos as well. The bundles sometimes encaged nuclei, often ran from one side of the cell towards the other, were not found in all cells, however, occurred regularly and will be an item of further investigation because their function is not yet elucidated.

REFERENCES

Bajer AS, Molè-Bajer J (1986) Reorganization of microtubules in endosperm cells and cell fragments of the higher plant Heamanthus in vivo. J. Cell Biol. 102: 263-281

Lang Selker JM, Green PB (1984) Organogenesis in Graptopetalum paraguayense E. Walter: shifts in orientation of cortical microtubule arrays are associated with periclinal divisions. Planta 160: 289-297

Van Lammeren AAM (1988) Structure and function of the microtubular cytoskeleton during endosperm development in wheat. Protoplasma (in press).

Van Lammeren AAM, Keijzer CJ, Willemse MTM, Kieft H (1985) Structure and function of the microtubular cytoskeleton during pollen development in Gasteria verrucosa (Mill.) H. Duval. Planta 165: 1-11.

Van Lammeren AAM, Kieft H (1987) Cell differentiation in the pericarp and endosperm of developing maize kernels (Zea mays L.) with special reference to the microtubular cytoskeleton. In: Embryogenesis in Zea mays L. A structural approach to maize caryopsis development in vivo and in vitro. Ph.D. Thesis, Agricultural University Wageningen, pp. 175.

Van Lammeren AAM, Kieft H, Provoost E, Schel JHN (1987) Immuno-gold labelling of tubulin in ultrathin cryosections of cultured carrot cells. Acta Bot. Neerl. 36: 125-132.

Willemse MTM, Van Lammeren AAM (1988) Structure and function of the microtubular cytoskeleton during megasporogenesis and embryo sac development in Gasteria verrucosa (Mill.) H. Duval. Sex. Plant Reprod. (in press)

Post-Pollination Phenomena and Embryo Development in the Oncidiinae (Orchidaceae)

S.C. Clifford and S.J. Owens,
Jodrell laboratory,
Royal Botanic Gardens, Kew,
Richmond, Surrey TW9 3DS,
United Kingdom

INTRODUCTION

This paper presents data on post-pollination phenomena and self incompatibility in species of the genera Oncidium, Lemboglossum, and Odontoglossum (Oncidiinae). The research is part of a broad study of breeding systems in the Orchidaceae ongoing at the Royal Botanic Gardens, Kew. Many species of the Orchidaceae, including species of the Oncidium alliance have been reported as being self incompatible (Scott 1865, East 1940, Charanastri and Kamemoto 1977).

Numerous post-pollination phenomena identified include stigma closure, changes in perianth, scent or nectar production, and ovary swelling. Genera in which such responses have been reported include species of Miltonia and Odontoglossum in the Oncidiinae, (Guignard 1886, Arditti 1979). Few data are presented on the sequence and timing of post-pollination phenomena. Several authors (Fitting 1909, 1910; Laibach and Maschmann 1933; Hubert and Maton 1939; Muller 1953) suggested that some or all of the observed responses are ellicited by substances held in the pollinia.

MATERIALS AND METHODS

All species used (Table 3) were from known, wild provenance and were grown at the Royal Botanic Gardens, Kew. All plants except those of Lemboglossum maculatum and Oncidium cheirophorum were kept at day temperature (DT) 24'C; night temperature (NT) 18.5'C; relative humidity (RH) 88%. L. maculatum and O. cheirophorum were kept at DT 17'C; NT 14'C; RH 80%. Daylength conditions were those prevailing in the UK.

Anatomy of the flower
Anatomical observations on the flower were made from material prepared by conventional wax embedding techniques (see e.g. Stevenson and Owens 1978). 5% glutaraldehyde was used instead of formalin propionic alcohol. for fixation. Sections (6-10 um.) were stained with safranin and alcian blue.

Pollinations
(a) Controlled selfs and crosses
   All pollinations were performed manually on open flowers. Pollinia from the required source species were placed into virgin stigmas. Subsequent observations and measurements of the post-pollination responses were recorded daily.
(b) Determination of ellicitors of the post-pollination responses. Experiments to find the causes of the post-pollination responses were made by pollinating virgin stigmas with:
   i Glass beads; ii Foreign (non-orchid) pollen; iii Foreign (orchid) pollen; iv

2mm cubes of 1% agar (Control); v Pollinia of L. maculatum (1 per cube) were incubated for 24h in the dark at 22'C. They were then removed and the cubes placed into the stigmas of L. maculatum and O. baueri. The pollinia which had been used for the extraction were used in parallel pollinations. Subsequent observations were made on a daily basis.

Pollen tube growth

Thick hand sections taken of pollinated stigmas and styles were mounted fresch in 0.1% aniline blue fluorochrome (Linskens and Esser 1957). Sections were examined using conventional UV microscopy (BP 350-450 exciter filter and LP 515 absorption filter).

Embryo sac development

Ovaries were removed at various stages of development, were fixed in formalin acetic alcohol (5:5:90), washed in water, cleared, and embryo sacs examined using differential interference contrast microscopy (Rudall and Linder in press).

Electron microscopy

Material for transmission electron microscopy was fixed and stained using conventional techniques (see e.g. Owens and Dickinson 1983).

RESULTS AND DISCUSSION

Individual flowers remain open for about 30 days in Oncidium species and up to 60 days in Lemboglossum.

Morphology and anatomy

The unit of pollination is the pollinium which comprises tetrads of binucleate pollen grains retained within a thick outer pollinium wall (Davis 1966; Brewbaker 1967). At anthesis, the stigma comprises a shallow depression filled with a clear viscous fluid in which single cells are often found floating. The stigma cavity remains open and wet (WN; Heslop-Harrison and Shivanna 1977) throughout the life of the open flower. There is no evidence to suggest a lack of stigma receptivity at any stage.

Morphologically, the stigma is variable in shape but anatomically it appears to be similar to that of Dendrobium speciosum (Calder and Slater 1985).

The style is fully continuous with the stigma and comprises a short, narrow, fluid-filled cylinder, running the length of the column. Though it enters directly into the ovary it branches at this point into 3 pairs of pollen tube guides, one pair associated with each placental ridge.

At anthesis, the ovary is trilocular with nucellar filaments bourne on each placental ridge. Following a compatible pollination, the archesporial cell, which has remained quiescent, acts directly as the megaspore mother cell (MMC; c. 23 days). The MMC undergoes meiosis to form a linear tetrad. Three cells of the tetrad degenerate while the persistent cell continues dividing to produce the mature bitegmic, tenuinucellar embryo sac (c. 40 days). This mode of embryo sac formation is a common feature of Orchidaceae (Swamy 1949, Davis 1966).

Post pollination responses

Four post pollination responses are indicated in Talbe 1. Two of these, stigma closure and perianth changes (lip fading, petal wilt, or colour changes), appear to be a part of a general response to pollination with pollen from within the tribe (Tables 1 and 2). Column and ovary swelling appear to depend upon the compatibility of the pollination (Tables 1 and 2). The pattern of response to a polliniun from a different species or genus varies according to how widely the taxa are separated.

Recent work confirmed that chemicals held in pollen walls play a critical role in the pollen-stigma interaction in plant species (see e.g. Clarke and Knox 1980; Dumas and Gaude 1982; Heslop-Harrison and Linskens 1984. Heslop-Harrison and Heslop-Harrison 1985).

Orchidaceae appear to be no exception . Fitting (1909) showed that subastances extracted from pollinia stimulate several post-pollination phenomena. Experiments reported below confirm this for Oncidiinae (Table 1). Glass beads and agar blocks placed on to the stigma have no stimulatory effect on post-pollination responses (Table 1). Agar blocks, however, which had pollinia placed on their surface for at least 24 hours prior to placing on the stigma, stimulate stigma closure and perianth changes (Table 1). It is highly likely that components from pollinia walls, leached into the agar blocks, effect the responses. The identity of the component(s) are unknown but the lack of a response with foreign (non-orchid) pollen suggests a difference betwenn Orchidaceae and other families.

Pollen tube growth and fertilization

Ovule development follows a compatible pollination. Fertilization occurs after 45-50 days following the stimulation of further pollen tube growth from the pollen tube guides to the micropyles. A minimum of 45,000 pollen tubes, estimated from transverse sections of pollen tube guides, are present in the ovary of O. excavatum. After incompatible matings, the response of the ovary may depend upon the species concerned. In O. excavatum ovular tissue degenerates within about 8 days while in L. bictoniense it is slower but the pollen tube guides are not visible in cross sections after about 8 days.

Self incompatibility (SI)

Evidence for the occurence of S I in 2 species of Lcmboglossum and 6 species of Oncidium is shown in Table 3. The conclusions relating to breeding system are supported by previous data (Scott 1865, Charanastri & Kamemoto 1977) for the species Oncidium baueri, O. maculatum and O. sphacelatum. Charanastri and Kamemoto (1975) record 68% of species in Oncidium showing self incompatibility.

In cases where incompatibility is expressed, the style is the main site for the reaction, with inhibited pollen tubes having been identified there (Scott, 1865 and this paper). A typical SI response appears to comprise:

A reduced level of pollen germination; a majority of germinated pollen has short tubes (10 pollen grains diameter) which barely penetrate the style. Pollen tube tips may be swollen or occluded, and may spiral; a minority of pollen tubes penetrate the style, and an even smaller proportion reach the ovary.

The post-pollination response in incompatible matings is initially similar to that in compatible matings although there appears to be a slower response (Table 2). Pollination followed by stigma closure in all cases (Table 2), predominantly by column swelling, and in about one third of replicates by ovary swelling. At about this time (7-8 days), the pedicel begins to turn brown, the flower abscicing at around 15 days (Table 2). Less than 20% of incompatible pollinations progress further with only 4.5% completing seed set, and many of these may be attributed to end of season effects. The various post-pollination responses are summarized in Fig. 1.

TABLE 1 The stimulation of post pollination responses in Oncidiinae. The effectiveness of pollens from various sources, of pollen leachates, and of inert material.

POLLINATION	STIGMA CLOSURE	PERIANTH CHANGES	COLUMN SWELLING	OVARY SWELLING
Orchid Pollinium				
i Intraspecific	+	+	+	+
ii Interspecific	+	+	+	+/-
iii Intergeneric	+	+	+/-	+/-
(within Oncidiinae)				
Agar Blocks				
i Impregnated	+	+	+/-	-
ii Not impregnated	-	-	-	-
iii Foreign pollen	-	-	-	-
(Non-orchid)				
iv Glass beads	-	-	-	-

TABLE 2 The timing in days (standard deviation) and percentage success of four post-pollination responses in compatible and incompatible pollinations.

POLLINATION	STIGMA CLOSURE/ %	COLUMN SWELL/ %	OVARY SWELL/ %	DATE DROPPED/ %
COMPATIBLE	1.76 (O.57) /100	2.47 (0.5) /98	6.1 (1.26) /83	16.5 (11.3) /24
INCOMPATIBLE	1.82 (0.9) /100	3.57 (1.4) /89	7.5 (2.75) /33	15.1 (4.7) /81

TABLE 3 The incidence of self incompatibility in species of Oncidiinae based on seed set and pollen tube growth studies

SPECIES	No. of Accns.	No reps.	Capsule develops	/ fails	Previous reports
L.bictoniense	2	16	0	16	SI
L.maculatum	2	11	2	10	SI Charanasri & Kamemoto (1977)
O.excavatum	4	20	2	18	(SI)
O.luridum	2	8	4	4	SC
O.maculatum	2	14	0	14	SI Charanasri & Kamemoto (1977)
O.sphacelatum	3	21	0	13	SI Scott (1865)
O.sphegiferum	1	11	6	5	SC

KEY:
   L. = *Lemboglossum*
   O. = *Oncidium*
   SI = Self incompatible
(SC) = low level of seed set, perhaps partially self compatible

Figure 1. A summary of post-pollination phenomena following compatible and incompatible pollinations in the Oncidiinae

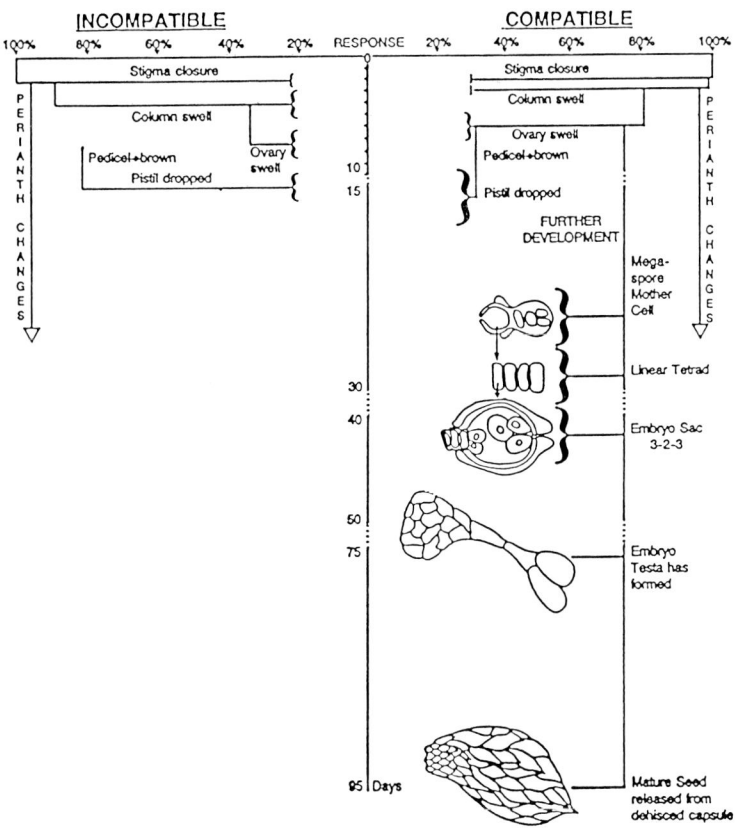

ACKNOWLEDGEMENTS

S.C.C. acknowledges receipt of a Kew Postgraduate Research Fellowship in collaboration with the department of Botany, University of Reading. The authors thank Professor H.G. Dickinson and Mrs. J. Stewart for useful discussion and encouragement.

LITERATURE CITED

Arditti J (1979) Aspects of the physiology of Orchids. Advances in Botanical Research 7:421-655

Brewbaker JL (1967) The distribution and phylogenetic significance of bi- and trinucleate pollen grains in the Angiosperms. American Journal of Botany 54:1069-1083

Calder DM, Slater AT (1985) The stigma of Dendrobium speciosum sm. (Orchidaceae): A new stigma type comprising detached cells within a mucilagenous matrix. Annals of Botany 55:297-307

Charanasri V, Kamemoto H (1977) Self incompatibility in the Oncidium alliance. Hawaii Orchid Journal VI(3):12-15

Clarke AE, Knox RB (1980) Plants and Immunity. Developmental and Comparative Immunology 3:571-589

Davis GL (1966) Systematic embryology of the Angiosperms. Orchidaceae, 194-195. John Wiley and Sons Inc New York

Dumas C, Gaude T (1982) Secretions et Biologie Florale II Leurs roles dans l'adhesion et la reconnaissance pollen-stigmate Donnees recentes hypotheses et notion d'immunite vegetale. Bulletin de la Societe Botanique de France. Actualites Botaniques:129, 89-101

East EM (1940) The distribution of self-sterility in flowering plants. Proceedings of the American Philosophical Society 82:449-518

Fitting H (1909) Die beenflussung der Orchideenbluten durch die bestaubung und durch andere umstande. Zeitschrift fur Botanik 1:1-86

Fitting H (1910) Weitere entwicklungsphysiologische untersuchingen an Orchideenbluten. Zeitschrift fur Botanik 2:225-266

Guignard L (1886) Sur la pollinisation et ses effets chez les Orchidees. Annales des Sciences Naturelles. Septieme serie. Botanique 4:202-240

Heslop-Harrison J, Heslop-Harrison Y (1985) Surfaces and secretions in the pollen-stigma interaction: a brief review. In Roberts K, Johnston AWB, Lloyd CW, Woolhouse HW (eds) The cell surface in plant growth and development. Journal of Cell Science supplement 2:287-300

Heslop-Harrison J, Linskens HF (1984) Cellular Interaction: a brief conspectus. In Linskens HF, Heslop-Harrison J (eds) Cellular Interactions. Encyclopedia of plant physiology 17:2-17

Heslop-Harrison Y, Shivanna KR (1977) The receptive surface of the Angiosperm stigma. Annals of Botany 41:1233-1258

Hubert B, Maton J (1939) The influence of synthetic growth controlling substances and other chemicals on post-floral phenomena in tropical Orchids. Biologisch Jaarboek 6:244-285

Laibach F, Maschmann E (1933) Ueber den wuchstoff der Orchideen pollineen. Jahrbuch fur Wissenschaftliche Botanik 78:399-430

Linskens HF, Esser K (1957) Uber eine spezifische Anfarbung der Pollenschlauche und die Zahl von Kallosepfropfen nach Selbstung und Fremdung. Naturwissenschaften 44:16

Muller R (1953) Zur quantitativen bestimmung von indolessigaure mittels papierchromatographie und papierelektrophorese. Beitrage zur Biologie Pflanzen 30:1-32

Owens SJ, Dickinson H.G (1983) Pollen wall development in Gibasis (Commelinaceae). Annals of Botany 51:1-15

Scott J (1865) On the individual sterility and cross impregnation of certain species of Oncidium. Journal of the Linnean Society 8:162-167

Stevenson, DW, Owens SJ (1978) Some aspects of the reproductive morphology of Gibasis venustula (Kunth) D. R. Hunt (Commelinaceae). Journal of the Linnean Society 77:157-175

Swamy BGL (1949) Embryological studies in the Orchidaceae I Gametophytes. American Midland Naturalist 41:184-201

Vij SP, Sharma M (1986) Embryo sac development in Orchidaceae In Vij SP (ed) Biology, Conservation and Culture of Orchids:31-48. Affiliated East-West Press Private Ltd, New Delhi

Polyembryony in *Oenothera*

Cornelia Harte and Renata Sniezko
Institut für Entwicklungsphysiologie
Universität zu Köln
Gyrhofstr. 17
D 5000 Köln 41
Fed.Rep. of Germany

Introduction

In **Oenothera** twin plants, two embryos in one seed, occur. They were observed in our own cultures and known to other geneticists working with this genus, like W. STUBBE and earlier RENNER (personal communication by STUBBE), but received no special attention. In the course of histological investigations on development of ovules, macrospore mother cells (MMC) and the embryo sac (ES), the female gametophyte, in **Oenothera** the occurrence of mature twin embryo sacs from two macrospores of one tetrad could be verified (NOHER de HALAC and HARTE 1975, SNIEZKO and HARTE 1984a, 1984b, 1986). On the other hand, ovules with two MMC were observed. Both become meiocytes and after meiosis two ES, one from each tetrad, can develop in one ovule. In the first case of twin ES, their genotypes are correlated to one another, the one being the complement of the other as far as the RENNER-complexes are concerned. In the second case of two ES in one ovule, developing from two MMC, the genotypes of the two egg cells are independent from one another. Both can have the same RENNER-complex. The question arises, how frequent a fertilisation of both egg cells in the two types of double embryo sacs in the ovules occurs and whether differences between different **Oenothera**-hybrids can be observed.

Genetics of the twin plants.

The genetic constitution of **Oenothera** species and hybrids is described by their RENNER-complexes. For the genetics of their progeny the competition between MMC of a tetrad in the ovules and between pollentubes has to be considered, as well as the lethals of some complexes.

Oe. elata H.B.K. (syn. Oe. hookeri Torr&Gray) is homozygous for the RENNER-complex **hhookeri**. In the species **Oe. biennis, albicans.rubens**, macrospore competition occurs and most viable ES contain the complex **albicans** in their nuclei. However, some ES can contain the RENNER-complex **rubens**. Pollengrains are only active when they contain the RENNER-complex **rubens** in the nuclei.

In the F_1 generation from the cross **Oe. biennis x elata**, the hybrids are **Oe. albicans.hhookeri**, and some **Oe. rubens.hhookeri**. From the reciprocal cross **Oe. elata x biennis** only the hybrid **Oe.hhookeri.rubens** is obtained. In the hybrid **Oe. albicans.hhookeri** with a ring of 14 chromosomes at meiosis, the macrospore competition leads to mostly **hhookeri-ES**, only some **albicans-ES** will occur (HARTE 1958 a, b, c). After self pollination of the F_1-plants, the segregation in the F_2-families allows conclusions about the frequency of the two RENNER-complexes in the ES of the F_1. The progeny of the hybrid **Oe. albicans.hhookeri** consists of mostly **hookeri**-type plants, only segregating for flower size, depending on crossing-over of the gene Co/co, and very rarely for flower colour, gene s/+s. The **albata**-type plants, **Oe. albicans.hhookeri**, indicate ES with the complex **albicans** and allow to estimate their frequency.

From the cross **Oe. suaveolens x elata**, two hybrids are obtained, **Oe. albicans.hhookeri** and **Oe. flavens.hhookeri**. The first behaves like the same combination of RENNER-complexes described above. The other can also be obtained from the reciprocal cross. In this hybrid the meiotic chromosome configuration is a ring of 4 chromosomes and five bivalents. As a consequence, at meiosis the complexes break into linkage groups and segregation and recombination is found in the progeny (HARTE 1948).

The hybrid **Oe.hhookeri.rubens**, with a lethal factor in **rubens** that prevents homozygosity for this complex in the progeny, most ES contain **hhookeri** and only some ES with **rubens** will occur. There must also be a strong competition between the two types of pollentubes. After self pollination the progeny consists mainly of plants of the **hookeri**-type, which are considered to be **Oe. hhookeri.hhookeri**, and others resembling the F_1. These types can be clearly distinguished. However, both are not identical with the reference types. The supposed **Oe. hhookeri.hhookeri** differs from the species **Oe. elata** in many quantitative characters as height, size of the leaves, flowers etc. and the same holds for the hybrid type. This segregation and the differences are observed in the following generations too, obtained by self pollination of the supposed **Oe. hhookeri.rubens**-plants. After self pollination of the **hookeri**-type plants, each F2-family is homogeneous, all plants resembling the parent. Taking into account the meiotic catenation of the chromosomes of the two complexes **hhookeri** and **rubens**, the segregation for quantitative characters can only be explained by crossing-over in several chromosomes during meiosis.

Segregating progenies were obtained by self pollination of plants of different genotypes from F1-families. The progenies of the later generations, F2 and F3, are obtained by self pollination of the plants corresponding in their combination of RENNER-complexes to the F1-generation (table 1).

Development of ovules with two embryo sacs.

According to the observations on macrospore competition, twin ES will always develop from macrospores at the opposite ends of the tetrad (NOHER de HALAC and HARTE 1975, SNIEZKO and HARTE 1984a, 1984b, 1986). They will originate from the two dyad cells of the first meiotic division and will be of different genotype. After fertilization of both egg cells, the twins can be distinguished by the corresponding phenotype. On the other hand, in twin tetrads from two MMC, the competition within each tetrad can lead to two mature ES of the same genotype in the ovule.

When in the progeny of a complex-heterozygote with high catenation of the chromosomes in meiosis and limited crossing-over the twins inherited the same RENNER-complex from the mother, then they must have originated from an ovule with two MMC. When they have different RENNER-complexes, however, a decission about their origin from two MMS or twin ES from one tetrad, is not possible for the individual case. Only statistical evaluation of the distribution of twin-types, based on the distribution of genotypes in the whole progeny, can reveal the occurrence of twins from sister ES and give an estimate of their frequency.

Observations on twin seedlings.

In the parents **Oe. biennis** and **Oe. elata** diploid twins occur with low frequency of less than 1%, both plants being always of the same phenotype. This gives proof of the presence of ovules with two ES in both parents. In **Oe. elata**, macrospore competition does not occur. The degeneration in the tetrad leaves only the micropylar macrospore to develop the ES (NOHER de HALAC and HARTE 1975). Here twin embryos can only develop from two ES originating from different MMC. In **Oe. biennis** macrospore competition is observed (RENNER 1921, NOHER de HALAC and HARTE 1975) and twin ES occur. From the observations on the parents it is obvious, that the genetic possibility for two ways for the origin of twin embryos is present in the material.

In **Oe. albicans.hhookeri** from **Oe. biennis** and **Oe. suaveolens**, resp., as female parents of the hybrid, the frequency of twins is the same in the three generations obtained by self pollination of plants with the appropriate genotype. In the families derived from the crosses with **Oe. suaveolens** only homogenous pairs, both plants of **hookeri**-phenotype, were observed. The actual numbers of pairs of twins are so small, that the difference between the observation 0 heterogenous pairs in 22 cases and the expectation of 3 heterogenous pairs to 19 homogeneous pairs is not statistically significant. This expectation is calculated from the sum of all **Oe. albicans.hhookeri**-families.

In all pairs of twins observed in the progeny of **Oe. flavens.hhookeri**, the plants were phenotypically different, having phenotypes corresponding to those found earlier in segregating F_2-progenies (HARTE 1948). These progenies can only be used for comparison of the frequency of twins but do not give information about their origin.

When plants with the RENNER-complexes **Oe. hhookeri.rubens**, irrespective of their origin, are self-pollinated, in the progeny two clearly distinguishable types were observed as described above. In most cases both twins had a phenotype corresponding to the **Oe. elata**-type. Some heterogenous pairs were observed, and a few pairs were found, in which both plants looked alike, corresponding to the phenotype of the F_1. The differences of the phenotypes from the reference types, parent and F_1, as described above do not interfere with the conclusions about the frequency and origin of the pairs of twins. In all generations of this hybrid, irrespective of the origin of the F_1, the frequency of twins is the same with an average of 2%. In the different hybrids, all containing the RENNER-complex **hhookeri**, the differences in the frequency of twins in the progeny are not statistically significant and vary around 2.3%.

The frequency of the homogenous pairs, both plants having the same genotype, is much higher than that of the heterogenous pairs. This means that in most cases the two ES developed from two MMC. However, the frequency of the heterogenous pairs, originating from

ES with different RENNER-complexes, is higher than is expected on the basis of the frequency of the two complexes in the whole progeny (Table 2). Therefore it can be concluded, that the egg cells in twin ES developing from one tetrad are both fertile. Irrespective of the origin of the two ES, their egg cells can be fertilized and give rise to normal, viable embryos.

In **Oe. elata** and the progenies from crosses with this species, haploids with the RENNER-complex **hhookeri** were observed earlier (HARTE 1973). When a pollentube entering the micropyle can trigger development of the embryo, haploids should be observed as partner of a diploid twin. However, no haploids were found. So for each ES a separate pollentube is necessary for triggering embryogenesis.

Summary.

In **Oenothera** twin seedlings are observed with considerable frequency. From histological observations it can be concluded that twins always are bizygotic, originating from fertilisation of the egg cells in two embryo sacs in one ovule. However, there are two pathways for the development of the double egg cells. They can be derived from two MMC, each with one macrospore developing to become an embryo sac, and most important for **Oenothera** from two developing macrospores from the poles of one meiotic tetrad.

Table 1.

Frequency of twin seedlings in the progeny of Oenothera-hybrids.

cross Oenothera	RENNER-complexes	generation	families	seeds	twins pairs	%
biennis x elata	albicans.hhook	F_1, F_2, F_3	15	2265	70	2.6
suaveolens x elata	albicans.hhook	F_1, F_2, F_3	6	942	22	2.3
suaveolens x elata	flavens.hhook	F_1	5			
			9	1897	41	2.2
elata x suaveolens	hhook.flavens	F_1	4			
elata x biennis	hhook.rubens	F_1, F_2, F_3	7			
			10	2184	85	2.0
biennis x elata	rubens.hhook	F_1, F_2	3			

Table 2. Genotypes of twin seedlings.

cross	RENNER-complexes	h+h	x+h	x+x	others*
Oe.biennis x elata	albicans.hhookeri	50	12	1	7
Oe.suaveolens x elata	albicans.hhookeri	22	0	0	0
Oe.suaveolens x elata and reciprocal cross	flavens.hhookeri	38+	0	0	3
Oe.elata x biennis and reciprocal cross	hhookeri.rubens	16	15	4	9

+ = all plants flava-type, both plants of a pair are different
* = the phenotype of one or both plants of the pair could not be
    identified because they died or were crippled.
h = plant of hookeri-type
x = plants not hookeri-type, either albata (with RENNER-complex
    albicans), or flava (with flavens) or laeta (with rubens), resp.

References

Harte C (1948) Cytologisch-genetische Untersuchungen an spaltenden
Oenothera-Bastarden. Z f induktive Abstammungs- und Vererbungslehre
82: 495-640
Harte C (1958) Untersuchungen über die Gonenkonkurrenz in der
Samenanlage bei Oenothera unter Verwendung der Letalfaktoren als
Markierungsgene. I. Die individuelle Komponente der Variabilität.
Z Vererbungslehre 89: 473-496
Harte C (1958) II. Die Umweltkomponente der Variabilität.
Z Vererbungslehre 89: 497-507
Harte C (1958) III. Die genetische Komponente der Variabilität.
Z Vererbungslehre 89: 715-728
Harte C (1973) Haploide Pflanzen bei Oenothera. Biol Zentralblatt
92: 361-364
Noher de Halac I, Harte C (1975) Female gametophyte competition in
relation to polarisation phenomena during megasporogenesis and
development of the embryo sac in the genus Oenothera. In Mulcahy D L
(ed) Gamete Competition in Plants and Animals. North Holland
Publishing Company, Amsterdam p 43-56.
Sniezko R, Harte C (1984) Polarity and competition between
megaspores in the ovule of Oenothera-hybrids. Pl Syst Evol 144:
83-97
Sniezko R, Harte C (1984) Callose pattern and polarization phenomena
in the ovules of hybrids between Oenothera hookeri and Oe.
suaveolens. Pl Syst Evol 147: 79-90
Sniezko R, Harte C (1986) Development of polarity in the ovules of
the F2-progeny of the Oenothera-hybrid albicans.haplo-hookeri. Pl
Syst Evol 154: 89-101
Renner O (1921) Heterogamie im weiblichen Geschlecht und
Embryosackentwicklung bei Oenotheren. Z f Botanik 13: 619-621

----

adress of second author:
UMCS, Instytut Biologii, Zaklad Anatomii i Cytologii Roslin
ul. Akademicka 19
20-033 Lublin , Poland

The Suspensor is a Major Route of Nutrients into Proembryo, Globular and Heart Stage *Phaseolus vulgaris* Embryos

Tom Brady and S.H. Combs*
Biology Department
Hamilton College
Clinton, NY 13323
USA

Embryos from proembryo through heart stage were examined for Prussian Blue deposition in both stained section preparations and in whole mounts which had been cleared with lactophenol. In all cases where Prussian Blue was observed in the embryo, it was also observed in the recurrent vascular traces which end in seed tissue adjacent to the suspensor. The precipitate could be seen in the cells surrounding the suspensor, in the vaculoes of the suspensor cells themselves, and in the embryo at the embryo-suspensor junction. The precipitate could often be seen in the endosperm with high concentrations in the area around the cotyledons; however, in only a very small percentage of the embryos examined, was there precipitate in the cotyledons adjacent to the endosperm indicating that, at least for the ions in question, there was little movement of material between the endosperm and the developing cotyledon, and that the major route for transport of material into the early Phaseolus embryo is through the suspensor.

INTRODUCTION

The role of the suspensor in the development of the angiosperm embryo has long been an enigma. A possible hypothesis is that the suspensor serves a two-fold function; first, to attach the developing embryo to the maternal seed tissue and second, to push the developing embryo into the nutrient-rich endosperm. The observation, by Schnepf and Nagl (1970) as well as by Clutter and Sussex (1968), of walled projections on the polytene suspensor cells of Phaseoleus led those authors and Gunning and Pate (1969) to hypothesize that the Phaseolus suspensor serves as a transfer organ, transporting material from the seed into the suspensor and possibly into the developing embryo. Work from the Sussex laboratory demonstrated that the suspensor is extremely active in both RNA and protein synthesis (Sussex et al., 1973). Alpi et al.

---

* Department of Biochemistry, Texas Tech University Health Sciences Center, Lubbock, Texas 79409 USA

(1975) found that the suspensor contains high levels of gibberellins early in development and that these decrease as the embryo matures. My laboratory has recently demonstrated that the attached suspensor is necessary for the maintenance of both protein synthesis and protein content (Walthall and Brady, 1986; Brady and Walthall, 1985) in heart-stage Phaseolus embryos in tissue culture; however, the suspensor may be, at least partially, replaced by physiological concentrations of gibberellin. This leads to the hypothesis that the suspensor may be either manufacturing and/or transporting gibberellin into the developing embryo.

In this study, we have sought to examine directly the hypothesis that the suspensor serves as a major route of transport of materials from the maternal plant through the seed into the developing embryo. In order to examine this hypothesis we used a modification of the Prussian Blue technique (Burbano, et al., 1976). Both ferric chloride and potassium ferrocyanide are readily soluble and are transported through the bean plant's vascular system; however, when the two come in contact, an insoluble blue precipitate, Prussian Blue, is formed, thus allowing one to directly observe the movement of solutes in a given system.

MATERIALS AND METHODS

P. vulgaris (var. Taylor's horticultural) were grown on light tables with a 16hr light 8hr dark cycle. When pods had reached the desired degree of maturity, a flap was cut in the central leaflet of the leaf above a given pod so as to include the midrib (Walbot, 1971). This flap was placed in a 10% solution of potassium ferrocyanide in a vial attached to a ring stand and the solution was allowed to flow through the vascular tissue of the leaf for one hour. The potassium ferrocyanide was chased with 1% sucrose for one hour. A 5% solution of ferric chloride was then fed through the leaf for one hour and again followed by a 1% sucrose chase for three hours. The pods were then picked. Pods under an inch long were processed in toto. Seeds from pods greater than one inch were removed and processed individually.

Sectioning - Pods or seeds were fixed by vacuum filtration with Navashin's Solution (Jensen, 1962), washed 3-4 hr in running water, dehydrated in TBA, infiltrated with paraffin at 56°C and sectioned. Paraffin was removed with xylene, the sections were rehydrated to 10%

ethanol/water stained with .01% filtered safranin, dehydrated in an ethanol series to xylene, and coverslipped with Permount.

Whole mounts - Pods or seeds were boiled in 10% ethanol to extract chlorophyll and then boiled for 10 min in lacto phenol (equal parts lactose, phenol, glycerin and water) to clear the tissue.

Photography - In order to more easily demonstrate the presence of Prussian Blue in thin sections, photographs were taken using a red Wratten series 25A filter.

RESULTS

In Phaseolus a single vascular trace enters the seed and this extends toward the chalazal end of the embryo sac. Two recurrent branches develop from this central trace and run on either side of the seed toward the area of the micropyle where the embryo is developing (Figs. 1 and 5). Prussian Blue is first observed in the vascular trace (Fig. 1). There are no connections of the vascular tissue with the embryo sac and thus dissolved materials move out of the vascular tissue into the parenchyma of the seed. Prussian Blue deposition is next seen most intensely in the area of the seed surrounding the suspensor (Figs. 3-5). This is followed by an accumulation of the precipitate in the endosperm (Figs 3-5). In Phaseolus the cellular endosperm is initially a fluid filled sac surrounding the embryo except where the embryo proper is attached to the suspensor (Fig. 5a). This sac and the embryo can be removed intact from the seed. Figure 5b shows heavy Prussian Blue deposition in the endosperm sac surrounding the embryo, however when the endosperm is removed from the embryo proper little or none of this precipitate is seen on the cotyledonary surfaces (Fig. 5c). In contrast, Prussian Blue can be seen in the area around the suspensor as well

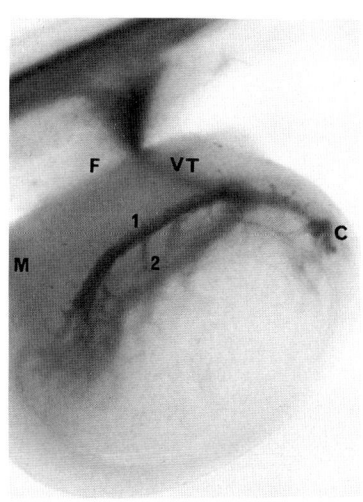

Fig. 1. Early cotyledonary stage of lactophenol cleared seed showing single vascular trace (VT) enters seed at funiculus (F) and terminates at the chalazal end (C). Two recurrent vascular traces (1,2) branch off midway down the VT and terminate at the micropylar end (M) in the region of the suspensor. Prussian Blue deposition (dark areas) defines the vascular traces. X 90

422

as in the wall projections between the integumentary tapetum and the suspensor, and then within the suspensor cells themselves (Figs. 2-4). Prussian Blue deposition is also seen at the junction of the suspensor and the embryo (Figs. 5a and 6) and in the region of the embryo, the future axis, which is attached to the suspensor.

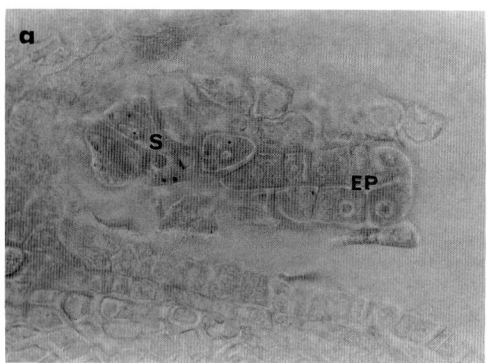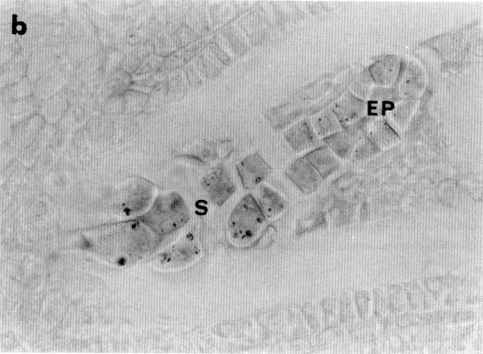

Fig. 2 a and b. Proembryos from successive seeds in same pod showing: a. Prussian Blue deposition in suspensor cells (S) and b. in both the suspensor (S) and the embryo proper (EP). X500

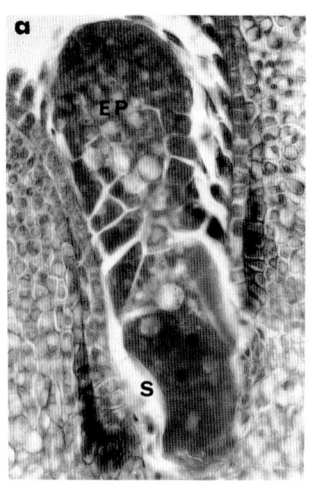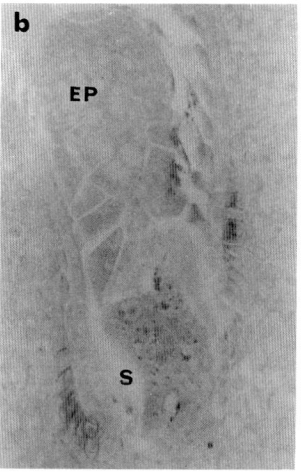

Fig. 3. Median longitudinal section of early globular embryo. a. Safranin stained. b. Safranin stained and photographed with red filter (Wratten 25A) to show Prussian Blue in integumentary tapetum (IT), endosperm adjacent to the embryo proper (EP) and suspensor (S). X250

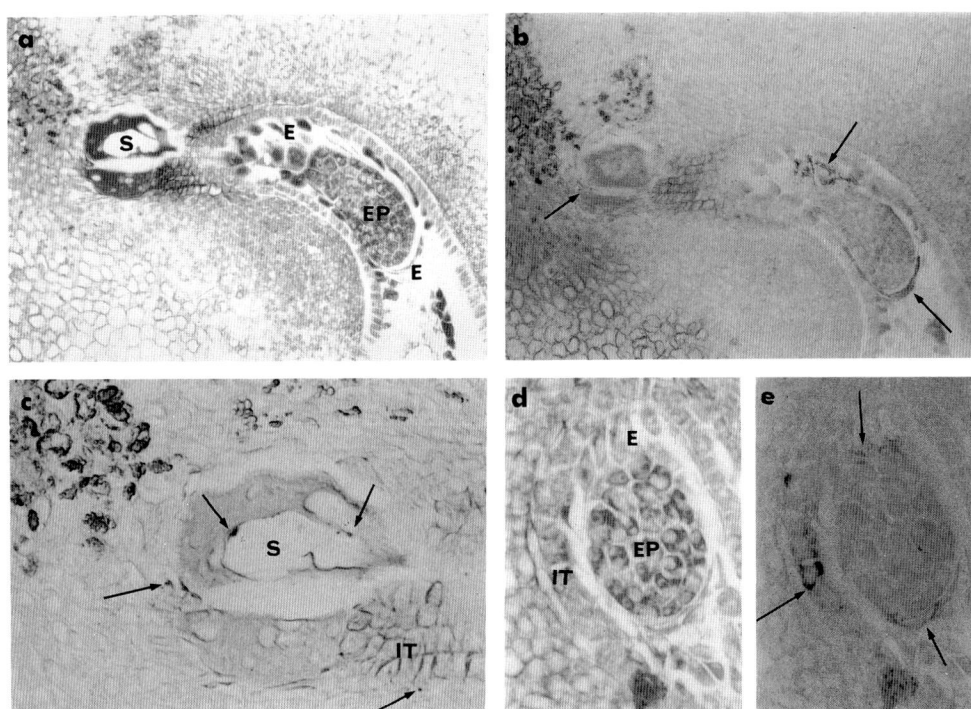

Fig. 4. Median longitudinal section of globular stage. a. Safranin
stained whole mount. X120 b. Red filter. PB deposition (→) in
suspensor (S), endosperm (E) adjacent to proembryo, and branched
parenchyma cells of seed coat. X120 c. Enlargement of basal cells
in a, red filter. PB in integumentary tapetum (IT), branched
parenchyma and suspensor cells. X280 d. Transverse section through
embryo proper (EP) of a, safranin stained and e. photographed with
red filter showing deposition in integument and endosperm. X250

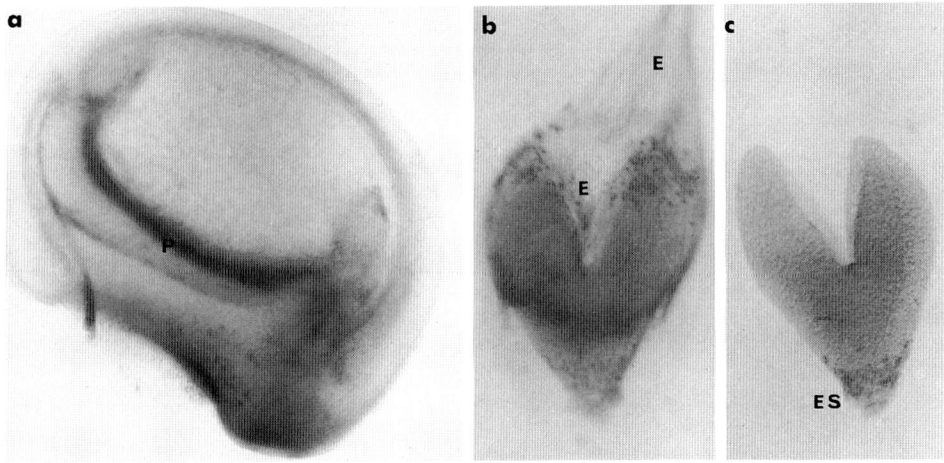

Fig. 5. a. Heart stage lactophenol cleared seed. PB deposition in
vascular traces, branched parenchyma cells, suspensor and endosperm
surrounding the embryo. X150 b. Dissected embryo with encapsulating
cellular endosperm showing PB deposition in endosperm cells. X250
c. Embryo with the endosperm removed. PB in embryo at the
embryo/suspensor junction and virtually no deposition in cotyledons
where they were in contact with the endosperm. X250

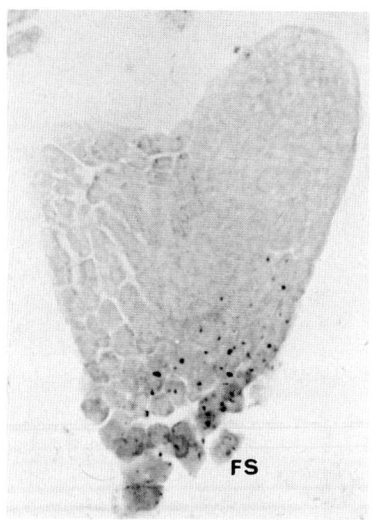

Fig. 6. Longitudinal section of heart stage embryo, photographed with red filter. There is substantial PB deposition at the endosperm/suspensor junction (ES) in the region of the future embryonic axis. x450

CONCLUSION

This study allows us to conclude that, at least up until the heart stage of embryo development in _Phaseolus vulgaris_, the suspensor is the major route of nutrients into the developing embryo proper.

BIBLIOGRAPHY

Alpi A, Tognoni F, D'Amato F (1975) Growth regulator levels in embryo and suspensor of _Phaseolus coccineus_ at two stages of development. Planta (Berlin) 127:153-162

Brady T, Walthall ED (1985) The effect of the suspensor and gibberellic acid on _Phaseolus vulgaris_ embryo protein content. Dev Biol 107:531-536

Burbano JL, Pizzolato TD, Morey PR, Berlin JD (1976) An application of the Prussian Blue technique to a light microsope study of water movement in transpiring leaves of cotton (_Gossypium hirsutum_ L.). J Exp Bot 27:134-144

Clutter ME, Sussex IM (1968) Ultrastructural development of bean embryos containing polytene chromosomes. J Cell Biol 39:26a

Gunning BES, Pate JS (1969) "Transfer cells": plant cells with wall ingrowths, specialized in the relation to short distance transport of solutes- their occurrence, structure and development. Protoplasma 68:107-133

Jensen WA (1962) Botanical Histochemistry, WH Freeman and Company, San Fransisco

Schnepf E, Nagl W (1970) Uber einige strukturbesonderheiten der suspensorzellen von _Phaseolus vulgaris_. Protoplasma 69:133-143

Sussex I, Clutter M, Walbot V, Brady T (1973) Biosynthetic activity of the suspensor of _Phaseolus coccineus_. Carylogia 25(Suppl): 261-272

Walbot V (1971) RNA metabolism during embryo development and germination of _Phaseolus vulgaris_. Dev Biol 26:369-379

Walthall ED, Brady T (1986) The effect of the suspensor and gibberellic acid on _Phaseolus vulgaris_ embryo protein synthesis. Cell Dif 18:37-44

"Structural Reserves" in Embryo Development

O.Erdelská
Institute of experimental
biology and ekology
Slovak Academy of Sciences
814 34 Bratislava
Czechoslovakia

Certain part of the structural elements formed during plant ontogenesis is not exploited. For various conditions partly different ways
of development are prepared in the organism, mostly relying on structural reserves. So for example the stem apex having been injured,
the stems may develop from axillary buds which, under normal conditions, remain only in the primordium stage. Very well known is also
the "oversize dimension" of the conductive tissues etc.

In embryogenesis, the development of "reserve" structures can be
traced already from the origin of sporogenous cells in the ovule
onward.

a/ In many species a multicellular archesporium is formed (Rosaceae,
   Asteraceae, Fagaceae, etc.). Usually, however, only one megaspo-
   rocyte continues to develop entering the meiosis.

b/ In tetradogenesis, usually four megaspores arise. In the majority of plant species, however, only one of them continues to develop, the other three degenerate. They develop only exceptionally
   (for example with the snowdrop) giving rise to several embryo
   sacs in one ovule.

c/ By the division of the funcional megaspore, an 8-nuclear (7-cel-
   led) embryo sac - female gametophyte - developes in the majority
   of angiosperms. Only one cell of the gametophyte (egg cell) has
   the function of a female gamete, the other ones (synergids, antipodals) are only potential gametes. At one side, they are not un-
   conditionally necessary for embryo formation, since species with
   an embryo sac without synergids (e.g. Plumbago), without antipodals (for example the types Allium, Oenothera) etc. also exist.
   On other hand, however, these cells may be potential gametes.
   Known are, for example embryos from synergids (Capsicum, and other
   ones).

d/ In the process of microsporogenesis, there usually arises an a-
   mount of pollen that many times exceeds the number of pollen

grains needed to fertilize the ovules. However, from the pollen grains or some of their developmental stages may originate the embryos in various types of androgenesis.

e/ Usually only from the apical cell of two-celled proembryo originates the embryo proper, and the basal cell gives rise to a temporarily functioning suspensor. The suspensor plays an auxiliary role in embryo development but it may also have an embryogenous potential, as proved by the model experiments of Haccius (1965 etc.). This, however, manifests itself when the developmental correlation of tissues in the embryo is out of order or also when the connection embryo-endosperm is discrupted.

In species with a delayed embryo differentiation (Paeonia, Tulipa and other ones) zygote division gives rise to a whole complex of cells with embryogenous potential. Under normal conditions, however only from one of them arises the embryo in the seed.

f/ Cotyledons (or the scutellum) are important organs of the embryos in seed plants. Their auxiliary character, however, is borne out by the fact that the embryo axis may continue its development in vitro even without them. In many species, however, they appeared as an adequate starting material for inducing the formation of additive embryos in vivo (Fergusson et al. 1979) or embryoids vitro.

g/ In course of the embryogenesis, the differentiation of organs, tissues and cells of the embryo axis passes off gradually. The excised embryo, however, may germinate before its differentiation has terminated. So, for example, the maize embryo is capable of germinating already at a time when its size attains only 1/10 of the mature embryo size. The wheat embryo may enter the post-embryonal development also before the formation of seminal root primordia is accomplished. From such prematurely germinating embryos miniature plants arise because the embryos have a smaller number of cells in the tissues. Embryos with a lower number of cells may arise, however, on the mother plant, too, if being cultivated under unfavourable temperature, moisture or nutrientconditions. A part of the embryo tissues thus represents a structural reserve that is not indispensibly needed for initial plant development, but may play an important role in ontogenesis and also in the practical utilization of cultivated plants.

h/ In addition to embryos from the zygote, asexually originating adventive embryos arise in the nucellus or in integuments of the

ovules (seeds) in some species. These too may be taken for a structural reserve of the embryogenesis.
Although the formation of structural reserves has its main biological significance for the preservation of the individuum or the species, structural reserves have appeared to by often utilizable also within breeding and cultivation programs. Known is, for example, the utilization of haploid embryos from synergids in breeding of Capsicum annuum, the utilization of adventive embryogenesis in the reproduction of citrus plants etc.

SUMMARY

Only a part of the initiated ovular cells participates directly in embryo development under normal conditions. Usually only one archesporial cell and one of four megaspores continue to develop. The embryo arises by the fertilization of one cell of the embryo sac. The other cells have an auxiliary character (function). The suspensor and the cotyledons or the scutellum may also be ranked among the structural reserve. A part of the tissue of the embryo axis too represent a structural reserve that is not decisive in the start of plant development but may become meaningful in later phases of the ontogenesis.
The formation of structural reserves is of major biological importance for preserving the individuum or the species. Structural reserves may be utilized with advantage, however, in breeding and cultivation programs (for example the utilization of haploid embryos from synergids, the adventive embryogenesis, etc).

REFERENCES

Fergusson J D, Mc Ewan J M, Card K A (1979) Hormonally induced polyembryos in wheat. Physiologia plantarum 45: 470 - 474

Haccius B (1965) Weitere Untersuchungen über Somatogenese aus den Suspensorzellen von Eranthis hiemalis: Embryonen. Planta 64: 219 - 224

Embryology and Taxonomy

Sporo-, Gametogenesis and Fertilization of *Escallonia* and *Brexia* with Comments on Their Taxonomy

O.P.Kamelina
Laboratory of Embryology
Komarov Botanical Institute
Prof Popov str 2
197022 Leningrad
USSR

Abstract

A comparative analysis of Escallonia and Brexia genera embryological
features has shown their significant differences and allowed to give
certain recommendations as to their taxonomy, affinity and position
in the phylogenetic system of plants.

Introduction

Relations between embryology and taxonomy are becoming increasingly
close and the data obtained by the comparative embryology are used
by the authors of the latest phylogenetic systems of flowering
plants. That is why the task to study non-investigated taxa and
first of all whose position in phylogenetic system and the affinities are not clarified is especially urgent for embryologists.

Embryological features which are stable and conservative often
allow to determine the absence of affinity in the taxa considered
by systematists to be closely related morphologically, e.g. Escallonia and Brexia. Both genera were studied in different systems and
taxonomic ranks as well as in various position, but usually they
belonged to the family of Escalloniaceae as separate genera.

The data on detailed investigations of male and female embryonic
structures of Escallonia and Brexia representatives are published
(Kamelina,1984,1985,1988). In this paper the data of the comparative
analysis of the embryological features are preceded by short embryological characteristics of these genera, most of the features being
described for the first time.

Observation

Escallonia is characterized by 4-locular anther with large connective. The development of the anther wall is centrifugal as in Dicotyledonous type but with additional divisions in the formed layers. The wall consists of the epidermis, the endothecium, two-three middle layers and of irregularly thickened tapetum. The latter is cellular, heteromorphic, with multilayer protuberance on the connective side. The fibrous thickennings are formed in the endothecium, in one or two remained middle layers and in the connective tissue attached to locula. Tanins are contained in epidermis. The multilayer sporogenic tissue is of convexed form. Tetrahedral and isobilateral tetrads of microspores are formed simultaneously. The mature pollen grains are two-celled.

The ovule of Escallonia is anatropous, unitegmic, with a short funiculus and tennuinucellar. It contains hypostaza, epistazalike cell group in the micropylar part of integument, integumentary tapetum surrounding the 2/3 of embryo sac from the chalaza side and tanin filled enlarged epidermis cells. The archesporium is unicellular; megasporogenesis has a linear megaspore tetrad at its end; the embryo sac is developed according to Polygonum type. The mature embryo sac is of a peculiar form: its rounded micropylar part is protruded from integumentary tapetum and has specific lateral protuberances of the central cell, the chalazal part is narrow, prolonged and tubularlike. The egg cell and two synergids are large. Calloze is accumulated forming a kind of a cap at the bazal end of synergids. Antipodes are presented by three small cells with dense cytoplasm. Polar nuclei fuse before fertilization. The cytoplasm of the egg cell and central cell contains some starch.

Porogamy and double fertilization are observed. The fusion of gamete nuclei takes place according to premitotic type. The endosperm is cellular with micropylar haustoria.

Brexia genus has 4-locular anther with a large connective. The anther wall is formed according to the basic type and consists of epidermis, endothecium, five-six middle layers and tapetum. Fibrous thickenings are formed in endothecium in two or three middle layers and in several connective tissue layers. The tapetum is cellular, multinuclear and morphologically homogeneous with orbicula. Microspore tetrads are simultaneously formed. The mature pollen grains are three-celled.

The ovule is anatropous with two integuments and crassinucellar. Parietal tissue is unilayer, integumentary tapetum is surrounding

the whole embryo sac. The archesporium is unicellular. The megaspore
tetrad is linear. The embryo sac is developed by Polygonum type. Its
elements are large and differentiated. Synergids have a filament apparatus. Polar nuclei are not fused prior fertilization. Three antipodal cells are situated in the narrow chalazal part of the embryo
sac surrounded by postament cells. Tanins are accumulated in some
anther and ovule tissues. The pollen tube intrudes the ovule poro-
gamically. The double fertilization is performed according to premitotic type. The endosperm is nuclear.

The comparison of the main embryological features allows us to
conclude that Brexia and Escallonia genera are different in many of
them. Thus, as to the features of male embryonic structures they are
as follows: the type of anther wall development (basic and dicotyledonous), the structure of the formed anther wall (7-9 and 5-6-lay-
ered with 4-6 and 2-3 middle layers accordingly); the structure of
tapetum (morphologically homogeneous and heteromorphic) and sporogenic tissues (oval and convexed); the mature anther wall (4-5-
layered and 3-layered); the mature pollen grains (three- and two-
celled). Such features as 4-locular longitudinally open anther with
a large connective, simultaneous type of tetrad microspore formation,
the accumulation of tanins and the formation of fibrous thickenings
in anther tissues are similar for both genera, though the two latter
are more spectacular in Brexia.

The common features of female embryonic structures are the following: the ovule type (anatropous for the both genera), the type
of megasporogenesis and gametogenesis (Polygonum type of embryo sac
development). However as to all other parameters Brexia and Escallonia genera are strictly distinguished. These differences are as
follows: a number of integuments (2 and 1), presence of parietal
layer in nucellus (crassinucellar and tenuinucellar ovules), a form
of micropyle (zigzag shape with short exostom and long endostom for
Brexia, while short and simple for Escallonia), the differentiation
of integumentary tapetum (either totaly surrounds the embryo sac or
only 2/3 of its length); the degree of vascularization (highly or
poorly developed); the presence of epistazalike tissue in micropylar part of Escallonia integument and its absence in Brexia. The
embryo sacs differ by form and morphology of their elements. Polar
nuclei of Brexia fuse at the time of fertilization, while those of
Escallonia fuse before it. The porogamy and the double fertilization of premitotic type are observed in both genera. Endosperm in

434

Legends

Some embryonic structures of Escallonia rubra (Figs.1-3,7,10) and
Brexia madagascariensis (Figs.4-6,8,9,11)

Fig.1 A part and a scheme (Fig.2) of the opening anther
Figs.3,6 Mature pollen grains
Fig.4 A part of anther and its scheme (Fig.5) with two-celled pol-
    len grains
Figs.7,9 Ovules
Fig.8 A mature embryo sac
Figs.10,11 Double fertilization
    (In Figs.2 and 5 strokes and dots show fibrous thickenings. The
    blackened parts of anthers and ovules in Figs.4,5,8,9 show the
    cells filled up with tanins)

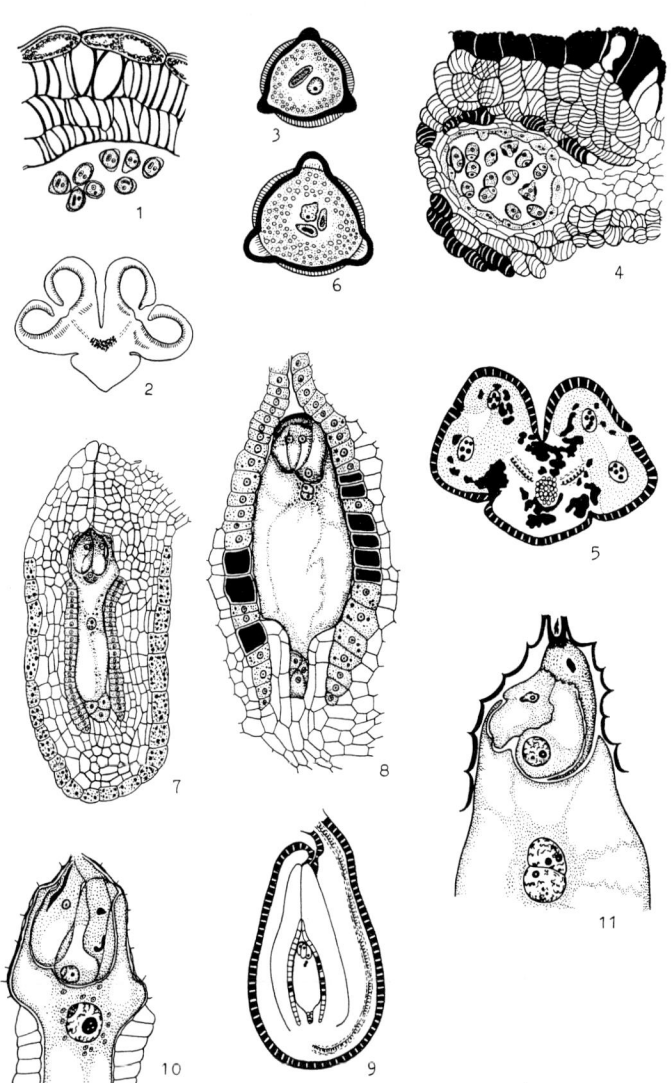

Brexia is of nuclear type, while that in Escallonia is cellular
with micropyllar haustoria.

Conclusion

Thus, a comparative analysis of both genera embryological features
has shown that they are sharply distinguished by most of their characteristics. It allows to make a conclusion that Escallonia and
Brexia genera are not closely related. The genus Brexia is distinguished by some features from other families it was formally related
to. It is worth to be considered as a separate family and should be
excluded from the family Escalloniaceae.

By their embryological features both genera are highly specialized but Escallonia has much higher number of advanced features
than Brexia.

The family Escalloniaceae by its embryological features (the
cellular heteromorphic tapetum, two-celled pollen grains, the anatropous with one integument tenuinucellar ovule, integumentary tapetum which does not surround the whole embryo sac, cellular endosperm with micropylar haustoria and some others) is related to the
most of Hydrangeales order families (in the scope of Takhtajan system, 1987).

The Brexiaceae family has more similar features with the Saxifragales order families such as, morphologically homogeneous cellular tapetum, three-celled pollen grains, two integuments in crassinucellar ovule, integumentary tapetum surrounding the whole embryo
sac and a nuclear endosperm. It is recommended to include Brexiaceae
into the order of Saxifragales as an independent family.

The present work once more demonstrates the importance of the
comparative-embryological method for phylogenetic plant taxonomy.

References

Kamelina OP (1984) To the embryology of the genus Escallonia (Escal-
    loniaceae) (in Russian with English summary). Bot Zh (Leningrad)
    69, 10: 1304-1316
Kamelina OP (1985) Family Escalloniaceae. In: Yakovlev MS (ed) Com-
    parative embryology of flowering plants Brunneliaceae-Tremandra-
    ceae (in Russian). "Nauka", Leningrad, p 9-14
Kamelina OP (1988) Embryology of the genus Brexia in connection
    with its systematic position (in Russian). Bot Zh (Leningrad)
    73, 3: 355-366
Takhtajan A (1987) System of Magnoliophyta (in Russian). "Nauka",
    Leningrad

Embryological Problems of the Family Rosaceae

R.Czapik
Department of Plant Cytology and Embryology
Jagellonian University
Grodzka 52
31-044 Kraków
Poland

Abstract

Contemporary knowledge of the Rosaceae embryology is based on data concerning ca one half of their genera, Still, a little is known about the embryological differences between the subfamilies or tribes and about the variability of embryological characters within the family. Thus is necessary to continue the routine work on inadequately known taxa and to do comparative-developmental observations of particular characters, paying attention to their inter- and intraspecific variations.

Key words: Rosaceae, embryological diagram, ovule, haustoria, secondary nucleus, apomixis.

Rosaceae belong to those families of the Angiospermae which are considered as embryologically well known. Such an opinion was expressed by Schnarf /1931/ in connection with three families only: Ranunculaceae, Rosaceae and Compositae. According to Davis /1966/ the higher degree of Rosaceae examination had been associated with the interest directed to hybrids and cultivated plants, the latter being rather common within the family. Apomixis is the second factor which has intensified the embryological research. In the list of apomicts by Khokhlov /1967/ Rosaceae came third in order of precedence after Gramineae and Compositae with regard to genera and species in which apomictic processes had been noted.

That is why embryological data with various degree of accuracy are known for more than one half of Rosaceae genera. This has been calculated from the monographs by Schnarf /1931/ and by Mandrik and Petrus /1985/ in which 77 names of embryologically investigeted genera were mentioned, compared with 124 genera in the classification of Hutchinson /1964/, which included also the taxa Neuradeae and Chrysobalaneae recognized as separate families by some of the systematists.

The embryological diagrams /characteristics/ of Rosaceae published by Schnarf /1931/, Davis /1966/, Poddubnaya-Arnoldi /1982/, Mandrik and Petrus /1985/ gave a uniform picture of the family. There has been, however, no documented elaboration dealing with the degree of its embryological uniformity. The systematists differed in proposed classification, which pointed to some unsolved problems of the mutual relation of the taxa. In this situation embryological arguments are welcome. They can hardly be expected from the generalized diagram in which differences and similarities between subfamilies or tribes are not marked. On the other hand, there were points in the diagrams which suggested the existance of differentiation, e.g. in the structure of the ovule, details of which are important for phylogenetic considerations. Nevertheless the comparative studies of particular characters were rare, a common deficiency of embryological investigations.

There was one comparative study within Rosaceae involving their four subfamilies: Spiraeoideae, Rosoideae, Pomoideae and Prunoideae published by Jacobsson-Stiasny /1914a/. It was based on the results of examination by various authors, mainly by Péchoutre /1902/, but the limited number of investigated species and of embryological characters prevented the authoress from drawing more reliable conclusion in several points. Nevertheless, the diagrams of the subfamilies showed some differences, e.g. the development of the obturator was noted only in Pomoideae, Prunoideae and partly in Spiraeoideae and not in Rosoideae; endosperm haustoria occurred in the members of Rosoideae, Pomoideae and Prunoideae but they were not found in Spiraeoideae. The investigations were not continued and often in the papers published in the following years no attention was paid to the several characters considered by Jacobsson-Stiasny /1914 a/.

The necessity of completing data concerning uninvestigated genera is obvious. In addition the examination should embrace as many embryological characters important from the phylogenetic point of view as possible. The lack of these data and their insufficient accuracy is an obstacle in the computational analysis. The need for planned studies on the morphological and anatomical characters was stressed by those who like Young and Watson /1970/, tried to use them in the taxonomical investigations.

In the phylogenetic considerations the structure of the ovule is taken into account in the first place /e.g. the type of the ovule, number of integuments, thickness of the nucellus/. The information we have in this subject, apart from some notes dealing with the embryology of particular species, are chiefly based on works by Péchoutre

/1902/ and Juel /1918/. On the whole, the data are inadequate to the needs. Let us consider as an example the type of the ovule. The distinction of three basic types: ortho-, ana-- and campylotropous is insufficient in Rosaceae. In the subfamilies Spiraeoideae and Prunoideae two types of ovules occur: ana- and hemitropous and the borderline between these types is not established. Another problem is that in Rosaceae crassinucellate ovule /an old embryological character/ is correlated with one or two integuments, free or with a tendency to concrescence. The insufficient data prevent us again from drawing a line between various integumentary forms or from following an evolutionary line leading to a single integument, a progressive embryological character. Because the type of an adult ovule may be misinterpreted comparative-developmental studies are necessary. Such a general research program for ovules was postulated by Bouman /1971/ who stressed that a similar final result of the development may be achieved in various ways.

To note new facts is important to pay attention to the possibility of occurrence of a given character in studied plants. <u>Haustoria</u> may serve as an example. Maheshwari /1963/ and Poddubnaya-Arnoldi /1982/ numbered them among characters important for phylogenetic and taxonomic studies. Davis /1966/ supposed that they might be an expression of convergent evolution, however, she believed that they were constant within taxa.

The diagrams of Rosaceae mentioned endosperm haustoria as a general character of the family /Poddubnaya-Arnoldi 1982/ or named the genera in which endosperm haustoria were found: Prunus /Schnarf 1931/ or Prunus and Rubus /Davis 1966, Mandrik and Petrus 1985/. This type of haustorium has been found by the present author in Potentilla Crantzii and P. arenaria. It was visible in the chalazal part of the embryo sac filled with nuclei of various size and shapes /Czapik 1961, 1962/. One can hardly suppose that it is limited to these three genera only. Still, little is known about its occurrence in Rosaceae as well as about its distribution within the above mentioned genera /Fig.1/.

Recently, a new type of haustorium - the embryo sac haustorium has been found in Dryas octopetala /Czapik 1986, 1987/. It was formed by the central cell before fertilization. In this type the growing haustorium leaves antipodals in situ, at the upper part of the grown up embryo sac. In such a situation the formation of the haustorium may be easily overlooked, especially in long embryo sacs with inconspicuous antipodals

The third type of the haustorium in Rosaceae, formed by the megaspores, was described in Potentilla heptaphylla by Rutishauser /1945/.
This type, probably rare, had not been observed in other species
either by Rutishauser himself or by other students of the genus.

One can expect, that all types of haustorium are formed in the
family more often than they are described. Jacobsson-Stiasny /1914 b/
supposed that there was a positive correlation between haustoria
and crassinucellate ovule. Rosaceae would be a good object for check-
ing this opinion.

Interesting from the theoretical point of view is the variability
connected with the time of the formation of the secondary nucleus.
The present author has observed it in 16 species of Rosoideae and
noted differences in the moment of the polar nuclei fusion /Czapik
1985 b/. The fusion may occur early before pollination, before fertilization, or simultaneously with the fusion of the sperm nucleus
with the polar nuclei. In some species or ovules the fusion does not
take place when fertilization is prevented, e.g. in Agrimonia eupa-
toria or is distinctly delayed e.g. in Rubus bellardii /Czapik 1983,
1985 b/. The variability of this process is characteristic of the
family, but the meaning of the differential behaviour of polar nuclei and the factors responsible for it are still unknown. The moment of the fusion has been methodically examined in Rosoideae only
and should be checked in the representatives of the remaining subfamilies in which the stage of unfused, but closely attached polar
nuclei, is commonly described or at least visible in the drawings of
embryo sacs /Fig.2/.

Apomictic studies have been dominationg in the embryological research of Rosaceae since 40´s. Elementary apomictic processes of all
types are common in the family and its taxa show various degrees of
the facultative apomixis up to the obligatory one. The studies are
very promising and undoubtedly they will be continued, following the
general apomictic problems. It seems that one of them is especially
worth drowing attention of the investigators: the occurrence of the
elementary apomictic processes or the full cycle of apomictic reproduction in diploids.

First examples of diploid apomicts: Potentilla argentea /Müntzing
1928/ and P. aurea /Shimotomai 1935/ were described in Rosaceae. One
may suppose that they are not unique. Recently, the present author
has found elementary apomictic processes in two diploid species:
Waldsteinia geoides /2n = 14/ and Sibbaldia procumbens /2n = 14/.
In Waldsteinia geoides /Czapik 1985 a/ apo- and diplosporous embryo

sacs developed, but their functional ability could not be checked.
In S. procumbens haploid parthenogenesis occurred in late autumnal
flower buds /Fig.3; unpubl./. It was not clear to what degree the
external conditions influenced the autonomous formation of embryos
and if the development of the seeds would be possible here. As the
apomictic processes in diploids are rare any further data are theoretically important.

Investigations applying TEM and SEM methods are in Rosaceae very
scarce. The present state of embryological examination of the family
is connected with the classical methods of the light microscope. It
may be, on a whole, better than of many other groups but it hardly
could be judged as satisfactory. Apart from filling the gaps at the
generic level it is necessary to check the embryological uniformity
of the family in planned studies. In addition, many important characters, at present neglected, should be observed including into the
observation their development and variations. The family is embryologically interesting and all the effords are worth being done to
have it properly examined.

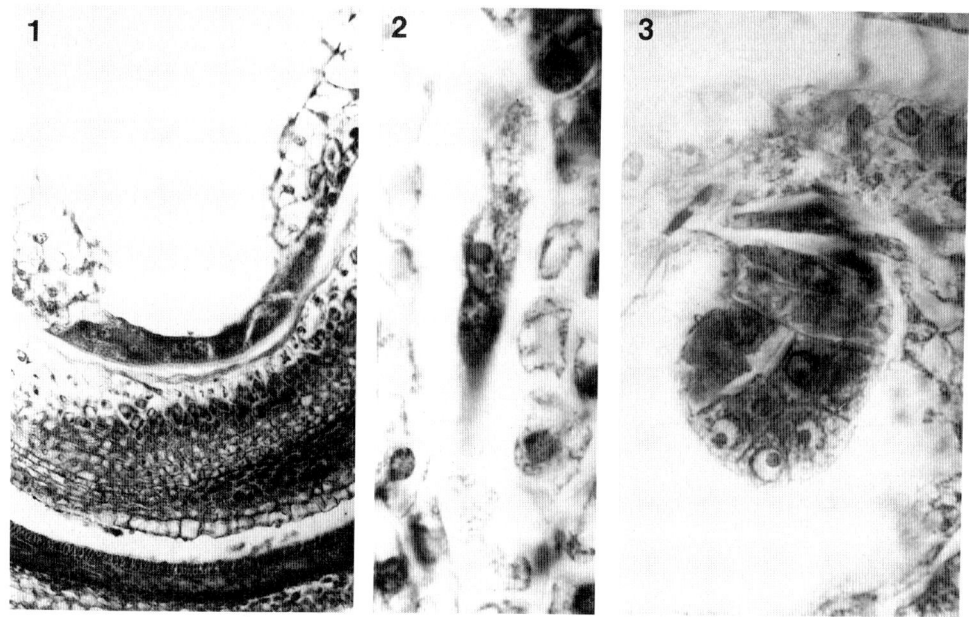

Figs 1 - 3. Rosaceae: 1. late stage of the endosperm haustorium
in Potentilla Crantzii; 2. polar nuclei attached one to another before fusion in Sibbaldia procumbens; 3. haploid parthenogenetic embryo from a closed flower of Sibbaldia procumbens /1: ×400; 2 and 3:
×750/.

442

References

Bouman F /1971/ The application of tegumentary studies to taxonomic and phylogenetic problems. Ber Dtsch Bot Ges 84: 169-177

Czapik R /1961/ Embryological studies in the genus Potentilla L. I. P. Crantzii. Acta Biol Cracov ser Bot 4: 97-119

Czapik R /1962/ Embryological studies in the genus Potentilla L. II. Potentilla arenaria. Acta Biol Cracov ser Bot 5: 29-42

Czapik R /1983/ The secondary nucleus in four species of the genus Rubus. Acta Biol Cracov ser Bot 25: 179-188

Czapik R /1985 a/ Apomictic embryo sacs in diploid Waldsteinia geoides Willd./Rosaceae/. Acta Biol Cracov ser Bot 27: 29-37

Czapik R /1985 b/ Secondary nucleus in Rosoideae. In: Willemse MTM, van Went JL /eds/ Sexual reproduction in seed plants, fern and mosses. Pudoc, Wageningen

Czapik R /1986/ Mono- and bisporic embryo sacs in Dryas octopetala L. /Rosaceae/. Acta Biol Cracov ser Bot 28: 31-38

Czapik R /1987/ Embryo sac haustorium in Dryas octopetala L. /Rosaceae/. Acta Soc Bot Pol 56: 209-214

Davis G /1966/ Systematic embryology of the angiosperms. John Willey New York

Hutchinson J /1964/ The genera of flowering plants /Angiospermae/. Dicotyledones, Vol I. Clarendon Press, Oxford

Jacobsson-Stiasny E /1914 a/ Versuch einer embryologisch-phylogenetischen Bearbeitung der Rosaceae. Sitz ber Kaiser Akad Wiss Wien Math-Natur Kl 123, Ab I, H VII: 763-800

Jacobsson-Stiasny E /1914 b/ Versuch einer phylogenetischen Verwertung der Endosperm- and Haustorialbildung bei den Angiospermen. Sitz ber Kaiserl Akad Wiss Wien Math-Natur Kl 123, Ab I, H V: 467-603

Juel HO /1918/ Beiträge zur Blütenanatomie und zur Systematik der Rosaceae. Kongl Sve Vet-akad Handl 58: 1-82

Khokhlov SS /1967/ Apomixis: classification and distribution /in Russian/. In: Dubinin HP /ed/ Uspekhi sovremennoy genetiki 1: 43-105. Nauka, Moskva

Maheshwari P /1963/ Embryology in relation to taxonomy. Vistas in Botany 4: 55-96

Mandrik VJu, Petrus JuJu /1985/ Family Rosaceae /in Russian/. In: Batygina TB, Yakovlev MS /eds/ Comparative embryology of flowering plants, Brunelliaceae-Tremandraceae. Nauka, Leningrad

Müntzing A /1928/ Pseudogamie in der Gattung Potentilla. Hereditas 18: 387-433

Péchoutre F /1902/ Contribution à l'étude du dévelopment l'ovule et la graine des Rosacées. Ann sc nat Bot 16: 1-158

Poddubnaya-Arnoldi VA /1982/ Cytoembryological characteristic of the angiosperms families /in Russian/. Nauka, Moskva

Rutishauser A /1945/ Zur Embryologie amphimiktischer Potentillen. Ber Schweiz Bot Ges 55: 19-32

Shimotomai N /1935/ Zur Kenntniss der Pseudogamie bei Potemtilla. Proc Imp Acad Tokyo 5: 338-339

Schnarf K /1931/ Vergleichende Embryologie der Angiospermen. Gebrüder Borntraeger, Berlin

Young DL, Watson L /1970/ The classification of dicotyledons, a study of the upper levels of the hierarchy. Austr j Bot 18: 387-433

Some Aspects of Reproductive Biology: Asexual Reproduction and Heterogenity of Seeds

T.B. BATYGINA

Department of Embryology
Komarov Botanical Institute
ul. Popova 2
197022 Leningrad
U.S.S.R.

Abstract

The work gives an analysis of the results of comparative and experimental studies of the phenomena of embryoidogenesis and defines its position within the system of reproduction and multiplication of angiosperms. The author introduces a new notion "embryoidogeny" to denote a specific mode of sporophyte formation, including into it reproductive (embryonic, nucellar, integumental) and vegetative (foliar, rhizogenic and cauligenous) embryoidogeny. Embryoidogenesis is a peculiar mode of the formation and development of a sporophyte (as compared to embyogenesis and gemmorhizogenesis) in homophazic (sporophyte -- sporophyte) reproduction. At the same time, so far as embryoid is a structural unit of reproduction, it appears to be possible to speak about the embryoidogenic type of asexual reproduction and multiplication. Reproduction and multiplication of flowering plants is represented by three types: embryogenic, embryoidogenic and gemmorhizogenic.

Results and Discussion

Maheshwari (1950) noted some essential differences of adventive embryos from sexual embryos. Steward et al. (1958) for the first time discovered an embryolike structure in culture in vitro of Daucus carota which became the object of general biological interest. Vasil and Hildebrandt (1966). Haccius, Lakshmanan (1969), Swamy and Krishnamurty (1981) most fully discussed the problem associated with differentiation of embryo-like structures, adventive buds and touched upon the so-called "foliar embryos" and came to different conclusions.

Investigation of the problems of morphogenesis in situ, in vivo and in vitro enabled us (Batygina, 1977; Batygina et al., 1978; Batygina 1984, 1987) not only to single out three modes of formation of the sporophyte embryogenesis, embryoidogenesis, organogenesis (gemmorhizogenesis), but also to establish their universality. Further analysis of the data available made it possible to more fully characterize the "embryoid" and get an idea of embryoidogenesis as of a specific, peculiar mode of sporophyte reproduction.

The embryoid is an embryo-like bipolar structure (with conjugated development of shoot and root apexes) - the rudiment of a new individual. It is formed asexually usually from one somatic cell, but can also originate from an "embryonal cellular complex", a meristematic focus of cells formed by the primary or secondary meristem. The embryoid does not usually have a common vascular system with the maternal organism. Typical of it is the formation of a new axis which connects the forming polarly arranged apexes of the shoot and the root. The embryoids differ from sexual and "apomictic" embryos (formed from gametophyte cells) not only by their origin, but also by a more slowed down and irregular (histologically unorganized) differentiation during the first stages of its development.

In connection with the elucidation of morphological nature and spread of embryoids, a new concept of asexual reproduction has been proposed by us (Batygina, 1987) according to which the embryoidogenic type of reproduction (along with the gemmorhizogenic type) is one of the basic types of asexual reproduction (Fig. 1).

Thus, embryoidogenesis is a peculiar mode of formation of the sporophyte (as compared to embryogenesis and gemmorhizogenesis) with homophase (sporophyte --- sporophyte) reproduction and multiplication. At the same time, since the embryoid is a structural unit of reproduction, it is probably, possible to speak about the embryoidogenic type of asexual reproduction and multiplication.

The phenomena of pseudoviviparity associated with the formation of the so-called "foliar buds" (="adventive", "brood" buds, as well as "foliar embryos", "embryo" by terminology of various authors) were repeatedly discussed in literature, Naylor (1932), Yarbrough (1932, 1934) showed on Bryophyllum that the "foliar embryo", the rudiment of the new individual is the result of endogenic differentiation of one meristematic focus of the primary meristem. It develops as a bipolar formation in which both polar structures (the shoot apex with the rudiments of the first leaves and the apex of the root) develop more or less simultaneously. We share the point of view of Yarbrough (1934) that such formations by the character of their development show more likeness with the embryos than with the adventive buds (gemmae). In his opinion, the "foliar embryos" differ from the "apogamic embryos" formed in seed only by their origin. As distinct from "foliar embryos" in Bryphyllum, McVeigh (1938) for the first time described the formation of the rudiments of new individuals from the foci of the secondary meristem (formed from the cells -derivatives of the leaf epiderm in Crassula multicava. She stresses that the primary root and the shoot apex in these embryos are formed exogenously from one focus of the secondary meristem. McVeigh points out that reproduction in various species of the family Crassulaceae is effected by different ways - the new individual can be formed either endogenously, from the focus of the primary meristem, or exogenously, from the locus of the secondary meristem. Thus, at least some of the formations which were earlier classified by many investigators as "foliar gemmae" or embryos formed on leaves, should be assigned to the category of embryoids.

The main critera for distinguishing between embryoids on one hand and adventive buds on the other, are not only the difference of their origin, but also the fact that embryoids are not the parts of the plant (they are self-dependent individuals of the subsequent generation) and as a rule, they are not connected with the maternal organism by a common conductive vascular system (Haccius, Hausner, 1975). Fragmentary data of the early works on the genesis of the so-called "foliar embryos" give possibility to interpret such structures in different ways. We do not exclude the presence of transitional forms, nor of two different ways of formation of the sporophyte in one and the same plant species.

The views of many botanists as to the relationship of sexual and asexual reproduction (mainly because of a different treatment of the sense of notions "apomixis" and "vegetative and asexual reproduction") are essentially different. The notion "apomixis" was applied by different authors in various senses - from a wide interpretation, as a synonym of asexual reproduction - to a narrow one -

"gametophytic apomixis" (Battaglia, 1963; Nogler, 1984; Grant, 1981).

In recent time the problem of meaning of such basic notions as "sexual process" and "sexual reproduction" has come to be disputable (Mogie, 1986; Dick, 1987).

At such extreme confusion of terms and notions relevant to the phenomena of reproduction, and as a result of great diversity of forms of gametophytic apomixis, the problem of classification of phenomena of sexual, apomictic and asexual reproduction acquires a special significance. We propose a classification of types and modes of reproduction in flowering plants where special attention is given to the embryoidogenic type of reproduction (Fig. 1).

It is known that the process of sexual reproduction does not, virtually, lead to the increase in the number of individuals; in connection with this, perhaps, it is more correct to compare sexual and asexual reproduction in discussing the ways (modes) of formation of new individuals. And the processes of reproduction proper which ensure the increase of the number of individuals of the new generation should be considered by means of other systems of comparison, for example, seed and vegetative reproduction, which means not the modes of reproduction, but the morphobiological systems of reproduction.

From our point of view, the terms "gamospermy" i.e. reproduction by means of seeds as a result of fusion of gametes and "apogamospermy" (="agamospermy" used by many authors) by means of seeds, but without fusion of gametes reflect all the diversity of the modes of seed reproduction, multiplication proper. The term "apogamospermy" includes both the phenomena of gametophytic apomixis (Nogler, 1984)and the phenomena of nucellar, integumental and "cleavage" "embryonic" – T.B.) embryony. The term "embryoidogeny" include not only the phenomena of reproductive (embryonal, nucellar, integumental) but also vegetative (foliar, rhizogenic and cauligenic) embryoidogeny. The term "apospermy" reflects the phenomena of vegetative reproduction represented by different modes (Levina, 1981).

The main conclusion to be drawn from this brief review is that at least some of the formations classified as "adventive buds" formed on leaves should, as we believe, be included into the category of embryoids, that is, referred to foliar embryoidogeny. Thus, in connection with this, the phenomena of pseudoviviparity are apparently represented not only by "gemmation", but also by embryoidogeny. It is to be noted that embryoids arise on different structures and at different stages of ontogenesis beginning with the earliest (zygote, embryo) up to the formation of reproductive structures (flower) both in situ, in vivo and in vitro. The seed has become that structural unit in which, in the process of evolution, a union of rudiments of different sporophytes – derivatives of both maternal (nucellus, integument) and daughter (zygote, embryo) generations can take place. Reproduction by means of embryoids effected in the seed structures essentially widens the diversity of seeds. Considering the embryoidogenic mode of reproduction it is apparently possible to single out the following categories of heterogeneity in seeds: seeds containing sexual embryo: seeds containing only embryoids originanting from the maternal sporophyte (in the case of integumental and nucellar embryoidogeny); seeds containing embryoids originating from the daughter sporophyte (embryonic embryogeny); seeds containing the sexual embryo and embryoids of different origin. In the complex system of forms of sexual and asexual reproduction of angiosperms reproduction by means of embryoids holds a unique position and, probably, is of different adaptive significance.

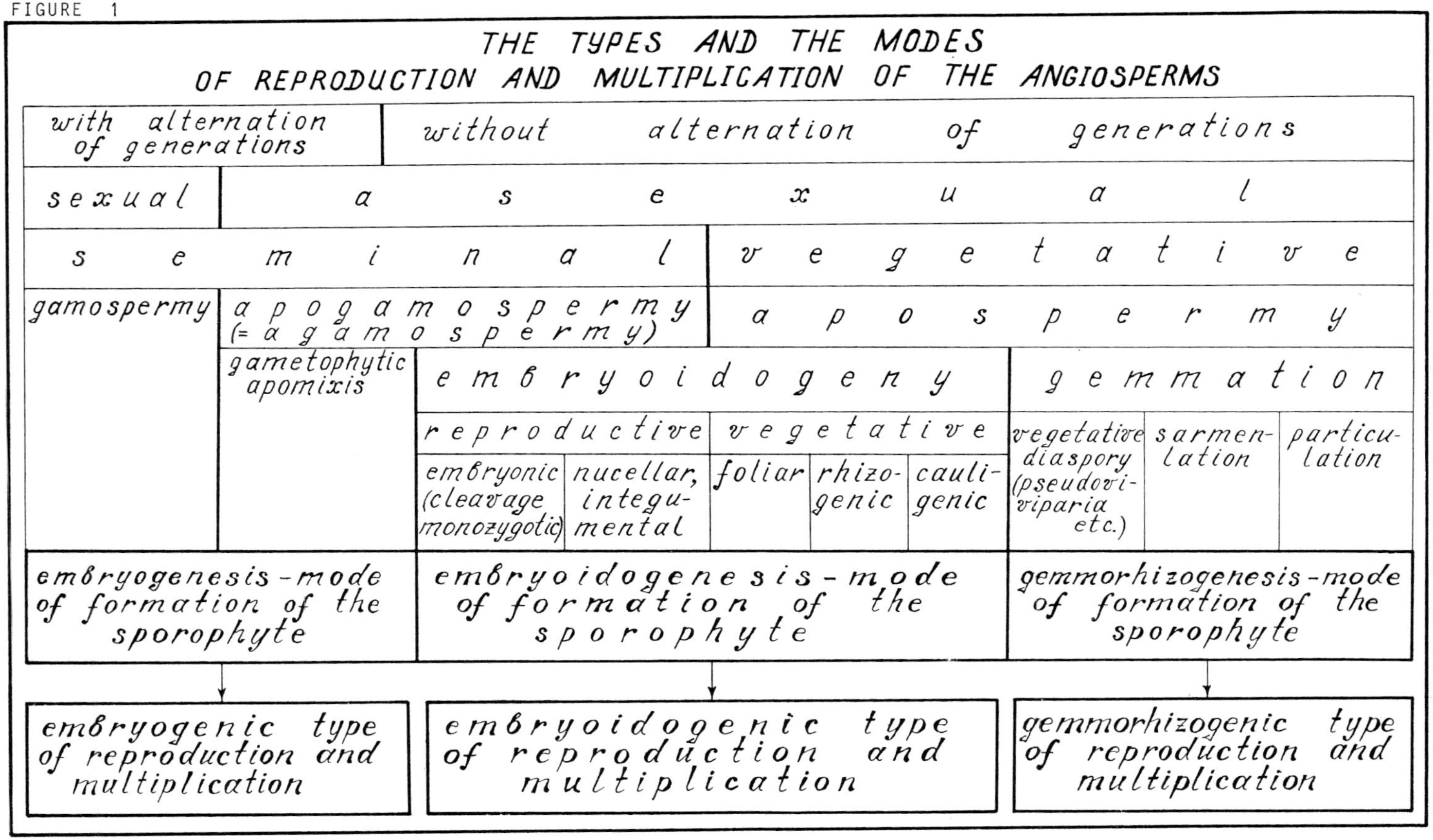

THE TYPES AND THE MODES
OF REPRODUCTION AND MULTIPLICATION OF THE ANGIOSPERMS

with alternation of generations		without alternation of generations		
sexual		a s e x u a l v e g e t a t i v e		
s e m i n a l		a p o s p e r m y		
	gamospermy	g a m o s p e r m y	g e m m a t i o n	

apogamospermy (= agamospermy)

gametophytic apomixis

e m b r y o i d o g e n y

reproductive — embryonic (cleavage monozygotic)

reproductive vegetative — embryonic nucellar, integu-mental; foliar, rhizo-genic, cauli-genic

vegetative — diaspory (pseudovivipary etc.), sarmen-lation, particu-lation

embryogenesis – mode of formation of the sporophyte

embryoidogenesis – mode of formation of the sporophyte

gemmorhizogenesis – mode of formation of the sporophyte

embryogenic type of reproduction and multiplication

embryoidogenic type of reproduction and multiplication

gemmorhizogenic type of reproduction and multiplication

FIGURE 1

References

Battaglia E (1963) Apomixis. In: Maheshwari P (ed) Recent advances in the embryology of Angiosperms. Intern Soc Plant Morphologists Depart Bot, Univ Delhi, p 221-264

Batygina TB (1977) Problems of morphogenesis in vivo and in vitro. In: Abstracts Indo-Soviet Symposium Embryology of crop plants. 23-26 Aug 1977. Univ Delhi, p 41-42

Batygina TB (1984) Problems of morphogenesis in situ, in vivo and in vitro. In: Novak FJ, Havel L, Dolezel J (eds) Proc Int Symp Plant Tissue and cell culture application to crop improvement. Olomouc Czechoslovakia. 24-29 Sept 1984. Czechoslovak Akad Sci, Prague, p 43-56

Batygina TB (1987) New concept of asexual reproduction in flowering plants. In: XIV Int Bot Congr. 24 July-1 Aug 1987. Berlin

Batygina TB et al (1978) Problems of morphogenesis in vivo and in vitro. Botanical Journal 63: 87-111

Dick MW (1987) Sexual reproduction: nuclear cycles and life-histories with particular reference to lower eukariotes. Biol J Lin Soc 30: 181-192

Grant V (1981) Plant specialization, 2n edn. Univ Press, New York

Haccius B (1965) Untersuchungen uber Somatogenese aus den Suspensorenzellen von Eranthis hiemalis Embryonen. Planta B 64: 219-224

Haccius B, Lakshmanan K (1969) Adventiv Embryonen Embryoide Adventiv-Knospen. Ein Beitrage zur Klarung der Begriffe. Osterr Bot Z 116: 145-158

Konar RN, Nataraja K (1965) Experimental studies in Ranunculus sceleratus L Development of embryo from the stem epidermis. Phytomorphology 15: 132-137

Konar RN, Thomas E, Street H (1972) Origin and structure of embryoids arising from epidermal cells of the stem of Ranunculus sceleratus L. J Cell Sci 11: 77-93

Levina RE (1981) Reproductive biology of Seed plants. Nauka, Moscow

Mogie M (1986) Automixis: its distribution and status. Biol J Lin Soc 28: 321-329

Naylor E (1932) The morphology of regeneration in Bryophyllum calycinum. Amer J Bot 19: 32-40

Nogler GA (1984) Gametophytic apomixis.In: Johri BM (ed) Embryology of Angiosperms. Springer, Berlin Heidelberg New York Tokio p 476-515

Steward FC, Mapes MO, Mears K (1958) Growth and organized development of cultured cells. II Organization in cultures grown from freely suspended cells. Amer J Bot 45: 705-708

Swamy B, Krishnamurty K (1981) On embryos and embryoids. Proc Ind Acad Sci Plant Sci 96: 401-414

Vasil JK, Hildebrandt AC (1966) Variations of morphogenetic behavior in plant tissue cultures II Petroselinum hortense. Amer J Bot 53: 9

Veigh H Mc (1938) Regeneration in Crassula multicava. Amer J Bot 25: 7-11

Yarbrough JA (1932) Anatomical and developmental studies of the foliar embryos of Bryophyllum calycinum. Amer J Bot 19: 443-453

Yarbrough JA (1934) History of leaf development in Bryophyllum calycinum. Amer J Bot 21: 467-482

Final Remarks

Present Status and Future Prospects of Sexual Reproduction Research in Higher Plants

HANS F. LINSKENS
Botanisch Laboratorium
Faculteit der Wiskunde en Natuurwetenschappen
Katholieke Universiteit
Toernooiveld
NL-6525 ED Nijmegen, The Netherlands

It is now one and a half century since the italian investigator, Giovanni Battista Amici (1824/1830), made the fundamental discovery, that germination of pollen grains on the stigma, thus forming pollen tubes which grow into the style and ovary, finally entering the ovules (F. Garbari, E. Pacini, eds: Plant Embryology, the tuscan contribution; Proceedings of the conference held in Pisa. Department of Botany; 25th October 1986, in honour of the biocentenary of the birthday of Giovan Battista Amici -1786/1863-.Pacini Editore, Pisa, 1987). Experimental plant embryology can be dated by the first embryo cultured on artificial medium by Hennig in 1904.

What significant progress has been made since that time!

The present state of fertilization research can be measured by the size increase of the Delhi textbook of embryology: Published in 1950 and written by the late Paul Maheshwari on 450 pages; the one-man book was transformed in 1963 under his editorship to a monograph written by 11 co-authors, followed in 1984 by the ambitious "Embryology of Angiosperms" with 20 authors on more than 800 pages. New generations of plant embryologists have not only extended the field by new methods, but also by increasing our insight to a point "that we are on the threshold of a new wave of discovery", as John Heslop-Harrison pointed out in his Foreword (to B.M. Johri, ed.: Embryology of Angiosperms; Springer, Berlin - Heidelberg - New York - Tokyo, 1984).

This symposium held in Siena for the second time, this time organized by Mauro Cresti, Paolo Gori and Ettore Pacini, gave an excellent overview of the present state of research going on in the field of sexual reproduction of higher plants. It will not be possible to summarize the fascinating new perspectives in a 15 minute talk. Instead, I will restrict the talk to some specific highlights, without mentioning the names of the related investigators.(*)

Some fields of embryology seem to be less active, although there is still some work going on. Clearly, the plant kingdom is so large that by far not all of the types of embryos and embryonic development have been described. However, new methods and more sophisticated instrumentation have opened new fields of research. During the recent years not only transmission and scanning electronmicroscopy but also tissue culture research have deepened the insight into the secrets of the sexual processes in plants. Also, more advanced techniques, like HPLC, NMR, GC/MS, micromanipulation, freeze substitution and immunological methods have helped a great deal.

(*) The details and references of the majority of the topics mentioned in this summary can be found in the abstracts and papers of this volume.

lf I were asked to summarize fields of plant reproduction research, which could be considered to be coming to a conclusion, one could consider the following:
- classification of embryo sac types,
- longevity and storage of pollen as a routine method for germplasm conservation,
- in vitro germination of pollen,
- production of haploids through anther culture techniques.

But I hesitate to go on. So many unsolved questions remain to be answered; an even for the topics mentioned, many species have to be listed, which are unknown for the special prerequisites they will deliver in the application of standard methods.

Pollen development

Pollen can be defined as a product of differential gene expresion. The various developmental stages during microsporogenesis can be characterized by specific transcription products.

There appears to be a considerable overlap in genes expressed in gametophytic and sporophytic tissues that offer the possibility to use pollen in discrimination and selection for resistance and tolerance to various environmental factors, which would be a great advantage in breeding programs.

The whole field of pollen development has received a lot of attention lately. But still, the molecular basis of the developmental switch from somatic cell division in the sporogenic tissue to meiotic division and sporogenesis is not well understood. Anther culture techniques are particularly suitable for solving some of the above questions at the molecular level.

Genetic analysis in yeasts has identified both activators and inhibitors of meiosis. It should attract the attention of plant embryologists to find similar genetic systems, and after that to identify the gene products which regulate meiotic induction. May be it is a protein kinase, as in yeast?

Disruption of the male gametophytic development from its tight control by sporophytic tissue leads to a reprogramming of pollen, frequently towards cell division and embryogenesis, a prerequisite for so many successful experiments for androgenesis.

In vitro production of mature, fully functional pollen is an indication that in vitro development can also be directed towards various intended developmental processes.

Synchronization of pollen development in the anther, especially those having very long anthers remains an open question. The spatial distribution of cytological processes along the anther is·continually changing in time. Recent analysis of lily data on the cytological processes suggest a wave form phenomenon, with a maximal growth rate at the peak and a basipetal move, which is still to be explained, either by chemical pulse, a mechanical trigger, as e.g. reduction of turgor, or by electrical impulses.

Plasmodial tapetal offer an unusual opportunity for the study of cell surface interaction and the behaviour of cytoskeleton elements and may also be useful for the elucidation of the synchronization trigger.

Pollen maturation

Pollen maturation during the period before the anthesis seems to be rather independent of the parent plant. Drastic dehydration before shedding is followed by further dehydration in the atmosphere; DNA repair in pollen, confirmed recently, must now be seen in relation to the effect of dehydration on generative cell DNA and its configuration.

A reinvestigation of pollen saccharides leads to the conclusion that these reserves in pollen grains, even if commonly considered "starch", are chemically different from one another. Starchless likely means that lower weight polysaccharide molecules are dispersed in the cytoplasm. It was found that phytic acid is common in many pollen species.

There is a natural variability within pollen for cold resistance. Pollen selection through storage or even temperature treatment during pollen development may take place. Selection for stress resistance can also take place during pollen development and pollen tube growth, but also among mature pollen.

Forced pollen shedding as a stress condition alters the pollen diameter, the starch content of the grains and the early seedling growth of the progeny. Cold and/or darkness lower viability. So stress reduced the mean size of pollen grains.

Future studies on pollen development, as well as pollen germination and fertilization, obviously can be used as parameters for environmental pollution, the effects of pesticide and herbicides, the effect of radiation, acid rain, ozone concentration in the atmosphere, and the presence of surfactants in precipitation.

Pollen wall studies

Numerous studies have been carried out on the early development of the pollen grain wall in different species. A general agreement has been reached regarding the formation of the primexine or exine template, whereas opinions differ over the order of appearance of the earliest elements of exine, depending on the taxon investigated. In general, formation of both the intine and the surface coating on and in the exine, as well as the attachment of allergens, take place during the final stage of pollen development.

The molecular structure and biosynthesis of sporopollenin has still not been elucidated. This biopolymer is the main constitutent of the exine structure and is extremely resistant to non-oxidative degradation. It is supposed that sporopollenin is a biopolymer of carotinoids and carotinoid esters, an asumption that is based on comparative studies on naturally occuring sporopollenin.

The open questions which are related to the chemical structure are:
- Does the sporopollenin-like material appear in plant metabolism only in association with reproductive structures?
- Are there species-specific differences in the chemical constitution of sporopollenin?
- Where does the genetic control of the pattern formation, for polarity, for site and number of apertures come from?
- What is the function of the exine sculpture? It is an adaptation for dispersal, or storage, or recognition?

There seems to be a fresh approach, using tracer experiments, which have shown that phenylpropane metabolism is involved in sporopollenin biosynthesis. These findings are confirmed by studies with inhibitors of carotenoid biosynthesis which demonstrated recently that an intact carotenoid metabolism is not a precondition for normal sporopollenin accumulation.

All these findings and questions point to an urgent need for investigatons of the carotenoid hypothesis of sporopollenin. Meiocytes, pollen and tapetum cells are the suitable material for the elucidation of the biosynthesis of sporopollenin.

Female gametophyte

Structure and organization of the female gametophyte has been investigated for an increasing number of species. However, the knowledge on the function and regulation of the processes in the embryo sac as a whole, and on the specific function of the compartments of the embryo sac, is still limited. The intensive studies of meiocytes, micro- and makrospores, their cytoplasmatic organization, the shifts in the ribosomal population, de- and re-differentiation provide physiologists with extensive field for future research.

It is now also possible to isolate whole embryo sacs using enzyme treatment and even liberate female gametes from the isolated embryo sacs. This technique will open new perspectives for the manipulation of isolated male and female gametes and studying the functions of the other cells of the embryo sac.

Embryologists have provided comprehensive information on the shape, size, position and development of the antipodial cells in a great number of angiosperms. But, their functions are less understood and research is restricted only to fixed material. These studies suggest that the antipodes actively function in absorption by means of their wall projections; their chief role seems to supply protein requirements for the coenocytic endosperm.

There are interesting indications that the suspensor manufactures and/or transports gibberellins, and thus is a major route of nutrients into the developing embryo.

Little is known about the biochemical interaction between synergids and pollen tubes. There apparently exists more than one relationship:
(a) The influence of synergids on the directional growth of pollen tubes by synthesis of chemotropical compounds has been discussed for a long time; up till now only indirect evidence has been presented.
(b) The opening of the pollen tube tip at the moment of approach to the egg apparatus could be another function; the effect of the oxygen tension as a trigger for the dissolution of the tube tip was demonstrated.
(c) Which initial gynoecial events are responsible for the stimulation for synergid degeneration?
(d) The function of the filiformous apparatus is still an open question. Does it contribute to:
    - the release of chemotropical substances by building up a gradient for the short distance orientation of the pollen tube tip into the embryo sac?
    - the absorption metabolism for the nutrition of the megasporophyte?
    - to the establishment of the egg polarity?
(e) Also, the relationship between the egg apparatus and the other ovule tissues is still future item.

Progamic phase

The relationship between the gametophytic and sporophytic generation on the physiological, biochemical and the ontogenetic level is by far not clear: How is the pollen tube growth related to sporophytic processes and responses?

There is new evidence for non-random fertilization: the fastest growing pollen tubes fertilize ovules in a specific region of the ovary, a phenomenon already known as selective fertilization. The amount of pollen used for pollination, as well as double pollination, influence seed set and sometimes seedling vigour. So a higher pollen load may promote pollen competition in favour of the fittest pollen producing fitter seedlings. Excess pollen may even function as mentor. Limited pollination apparently leads to the greatest genetic variability.

Problems could arise because pollen grains, with new genomic construction, have a reduced competitive ability. Unreduced gametes could be a potential for employing sexual polyploidization.

Gynogenesis offers the possibility to produce haploid embryos. The study of apomixis should attract more research attention because of the growing importance in practical breeding programs.

More research should also be directed to understanding the mechanisms of chemotropic growth of pollen tubes in vivo. New results presented during this meeting suggest that ethylen may be one of the chemically possible stimuli. But the question of pollen tube guidance and target finding has raised new discussion since the widely accepted interpretation of pollen tube orientation from in-vitro experiments differs from in-vivo studies of pollen tubes after certain intra- and interspecific pollinations. It seems possible that during the course of the progamic phase several different orientation mechanisms function: Mechanical and tactile guidance may have a greater importance than chemical gradients. So the chemical gradient theory is again open for question.

Another unsolved problem is the transfer of various materials from the pistil into the pollen tube. Also the general translocation processes within the flower during flower development and the later changes due to pollination did not attract further interest of investigators.

Evidence is given for the exclusive location of lectin activity in reproductive organs of plants, suggesting a possible involvement in pollen-pistil interaction and in incompatibility responses.

Sperm cells

Isolation of generative cells and sperm cells appears to be a routine procedure, even from trinucleate pollen. Further manipulation of the fertilization process can be forseen, since injection of sperm cells into synergids or egg cells of in vitro cultured ovules using micromanipulators has been successful; sperm cells are dimorphic, at least in some of the more extensively investigated species. Those isolations may also help to understand preferential fertilization of plastid-rich sperms with egg cells.

The isolation of protoplasts from pollen grains, pollen tubes and egg cells seems to be a promising method for the investigation of cell wall components, cell wall regeneration, organelle development and phytin storage bodies.

The relationship between the vegetative nucleus and the generative cell or sperm cells in the pollen tube is not very well understood. The former may have a physiologically active role in mitosis of the generative cell and sperm formation.

Investigations of the cytoskeleton in pollen tubes have made great progress. Organelle movement and the traffic of fibrillar elements could be followed by differential interference microscopy, using continuous recordings made by video recorder, and the application of fluorochromes. Movement of the vegetative nucleus, generative cell and the gametes seems to be associated with the actine cytoskeleton. Bundles of microtubules are located in the cortical plasma of the generative cell and extend into the tail-like part of the cell. They appear to be primarily engaged in shaping the cell.

There is some evidence for the presence of kinesin in pollen tubes, a protein known to be involved in the movement of vesicles along microtubules in animal cells.

Application of tissue and cell culture techniques marked the entrance of plant reproduction into the area of biotechnology. Somatic embryogenesis offers the bypass of sexual fertilization. Embryogenesis of plants from immature microspores and pollen led to a technology which delivered microspore-derived plants from more than 200 species from 83 genera in 37 families. Nevertheless, progress is a bit frustrating and erratic since very few definitive recommendations about the successful culture conditions for an untried species can be made. More research on the culture-induced changes in the gametophytic development has to be done, especially since more knowledge on gene activity and gene regulation is how becoming available.

Thus, there is hope that this knowledge can soon be applied to the transition from gametophytic to the sporophytic stage.

Fertilization barriers

The way in which products of identical S-alleles carried in pollen and style, and isolated from many species interact to cause arrest of pollen tube growth within the style is still not understood. There is some evidence that adenyl cyclase plays a role in th recognition events involved in incompatibility. Signal molecules from pollen binding to stigmatic target cells could activate this enzyme, and in turn produce cyclic AMP as a second messenger.

Strange enough, the problems of cross incompatibility between species, - or as we prefer to say: of incongruity, because the phenomena between species are evidently not directed by the S - gene system, - have not yet found investigators to tackle the problem with more sophisticated methods than just looking for the behaviour of the pollen tubes.

Also the biochemical investigation of the incompatibility mechanism in trimorphic heterostylic species is still in the beginning stage. Male sterility is an important field of research and the recently published 1000 paged monograph attests to its importance. Nevertheless, the practical application of gametocides for many crops has still to be solved.

Final remarks

It is now 25 years since plant embryologists and plant sexologists came together for the first time in a special international meeting; this took place in 1963 at Nijmegen. Only a few participants of the present day symposium will remember the gathering which took place a quarter of a century ago. Most of the former participants left science or faded away. At that time, the center of discussion concerned the question whether or not more than one pollen tube could participante in the formation of one zygote, a controversy resulting from the Mitschurin-Lyssenko biology. If one looks in the proceeding volume of this first meeting, which was published under the title "Pollen Physiology and Fertilization", one realizes the great progress that has been made since then. Especially, the important contribution of electron microscopy has to be mentioned.

Some 10 years ago Heslop-Harrison (J. Heslop-Harrison: The forgotten generation; Some thoughts on the genetics and physiology of angiosperm gametophytes. In: D.R. Davies and D.A. Hopwood, eds., Proceedings of the 4th John Innes Symposium, John Institute, Norwhich, 1979) characterized the gametophytic part of the angiosperm life cycle as the "forgotten generation". Times changed and the forgotten generation became the center of interest of many biologists assembled here.

If some of the present participants meet again another quarter century later, more progress will be made and many of the problems I mentioned today will be solved. Some answers to questions provided by former generations will be rediscovered and resolved again, likely with more sophisticated methods.

I agree with Pacini (New Techniques and Recent Findings in Embryological Researches. In: Atti Soc. Tosc. Nat., Mem., Ser. B, 94, 189-201, 1987) who stressed that embryological research in recent years has made significant advances, especially in the biology of pollen. This is certainly due to the fact that the study of the male gametophyte is easier than the female and, for which the pathway and the means of spermatic nuclei from the synergids to the egg cell and proendospermic cells have only been observed in a few species. From the observations published, it must be concluded that even the female gametes to be fertilized are predetermined.

If my calculation is correct, and the meetings and conferences organized on personal initiative of european and american plant sexologists are included, than the next meeting will be the 16th, - not included the numerous embryological meetings in India, Australia and New Zealand. This fact demonstrates not only the increasing interest in plant embryology as part of the plant development biology but also that sexual plant reproduction is becoming more and more recognized as a fundamental body of knowledge for plant breeding, cell genetics and plant biotechnology. Since we now have a special journal (*) which will cover the whole field of sexual reproductive processes, normal and deviant, it demonstrates the recognition of our field. The research in the field of sexual plant reproduction has essentially a transdisciplinary character, and a great future.

Acknowledgement

I thank Mauro Cresti, Jack Van Went and Professor John Stoffolano for reading and critizising the manuscript.

(*) "Sexual Plant Reproduction", Springer International, Berlin - Heidelberg - New York - Tokyo. Published quarterly. First issue came out March 1988.

Abstract of Poster Presentations

Microsporogenesis and Pollen Behaviour in Jojoba [*Simmondsia chinensis* (Link) Schneider]

ARONNE G.

Istituto di Botanica, Facoltà di Agraria
Università di Napoli
Portici, Napoli
Italy

Plants of Simmondsia chinensis were grown in the Botanical Garden of Portici (Napoli). Flowering occurred during February, March and April 1988. In this period temperature ranged 2° to 25°C whereas according to Dunstone and Begg (CSIRO, Camberra, Australia, 1980) the optimum temperature for pollen germination is 27°C. Since the possibility of growing this species in our region has been considered for oil production, we studied its floral biology. Microsporogenesis, pollen behaviour and in vitro germination requirements have been analyzed.

During microsporogenesis, callose deposition was studied by fluorescence method; cytomixis and meiotic irregularities were often observed. Tetrads differentiate simultaneously and pollen grains show different shape and size; when treated with aniline-blue they present different fluorescence intensity. The pollen grain is tricolporate with a granulate exine layer. Problems of harmomegathy were considered: hydration was immediate in water while took about 1 h in 65% and 5 to 6 h in 80% sucrose solution. DAPI treatment (Vergne et al., Stain Technol.,62: 299-302, 1987) shows the presence of three nuclei and two of these are characterized by a more intense fluorescence.

Germination was observed by hanging drop technique. The pollen was sampled from recently opened anthers. Overall it showed low percent germination and a complete loss of germinability after two days. In February the highest percent germination (60%) was obtained in a water solution containing 80% sucrose, 100 ppm boric acid and 300 ppm calcium nitrate. Differently in April the best results (only 30%) were obtained with 20% sucrose, 100 ppm boric acid, 300 ppm calcium chloride and 100 ppm potassium nitrate. In April, however, there was a large failure in the elongation of the pollen tubes.

The relationship between germinability and the fluorochromatic reaction, with fluorescein diacetate (Shivanna and Heslop-Harrison, Ann. Bot. 47: 759-770, 1981 and quoted literature) as a general test for pollen viability, is also studied.

The results of this work suggest that the cytomixis could be caused by environmental stresses: the rapid loss of germinability is probably connected with the presence of three nuclei in the pollen grain (Brewbaker, J. Hered, 48: 217-277, 1957), while diversity among grains could be reported to irregular meiosis.

The problem of pollen germination is worthy of further studies before growing Jojoba in our region.

A Structural Investigation of the Ovule in Sugar Beet, *Beta vulgaris*: the Degenerated Synergid and the Micropylar Nucellus

Lone Bruun & P. Olesen[1]
Botanical Laboratory
University of Copenhagen
14o Gothersgade
DK-1123 Copenhagen K
Denmark

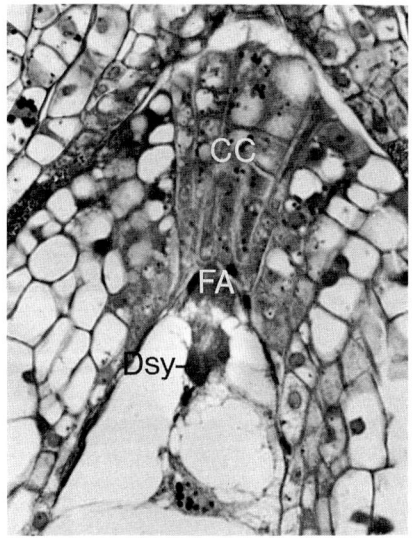

Fig. 1: The degenerated synergid (Dsy) and the columnar cells (CC) X 475.

As part of a comprehensive reseach programme on the reproduction of sugar beet, the mature embryo sac and the micropylar nucellus have been investigated by light- and electron microscopy (Materials and Methods, see Bruun 1987).

One of the synergids degenerates during ovule maturation before pollination (Bruun 1987). The cytoplasm stains intensely with aniline blue black (fig. 1). The filiform apparatus (FA) persists after degeneration of the synergid (fig. 1).

The columnar cells of the micropylar nucellus is located exactly in the elongation axis of the embryo sac towards the micropyle. The columnar cells are surrounded by a very

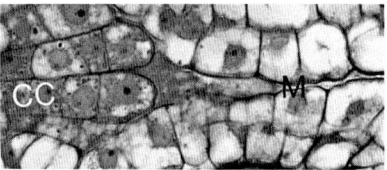

Fig. 2: PAS+ substance in the micropyle (M) X 785.

strong PAS+ cell wall (fig. 1). In the micropyle a PAS+ substance is often observed (fig. 2).

In sugar beet the PAS+ cell walls of the columnar cells, the PAS+ substance in the micropyle and the early degeneration of the synergid (Bruun & Olesen 1988) may all be events involved in the precise guidance of the pollen tube(s) towards the egg apparatus.

Bruun,L.(1987) The mature embryo sac of the sugar beet, Beta vulgaris: A structural investigation. Nord. J. Bot. 7: 543-551

Bruun,L. & Olesen,P.(1988) A structural investigation of the ovule in sugar beet, Beta vulgaris: The micropylar nucellus. Nord. J. Bot. (in press).

1: Biotechnology Section, A/S De Danske Sukkerfabrikker, P.O.Box 17, DK-1oo1 Copenhagen K, Denmark.

Embryo Stratification and Destiny of Embryonic Territories in Some *Hypericum*

M. BUGNICOURT and J. PARE
Laboratoire d'Embryologie végétale   Faculté des Sciences
33, rue Saint-Leu   80039 AMIENS CEDEX   FRANCE

For 32 species among the flora of *Hypericum* of France* we have clearly established : the diagram of filiation of the blastomeres (to be a direct descendant), the rules of segmentation of the proembryo (embryotectonic), the hypophysial generation and the organisation of the quiescent center, the appearence of the embryogenous potentialities (embryogeny).

*H. Androsaemum, H. Ascyron, H. barbatum, H. calycinum, H. Coris,
H. corsicum, H. Desetangsii, H. elatum, H. empetrifolium, H. fragile,
H. Helodes, H. hircinum, H. hirsutum, H. Hookerianum, H. humifusum,
H. linarioides, H. Metroi, H. montanum, H. mysorense, H. nummularium,
H. olympicum, H. orientale, H. orientale var. ptarmicaefolium, H. patulum,
H. perforatum, H. polyphyllum, H. pulchrum, H. quadrangulum, H. repens,
H. Richeri, H. rumeliacum, H. tetrapterum, H. tomentosum.*

This study is marked by a dominant feature : the constancy of development and of the rules of segmentation (Megarchetype IV ; Period II ; Série A' ; Group 9). The *Hypericum* proembryo is divided very early into three fundamental regions : superior level or embryonic globule, median level or hypophysial, inferior level or suspensor. Each level is composed of two sub-levels : superior layer, cotyledonary and hypocotyledonary regions ; median layer, quiescent center and root cap ; inferior layer, suspensor issued from ca (terminal cell of bicelled embryo) and cb (basal cell of bicelled embryo). The segmentation of the hypophysial level is fundamental in all cases for the continuation of the embryogenesis.

The standard embryo of *Hypericum* corresponds to the following description : 1) cotyledonary sub-level with only one layer until the evolution of the axial symetry, 2) hypocotyledonary sub-level differenciated very early into cortex and central cylinder, 3) hypophysial level, differenciated at the function of the proembryo and the suspensor, divided by the formation of horizontal wall shaped like a watch glass (whose periphery touches the lower walls of the dermatogen of the hypocotyledonary region), 4) regular suspensor, filamentous, ten-celled.

Some species show minor adaptations to this description ; three species : *H. perforatum, H. calycinum, H. Hookerianum* are exceptions to the rule which present in exceptional or experimental conditions profoundly unstable development. In this special case the suspensor passes the first modifications and then possibly the hypophysial level and finally (and optionally) the embryonic globule. These disorders envolve most often in two directions : 1) adjustement which orders normal embryogenesis, 2) anarchy which desorganizes ends embryogenesis.

The Stylar Transmitting Tissue of *Trimezia fosteriana* (Steym), Iridaceae

Per-Arne Bystedt

Department of Botany

University of Stockholm

106 91 Stockholm Sweden

The style of Trimezia fosteriana is hollow and the canal is lined with secretory cells of the transmitting tissue. This is in accordance with most other monocotyledons (Kronestedt & Walles 1986). The top and middle part of the stylar canal is lined with one layer of secretory cells. The secretion fills the lumen of the canal. Remnants from disintegrated cells are sometimes present in the secretion. The secretory process removes the cuticle from the cell walls. The cuticle in the middle part of the style remains undamaged and covers the secretion. Pollen tubes grow in the secretion between the cell wall and the cuticle. The transmitting cells and the cuticle appear unchanged after pollen tube growth. The basal (0.5 mm) part of the stylar canal divides into three narrow canals. The epithelial cells vary in size and shape. Some cells are small with a dome-shaped wall facing the canal. Other cells are larger, slightly papillae-shaped and protrude into the lumen of the canal. A cross section can give the impression of large intercellulars between the cells. Cuticle remnants in the secreted product between the protruding parts of the epithelial cells reveal the true situation. Work in progress is concentrated on the histochemistry and ontogeny of unpollinated and pollinated pistils.

Kronestedt E, Walles B (1986) Anatomy of the
    Strelitzia reginae flower.
    Nord. J. Bot. 6 : 307-320

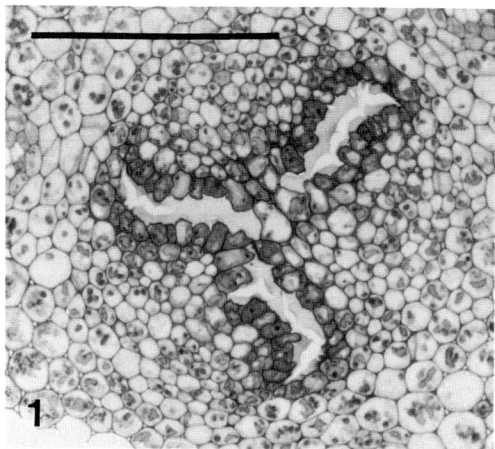

Fig 1. Cross section of the middle part of the style.
Bar =100 µm.

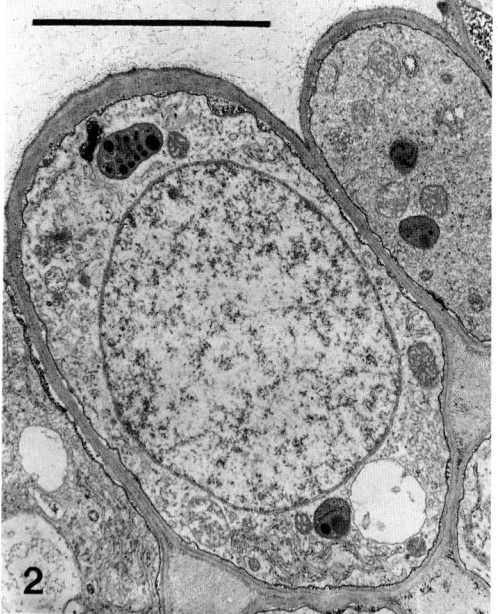

Fig 2. Cross section through the same part as in Fig 1. Bar= 10 µm.

Effects of Different PH on in Vitro Germination of *Pinus pinaster* Aiton and *P. pinea* L. Pollen Grains (*)

G. Cela Renzoni, L. Viegi
Dipartimento di Scienze Botaniche
Università di Pisa
Italia

The effects of different pH values (6.17, 5.50, 4.54, 3.57 in deionized water and 6.36, 5.43, 4.51 and 3.58 in deionized water plus 9% sucrose) on in vitro germination and pollen tube growth in pollen of <u>Pinus pinaster</u> Aiton and <u>P. pinea</u> L. from S. Rossore (Pisa, Italy) were investigated. Standard conditions were the same as in previous in vitro experiments (CELA RENZONI et al., 1983, 1986).

Our results indicate that pollen behaviour was significantly affected by pH 3.5 in both species.

The germination percentage was negatively influenced by pH 3.5, in agreement with TANAKA (1955), WOLTERS et al. (1987). The addition of sucrose had a favourable effect on germination rate, though less significantly in <u>P. pinea</u>.

Pollen tube growth in both species proved similar with all media, except for the pH 3.5, in which elongation was considerably less than in the other media. A marked acceleration of tube growth was also obtained with the addition of sucrose.

We hypothesized a negative effect of acidity on the presence of bilateral pollen tubes only in <u>P. pinaster</u>. The addition of sucrose had a significant influence on the number of bilateral tubes, almost doubled in all the pHs in both species.

The positions of the tube nucleus in the different pollen tube types were also considered in relation to previous reports (TANAKA, 1956).

REFERENCES

Cela Renzoni G, Viegi L, Stefani A (1983) Germinazione in vitro di polline di <u>Pinus pinea</u> L. di diversa provenienza. Giorn Bot Ital 117 (Suppl 1): 192-193.

Cela Renzoni G, Viegi L, Stefani A (1986) Effects of different media components on the germination of <u>Pinus pinaster</u> and <u>P. pinea</u> pollen. In: Cresti M, Dallai R (eds) Biology of Reproduction and Cell Motility in Plants and Animals. University, Siena p 163-168

Tanaka K (1955) The pollen germination and pollen tube development in <u>Pinus densiflora</u> Sieb. et Zucc. I. The effect of storage, temperature and sugars. Sci Rep Tohoku Univ (Biol) 21: 185-198

Tanaka K (1956) The pollen germination and pollen tube development in <u>Pinus densiflora</u> Sieb. et Zucc. II. The tube growth and tube nucleus. Sci Rep Tohoku Univ (Biol) 22: 219-224

Wolters J H B, Martens M J M (1987) Effects of air pollutants on pollen. Bot Rew 53: 372-414

(*) Supported by Italian M.P.I.

Pollen Grain Germination and Pollen Tube Growth of Chinese Giant Lily by SEM, EDAX and CTC Fluorescence

Chen Fang
Biology Department
Sichuan University
Chengdu, China

Pollen grain germination and pollen tube growth of Chinese giant lily (Cardiocrinum giganteum) have been investigated as follows: scanning electron microscopy (SEM) morphology of pollen germination in vivo, and calcium distribution in germinating pollen grains and growing pollen tubes in vitro with energy-dispersive X-ray analysis (EDAX) and chlorotetracycline (CTC) fluorescence. Research materials were mainly collected from the plants wild-grown at Mount Emei in Sichuan. Self- and cross-pollination was made by hand in the open, and pollen grains collected were germinated in vitro as discribed commonly. Samples were withdrawn at the required intervals and prepared for SEM, EDAX and CTC fluorescence. SEM observation revealed that pollen grains start to germinate on stigmas 30 min after pollination, and there are linked secretions between germinating grains and between pollen and stigma as well as secretion granules on the exine surfaces and at the germinal furrows. For EDAX, measured along the pollen tube axis from the tip to the base, the tubes show a maximum of calcium concentrition at the tip region. Behind the region of about 5-10 um from the apex, the calcium concentritions fall remarkably and remain relative constant in the rest of the tube. The germinating grains show higher calcium amounts at the exine surfaces, especially at the germinal furrow region having a lot of calciumrich bodies. Likewise, CTC fluorescence intensity shows the highest fluorescence at the growing tube tip, getting weakly in the basipetal direction, even disappeared finally. These findings seem to reveal a calcium distribution gradient declining from the tip to the older parts of the pollen tube, which might be considered as evidence for a calcium gradient playing an important role in the pollen tube tip growth.

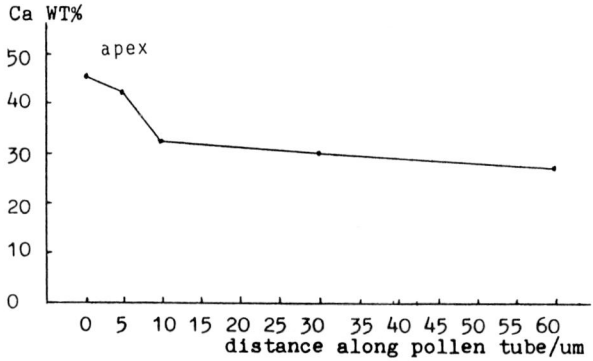

Fig.1. The tip-to-base calcium gradient in the pollen tube of Cardiocrinum giganteum measured by EDX.

The Development and Ultrastructure of Pollen in *Frittillaria thunbergii* MIQ

Z.K. Chen, F.H. Wang and F. Zhou

Institute of Botany
Academia Sinica
Beijing 100044, China

Fritillaria thunbergii Miq used in this study was collected from January to February, 1986 in Hangzhou, Zhejiang, China. In the middle of January, the sporogenous cells are found in anther. In the meantime, the cells in the middle layer of the anther wall are rich in amyloplasts which may be a product of dedifferentiation of the chloroplasts. By the end of January, the microspore mother cells are formed and a large number of mitochondria and plastids are found in their cytoplasm.

In early February the meiosis takes place in the microspore mother cells. At this moment, the prominent peritapetal membrane is formed in outer tangential wall of the tapetal cell and the orbicles are mainly distributed near the inner tangential wall. In the microspore stage, the cytoplasm of the tapetal cell contains a lot of plastids which possess two kinds of osmiophilic substances with different electron density. When the tapetal cells became disintegrated these plastids together with the periplasmodia are progressively extruded from tapetal cell into anther locule.

From the stage of meiosis I to vacuolate period of the microspore formation, the cytoplasm contains a large number of mitochondria, plastids and rough endoplasmic reticulum which is usually arranged parallel to the plasmalemma. At late vacuolate period of the microspore, their most striking feature is that the amyloplasts and lipid bodies become the main contents of cytoplasmic organelles and almost occupy the whole space of the cytoplasm except for the nucleus (Fig. 1).

During the maturation of pollen grains, the remains of primexine can be seen in the arcade between columellae. The vegetative nucleus usually contains two nucleoli and plastids are absent in the generative cell. So the model of the plastid genetics conforms to the Lycopersicum type.

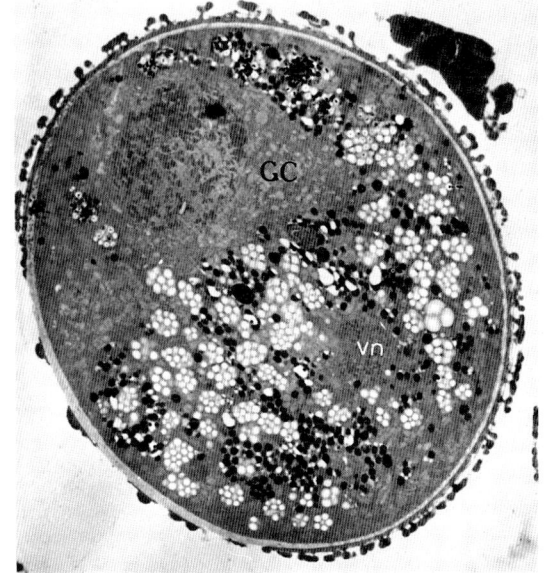

Effects of Acidity on in Vitro Pollen Germination and Pollen Tube Growth of *Nicotiana tabacum*

D. DANI, M. CRESTI and F. CIAMPOLINI
Department of Environmental Biology
University of Siena
Via Mattioli 4
53100 SIENA - ITALY

Fresh pollen of <u>Nicotiana tabacum</u> was cultured on Brewbaker and Kwack medium at pH 5,6 (control), 4.5, 4, 3.8, 3.6, 3.4 and 3.2, adjusted with H_2SO_4. Germination and pollen tube length of 100/150 grains was measured after 3/5 hrs of culturing. For TEM a standard glutaraldehyde/OsO_4 technique and embedding Spurr's resin was used. Microtubules were studied using monoclonal mouse anti-tubulin and rabbit anti-mouse antibody conjugated with fluorescein-isothiocyanate. Microfilaments were studied with labeling by phalloidin.

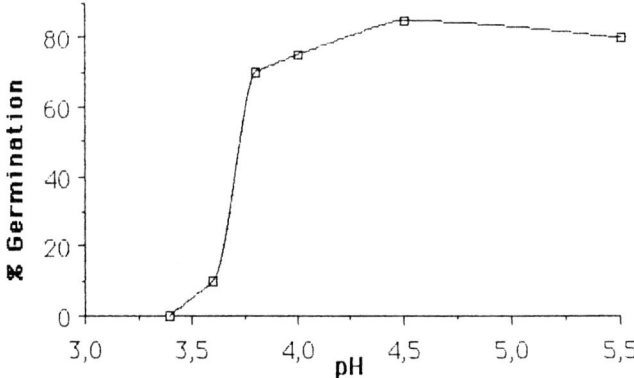

Percentage of germination at different pH.

The highest percentage of germination is found at pH 4.5. With lowering pH both pollen germination and length decrease till pH 3.4 when pollen germination is completely blocked. At pH 3.8 and 3.6 we observed aberrant phenomena as pollen tube biforcation and grains with 2 or 3 tubes. At pH 3.4 some pollen grains start germination but they burst before a real tube is formed. At pH 3.8 and 3.6 the pollen tubes show undulating cell walls, aberrant endoplasmic reticulum and mitochondria, and reduced electron-density of the cytoplasm. Fluorescence of microtubules and microfilaments is strongest at pH 4.5. It diminishes progressively when the pH is lowered, indicating a progressive breakdown or instability of these structures. Our observations show that low pH of the culture medium have a sincere effect on both pollen germination and tube growth in vitro.

A special thank to Dr. G. Cai for his help in the preparation of the paper.

Male Research Programme in *Brassica*

S. Detchepare, P. Heizmann, M. Murgia*, M. Cresti* and C. Dumas

Laboratoire de Reconnaissance Cellulaire et Amélioration des Plantes
Université Claude Bernard, LYON I. Bat. 741
43 Bd du 11 novembre 1918
F-69622 Villeurbanne Cedex
FRANCE

*Dipartimento di Biologia Ambientale
 Università di Siena
 Via P.A. Mattioli 4
 53100 Siena
 ITALY

In order to study male gametophyte development in Brassica oleracea L., the various stages of microspores and pollen were assessed in fresh anthers with the use of fluorescent dyes and cytological observations (Vergne et al. 1987).

On one hand, an ultrastructural study is currently in progress in order to investigate the male germ unit formation during the second pollen mitosis.

On the other hand, developmental variations in SDS-PAGE protein patterns of anthers and pollen grains were analysed from the late vacuolate microspore stage to the end of the pollen maturation period. Total protein staining and concanavalin A-binding glycoprotein detection showed that a specific set of developmental polypeptides appears during the tricellular pollen stages.

Protein synthesis was studied by (^{35}S)-methionine incorporation, SDS-PAGE, and autoradiography. Two periods in the protein synthetic activity were detected: the first one corresponding to the microspore and bicellular pollen stages; the second one corresponding to the mid, late, and mature tricellular pollen stages. During this second period, new polypeptides are synthetized and most of them are correlated with the developmental polypeptides that appear in total and Con A-binding protein patterns. These results suggest the occurence of a metabolic reorientation after the second pollen mitosis, at the time of sperm cell maturation.

Vergne P., Delvallée I. and Dumas C., 1987 – Rapid assesment of microspore and pollen development stage in wheat and maize using DAPI and membrane permeabilization. Stain Technology, 62: 299-304.

Interspecific Incompatibility Between *Brassica napus* and *B. oleracea*

S. Dharmaratne & T. Hodgkin
Scottish Crop Research Institute
Invergowrie
Dundee   DD2 5DA
United Kingdon

Brassica napus is an amphidiploid species which may be resynthesized by inter-crossing the parental species, B. campestris and B. oleracea. While the two parent species are self-incompatible, each possessing a single-locus sporophytic self-incompatibility system, B. napus is usually self-compatible. Test crosses between cultivars of each of these three species showed that pollen from B. oleracea usually failed to penetrate the stigmatic surface of B. napus pistils although all other combinations of interspecific crosses were compatible with respect to pollen-tube penetration. Three out of four synthetic B. napus lines, obtained by embryo culture following interspecific pollinations of B. campestris with B. oleracea, were also incompatible when pollinated with B. oleracea pollen although the parental lines used were reciprocally cross-compatible. However, the fourth synthetic was reciprocally compatible with B. oleracea. The inter-specific incompatibility could be overcome by bud pollination or by treatment with cycloheximide as is the case for intra-specific incompatibility in Brassica species.

Embryology of Barley III: Synergids and Pollen Tube Before and After Fertilization

Kirsten Engell
Botanical Laboratory,
University of Copenhagen
Gothersgade 140,
DK-1123 Copenhagen K.

Abstract for Poster.

This study is a part of an investigation of details in the embryology of Hordeum vulgare cv. Bomi. Earlier investigations (Engell 1988a, 1988b) have been made on the fertilization and the embryo formation from the time of pollination and the next 3-4 days followed in details every minutes or hours, based on light microscopy (LM) studies of serial sections. This is a preliminary report of an investigation of the synergids before and after fertilization observed with LM as well as transmisssion electron microscopy (TEM). 48 hours before the stigma is receptive for pollen grains the synergids are cells of equal size, pyriformed and with a characteristically hooked shape. The persistent synergid preserves this shape more than 26-28 hours after pollination but the other synergid begins to degenerate between 20-0 hours before the pollination takes place. The nucleus disappears, the membranes of the organelles are more sensible for staining and the organelles are difficult to recognize. Thus in Hordeum vulgare cv. Bomi one of the synergids is degenerated before pollination, probably a necessity because of the short time interval (45 minutes) between pollination and fertilization. The cytoplasm of the degenerated synergid is pressed towards the periphery when the pollen tube contents is disharged in this cell. Because of rod-shaped starch grains only in the pollen tube it is easy to distinquish between pollen tube- and degenerated synergid contents. The egg cell as well as the central cell has many sphaerical starch grains in their plastids, whereas the persistent synergid has lots of plastids without starch grains. Remnants of the persistent synergid can still be seen as late as 50 hours after pollination.

Figs 1-2. Hordeum vulgare cv. Bomi. 50 minutes after pollination. (LM).

Fig.1 The zygotic nucleus with materials from a sperm nucleus (sn) and the egg nucleus (en). Degenerated synergid (Ds) with starch grains from pollen tube. - Fig.2 The intact persistent synergid (Ps), the degenerated synergid and the egg cell. X 800

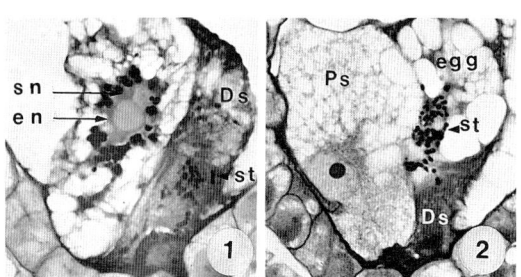

Literature:

Engell K (1988) Embryology of Barley I: Time course and analysis of controlled fertilization and early embryo formation based on serial sections. (in prep.)
Engell K (1988) Embryology in Barley III: Synergids and egg cell, zygote and early embryo formation. (in press).

Rapid, Even Distribution of Pollen Populations on Semi-Solid Media Using a Modified Replica Plate Technique

WILLIAM A. FEDER and JANE S. MIKA

Suburban Experiment Station
University of Massachusetts, Waltham
240 Beaver Street, Waltham, MA.
U.S.A.

Pollen populations are proving to be very useful tools for a wide variety of studies, including the measurement of biotic and abiotic stresses. A majority of these studies require fresh pollen which readily germinate on appropriate selected artificial media. A second requirement is the even distribution of single pollen grains on the medium so that numbers of pollen grains, rate of germination, and tube lengths are easily measured. Brush and other gross dusting techniques produce clumpy, uneven population distribution on semi solid media. This poster describes a simple technique which produces even population distribution.

Anthers are removed from fresh flowers and put into disposable test tubes containing 5 ml of the appropriate liquid growth medium. Tube is shaken vigorously and the anthers are removed, leaving a pollen suspension. The suspension is transported to the laboratory in a cool box. The tube is vigorously shaken for 30 sec. and 0.2-Q5 ml suspension in withdrawn and placed on 2.54 cm filter paper disc resting on a sintered glass suction platform. Excess fluid is immediately drawn off using a suction pump. The disc is then removed, inverted and pressed pollen side down onto the surface of semi-solid medium in a Petri Dish. After assuring contact of the entire disc with the medium surface,the disc is removed and discarded. The pollen now adheres to the media surface and is evenly distributed on that surface with a minimum number of clumps and piles. The process insures the freshness of the pollen, requires no dehydrating step, and produces a working pollen distribution on the medium surface which is viable, easily quantified, and ready for treatment, study, and observation.

Pollen Polysaccharide Reserves in Some Plants of Economic Interest

G.G. FRANCHI and E. PACINI*

Sections of Pharmaceutical Botany and *of Botany
University of Siena
Via Mattioli 4, 53100 Siena, ITALY

A recent paper by Baker and Baker (1979) on starch reserves in Angiosperm pollen grains only reports finding black and dark blue starch by the IKI test. However other types of insoluble polysaccharides, made not only by amylose and amylopectin, are present: they are composed mainly of amylopectin and dextrin, which stain from dark red to brown with the IKI test. Starch was detected in the shed pollen of 152 plant species and cultivars of economic interest. The cytological detection of starch was effected microscopically either by 1) staining with iodine/potassium iodide (IKI) or 2) observing by polarization microscopy. The occurrence of polysaccharides dispersed in the cytoplasm was detected by PAS preceeded by aldehyde blockade. Plants belong to the following families: Cucurbitaceae, Graminaceae, Labiatae, Leguminosae, Liliaceae, Oleaceae, Rosaceae, Rutaceae, Solanaceae.

A true starch (blue or black with the IKI test and polarizable) was found in only six of the examined specimens. The absence of starch does not mean that there are no polysaccharides. Pollen grains without starch stain intensively and uniformly with PAS. This means that polysaccharides are dispersed in the cytoplasm (cf. Ciampolini et al., 1982, where ripe pollen grains were observed with TEM after PATAg test). On the contrary starchy pollen grains never show a PAS positive cytoplasm.

Starchy and starchless grains may occur contemporaneously in the same anther, as in some cultivars of Olea europaea, Amygdalus persica, Cydonia oblonga and Pyrus communis. In two Citrus species some pollen grains have brown starch according to the IKI test, and other grains have black starch. This variability in viable grains seems to be due to an extreme asynchrony of pollen development perhaps linked to prolonged vegetative reproduction. Some species and cultivars also have a high percentage of unviable pollen; if they have starch grains at the cytoplasm degeneration stage, this starch persists even though it may not be present in the ripe viable pollen.

The occurrence of starch and its physical and chemical features are useful data in the study of several aspects of pollen biology. The type of reserve is a function of pollination type and vectors (Stanley and Linskens, 1974) and of the nutritive relationships between pollen tube and pistil.

References

Baker H.G. and Baker I., 1979 - Starch in Angiosperm pollen grains and its evolutionary significance. Amer. J. Bot. 66: 591-600.

Ciampolini F., Cresti M. and Kapil R.N., 1982 - Germination of cherry pollen grains in vitro - an ultrastructural study. Phytomorphology 32: 364-373.

Stanley R.G. and Linskens H.F., 1974 - Pollen. Biology, biochemistry, management. Springer Verlag, Berlin.

Some Parallelisms in Pollen Wall Configuration Between Orchidaceae and Annonaceae

M. Hesse

Institute of Botany and Botanical Garden
The University of Vienna
Rennweg 14
A - 1o3o Wien
Austria

The "typical" angiosperm pollen wall consists of a tectum with well developed columellae,a thin foot-layer ("basal layer"),and an endexine;furthermore of a thin,homogeneous intine except,of course,in the aperture regions. A quite different sporoderm type is found in some families as a matter of convergent or parallel evolution. Striking convergences with at least a dozen peculiar features of the respective pollen wall configuration we find in the by no means related Lauraceae and Zingiberaceae (STONE 1987) or in Annonaceae and Orchidaceae (BURNS-BALOGH & HESSE 1988, LE THOMAS et al. 1986, WAHA 1987): 1.A tendency towards exine reduction in proximal sporoderms of compound pollen (permanent tetrads or polyads) is common in orchids and occurs sometimes in Annonaceae. 2.The exine may consist of a "typical" exine described above (a rather rare situation both in orchids and Annonaceae), 3.A smooth exine is found in Auxopus,Lecanorchis;Piptostigma. 4.The basal layer is slightly lamellated in Lecanorchis,Phragmipedium and Vanilla,but often in the Annonaceae. 5.The basal layer is bi-layered in Pogonia;Annona. 6.The basal layer is extremely lamellated in Phragm. x cardinale;Uvaria. 7.Beneath the tectum we find only exine "stalactites" and no basal layer (Arethusa,Cleistes,Cypripedium,Epipogium,Pogonia,Stereosandra; Polyceratocarpus). 8.The exine consists only of sporopollenin globules within the intine (Acianthus,Arethusa,Cypripedium,Galeola,Habenaria,Stereosandra;Guatteria). 9.The exine consists only of sporopollenin elements on top of the intine (Auxopus, Cleistes,Corybas,Paphiopedilum,Triphora;Duguetia,Pachypodanthium). 1o.The germination area is indicated only by thickened intine regions (Lecanorchis;Polyalthia). 11.Before germination the intine protrudes enormously through the aperture region (Isotria;Anaxagorea). 12.Last not least the formation of polyads itself in Annonaceae. All these parallelisms rather depend on similar ecophysiological demands,which are far from being sufficiently known at time,and not on systematic relationships. But beside this,is there some atavism in orchids,"remembering" the ancient,primitive status of the early dicot angiosperms ?

BURNS-BALOGH P, HESSE M (1988) Pollen morphology of the cypripedioid orchids.
    Plant Syst Evol 158: 165-182
LE THOMAS A, MORAWETZ W, WAHA M (1986) Pollen of palaeo- and neotropical Annonaceae.
    In BLACKMORE S, FERGUSON (eds) Pollen and Spores,Form and Function. Linnean
    Soc. Symp. Series 12: 375- 388. Academic Press London
STONE DE (1987) Developmental evidence for the convergence of Sassafras (Laurales)
    and Heliconia (Zingiberales) pollen. Grana 26: 179-191
WAHA M (1987) Different origins of fragile exines within the Annonaceae. Plant
    Syst Evol 158: 23-27

Placental Pollination of *Lilium longiflorum* to Study Pollen Tube Growth

J. Janson

Department of Plant Cytology and Morphology

Agricultural University

Arboretumlaan 4

6703 BD Wageningen

The Netherlands

To overcome self-incompatibility and incongruity in <u>Lilium</u> <u>longiflorum</u> breeding programs, the cut-style pollination can be used. With this method, pollen tubes grow along the placenta but rarely enter the micropyles. Accordingly, the seedset is low, even if compatible pollen is used. To study the phenomenon of the pollen tube not entering the ovule, placental pollination was carried out as a model system.

Ovaries were sterilized and longitudinally cut in 6 sectors, each having a placenta with a row of ovules. These were placed on different media : either a supplemented B5 medium, a Brewbaker and Kwack medium with 12 % sucrose (BK) or an agar plate containing 6 % mannitol. Compatible and incompatible pollen grains, pollen tubes isolated from the style and pregerminated pollen grains were placed on the placenta. The flowers delivering the placentas were of variable ages and half of the stigmas were pre-pollinated and the placentas isolated (and pollinated) before the pollen tubes coming from the stigma had reached their ovules. One week after culture the ovules were cleared and the pollen tubes were coloured with 1 % aniline blue in the clearing solution. SEM preparations were made of an experiment using compatible pollen grains.

The number of pollen tubes formed when pollen grains are placed on the placenta depends upon the medium used. On the B5 medium and the mannitol plate a bursting of the pollen grains or tubes is observed more often in comparison with the BK medium. A part of the pollen tubes grow in between the ovules, over the inner integument, and occasionally over the outer integument. On the BK medium an abundant pollen tube growth is observed on the inner integument in contrast with ovules on the B5 medium and the mannitol. Penetration of the ovule is not often observed (less than 5 %), but appears on all media. The different kinds of pollen grains and tubes, varying the age of the placenta and the medium do not improve the low percentage of ovules with a pollen tube growing into the micropyle. Pre-pollinating the stigma might have some influence. On all tested media, the percentage of ovules with a pollen tube in the micropyle is very low, which was also observed after cut-style pollination. This can be a consequence of the short length of the pollen tube if compared with pollination on the stigma and/or that the ovule is not receptive or that a pollination signal is missing.

Visualization of Microfilaments in Cytoplasm and Spindles of Meiocytes, Microspores and Pollen of *Gasteria verrucosa*

A.A.M. van Lammeren, J. Bednara*, M.T.M. Willemse

Department of Plant Cytology and Morphology

Agricultural University Wageningen

Arboretumlaan 4

6703 BD Wageningen

The Netherlands

Plant cells exhibit two types of cytoskeletons, the microtubular and the microfilamental network. In developing pollen of <u>Gasteria</u> the microtubular (MT) skeleton was investigated (Van Lammeren et al. 1985**) Here we present observations on the actin skeleton before, during and after meiosis.

Cells were removed from anthers, sticked onto poly-lysine-coated slides, incubated in extraction buffer (Pipes buffer containing Nonidet and DMSO), stained with Phalloidin-rhodamin in Pipes buffer, washed, embedded in Mowiol and Citifluor and observed with epifluorescence microscopy. Exine was permeabilized (4-methylmorphaline N-oxide monohydrate).

Microfilaments (MFs) were detected throughout all stages of pollen development (<u>Fig. 1</u>). At prophase bundles were observed in the cytoplasm (<u>a</u>). During metaphase I and II the bundles in the cell cytoplasm persisted but additional MFs coaligned the MTs of the spindles (<u>b</u>). After cold treatment the MFs in spindle disappeared. After pre-treatment with cytochalasin B only the spindle MFs persisted (<u>d</u>). Anaphase stages showed coalignement of MFs with MTs running from pole to pole and from kinetochores to poles. The latter decreased in size as the chromosomes moved pole ward. At telophase I and II MFs radiated from the poles. Dyads (<u>c</u>) young microspores (<u>e</u>) exhibited cytoplasmic MFs. In pollen grains a cortical network of exclusively MFs was observed (<u>f</u>).

Microtubules and MFs act together in the spindle and during positioning of the nucleus.

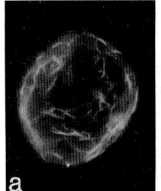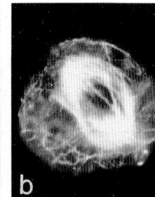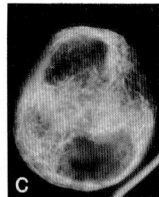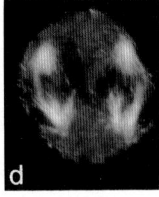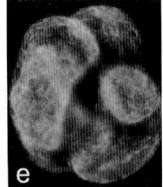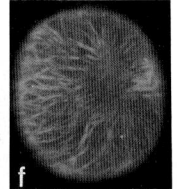

Fig. 1. Light micrographs of Rhodamin-phalliodin labelled actin in developing pollen of <u>Gasteria</u>.

** Van Lammeren AAM, Keijzer CJ, Willemse MTM, Kieft H. Planta 165, 1–11
* Pant Anatomy and Cytology Department, Maria Curie Slodowska University Lublin, Akademicka 19, 2033 Poland

Structure of the Microtubular Cytoskeleton in Callus and Developing Somatic Embryos of Carrot (*Daucus carota* L.)

A.A.M. van Lammeren, E. Provoost, J.H.N. Schel

Department of Plant Cytology and Morphology

Agricultural University

Arboretumlaa 4

6703 BD Wageningen

The Netherlands

Embryos from carrot suspension cultures were observed by transmission and scanning electron microscopy. Also, semithin sections of polyethylene glycol (PEG) embedded embryos were labelled for tubulin with antibodies and FITC to investigate the cytoskeleton by light microscopy.

With electron microscopy bundles of microtubules were demonstrated close to the cell wall: the cortical microtubules (see also Van Lammeren et al., 1987). The 3-dimensional orientation was, however, difficult to recognize. In 3-5 μm sections of polyethylene glycol embedded embryos preprophase bands, spindle tubules and cortical microtubules were found. At prophase a scaffold of microtubules was observed surrounding the nucleus. During metaphase only spindle tubules were observed. Cortical microtubules were randomly orientated in the subepidermal and central cells of globular and heart-shaped embryos. About parallel bundles were frequently observed in the protoderm then. Such bundles were also demonstrated in cortex cells of torpedo-shaped embryos where they were often thicker and showed a stricter arrangement perpendicular to the length axis of the cells. Isodiametric cells of the root tip and cotyledons exhibited randomly patterning of microtubules preferentially. The results indicate that cell morphogenesis and the arrangement of the cytoskeleton are correlated.

Van Lammeren AAM, Kieft H, Provoost E, Schel JHN (1987) Immuno-gold labelling of tubulin in ultrathin cryosections of cultured carrot cells. Acta Bot. Neerl. 36(2): 125-132.

Current Research and Use of True Potato Seed in Developing Countries

Malagamba, J.P., and Accatino, P.
International Potato Center (CIP), Lima, Peru

One of the main advantages of sexually propagating potatoes, i.e. using the true seed (TPS), is the flexibility offered by the different systems by which TPS can be utilized in various climates and farming conditions of the tropics. In many of those conditions TPS use not only reduces considerable the costs of the planting material, but also allows the farmer to grow a healthy crop at anytime as required by his farming system. Other valuable characteristics of TPS are the simple storage and transport required from the production site to the farmer.

Research on TPS at the International Potato Center (CIP) started in 1977. Efforts have concentrated since then in the obtaining of improved genetic materials and techniques. Progenies recently developed give tuber yields over 40 t/ha and meet acceptability standards of many growing areas of the tropics. Several alternative systems of using TPS have also developed, e.g. for producing potatoes for consumption directly in one season by transplanting seedlings to the field or as the base for producing seed tubers (seedling tubers) for planting the following seasons (Malagamba, et al., 1984).

Several countries have shown fast progress: Philippines: The use of a locally developed hybrid progeny has proved as a good alternative in farmers' fields of two regions, Mindanao and Negros, (Vander Zaag et al., 1986). In Mindanao, yields ranging from 30 to 45 t/ha were obtained by farmers using seedling tubers produced at the farmers' cooperative field. Vietnam: During the 1985-86 season, over 20 kg of open-pollinated TPS were collected and distributed to 35 cooperatives in 10 provinces. Seedlings were transplanted into 50 ha and produced over 3 million seedling tubers per hectare. Yields, when using those seedling tubers were higher by 50% than the local variety (Vander Zaag et al., 1986). The program grew further last year. Sri Lanka: One of the early adopters of TPS for providing low-cost planting materials to many small farmers (Upadhya et al., 1986). Higher yields observed when using hybrid seed have further expanded prospects for adoption in several areas of the country. Bangladesh: On-farm research and demonstrations to farmers in various potato growing areas have shown the advantages of using TPS. Seedling tubers of several hybrids had higher yields than imported seed tubers of commercial varieties (Sikka et al., 1986). Egypt: A strong research program for adapting TPS technology to the local conditions was developed the last few years with increasing number of farmers involved. Chile: A more specialized operation, production of TPS, was undertaken by national scientist in collaboration with CIP. About 30 kg of hybrid TPS of six selected families for distribution to scientists throughout the world were produced in early 1988.

References

Malagamba, P., and A. Monares and D. Horton. 1984. Design and evaluation of different systems of potato production from true seed. pp. 315-316. In: Winiger, F.A. and Stockli (eds.) 9th Triennial Conf. Eur. Assoc. Potato Res. Interlaken, Switzerland.

Sadik, S., C. Engels, A. El-Bedewy, A. Sharara, and A. El-Imrity. 1986. Regional Research Progress Report (Internal). CIP.

Sikka, L.C., H. Kabir, H. Sarker, E.H. Choudhury, and A. Quasem. 1986. Regional Research Progress Report (Internal). CIP.

Upadhya, M.D., H.O. Agrowal, S.N. Bhargava, P.C. Panday, M.S. Kadian, and K.C. Thakur. 1986. Regional Research Progress Report (Internal). CIP.

Vander Zaag, P., B. Fernandez, and B. Susana. 1986. Regional Research Progress Report (Internal). CIP.

Induction of Hysteranthy in the Saffron Crocus (*Crocus sativus* L.)

M. Negbi, Ora Plessner, Meira Ziv and D. Basker[*],

Department of Agricultural Botany

The Faculty of Agriculture

The Hebrew University of Jerusalem

P.O.Box 12, Rehovot 76 100, Israel

The saffron crocus (Crocus sativus L.), a sterile cultivated geophyte, is propagated by annual replacement corms for the red stigmatic lobes that constitute, after drying, the saffron spice. C. sativus is a sub-hysteranthous species, i.e., it blooms in autumn shortly after planting, before, with or after leaf appearance. The remainder of its growing season constitute of initiating, filling and maturing the daughter corms which can be lifted at the beginning of summer. Cultivation is still performed by traditional methods in the countries where it is grown: corm planting, flower harvesting, stigma separation and corm lifting are all carried out manually. These labour intensive practices greatly contribute to the high price at which the spice is sold.

The study of the physiological background of the developmental processes in C. sativus was aimed at improving the cultivation methods. It is demonstrated that a controlled temperature regime during corm storage affects flowering and production of daughter corms. Moreover, the hysteranthous habit is promotable by specific combination of controlled environmental corm storage and planting conditions. The application of these methods can result in mechanical flower harvesting to replace manual hand picking.

---

[*]Food Science Department, Agricultural Reaserch Organization, P.O.Box 6, Bet Dagan, Israel

Quality of *Zea Mays* Pollen Grains

P. ROECKEL, C. DIGONNET, B. FENNET and C. DUMAS

Reconnaissance Cellulaire et Amélioration des Plantes
UCB-LYON I
43 Bd du 11 Nov. 1918
69622 Villeurbanne Cedex
FRANCE

Pollen quality is a parameter which is rather difficult to estimate. In maize, pollen quality has been monitored using cytological (fluorochromatic reaction), (Heslop-Harrison et al. 1984) physiological (water content measurement and in vitro germination), biochemical (ATP assay with luciferine-luciferase reaction) (Matthys-Rochon E. et al. 1988) and biophysical techniques (Nuclear Magnetic Resonance of ^{31}P). All these tests have been performed every 30 min. on samples from the same pollen population which has been submitted to natural ageing process.

At anthesis, the maize pollen water content is about 60% of the fresh weight. During ageing, pollen undergoes a gradual dehydration which is correlated to decreases of FCR score, in vitro germination rate and ATP content.

This pluridisciplinary approach allows a good estimation of pollen quality. From the data, the water content measurement is the best test currently available for species which display hydrated pollen grain at anthesis.

References

Heslop-Harrison, J., Heslop-Harrison Y., Shivanna K.R. (1984) — The evaluation of pollen quality and a further appraisal of the fluorochromatic (FCR) test procedure. Theor. Appl. Genet. **67**: 367–375.

Matthys-Rochon E., Roeckel P., Detchepare S., Wagner V., Vergne P., Dupuis I. (1988) — Plant sperm cells and potential biotechnological prospects. Proceedings of the international conference on research in Plant Sciences and its relevance to future. Dehli, India (In Press).

Male Sterility in Flowers of *Rosmarinus officinalis*

L. Roiz and R. Dulberger
Department of Botany
The George S. Wise Faculty of Life Sciences
Tel Aviv University
Tel Aviv, 69978
ISRAEL

Rosmarinus officinalis L. is a shrub native to the Mediterranean area, but occurring mainly in southern Europe. In Israel it is widespread as a cultivated ornamental. Plants examined from Tel Aviv and Jerusalem were found to bear sexually polymorphic flowers, the polymorphism expressed in morphological traits and pollen stainability.

In hermaphrodite flowers (MF) the stamens were exserted from the corolla tube; the anthers contained up to 20% non-stainable pollen grains. Male-sterile flowers (MS) were markedly smaller than the hermaphrodite ones and their stamens were short and hidden within the corolla tube. The aborted anthers were either totally shrivelled or did not dehisce and contained 80-100% non-staining grains. Hermaphrodite and female forms were linked by a series of intermediary forms (INT) intergrading in so far as length of stamen filaments and degree of anther abortion. In such partially male-sterile flowers the percentage of non-stainable pollen grains varied from 20-80%.

There was an inverse correlation between the length of the stamen filaments and the percentage of aborted pollen grains.

Out of 46 plants examined three had >90% MF flowers, one had >90% MS flowers and 42 had MF and/or MS flowers mixed in varying proportions with INT flowers.

In a sample of 105 flowers (11 plants) collected randomly and examined for pollen abortion, only 25% were of the MF type.

Pollen abortion percentages varied considerably in flowers of individual racemes. Neither was there any definite pattern in the distribution of flowers of different sexual types on racemes along the main axis. The degree of male-sterility is apparently determined at each developmental site.

Seed set obtained from controlled hand-pollinations indicated that the plants are self-compatible.

In rosemary, it is possible that male sterility and protandry promote outcrossing, while allowing selfing. MS and INT flowers may also enhance the attractiveness of the reproductive shoots, while saving energy necessary for pollen production.

The Embryo Sac in Restionaceae: A Systematic Survey

Paula Rudall

Jodrell Laboratory,

Royal Botanic Gardens, Kew

Richmond, Surrey TW9 3DS.

United Kingdom.

Microscopical data have been extensively utilised in the systematics of the order Poales, (the grasses and allied families), largely due to a relative shortage of convenient macromorphological characters, as the flowers and leaves are often very reduced. In particular, the occurrence of proliferated antipodals in mature Restionaceae embryo sacs has been taken to indicate a relationship with Poaceae, where this condition (the 'Poaceae variant'; Anton & Cocucci, 1984) is nearly universal. This survey was undertaken to examine this character in the context of the systematics of the order, currently under review, and to provide a detailed survey of the megagametophyte in Restionaceae using modern techniques of light microscopy. Previously mainly African Cape genera had been examined, representing a single, probably specialised, subfamily (Johnson & Briggs, 1981); however Restionaceae also occurs widely in Australasia, and there is a single genus in the New World (not included in this survey). In general proliferation of antipodals occurs in South African taxa, but not in Australasian taxa, and since the character occurs sporadically it cannot be used to support the view that the two families are sister groups (Rudall & Linder, in press). The mature megagametophyte in Restionaceae characteristically has a well developed hypostase, a thick nucellus that is uniseriate at the micropylar end, and copious starch in the mature embryo sac. The starch granules are most clearly visible using differential interference contrast optics with cleared ovules, since they are refractive, and also the chloral hydrate in Herr's clearing fluid may cause them to swell slightly. Embryo sac development is of the Polygonum type, which is fundamental to the Monocotyledons and almost universal in the Poales.

Anton A M, Cocucci A E (1984) The grass megagametophyte and its possible phylogenetic implications. Plant Sytematics and Evolution 146: 117-121

Johnson L A S, Briggs B (1981) Three old southern families - Myrtaceae, Proteaceae and Restionaceae. In: A. Keast (ed). Ecological Biogeography of Australia, pp 429-469. Junk, The Hague

Rudall P, Linder H P (In press) The megagametophyte and nucellus in Restionaceae and Flagellariaceae. American Journal of Botany

Morphogenetic Processes in Wheat Anther Culture

L. SAGI and E. SZAKACS

Department of Genetics
Agricultural Research Institute
P.O. Box 19
H-2462 Martonvasar
HUNGARY

Anther culture is an efficient method for producing haploids, dihaploids, alien addition and substitution lines in wheat either for practical or basic research. Although wheat is a relatively good subject for anther culture it is desirable to increase the green plant production rate to fulfill the above purposes.

Pollen embryos induced by anther culture were subcultured to produce a high amount of regenerating material. Adventitious secondary embryos were developed on embryogenic calli selected during subculture. In the present study differentiation of primary and secondary pollen embryos was observed to demonstrate if secondary embryos have similar morphogenetic characters to those of primary embryos.

Anthers of Benoist variety (Triticum aestivum L.) containing microspores in the uninucleate stage were cultured on Potato - 2 medium. After 1 month induced primary pollen embryos were subcultured each 4 weeks on Murashige-Skoog medium with decreasing concentrations of 2,4 - D. Primary embryos wer transferred for shoot induction onto 190-2 medium in parallel (Zhuang and Jia, 1983).

Both kinds of embryos produced internal or peripheral meristematic regions in the first 4 days of differentiation. They developed into leaf primordia and shoot apeci in 8-10 days in case of primary pollen embryos and in 15-20 days for secondary ones. This delay might be due to different media used for differentiation.

Secondary embryos showed normal bipolar structure and developed into green plants on Murashige-Skoog medium with decreasing concentrations of 2,4 - D, as well as on 190-2 medium.

Regenerated plantlets from secondary embryos showed both diploid ($2n = 42$) and haploid ($2n = 21$) chromosome numbers demonstrating that no somatic embryogenesis had taken place.

It is concluded that secondary embryos can also be induced on the haploid level and that they have a similar differentiation pattern as primary pollen embryos.

Zhuang JJ, Jia X (1983) Increasing differentiation frequencies in wheat pollen callus. In: Cell and Tissue Culture Techniques for Cereal Crop Improvement, Science Press Peking, p. 431.

Structure and Development of the Pollen Wall in *Larix leptolepsis*

C. SAID

Reconnaissance Cellulaire et Amélioration des Plantes
UCB-LYON I
43 Bd du 11 Nov. 1918
69622 Villeurbanne Cedex
FRANCE

Microsporogenis in Laricina has been studied at the light microscopical level by Eriksson (1982) and Hall (1982). Our work presents an ultrastructural study of the pollen wall and of its development in a Gymnosperm: Larix leptolepis. The exine is thin and divided in a granular ectexine and a lamellar endexine; the intine is thick and microfibrillar in its inner part. The formation of the sporoderm begins as the spores are enclosed within the callose envelope of the tetrad. The exine template, which appears between the plasma membrane of the microspore and the callose wall, is divided early in two parts: an outermost layer (future ectexine) and an innermost layer (future endexine). Only the formation of the endexine involves deposition of sporopollenin on trilaminar tapetum.

After the release of the microspores from the tetrad, the tapetum degenerates in a periplasmodium, characterized by the occurence of orbicles. The periplasmodium remains in contact with the developing exine and may convey sporophytic material into it. The intine is the last layer formed by material of gametophytic origin.

Pollen wall development in Larix leptolepis and in Angiosperms appears identical in many important aspects, particularly in the double origin of its layers.

References

Eriksson G. (1968) Meiosis and pollen formation in Larix. Roy. Coll. For Stockholm.

Hall J.P. (1977) Microsporogenesis in Laricina. Can. J. Bot., **60**: 797–805.

Nucleic Acid and Protein Synthesis and Release During Development of Male Gametophyte (MG) of *Clivia nobilis* in Vitro

Tang Pei-hua and Zhu Ying-min
Institute of Botany
Academia Sinica
Beijing 100044, China

Synthesis of RNA and protein during development of MG and release of newly synthesized RNA (NSRNA) and protein (NSP) from germinating pollen (GP) were observed and proved by using autoradiography labeled with ^{3}H-uridine, ^{3}H-leucine and treatment with inhibitors (Actinomycin D and Cycloheximide). The dynamics of protein synthesis is similar to the same of RNA. Both of them have three peaks and two intermissions. The 1st peak is in 9-12 days before dehiscence of anthers (BDA). The 2nd is in 5-7 days BDA and intermitted from 48th h BDA. The 3rd begins from the 1st h after culture (AC), decreases at 6th h AC and stops before 20th h AC. The dynamics of release NSRNA and NSP is as follows: 1 h after labeling (AL) NSP was full of pollen (Fig.1). 1.5 h AL NSP concentrated near germinating pore (Fig.2). 2 h AL NSP transported to pollen tube (PT)(Fig.3). 5 h AL it diffused out (Fig.4). At 1.5 h AL NSRNA began to diffuse into protoplasm of GP (Fig.5). 2 h AL it transported to PT (Fig.6). 3 h AL it more and more transported into PT (Fig.7). 4 h AL it concentrated in the tip of PT and began to diffuse out (Figs.8,9). 5 h AL the most of NSRNA was out of PT (Fig.10).

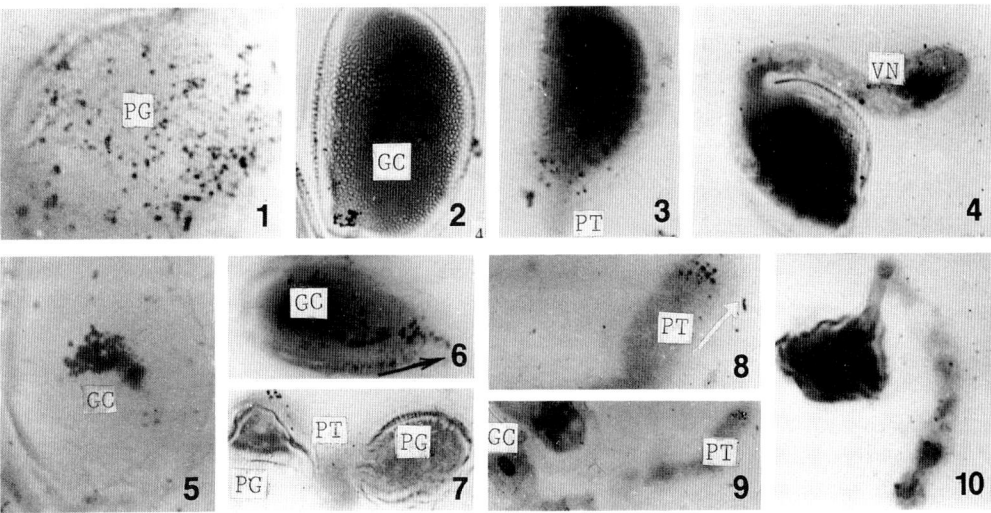

Genotypic and Environmental Variation in Production of 2n-Gametes in *Lilium*

Jaap M. van Tuyl and Pauline Stekelenburg

Institute for horticultural plant breeding (IVT)

Mansholtlaan 15

6708 PA Wageningen

The Netherlands

The production of 2n-gametes (meiotic polyploidization) occurs in many species, because of irregularities in meiotic division. The frequency of 2n-gametes is naturally extremely low, but can be influenced by environmental conditions like high temperatures and by selecting the most suitable genotypes. In our lily breeding research programme 2n-gamete producers were traced by crossing a series of diploid Asiatic hybrids with tetraploid parents. Over 500 of these (4x x 2x) and (2x x 4x) crosses resulted in a dozen tetraploid seedlings. One genotype 'Connecticut King' appeared especially responsible for this result. Therefore this hybrid was used to study the influence of temperature on the production of 2n-gametes. In the temperature range going from 14°, 18°, 22° to 26°C an increasing percentage dyads in the pollen mother cells was found, indicating the formation of 2n-gametes.

Wide interspecific lily hybrids are usually completely sterile. In rare cases, however, some fertile pollen can be detected. In a group of more than 50 embryo cultured hybrids of the cross **Lilium longiflorum** x **Lilium candidum** only one hybrid showed a pollen fertility of 25%. Meiotic studies revealed that in this case 100% 2n-pollen was formed. Comparable cases were found in the interspecific hybrids "Shikayama" x **L.henryi** and **L.auratum** x **L.henryi** ; used as pollen parents, these produced triploid progenies. Backcrossing these triploids with **L.auratum** x **L.henryi** gave a number of aneuploids with a chromosome number between 36 and 48. In contrast to the wide interspecific hybrids, seedlings from the interspecific cross of the Asiatic hybrid "Enchantment" and the related **Lilium pumilum**, produced fertile pollen. Meiotic studies of several of these hybrids showed, that not only haploid pollen was formed but also relatively high percentages 2n and also some, possibly triploid, (giant)pollen. The application of meiotic polyploidization in the breeding of lilies can be of great importance for further developments in lily culture.

Time and Space Concordance of the Growth, Differentiation and Metabolism in the Developing Seed of *Nelumbo nucifera* Gaertn

V.E.Vasilyeva, G.E. Tytova, I.P. Ermakov[x], N.M. Morozova &
K.A. Chochya
Department of Embryology
Komarov Botanical Institute
ul.Popova 2
197022 Leningrad
USSR

A certain sequence of stages of development of the ovule and seed is possible only in the presence of regulatory mechanisms which ensure coordination of growth, differentiation and metabolism both in space and in time. The present investigation shows a number of morphogenetic and morphophysiological correlations in the ovule development referring to the differentiation of its structural elements and distribution of starch at the stages before and after fertilization.

Disruption of coordination caused by the absence of fertilization resulted in a significant decrease of sucrose content almost up to its complete disappearence in the ovule, ovary and receptacle.

During the seed formation (1-15 d.a.p.) the dynamics of total nucleic acid content in the embryo and surrounding tissues was of the same pulsating character as the accumulation of soluble carbohydrates and growth rate. At this there were noted certain zones in the endosperm, the nucellus, the chalazal part differing both in the content of total nucleic acids and morphology. The main NA maxima coincided in time with the stages of initiation of the cotyledon, differentiation of the primordia of the 2-nd and 3-d plumule leaves. It is interesting to note that the NA maxima in the reproductive structures are manifested almost simultaneously during the first stages of seed formation. However beginning with stage of the embryo autonomy this concordance decreases.

Thus, the seed during its first stages of formation functions as a more rigidly coordinated system. On reaching the embryo autonomy new mechanisms of regulation appear to be activated in the embryo, which are responsible for the embryo passing on to a self-dependent way of development.

---

[x]Department of Biology, Moscow State University, Moscow, USSR

Study of the Male Developmental Programme at the Protein Level in Three Crop Species

P. Vergne, S. Detchepare, I. Delvallée and C. Dumas

Reconnaissance Cellulaire et Amelioration des Plantes
Université Claude Bernard LYON I
43, Bd. du 11 Novembre 1918
69622 Villeurbanne Cedex
FRANCE

Various electrophoretical techniques were used to monitor pollen development in three crop species: wheat (Triticum aestivum), maize (Zea mays) and kale (Brassica oleracea). These studies were performed either on whole anthers or isolated microspores and pollen grains. Jointly with protein analysis, careful cytological controls of the developmental stages as well as the viability of pollen samples were done using fluorescent probes, of fluorescein diacetate (Heslop-Harrison et al. 1984) in combinations with DAPI staining (Vergne et al. 1987).

This methodology allowed us to characterize several important modifications in protein patterns during pollen differentiation. In particular, changes occur soon after the first pollen grain mitosis. In addition, a set of new proteins is synthesized during pollen maturation and sperm cell differentiation. In the three species, mature pollen displays very specific patterns, and the detected pollen-specific proteins might be related to the ability to germinate and to achieve fertilization.

These modifications in protein synthesis and patterns during late pollen development are likely to result from haploid gene expression.

References

Heslop-Harrison J., Heslop-Harrison Y., Shivanna K.R. (1984) The evaluation of pollen quality and a further appraisal of the fluorochromatic (FCR) Test procedure. Theor. Appl. Genet., **67**: 367–375.

Vergne P., Delvallée I. and Dumas C. (1987) Rapid assessment of microspore and pollen development stage in wheat and maize using DAPI and membrane permeabilization. Stain Technol. **62**: 299–304.

Embryology of Plantaginaceae

F. VIGNON and J. VIGNON
Laboratoire d'Embryologie végétale,   Faculté des Sciences
33, rue Saint-Leu   80039 AMIENS CEDEX   FRANCE

The scarce embryonical observations about *Plantaginaceae*, realised by R. SOUEGES, are old. Thus, we are inciting to be interested in this family. Among the *Plantaginaceae* which are studying, we propose some stages of embryo and endosperm developments for three species adapted to different natural environments : *Plantago major* L. (all not halophilic grounds), *Plantago coronopus* L. (sandy grounds) and *Littorella uniflora* (L.) Ascherson (sandy grounds flooded during a long time).

The embryo development for these species conforms to the *Myosurus* type such *Plantago lanceolata* L.. From the bicelled proembryo, the apical cell (ca), the first division of which is vertical, produces the levels l and l' ; while the basal cell cb, partitioned transverly (m and ci) is at the origin of the hypophysis (h) and the suspensor (s). The level l contributes to the stem tip (pvt) and the cotyledonary zone (pco) ; the level l' produces the hypocotyledonary region (phy). The cell at the origin of hypophysis (h) is grand-daughter of m. At first partition of this one is concave and results in two super-imposed cells. The lower cell will form the rot cap. The suspensor is filamentous, constitued by a linear suit of cells derived from transversal partitions of the lower half of m and of all of ci. This long suspensor is rarely visible because the ovule is very curved.

The endosperm is an helobial type, but, for the selected exemples, the first partition of the primary endosperm nucleus produces two chambers, micropylar and chalazal, more or less equals. Next, the micropylar chamber is divided in two juxtaposed cells, and later, in four cells super-imposed two by two. Then, it appears two four-celled layers while the split of the micropylar chamber nucleus is not followed by a cell partition. From this time, we can locate three zones : - a chalazal cell seems to be always binuclear and becomes a very long chalazal haustorium (ch) scarcely visible entirely ; - a median zone with a great many cytokinesis resulting in small cells, produces the endosperm proper where the globular embryo is developing ; - the lowest four cells do not divide but continue to grow and become four large micropylar haustoriums (mh).

The embryo development seems to begin when the three zones produced by endosperm nucleus, are very established. You will observe that in these *Plantaginaceae*, the embryo development is like the one of dialypetalous species (*Myosurus*, *Capsella*, *Reseda*; *Ruta* ...), while the endosperm development is a type principally known in gamopetalous species (*Scrofulariaceae*, *Orobanchaceae* ...).

Use of Anther Culture in Incompatibility Studies

Gilles Vincent* and Mario Cappadocia**

*Jardin botanique de Montréal and ** Institut botanique de l'Université de Montréal,
4101 est, rue Sherbrooke,
Montréal, Qc, H1X 2B2
Canada

The problem taken into consideration in our research program is the mechanism(s) of generation of new S-alleles at the S-locus, in species characterized by self-incompatibility of the gametophytic type. A number of theories have been proposed to explain why obligate inbreeding has often resulted in the appearence of new S-alleles while all the efforts to detect new S-alleles after mutagenic treatments have failed. All these theories have recognized, at various degree, the dependence of the phenomenon on the general level of homozygosity. Our objectives are therefore: i) the production of perfectly homozygous plants in self-incompatible species via anther culture ii) the use of molecular tools in order to analyse the S-locus in plants already characterized genetically. This analytical approach and the use of fully homozygous plants should help in clarifying the incompatibility phenomena and in better understanding the functioning of the S-gene(s). Recently we produced haploids in self-incompatible Solanum chacoense and were able to analyse few doubled haploids and their progeny. The results obtained indicate that the genetic background of the regenerated homozygous plants affects the proper functioning of the S-locus and that this status of disturbance is also maintained in the following generation. Our analysis, however, was conducted on a very limited number of individuals and, for the more, no appropriate tester stocks (i.e. genotypes containing known S alleles) were available at that time in S. chacoense. We have now produced more monohaploids, diploids and tetraploids in two additional clones of S. chacoense and initiated the analysis of the androgenetic individuals. We have also generated a number of suitable tester stocks.

Sperm Cells in Apiaceae

M. Weber
Institute of Botany
University of Vienna
Rennweg 14
1030 Vienna
Austria

Two types of sperm cells are found in mature pollen grains of *Apiaceae*. For example *Eryngium campestre* (Fig.1) shows the plasma-rich sperm cell type having numerous organelles, mainly mitochondria and dictyosomes. The mitochondria (m) are smaller than those in the vegetative cytoplasm (M). The nucleus is elongated and slightly sickle-shaped. The folded sperm cell wall consists of two plasma membranes, which form broad projections on both poles of the cell. Inbetween the plasma membranes there is a deposition of wall material.

The second type of sperm cells is observed in *Orlaya daucoides* (Fig.2). Here the sperm cells have very little plasma and lack organelles. However, in a few cases unidentified inclusions are seen besides the spindle-shaped nucleus. The cell is limited by two plasma membranes, forming long slender projections (arrows) on both poles of the spindle-shaped cell.

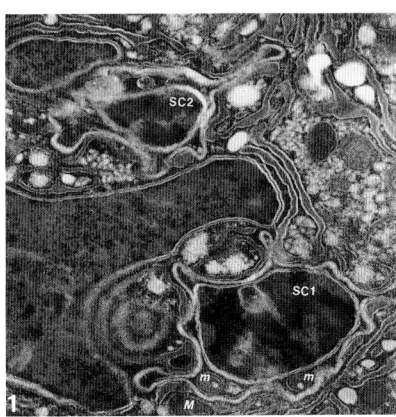
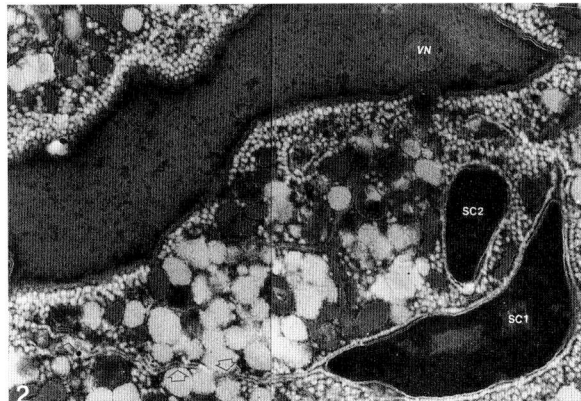

Author Index

Subject Index